Health Effects of Municipal Waste Incineration

Editors
Holly A. Hattemer-Frey, M.S.
Senior Risk Assessment Specialist
Advanced Sciences, Inc.
Oak Ridge, Tennessee

and

Curtis Travis, Ph.D.
Director
Office of Risk Analysis
Oak Ridge National Laboratory
Oak Ridge, Tennessee

CRC Press
Boca Raton Ann Arbor Boston

Library of Congress Cataloging-in-Publication Data

Health effects of municipal waste incineration / editors, Holly A. Hattemer-Frey and Curtis Travis
p. cm.
Includes bibliographical references and index.
ISBN 0-8493-4933-8
1. Hazardous wastes--Incineration--Health aspects--United States. 2. Health risk assessment--United States.
I. Hattemer-Frey, Holly A. II. Travis, C. C.
RA578.H38H43 1991
363.72'8--dc20 90-15013 CIP

Direct all inquires to CRC Press, Inc., 2000 Corporate Blvd., N.W., Boca Raton, Florida, 33431.

International Standard Book Number 0-8493-4933-8

Library of Congress Card Number 90-15013
Printed in the United States

PREFACE

A serious problem confronting industrialized countries is disposal of the large quantities of municipal solid waste (MSW) being generated by municipalities. It is estimated that within 10 years half of all municipalities will lack adequate landfill space.[1] Consequently, numerous cities and counties are considering MSW combustion as an alternative waste management option.[2] Currently, 111 MSW incinerators are operating in the U.S., 210 facilities are in the planning/construction stage, and about 300 facilities are expected to be in operation by the year 2000.[1]

There has been considerable local opposition to the construction of MSW incinerators because of their feared potential health and environmental impact. In this atmosphere of intense debate, decision makers are being presented with massive amounts of information and pressured to address public skepticism and concern about the safety of these facilities. It has become a formidable challenge for any one individual to critically evaluate the myriad of information available. This book attempts such an evaluation.

Although the acceptance of an MSW incineration facility must be based on community needs, health risks, costs, and waste management benefits, the evaluation of potential risks and benefits should be based on the best available evidence and be made in a rational, logical manner. The evidence presented in this book should not be viewed as either support for or criticism against MSW incineration as a waste management tool, but is intended to provide a perspective on some basic issues concerning MSW incineration. This book represents a compilation of chapters written by experienced individuals in the areas of emissions estimation, deposition modeling, risk assessment, indirect exposures, uncertainty analysis, and risk communication.

Estimating the magnitude of pollutant emissions from MSW incinerations and their potential health impact has become an increasing concern, especially as more facilities are being proposed and built. Chapter 1 presents data on pollutant emissions from MSW incinerations and discusses some of the statistical properties associated with such emissions. Emissions estimates, obtained during short-term stack tests, are important because they are used to predict air concentrations and the extent of human exposure to facility-emitted pollutant via inhalation.

Chapter 2 describes the methods and results of the Environmental Protection Agency's (EPA) regualtory analysis of MSW incinerator emissions, and the Agency's recent determination that further federal regulation of these emissions is warranted.

Chapter 3 provides an overview and a description of the types of air quality models available and how they can be used to estimate air concentrations and deposition rates in the vicinity of an existing or a new MSW incinerator. Deposition rates are important because they define the rate at which airborne pollutants will settle onto water, soil, and plant surfaces. Hence, quantifying deposition rates serves as the basis for estimating the extent of human exposure through the food chain.

Chapter 4 quantifies the extent of human exposure to a representative organic (dioxin) and a representative metal (cadmium) through the terrestrial food chain pathway from an arbitrary MSW incinerator. The multimedia partioning model used estimates the extent of daily intake that results from ingestion of contaminated beef, beef liver, milk and dairy products, chicken, pork, lamb, eggs, fruits, vegetables, and grains.

Chapter 5 identifies the major agricultural food chain sources of human exposure to dioxin emitted from a rural MSW incinerator. An environmental transport approach was used to model both the environmental fate of dioxin and its disposition in common farm livestock.

Recent studies indicate that the deposition of atmospherically emitted pollutants could result in additional, indirect avenues of exposure for humans and other organisms. Therefore, a methodology was developed by the EPA to examine both direct and indirect exposures and risks from atmospheric emission sources. Chapter 6 uses this methodology to study potential risks to human health and the environment from MSW incinerators.

In late 1987, the EPA proposed a revision in its cancer potency estimate for dioxin. Adoption of the new potency estimate for dioxin would dramatically reduce health risk estimates for MSW incinerators. Chapter 7 discusses the EPAs recent reevaluation of the cnacer potency estimate for dioxin and the policy implications associated with this unprecedented action.

Studies of the risks posed by MSW incinerator emissions often conclude that exposures to facility-emitted polutants through the food chain are likely to be more significant than exposures via direct inhalation. For this reason, predictions of PCDD and PCDF concentrations in plants, milk, beef, and other food items are often necessary to conduct a full evaluation of the health risks associated with an MSW incinerator. Chapter 8 presents a general approach for estimating the possible extent of bioaccumulation of dioxins and furans in cow milk and beef.

The health risk assessment process, as applied to MSW incineration facilities, has evolved largely in response to concerns raised during the regulatory review process. Chapter 9 outlines the evolution of health risk assessments for MSW incinerators in California and describes the methodology used in and technical issues inherent to health risk assessments for this type of facility using a case study from a California facility.

Health risk assessments should not only characterize the potential risk to affected populations, but should also express risk estimates in such a way that the assumptions made to derive those estimates and the uncertainties associated with such estimates are evident to the decision maker. Chapter 10 discusses the sources of uncertainty on the decision-making process, and methods available to explicitly evaluate uncertainties in the risk assessment process.

Chapter 11 uses a probabilistic simulation technique to perform a sensitivity and uncertainty analysis of the multimedia (food chain) impacts of MSW incineration. The distributions generated from this analysis reflect the variability (uncertainty) in the input parameters used in the multimedia partitioning model.

Public concern about possible adverse effects to human health and the environment from exposure dioxins and furnas emitted from MSW incinerators present the need to assess and compare the toxicity of these compounds collectively rather than focusing on a few, well-studied compounds. Chapter 12 evaluates several different weighing schemes currently used in health risk assessments and discusses the need for a common, numerical expression of dioxin and furan toxicity to be used in all quantitative risk assessments for MSW incinerators.

Health-related issues associated with ash management remain unsolved. Chapter 13 defines the characteristics of ash generated from MSW combustion, discusses the potential for pollutants to leach from the ash and contaminate the environment, explains the different types of leaching tests currently used, and discusses some important current issues associated with ash management.

The State of Michigan's Toxic Substance Control Commission conducted an in-depth review of the policies and regulations governing the disposal of resides generated from MSW incineration. Chapter 14 presents the conclusions and recommendations of the Committee.

Chapter 15 presents the authors' views concerning the lack of comprehensive analysis needed to accurately assess the risks associated with MSW incineration and to develop appropriate guidelines for the use of this technology. The authors discuss several issues including the need to consider an integrated waste management strategy and the need to focus attention on source-based strategies to the amount of MSW generated initially.

To facilitate the comprehensive, concurrrent evaluation of various waste disposal choices, the Institute for Energy and Environmental Research developed a comprehensive and systematic approach for evaluating the costs, benefits, and environmental impact of differing MSW disposal systems to provide public officials with a detailed methodology for analyzing, comparing, and ranking MSW management alternatives. Chapter 16 presents the results of applying this approach to the City of Cologne, West Germany. The comparison of several waste management options, including information on costs, environmental impact, energy savings, and future landfill space requirements, are presented in that chapter.

Integrated MSW management has gained recent appeal as the waste disposal crisis grows more acute around the country. Many state and local governments have discarded their search for a single disposal alternative, such as MSW incineration and have developed a hierarchy of disposal alternatives, with waste reduction and recycling at the forefront. Chapter 17 explores some of the factors involved in establishing feasible community recycling goals, evaluates the performance of various recycling technologies, and estimates their maximum potential for reducing the amount of MSW that must be disposed of in a landfill.

Chapter 18 provides a perspective on MSW incineration as a source of human exposure to the highly toxic dioxins and furans. In these two chapters, the authors evaluate the following key issues: (1) Are MSW incinerators the major source of dioxin input into the environment? (2) Are environmental concentrations of PCDD/PCDFs around operating MSW incinerators substantially elevated? and (3) Are MSW incinerators the major source of human exposure to PCDDs and PCDFs?

The fact that dioxin has been measured in virtually every media indicates that dioxin emissions are widespread. Chapter 19 estimates the magnitude of dioxin emissions into the U.S. environment and provides data on the relative contribution of four known dioxin sources, MSW incinerators, hospital waste incinerators, motor vehicles, and residential wood combustion, to total dioxin input.

The remaining chapters address issues concerning risk communication and public opposition to MSW incinceration. The probability of risk is increasingly estimated using health risk assessment. To the public, however, the very act of undertaking a "risk" assessment seems to legitimize their initial concerns about the seriousness of the threat. Chapter 20 introduces an approach that incorporates evaluation of alternative risks, examination of the potential for catastrophic circumstances, and discussions of ways that enable those affected to control the risks in some meaningful way.

When an MSW incinerator is proposed as part of an integrated waste management plan, public opposition swells. Most citizens do not want these facilities located in their neighborhood. Chapter 21 addresses some of the public's concerns associated with MSW incineration.

As an increasing number of MSW incineration projects are being planned and developed, a concomitant increase in organized public opposition has also arisen. Chapter 22 discusses some of the reasons behind this public opposition and proposes ways to effectively address public concerns about potential health risks associated with emissions from these facilities. Several case studies illustrating these concepts are also presented.

It was our intent that the presentation and interpretation of information presented herein be useful to the scientist, the layperson, those individuals whose incineration-related activities are regulated, and to those who make regulatory decisions. While much remains to be learned about the human health risks associated with MSW incineration, it is our goal that the material presented in this book stimulate further research and encourage readers to formulate opinions about some of the fundamental issues affecting the management of municipal solid waste.

REFERENCES

1. **U.S. Environmental Protection Agency**, Municipal Waste Combustion Study: Characterization of the Municipal Waste Combustion Industry, EPA/530-SW-87-021H, Pollutant Assessment Branch, Research Triangle Park, NC, 1987.
2. **Petty, J. and Rodgers, P.,** Source, distribution, and fate workshop summary, in Dioxins in the Environment, Kamrin, M. A. and Rodgers, P. W., Eds., Hemisphere Publishing, Washington, D.C., 1985, 299.

THE EDITORS

Holly A. Hattemer-Frey, M.S., is a Risk Assessment Specialist for the Advanced Sciences, Inc. (ASI) Corporation in Oak Ridge, TN, where she manages a wide range of risk assessment projects, including remedial investigation/feasibility studies (RI/FS) for various DOE facilities.

Ms. Hattemer-Frey graduated cum laude from Ohio Wesleyan University, Delaware, OH, in 1983 with a B.A. in Zoology. She received an academic fellowship to pursue graduate work in Environmental Sciences. She received a Master of Environmental Sciences with an emphasis in Applied Ecology in 1986 from Miami University, Oxford, OH. She was an Information Specialist with the Carbon Dioxide Information and Analysis Center, Oak Ridge National Laboratory, where she researched, wrote, and edited a publication on climatic indices to help meet the information needs of CO_2-climate researchers evaluating the effects of increased atmospheric CO_2 levels.

During the $3^1/2$ years as an Environmental Scientist with the Office of Risk Analysis, Oak Ridge National Laboratory, she provided technical support in the areas of risk analysis and human exposure assessment, authored or coauthored more than 35 papers and technical reports, and gave invited presentations on diverse risk assessment topics. Her research interests include applying environmental partitioning models to evaluate the extent of human exposure to organics through the food chain, investigating the uptake of organics by aerial plant parts as a potential pathway of vegetative contamination and human exposure, quantifying human exposure to dioxin from municipal waste incineration and background environmental contamination, evaluating potential human impact of deliberate environmental introductions of genetically altered organisms, and applying pharmacokinetics and pharmacodynamics to improve the risk assessment process.

Ms. Hattemer-Frey is a member of the Society for Risk Analysis and has served on the Science Advisory Panel for the Montgomery County, Maryland, Power Plant and Municipal Solid Waste Incincerator Research Project (1988), as a consultant on the review of the Preliminary Assessment of Risks from Waste Incineration in the Commissioners/Cherry Street Area of Toronto, Canada (1988), and was a National Academy of Sciences Consultant to the Committee on Animals of Monitors of Environmental Hazards (1989). She was invited to participate in the Environmental Protection Agency's Workshop on Municipal Solid Waste Technology, performed a technical review of the incineration portion of the Environmental Protection Agency's Sludge Technical Regulations for the American Association of Sewerage Agencies (1989), and organized and edited a special issue of *Risk Analysis* on municipal waste incineration.

Curtis C. Travis, Ph.D., has been the Director of Oak Ridge National Laboratory's Office of Risk Analysis since 1984 and is responsible for coordinating all risk assessment activities at the Laboratory.

Dr. Travis obtained his Ph.D. in 1971 from the University of California at Davis in Applied Mathematics. He received an M.A. in Biomathematics in 1967 and a B.S. in Mathematics and Physics in 1966 from the California State University at Fresno. He served as a Research Engineer at the Jet Propulsion Laboratory, California Institute of Technology, working on computer analysis of space flight simulation. He was also an Assistant Professor of Mathematics at Vanderbilt University (1971—1974) and at the University of Tennessee, Knoxville (1974—1976). In 1976, he joined the Research Staff of the Environmental Sciences Division, Oak Ridge National Laboratory, where he performed radiological risk assessments and conducted research in theoretical population ecology. As Leader of the Exposure Analysis Group (1977—1984), he was responsible for developing environmental transport and exposure models for toxic chemicals and became nationally recognized for his efforts in multimedia modeling.

Dr. Travis is President-Elect of the Society for Risk Analysis and will serve in that capacity from 1990 to 1991. He continues to function as Editor-in-Chief of *Risk Analysis: An International Journal*, which he has done since 1983. He is a member of the Science Advisory Board, National Center for Toxicological Research (1986—present), National Academy of Sciences Committee on Biomarkers for Immunotoxicology (1989), the National Academy of Sciences Board on Radioactive Waste Management Review of TRU Waste at Idaho National Laboratory (1989—1991), Chairman of the Office of Science and Technology Policy Task Force on Risk Analysis (1985—present), and Course Director for the Society of Risk Analysis course on carcinogen risk analysis. He has also served as Director of the NATO Advanced Study Institute for Biologically Based Methods in Cancer Risk Assessment (1988) and as a member of the Environmental Protection Agency Review Panel for Cancer Risk Assessment Guideline Revision (1989). He is the Technical Lead on the Office of Science and Technology Policy's project to evaluate the scientific bases of assumptions used by federal agencies in assessing risk.

Dr. Travis has authored more than 180 papers and technical reports, has been the author or coauthor of 7 books, and drafted the Guidance Document for Health Assessments at Superfund Sites for the Agency for Toxic Substance and Disease Registry (ATSDR). He has delivered more than 50 invited presentations on the subject of risk assessment since 1986. His current research interests include multimedia partitioning of chemicals, physiological-based pharmacokinetic and pharmacodynamic modeling, and risk assessment for biotechnology and the incineration of municipal and hazardous wastes.

CONTRIBUTORS

Denise Alston-Guiden
Associate Engineer
Air Quality Management Department
ROY F. WESTON, INC.
West Chester, Pennsylvania

Michael T. Alberts
Staff Engineer
Department of Environmental Services
Radian Corporation
Sacramento, California

Greg D. Belcher
Martin Marietta Energy Systems
Oak Ridge, Tennessee

Randall J. F. Bruins
Environmental Scientist
Environmental Criteria and Assessment Office
U.S. EPA
Cincinnati, Ohio

Paul Chrostowski, Ph.D.
Vice President
Clement Associates, Inc.
Fairfax, Virginia

David H. Cleverly
Environmental Scientist
Office of Technology Transfer and Regulatory Supprt
U.S. EPA
Washington, D.C.

Richard J. Cook, Ph.D.
Professor of Chemistry
Kalamazoo College
Kalamazoo, Michigan

Gaynor Dawson, Ph.D.
Chemical Engineer
ICFKE
Richmond, Virginia

Richard Denison, Ph.D.
Staff Scientist
Toxic Chemicals Program
Environmental Defense Fund
Washington, D.C.

R. Wyatt Dietrich
Senior Scientist
Department of Environmental Services
Radian Corporation
Sacramento, California

Larry Fradkin
Office of Senior Official for Research and Development
U.S. EPA
Cincinatti, Ohio

Bernd Franke
Institute for Energy and Environmental Research
Takoma Park, Maryland

Michael Gough, Ph.D.
Senior Fellow
Center for Risk Management
Resources for the Future
Washington, D.C.

Holly A. Hattemer-Frey
Advanced Sciences, Inc.
Oak Ridge, Tennessee

Albert Iannacone
Environmental Scientist
Environmental Sciences Division
NSI Technology Services Corporation
Kansas City, Kansas

Annie M. Jarabek
Environmental Scientist
Office of Research and Development
U.S. EPA
Cincinatti, Ohio

Kay H. Jones, Ph.D.
Zephyr Environmental Consulting
Seattle, Washington

Ray Kapahi
Vice President
Ecoserve Environmental Services
Pittsburg, California

Robert G. Kellam
Chief
Pollutant Assessment Branch
U.S. EPA
Durham, North Carolina

Carolyn S. Konheim
President
Konheim & Ketcham, Inc.
Brooklyn, NewYork

John A. Lowe
Industrial Hygienist
Department of Environmental Sciences
Radian Corporation
Sacramento, California

Michael Marchlik
Senior Consultant
Safety and Environmental Health Section
Arthur D. Little, Inc.
Cambridge, Massachusetts

Patricia M. McGinnis, Ph.D.
Managing Toxicologist
Chemical Hazard Assessment Division
Syracuse Research Corporation
Cincinatti, Ohio

Rayburn Morrison
Chemical Engineer
Office of Air Quality Planning and Standards
U.S. EPA
Durham, North Carolina

Charles Papke
Resource Management Associates
Napa, California

Norman C. Possiel
Meteorologist
Office of Air Quality Planning and Standards
U.S. EPA
Durham, North Carolina

Brenda L. Riddle
Environmental Engineer
Emission Standards Division
U.S. EPA
Research Triangle Park, North Carolina

Shawn L. Sager
Senior Associate Chemist
Clement Associates, Inc.
San Francisco, California

Ellen K. Silbergeld, Ph.D.
Senior Scientist
Toxic Chemicals Program
Environmental Defense Fund
Washington, D.C.

Paul C. Siebert
Section Manager
Air Pollution Control and Permitting
ROY F. WESTON, INC.
West Chester, Pennsylvania

Jeffrey B. Stevens, Ph.D.
PACE Laboratories, Inc.
Minneapolis, Minnesota

Curtis C. Travis, Ph.D.
Director, Office of Risk Analysis
Health and Safety Research Division
Oak Ridge National Laboratory
Oak Ridge, Tennessee

Lillie Craig Trimble
Senior Partner
Bluffs Specialties
Kansas City, Missouri

Duncan Turnbull, D.Phil.
Senior Science Advisor
ENVIRON Corporation
Arlington, Virginia

James F. Walsh
ROY F. WESTON, INC.
Bannockburn, Illinois

Stephen T. Washburn
Project Manager
ENVIRON Corporation
Princeton, New Jersey

TABLE OF CONTENTS

Chapter 1

AN ANALYSIS OF WORLDWIDE RESOURCE RECOVERY EMISSIONS DATA AND THE IMPLICATIONS FOR RISK ASSESSMENT

Paul C. Siebert, Denise Alston-Guiden, and Kay H. Jones

TABLE OF CONTENTS

I. INTRODUCTION

Emissions from the combustion of municipal solid waste (MSW) and their potential health impacts have become an increasing concern, especially as more resource recovery incineration facilities are planned and built. In particular, emissions of polychlorinated dibenzo-*p*-dioxins (PCDDs), polychlorinated dibenzofurans (PCDFs), and trace metals have received much attention because of their high toxicity, despite the relatively low quantities emitted. The mechanisms involved in the formation and possible control of PCDDs and PCDFs can only be hypothesized at this time. Conversely, metallic compounds result from the various oxidation-reduction reactions that occur within the combustion chamber.[1] Depending on the vaporization and condensation temperatures of the compounds and the temperature, oxygen, and carbon monoxide profiles within the incinerator, different amounts and forms of metallic compounds can be formed and adsorb onto particles.

Limited emissions test data are available because of the high cost of analyzing organic compounds and trace metals. Hence, an appropriate evaluation of available emissions data is crucial to producing accurate risk assessments for resource recovery facilities. Emissions data are also used in environmental impact analyses, as the basis for permit limitations, and to verify hypotheses regarding the effectiveness of various air pollution control technology options.

ROY F. WESTON, INC. (hereafter known as WESTON) has attempted to collect the best data available for estimating emission factors for various pollutants of concern released from resource recovery facilities. Although this database has over 500 citations, only a small number of those citations are facility test reports, which are the preferred data source for emission factor development. Facility test reports are the preferred data source, since secondary sources (e.g., journal articles and conference papers) frequently do not include all the data necessary to ensure a meaningful comparison. Primary data sources typically include all of the information necessary to standardize reported emissions rates into a common basis for analysis. WESTON has developed a computer program into which new emissions test data can be entered and conversions to standardized units calculated. Although the number of emission test reports currently included in WESTON database is still limited, more data are used to estimate resource recovery emissions than are currently being used in other risk assessments. This larger test sample size allows various statistical analyses to be performed and makes the overall emissions estimate less susceptible to influence by an atypical facility test.

The current WESTON database, called the Roy F. Weston, Inc. Resource Recovery Emissions Data System, includes data from 58 facilities in the U.S., Canada, and Europe, covering a wide range of fuel types, combustion systems, throughputs, and air pollution control technologies.[2-72] While the WESTON database quantifies emissions of organics, trace metals, acid gases, and criteria pollutants, this chapter focuses on PCDD, PCDF, and trace metal emissions. In addition, this chapter discusses some of the statistical properties associated with such emissions and their implications for improving health risk assessments.

II. ANALYSIS OF EMISSIONS DATA

Although data on PCDD/PCDF emissions are increasing steadily, information on some metals, other organics, and even the criteria or regulated pollutants is limited. Therefore, emissions data obtained from different facilities were used for each category of pollutant, so that reasonable emissions estimates could be made. Facility test reports were used as the primary data source for most facilities considered; however, for some European facilities, it was necessary to use data summaries presented in other documents.

The pollution control equipment used and differences in temperature, oxygen, and carbon monoxide profiles within the facility may affect pollutant emissions. Incinerator characteristics, such as whether the facility recovers heat, facility design (e.g., modular, which allows off-site construction vs. nonmodular, on-site construction), the type of combustion system installed (e.g., single or two-stage), the amount of excess air available during combustion (e.g., excess or starved air system), and the type of fuel burned (e.g., massburn vs. refuse-derived fuel — RDF) affect these profiles and may also affect emissions. Differences in waste composition affect metal and acid gas emissions the most, since metals as well as sulfur, chlorine, and fluorine originate in the waste and are not formed during combustion.

Emissions data tend to be log-normally distributed (i.e., the logarithms of the data are normally distributed). Hence, expected emissions were derived by assuming a log-normal distribution. As shown in Figure 1, log-probability plots of PCDD/PCDF emissions (expressed in toxic equivalent [TEFs] using the U.S. EPA method[57]) from selected facilities with high efficiency particulate control devices appear linear for the three data sets evaluated: (1) facilities with acid gas control, (2) facilities without acid gas control, and (3) all facilities. The log-normality of the data available in 1988 for these three data sets was demonstrated using the Shapiro-Wilk statistic and a univariate analysis (W: log normal greater than 0.9 and test statistic for rejection of log-normality greater than the significance level of 0.05).[63] Therefore, log-normality was also assumed for metals and other pollutants evaluated, and statistical analyses of emissions data were performed on the logarithms of the data points.

Statistical analyses confirmed the hypothesis that facilities with and without heat recovery are distinct. A simple linear regression analysis of the logarithm of total PCDD emissions using the 1988 database vs. a dummy variable (representing whether the facility recovers heat) showed that emissions from the two types of facilities are significantly different ($R^2 = 0.494$ and test statistic probability that emissions from the two types of facilities are significantly different was greater than 99%).[63] A *t*-test also confirmed that the two facility types are significantly different. (The probability of the means being different was 99% with unequal or equal variances, while the probability of the variances being unequal was 60% for the 1988 data.[63] Using the current database, the probability of the means being different was 99% with unequal or equal variances, while the probability of the variances being unequal was 87%.)

A statistical analysis of the current database has not confirmed the validity of the distinction in emissions between massburn and RDF facilities. A *t*-test was performed on the logarithm of total PCDD emissions. The probability of the means being different was only about 30%, while the probability of the variances being unequal was 48%.

Because of the limited data available for RDF facilities and the potential differences in combustion and emissions for the two fuel types, emissions data for RDF and massburn facilities have been segregated. The validity of treating facilities with and without acid gas control as two distinct data sets was also assessed statistically for PCDD and PCDF emissions. A simple linear regression of the logarithm of PCDD and PCDF toxic equivalent emissions (expressed in TEFs) from the 1988 database and a dummy variable (indicating whether the facility did or did not have acid gas controls) showed with 88% confidence that the two sets were significantly different ($R^2 = 0.247$ and test statistic probability that emissions from the two types of facilities were significantly different was greater than 88%).[63] Again, a *t*-test demonstrated that the two data sets were significantly different. (The probability of the means being different was 82% with unequal variances and 88% with equal variances, while the probability of the variances being unequal was 94% for the 1988 database. For the database presented here, the probability of the means

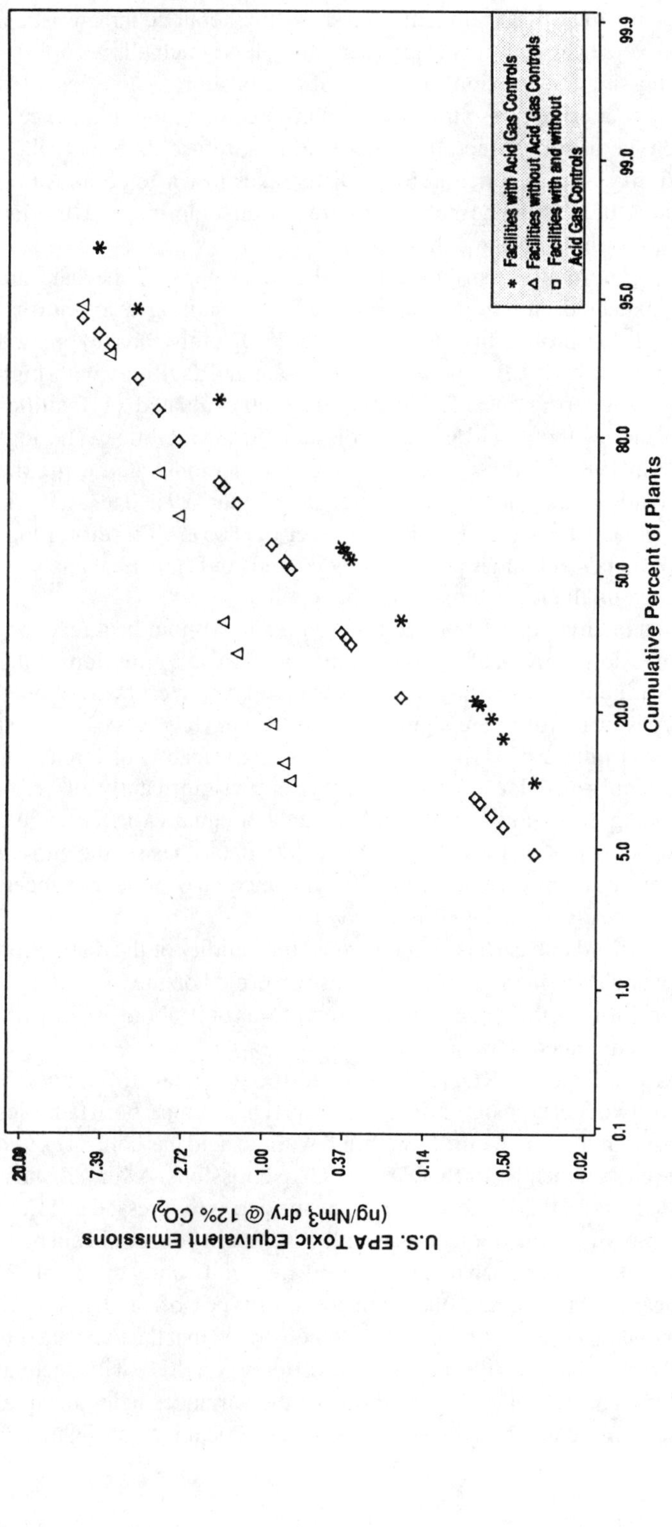

FIGURE 1. Log normal probability plot of dioxin/furan emissions (expressed in toxic equivalents for selected massburn facilities.

being different was 99% with unequal variances and 98% with equal variances, while the probability of the variances being unequal was 85%.)

Emissions test results from nonmodular (excess air) massburn facilities with heat recovery, standard operation during testing, and acceptable testing procedures were primarily considered. For a few metals and organics, however, so little data were available that modular facilities were also considered. Standardized results from individual emissions test runs were averaged to calculate one emission value for each pollutant measured at a given facility. Facility emission values were calculated by taking arithmetic averages of individual emission test run results in accordance with standard emission testing procedure.

The statistical analyses used to develop one emission factor for each compound emitted from like facilities varied depending on the number of data points that were deemed acceptable for inclusion. The geometric mean was used as the best estimate of long-term emissions, regardless of the number of data points. (Using the geometric mean as the best estimate of long-term emissions is equivalent to assuming that the distribution describing emissions from all facilities in the population taken as a whole also describes emissions from any one facility in the same population as they vary over time.) Upper- and lower-bound emission factors were estimated from: (1) the 95% confidence intervals for data sets with seven or more points; (2) the geometric mean plus or minus the geometric standard deviation for sets with five or six points; and (3) the maximum and minimum value reported for sets with four or fewer points. Based on the same assumption regarding the distribution of emissions from one facility over time, the upper-bound value (calculated from tests at a population of facilities) was considered an appropriate estimate of short-term emissions from any one facility.

III. ANALYSIS OF DIOXIN/FURAN EMISSIONS DATA

A. SELECTION OF FACILITIES

Table 1 lists all facilities for which PCDD emissions data are available and their respective total values (column A). Facilities included in this list, however, use diverse combustion and air pollution control technologies. The data were screened to include only those values representing the most common technology type, massburn with excess air, according to the following steps:

- All facilities without heat recovery were eliminated, leaving those in column B of Table 1.
- Facilities that experienced abnormal conditions during test periods were eliminated, leaving those in column C.
- Facilities that burn RDF rather than a massburn of MSW were eliminated, leaving those in column D.
- Massburn facilities with a furnace capacity of less than 50 tons/d were eliminated, leaving those in column E. (These smaller modular facilities usually are two-stage starved air systems in contrast with the larger single-stage, excess air systems. Although the Pittsfield facility is a modular facility, it was retained, because it is a large capacity, excess air facility.)

The following sections explain these criteria in more detail.

1. Facilities with Heat Recovery

Combustion parameters, such as temperature and retention time within the furnace, can substantially affect the formation of PCDDs, PCDFs, and their precursors. These combustion parameters are inherently different in facilities that recover heat compared with facilities without heat recovery, especially since facilities without heat recovery tend to be older and less efficient.[62] Hence, the following facilities were excluded from our analysis, because they lacked

TABLE 1
Worldwide PCDD Emissions Data by Facility Type[a] (Tetra through Octa Homologs) (ng/Nm^3 @12% CO_2, dry)

Facility	Ref. No.	Country	A All Data	B Plants with Heat Recovery	C Best Data (Normal Operating & Testing)	D Mass Burn Facilities	E Mass Burn Excess Air Facilities > 50 TPD	F Facilities with Acid Gas Controls
Philadelphia,N.W., PA	51	USA	589	-----	-----	-----	-----	-----
Toronto	4	Canada	612	-----	-----	-----	-----	-----
Italy 6	3	Italy	675	-----	-----	-----	-----	-----
Italy 5	3	Italy	746	-----	-----	-----	-----	-----
Niagara RDF, NY	28	USA	853	853	853	-----	-----	-----
Zaanstad	43	Netherlands	1,294	-----	-----	-----	-----	-----
Brasschaat	7	Belgium	3,383	-----	-----	-----	-----	-----
Hamilton/Wentworth SWARU	14,25	Canada	4,259	4,259	-----	-----	-----	-----
Herelbeke	7	Belgium	4,657	-----	-----	-----	-----	-----
Italy 4	3	Italy	5,003	-----	-----	-----	-----	-----
Italy 3	3	Italy	8,622	-----	-----	-----	-----	-----
h Hampton, VA	20,72	USA	11,188	11,188	-----	-----	-----	-----
Italy 2	3	Italy	33,047	-----	-----	-----	-----	-----

a Based on data collected as of October 1, 1989.

b Results shown are based on an average of six sets of tests (12 total). Pilot study of various control configurations conducted by Environment Canada & Flakt on a gas slipstream of about 6 to 9 percent of the flow from one of four 250 ton/day units of a massburn facility (i.e. pilot unit equivalent to 23 TPD maximum).

c Measured values include tetra through hexa homologues

d Measured values do not include the penta homologues.

e Measured values include tri through octa homologues.

f Tests performed during study included changing combustion temperature and burnrate; results shown are for tests performed during normal operating conditions.

g Values shown are for tests performed when natural gas was not used.

h The name of the facility was not supplied in the test reports, but were taken from the U.S. EPA Municipal Waste Combustion Study (76).

heat recovery (Table 1): Italy 1 through 6, Philadelphia Northwest, Valmadrera, Toronto, Brasschat, Herelbeke, and Zaanstad[3,4,7,18,43,51] leaving those in column C. A comparison of emissions values for facilities listed in columns A and B of Table 1 show that facilities with newer, more sophisticated resource recovery technology have lower emissions of PCDDs than older, less efficient facilities. As noted previously this difference in emissions has been statistically demonstrated.

2. Abnormal Facility Operation during Testing

Facilities that experienced abnormal conditions during any testing program were also discarded, since inclusion of these types of data could bias overall emissions estimates. For example, the Montreal test data[15] were eliminated, because extremely low PCDD and PCDF emissions were reported. Such low emissions factors may have been due to a reduced load during testing.[62] The Hamilton, Ontario, facility[14,25] experienced unsteady burning conditions due to inconsistencies in the refuse preparation and feeding system.[62] The situation was further aggravated by higher than usual combustion temperatures and the deteriorated condition of the electrostatic precipitator (ESP). Test results from the Hampton, VA, facility[20,72] were discarded due to low combustion temperatures during testing.

3. Type of Fuel Burned

The design of RDF facilities is substantially different from the design of massburn facilities. RDF is fuel that has been shredded and processed into pellets. Table 1 shows that PCDD emissions from RDF facilities generally tend to be higher than emissions from massburn facilities. As discussed earlier, this difference has not been shown to be statistically significant. However, to ensure the suitability of the database for massburn facilities, the Orrington, ME; Lawrence, MA; Pigeon Point, DE; Biddeford, ME; Eksjo, Sweden; Albany, NY; and Niagara, NY, facilities[12,28,30,49,50,52,70] were deleted leaving the facilities in column D of Table 1.

4. Facility Design

Smaller facilities with a furnace capacity of less than 50 tons/d are typically modular, starved air, two-stage systems. The first stage of these units has less than stoichiometric amounts of air available. (Stoichiometric refers to the exact amount of air necessary for complete combustion.) In contrast, larger massburn units usually are single-stage combustors with 1.5 to 2.5 times the stoichiometric amounts of air available (i.e., 50 to 150% excess air available). Since the operating parameters and combustion processes of modular facilities are different from the processes of other massburn systems, it is likely that their emissions are also different. Five facilities, Prince Edward Island (PEI), Canada; Alexandria, MN; Oneida County, NY; Oswego County, NY; and Cattaraugus County, NY,[16,22,27,29,32] were eliminated, since they are modular plants, leaving the facilities in column E of Table 1.

Also although the Quebec City MSW incinerator, located near Quebec City, Quebec, Canada, is a four-unit, massburn, refractory-lined facility with an overall rated capacity of 1000 tons/d, it was not used to determine the dioxin/furan emission factors. These data[17] were not used for the following reasons:

- The results were based on a pilot study to determine the relative collection efficiency of a dry/dry scrubber system at various flue gas temperatures and a wet/dry scrubber system that recycled part of the lime used to reduce acid gases.
- Only a slipstream from incinerator number three flue gas was used. The flue gas that entered the air pollution control system was estimated to be equivalent to that of the gases generated by combusting approximately 23 tons/d maximum. Because of this estimated throughput, the Quebec City combustor was considered to be more similar to a modular facility.

Although the Pittsfield, MA, facility is also a modular facility, it was retained because it is a large-capacity, excess air facility.

B. CALCULATION OF DIOXIN/FURAN EMISSION FACTORS

Since the PCDD emissions listed in column E of Table 1 represented a variety of units and operating conditions, the data were converted to standard units and conditions. PCDD and PCDF emissions were converted to nanograms per normal cubic meter (ng/Nm^3) on a dry basis and corrected to 12% carbon dioxide (CO_2). To convert the emissions data to these units required correcting for temperature, moisture, and excess air (%CO_2) content in the flue gas, as well as converting to metric units. These conversions are done automatically by WESTON's computerized resource recovery database.

Tables 2 and 3 summarize standardized average emission factors for the tetra- through octachlorinated homologs of dioxins and furans from facilities with and without acid gas control, respectively. The term homolog of dioxin or furan refers to a group of structurally related compounds that have the same number of chlorine atoms attached to the benzene rings. For example, tetrachlorinated compounds have four chlorines, while octachlorinated compounds have eight. Isomers are chemical compounds that belong to the same homolog. They have the same number of chlorine and other atoms but differ in arrangement of those atoms and, hence, in some of their chemical properties. For example, 2,3,7,8-T_4CDD and 1,4,6,9-T_4CDD are isomers of the tetrachlorinated dioxin (T_4CDD) homolog. Congeners are members of the same general class of compounds. For example, 2,3,7,8-T_4CDD and 1,2,3,4,7,8-H_6CDD are both congeners of dioxin.[57]

Only emissions data for the selected massburn, excess air facilities were considered. In addition, emission rates expressed in TEFs calculated according to the U.S. EPA,[57] Eadon,[73] California,[74] Ontario,[75] and New York[68] weighting schemes are presented for each facility. Toxic equivalency weighting schemes express emissions of PCDD/PCDF homologs in terms of their assumed toxicity relative to the known toxicity of the most toxic (and most researched) congener, 2,3,7,8-T_4CDD.[57]

If a particular facility test report did not include data for a given congener, emissions data from other facilities that did measure that congener were used to estimate the toxic equivalent emission factors for facilities without measured data. The method used to estimate emissions of congeners that were not measured or reported at a given facility was to ratio emissions of the various congeners for which measured data were available. T_4CDD emissions are almost always measured, which allowed the ratio of each other homolog to the T_4CDD homolog to be calculated for each facility for which data was reported. Then an overall expected ratio of the other homologs to T_4CDD was calculated by computing the average of the ratios developed for individual facilities. Similarly, since the 2,3,7,8-isomer is presumed to be the most toxic isomer of each homolog, the ratio of 2,3,7,8-isomer to total homolog emissions were quantified for each facility for which the appropriate emissions data were reported. These ratios were averaged (using the geometric mean) to obtain the expected ratio of 2,3,7,8-to-homolog emissions. Thus, emissions of a 2,3,7,8-isomer could be estimated from measured emissions of the entire homolog, or, if homolog emissions were not reported, both homolog emissions and, subsequently, the 2,3,7,8 congener emissions, could be estimated from measured emissions of T_4CDD.

Thus, using measured data and estimated values where needed, total PCDD and PCDF emissions for the tetra- through octa- homologs were calculated for each of the selected facilities. Toxic equivalent emissions were also calculated for each facility. These values are presented in Tables 2 and 3 for facilities with and without acid gas control, respectively. The data presented in Tables 2 and 3 allow expected emissions of PCDDs and PCDFs for any massburn excess air facility to be calculated.

Tables 4 and 5 present the geometric mean emission rate and upper and lower 95% confidence

TABLE 2
Comparison of Dioxin and Furan Emissions from Selected Facilities with Acid Gas Controls
($\mu g/Nm^3$, dry @12% CO_2)

Species	Wurzburg (35)	Hogdalen (39)	Marion (40)	Commerce (10)	Stellinger Moor (8)	Stapelfeld (8)	Borsigstrasse (8)	Bristol (37)	Stanislaus (68)	Millbury (67)	Bridgeport (69)	Jackson Co (66)
2,3,7,8-T4CDD	0.018	0.004	0.022	0.011	0.740	0.140	0.340	0.020	0.010	0.061	0.317	0.020
Other T4CDD	1.892	0.247	0.209	0.056	20.860	8.490	41.680	0.328	0.438	0.899	0.046	0.062
2,3,7,8-P5CDD	0.200	0.040	0.011	0.009	1.890 *a*	0.755 *a*	3.677 *a*	0.013	0.009	0.101	0.009	0.028
Other P5CDD	2.340	0.396	0.074	0.140	20.444 *b*	8.168 *b*	39.771 *b*	0.392	0.085	0.853	0.084	0.103
2,3,7,8-H6CDD	0.390	0.449	0.014	0.020	6.620 *a*	2.645 *a*	12.878 *a*	0.110	0.057	0.596	0.039	0.111
Other H6CDD	2.790	0.712	0.078	0.073	26.432 *b*	10.561 *b*	51.421 *b*	0.958	0.158	1.330	0.016	0.170
2,3,7,8-H7CDD	2.200	0.204 *a*	0.108	0.037	17.593 *a*	7.029 *a*	34.225 *a*	0.474	0.088	0.950	0.075	0.175
Other H7CDD	2.100	0.175 *b*	0.092	0.018	15.097 *b*	6.032 *b*	29.369 *b*	0.578	0.097	0.789	0.024	0.077
Total O8CDD	10.170	0.501 *b*	0.841	0.190	16.480	16.480	21.170	1.219	0.080	4.150	0.401	0.214
Total PCDD	22.100	2.727	1.449	0.554	126.157	60.300	234.532	4.092	1.019	9.729	1.013	0.959
2,3,7,8-T4CDF	0.250	0.053	0.234	0.045	4.570	1.780	5.140	0.791	0.149	0.526	0.006	0.063
Other T4CDF	9.350	0.910	0.249	0.911	139.960	50.560	106.280	4.397	3.252	12.600	0.038	0.163
2,3,7,8-P5CDF	1.460	0.106	0.036	0.100	15.580 *a*	6.225 *a*	30.308 *a*	0.482	0.275	1.042	0.031	0.038
Other P5CDF	7.800	0.923	0.017	0.211	53.957 *b*	21.558 *b*	104.966 *b*	2.167	1.694	5.970	0.061	0.088
2,3,7,8-H6CDF	1.610	0.277	0.003	0.071	18.269 *a*	7.299 *a*	35.541 *a*	0.688	0.291	2.561	0.131	0.075
Other H6CDF	4.430	0.923	0.000	0.002	20.522 *b*	8.199 *b*	39.923 *b*	0.751	0.034	3.540	0.003	0.028
2,3,7,8-H7CDF	1.770	0.260 *a*	0.005	0.037	22.404 *a*	8.951 *a*	43.584 *a*	0.638	0.189	2.083	0.312	0.117
Other H7CDF	0.310	0.057 *b*	0.000	0.000	4.940 *b*	1.974 *b*	9.610 *b*	0.258	0.000	1.150	0.001	0.037
Total O8CDF	0.880	0.112 *b*	0.050	0.016	2.270	2.980	4.830	0.192	0.019	2.550	1.475	0.159
Total PCDF	27.860	3.621	0.594	1.393	282.472	109.526	380.183	10.364	5.902	32.022	2.058	0.768
c Total PCDD & PCDF	49.950	NA	2.043	1.944	496.730 *d*	219.000 *d*	532.690 *d*	14.456	6.914	43.443	6.142	3.453
Toxic Equivalent:												
EPA	0.374	0.067	0.058	0.034	4.705	1.716	7.524	0.178	0.071	0.354	0.330	0.050
Eadon	0.810	0.112	0.123	0.069	9.661	3.689	16.457	0.463	0.162	0.723	0.341	0.085
New York	0.909	0.125	0.126	0.075	10.706	4.092	18.003	0.492	0.183	0.800	0.343	0.087
California	2.107	0.238	0.307	0.170	24.726	9.678	43.252	1.363	0.460	1.916	0.379	0.162
Ontario	10.707	1.289	0.312	0.677	118.000	44.289	143.593	4.252	2.765	11.087	0.418	0.252

a *Indicates 2,3,7,8-isomer specific emissions are unknown (not analyzed or given); ratios of 2,3,7,8 isomer to homologue from WESTON's worldwide database were used to estimate 2,3,7,8 isomer values (for facilities w/ & w/o acid gas controls).*

b *Indicates homologue specific emissions are unknown (not analyzed or given); ratios of homologue to T4CDD from WESTON's worldwide database were used to estimate homologue values (for facilities w/ & w/o acid gas controls).*

c *Total PCDD & PCDF are from actual test reports.*

d *Total PCDD & PCDF includes tri through octa homologues.*

TABLE 3
Comparison of Dioxin and Furan Emissions from Selected Facilities without Acid Gas Controls
(ng/Nm^3 @12% CO_2, dry)

Species	Pittsfield (53)	Zurich (34)	Tulsa (41)	Saugus (48)	Chicago (21)	North Andover (47)	Westchester (31)	Pinellas (13)	Alexandria (36)
2,3,7,8-T4CDD	0.585	0.242	0.101	1.860	0.588	1.073	0.279	0.163	0.054
Other T4CDD	1.664	5.809	1.512	32.375	8.610	13.696	2.572	3.377	2.733
2,3,7,8-P5CDD	0.283 a	1.408 a	0.189	3.667	0.805 a	2.119	0.733	0.332	0.129
Other P5CDD	3.066	15.231	2.799	34.415	8.706 b	19.728	2.101	4.608	2.222
2,3,7,8-H6CDD	1.094 a	7.544 a	0.524	5.438	4.705 a	3.226	0.455	2.786	0.661
Other H6CDD	4.369	30.122	4.571	31.696	18.786	15.034	3.476	7.914	2.286
2,3,7,8-H7CDD	2.785 a	19.620 a	2.189	17.328 a	5.891 a	20.951 a	2.994 a	6.310	1.373
Other H7CDD	2.390	16.836	2.229	14.869	5.055	17.979	2.569	6.690	1.286
Total O8CDD	20.106	74.273	4.810	40.031	3.683	28.593	8.844	16.600	2.788
Total PCDD	36.342	171.085	18.924	181.678	56.829	122.399	24.023	48.780	13.532
2,3,7,8-T4CDF	1.070 a	2.752 a	2.905	25.042	10.609 a	21.597	2.159	0.944	3.501
Other T4CDF	12.052	30.981	4.403	169.909	119.454	63.687	27.910	20.456	22.324
2,3,7,8-P5CDF	3.532 a	9.252 a	1.382	17.547	6.635 a	16.500	6.327	2.027	1.706
Other P5CDF	12.233	32.044	1.953	96.196	22.977 b	36.271	11.188	9.773	10.109
2,3,7,8-H6CDF	5.959 a	13.322 a	1.437	22.663	41.465 a	9.524	1.951	5.434	2.069
Other H6CDF	6.694	14.965	0.384	52.266	46.578	6.422	15.731	6.366	0.734
2,3,7,8-H7CDF	16.545 a	15.369 a	1.631	31.832 a	8.856 a	43.249 a	8.677 a	6.519	1.375
Other H7CDF	3.648	3.389	0.719	7.019	1.953	9.536	1.913	1.831	0.587
Total O8CDF	11.544	12.404	0.706	19.032	0.890	44.620	0.386	1.460	0.566
Total PCDF	73.277	134.478	15.520	441.505	259.415	251.406	76.242	54.810	42.971
c Total PCDD & PCDF	109.600	305.563	34.452	623.183	277.122	373.805	100.266	103.590	56.503
Toxic Equivalent:									
EPA	1.368	2.828	0.701	9.226	3.617	6.573	1.622	0.896	0.761
Eadon	2.480	5.971	1.735	19.971	7.639	15.956	3.846	1.613	1.942
New York	2.621	6.416	1.802	21.502	8.265	16.660	4.025	1.800	2.092
California	6.262	15.330	4.750	50.434	20.464	43.597	9.921	4.097	5.554
Ontario	17.446	46.626	6.496	172.256	92.833	76.760	26.703	19.754	19.758

a Indicates 2,3,7,8-isomer specific emissions are unknown (not analyzed or given); ratios of 2,3,7,8 isomer to homologue from WESTON's worldwide database were used to estimate 2,3,7,8 isomer values (for facilities w/ & w/o acid gas controls).

b Indicates homologue specific emissions are unknown (not analyzed or given); ratios of homologue to T4CDD from WESTON's worldwide database were used to estimate homologue values (for facilities w/ & w/o acid gas controls).

c Total PCDD & PCDF are from actual test reports.

d Total PCDD & PCDF includes tri through octa homologues.

TABLE 4
Summary of Expected PCDD and PCDF Emissions for Facilities with Acid Gas Controls[a]
(ng/Nm^3 @ 12% CO_2, dry)

	Pollutant	Geometric Mean	Upper 95% Confidence Interval	Lower 95% Confidence Interval
	2,3,7,8-T4CDD	0.04	0.12	0.02
	T4CDDs	0.89	3.40	0.23
b	PCDDs	6.23	23.51	1.65
	T4CDFs	3.77	18.72	0.76
b	PCDFs	10.97	45.23	2.66
	EPA TEF	0.28	0.88	0.09
	Eadon TEF	0.55	1.78	0.17
	New York TEF	0.59	1.94	0.18
	California TEF	1.28	4.40	0.37
	Ontario TEF	4.20	17.58	1.00

a Assuming a log normal distribution.
b Tetra through octa homologues only.

TABLE 5
Summary of Expected PCDD and PCDF Emissions for Facilities without Acid Gas Controls[a]
(ng/Nm^3 @12% CO_2, dry)

	Pollutant	Geometric Mean	Upper 95% Confidence Interval	Lower 95% Confidence Interval
	2,3,7,8-T4CDD	0.33	0.78	0.14
	T4CDDs	3.84	9.69	1.53
b	PCDDs	51.06	106.31	24.52
	T4CDFs	27.66	60.93	12.56
b	PCDFs	97.43	217.79	43.58
	EPA TEF	2.06	4.24	1.00
	Eadon TEF	4.50	9.34	2.17
	New York TEF	4.80	9.95	2.32
	California TEF	11.86	24.51	5.74
	Ontario TEF	34.41	75.05	15.77

a Assuming a log normal distribution.
b Tetra through octa homologues only.

intervals (assuming a log normal distribution) for facilities with and without acid gas control, respectively. Emission rates are presented for the following five groups of PCDDs and PCDFs:

- The 2,3,7,8 congener of the tetrachlorinated dibenzo-*p*-dioxin (2,3,7,8-T_4CDD), which has four chlorine atoms in the 2,3,7, and 8 positions on the benzene rings.

- Total tetrachlorinated dibenzo-*p*-dioxins (TCDDs), consisting of all dioxin congeners with four chlorine atoms attached to the benzene rings.
- Total PCDDs (tetra- through octa- homolog), or all dioxin congeners with four to eight chlorine atoms.
- Total tetrachlorinated dibenozofurans (TCDFs), or all furan congeners with four chlorine atoms on the benzene rings.
- Total PCDFs (tetra- through octa- homologs), or all furan congeners with four to eight chlorine atoms attached to the benzene rings.

In addition, emission rates are given for four toxic equivalent measures. The 95% confidence intervals of the mean are the best statistical predictor of the range within which there is 95% confidence that the true mean emission rate of the sample facilities would be expected to fall and the range within which 95% of the results of new tests done at similar facilities would be expected to occur.

Due to the small sample sizes involved, extreme high or low values may significantly influence results and may lie outside the 95% confidence intervals. The distance outside the confidence intervals that an extreme value may lie can appear especially large when the confidence intervals and extreme values are reported as untransformed numbers (such as in Tables 2, 3, 4, and 5) rather than as log transformed values (such as in Figure 1). Nevertheless, confidence intervals based on a log-normal distribution are the statistically correct measure of variability. Assuming a normal distribution generally results in higher mean emission factors as well as broader confidence intervals. Such an assumption would also bias (overpredict) exposure estimates. For risk assessment purposes, the geometric mean value should represent expected emissions from a facility over the long-term. However, when setting permit limitations for new facilities, the upper 95% confidence interval is more representative of emissions that can be attained during any given emission test.

C. STATISTICAL ANALYSES

Statistical analyses were performed to determine the effect of facility design and operation on emissions of dioxins and furans. Since the current set of massburn (excess air) facility emission test data is small, some results are inconclusive but do give indications of possible relationships. Table 6 presents the U.S. EPA toxic equivalent emission rates for selected facilities along with several combustion process and air pollution control parameters, including capacity, flue gas temperature, and the pollution control systems used. Preliminary analyses of the effects of these parameters and other factors on emissions are briefly outlined in the following sections.

1. Application of Results in Risk Assessments

Emission data are used in risk assessments to estimate inhalation and food chain exposures in which local terrain and topographical features influence the deposition of pollutants onto plant, water, and soil surfaces. Table 7 and Figure 2 show the relationship between plant capacity, local meteorology, and estimated inhalation exposure risk for the maximally exposed individual. These data show that except in areas with unique topographical conditions, such as the Rutland, VT, site, the expected upper-limit individual risk of developing cancer from exposure to dioxins and furans via inhalation of emissions from resource recovery facilities is less than one chance per million.

2. Type of Acid Gas Control Equipment Used

The effect of installing an acid gas control system (i.e., a scrubber) on dioxin and furan emissions remains speculative. There are still too few data available to conclusively determine the correlation between pollution control equipment used and dioxin/furan emissions. The data

TABLE 6
Facilities Included in the Development of Dioxin and Furan Emission Factors

Facility (Ref. No.)	Start-up Date/ Test Date	(No. of units) /Total Capacity (tons/day)	Stack Temperature (deg. F)	Type of Boiler/ Grate Systems	Pollution Control Systems	Front-half Particulate Loading (gr/dscf @12% CO2)	U.S.EPA Toxic Equivalence Emissions of Dioxin & Furans [a] (ng/Nm^3 @12% CO2,dry)
b Commerce (10)	1987/1988	(1) / 380	270	Foster-Wheeler water-walled/Detroit Stoker	SNCR, spray dryer & fabric filter	0.005	0.034
Jackson, Co., Michigan (66)	1988/1988	(2) / 200	289		spray dryer & fabric filter	0.001	0.050
c,d Stockholm-Hogdalen (39)	1986/1986	(1) / 397	343	Martin water-walled	dry injection & fabric filter	0.001	0.067
Marion (40)	1986/1986	(2) / 550	257	Martin water-walled	spray dryer & fabric filter	0.011	0.058
b Stanislaus (68)	1988/1988	(2) / 800	288	Martin water-walled	SNCR, spray dryer & fabric filter	0.004	0.071
Bristol (37)	1986/1988	(2) / 650	284	Martin water-walled	spray dryer & fabric filter	0.010	0.178
Bridgeport (69)	1988/1988	(3) / 2250			spray dryer & fabric filter	NA	0.330
Millbury (67)	1988/1988	(2) / 1500	275	Foster-Wheeler water-walled/Detroit Stoker	spray dryer & ESP	0.009	0.354
Wurzburg (35)	1984/1985	(2) / 660	365	Martin water-walled	dry injection & fabric filter	0.004	0.374
Tulsa County (41)	1986/1986	(2) / 750	355	Martin water-walled	ESP	0.009	0.701
Alexandria/ Arlington (36)	1987/1987	(3) / 975	342	Martin water-walled	ESP	0.027	0.761

a U.S. EPA and other toxic equivalency weighting schemes weight the emissions of the various PCDD and PCDF congeners by their assumed toxicity relative to 2,3,7,8-T4CDD.
b Selective Non-Catalytic Reduction (SNCR) is used to reduce NOx emissions.
c Some 2,3,7,8-isomer specific emissions are unknown (not analyzed or not given). Ratios of the 2,3,7,8 isomer to total homologue emissions from WESTON's worldwide database were used to estimate these 2,3,7,8 isomer values.
d Some homologue specific emissions are unknown (not analyzed or not given). Ratios of the homologue to T4CDD emissions from WESTON's worldwide database were used to estimate these homologue values.

TABLE 6 (continued)
Facilities Included in the Development of Dioxin and Furan Emission Factors

	Facility (Ref. No.)	Start-up Date/ Test Date	(No. of units) /Total Capacity (tons/day)	Stack Temperature (deg. F)	Type of Boiler/ Grate Systems	Pollution Control Systems	Front-half Particulate Loading (gr/dscf @12% CO2)	U.S.EPA Toxic Equivalence Emissions of Dioxin & Furans a (ng/Nm^3 @12% CO2,dry)
	Pinellas (13)	1983/1987	(3) /3,000	514	Martin water-walled	ESP	0.002	0.896
c	Pittsfield (53)	1981/1986	(2) / 240	407	Vicon	electrified granular bed filter	0.052	1.368
c	Westchester (31)	1985/1985	(3) /2,250	460	Foster-Wheeler water-walled/Von-Roll	ESP	0.043	1.622
c,d	Stapelfeld (8)	1979/1984	(2) /1,110	NA	Steinmuller	ESP & wet scrubber	0.005	1.716
c	Zurich (34)	1978/1981	(2) / 563	475	Martin water-walled	ESP	NA	2.828
c,d	Chicago, N.W. (21)	1970/1980	(4) /1,600	449	Martin water-walled	ESP	NA	3.617
c,d	Stellinger Moor (8)	1972/1984	(?) / 172/unit	NA	Martin water-walled	dry scrubber & ESP	NA	4.705
c,d	Borsigstrasse (8)	1967/1984	(?) / 423/unit	NA	Martin water-walled	dry scrubber & ESP	NA	7.524
c	North Andover (47)	1985/1986	(2) /1,500	570	Martin water-walled	ESP	0.010	6.573
c	Saugus (48)	1975/1986	(2) /1,500	521	water-walled/ Von-Roll	ESP	0.015	9.226

a U.S. EPA and other toxic equivalency weighting schemes weight the emissions of the various PCDD and PCDF congeners by their assumed toxicity relative to 2,3,7,8-T4CDD.
b Selective Non-Catalytic Reduction (SNCR) is used to reduce NOx emissions.
c Some 2,3,7,8-isomer specific emissions are unknown (not analyzed or not given). Ratios of the 2,3,7,8 isomer to total homologue emissions from WESTON's worldwide database were used to estimate these 2,3,7,8 isomer values.
d Some homologue specific emissions are unknown (not analyzed or not given). Ratios of the homologue to T4CDD emissions from WESTON's worldwide database were used to estimate these homologue values.

TABLE 7
Relationship between Plant Capacity, Meteorology, and Estimated Modeled Maximum Individual Inhalation Exposure Risks for Resource Recovery Plants[79]

Plant	Plant Capacity (tons/day)	Status[a]	Annual Dilution Factor (X/Q) ((ug/m^3)/(gm/sec))	Inhalation Exposure Risk[b] (chances/million)
Philadelphia, PA	2250	UC	0.044	0.246
Delaware Co., PA	2688	PR	0.033	0.220
Montgomery Co., PA	1200	PR	0.035	0.104
Montgomery Co., MD	1800	UC	0.032	0.143
Pennsauken, NJ	500	PR	0.031	0.038
Camden, NJ	1400	PR	0.012	0.042
Gloucester Co., NJ	575	PR	0.068	0.097
Pigeon Point, DE	810	PR	0.151	0.304
Johnston, RI	750	UC	0.055	0.102
Detroit, MI	3000	C	0.049	0.365
Portland, ME	400	C	0.230	0.229
Los Angeles, CA	1600	PR	0.130	0.517
Rutland, VT	300	C	2.000	1.490
Ocean City, NJ	400	UC	0.088	0.087
Hudson Falls, NY	400	PRC	0.280	0.278
Dutchess Co., NY	400	PRC	0.095	0.094

[a] UC = Under Consideration
PR = Permit Review
PRC = Permit Review Completed
C = Under Construction

[b] Using the upper 95% confidence interval U.S. EPA toxic equivalent emission factor for all mass burn facilities with and without acid gas controls.

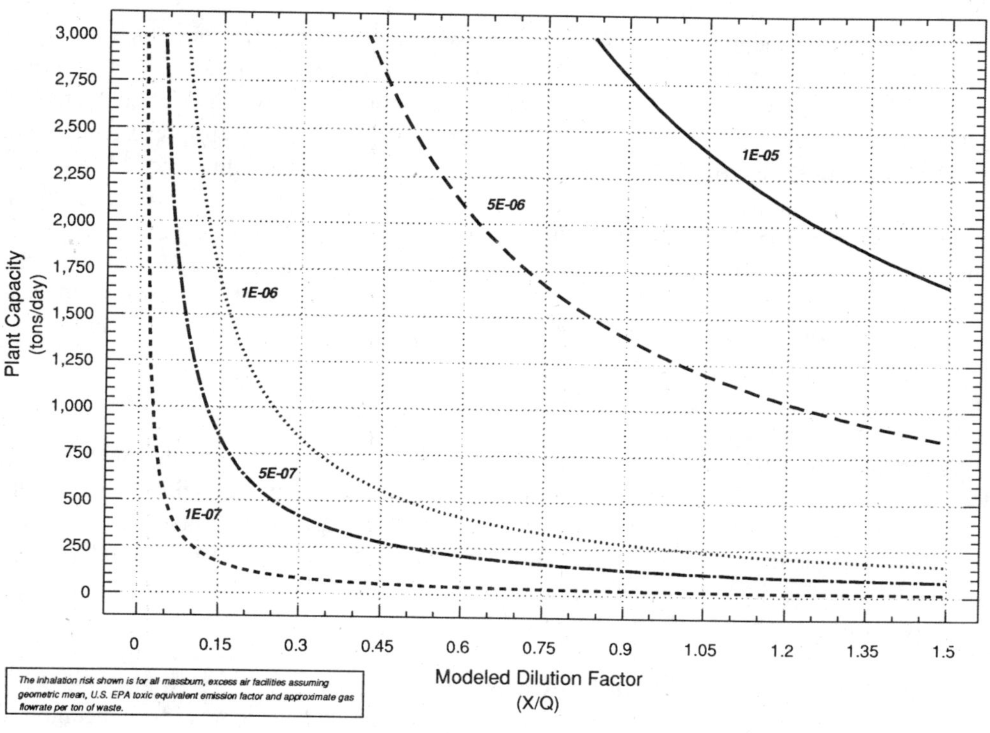

FIGURE 2. Individual inhalation risk vs. plant capacity and atmospheric dilution factors.[79]

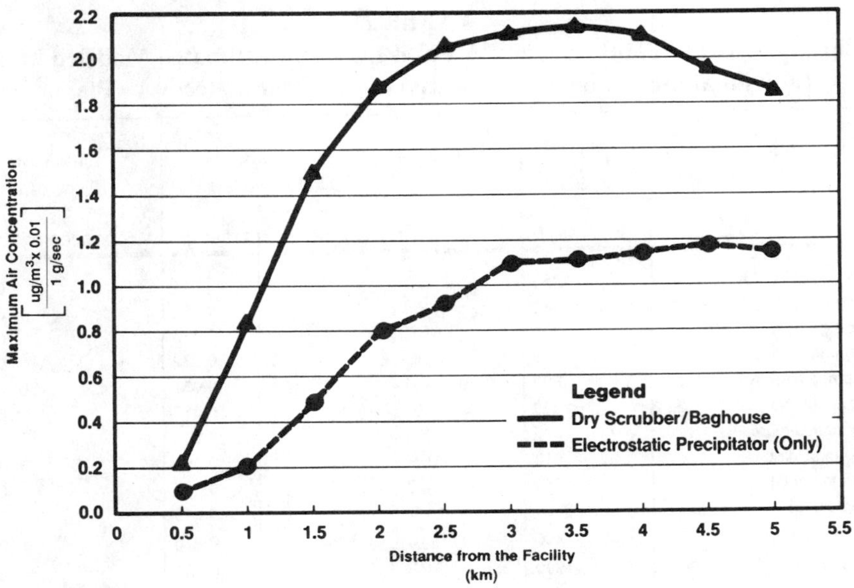

FIGURE 3. Maximum predicted normalized annual air concentrations (for a unit emission rate of 1 gram per second) at sites downward from a hypothetical resource recovery facility as a function of the air pollution control system used (stack height = 200 ft).[79]

presented in Figure 1 and Tables 4 and 5, however, indicate that geometric mean PCDD/PCDF emissions (expressed in TEFs) are 86% lower for facilities with scrubbers. On the other hand, air concentrations of PCDDs and PCDFs (and subsequent health impacts) may not be reduced by scrubbing since stack temperatures decrease when scrubbers are added. Exit gas temperatures for facilities with a scrubber/baghouse system vs. facilities with an ESP alone vary by as much as 200°F.

Figure 3 depicts the maximum annual ambient air concentration for a unit emission rate at sites downwind of a hypothetical 1400 ton/d resource recovery facility located in suburban Washington, D.C. The scrubber/baghouse facility was assumed to have an exit temperature of 220°F, exit velocity of 55 ft/s, and a stack diameter of 11 ft. The ESP facility was assumed to have an exit temperature of 400°F, exit velocity of 58 ft/s, and a stack diameter of 12 ft. Figure 3 shows that maximum predicted air concentrations for the same emission rate would be almost double for the scrubbed facility. Thus, individual inhalation risks resulting exposure to dioxins and furans emitted from resource recovery facilities could be reduced by about 76% with the application of dry scrubbers. Deciding whether such a reduction in risks (which are already less than 1 in 1 million in most cases) justifies the added cost of installing scrubbers on all modern facilities is beyond the scope of this chapter.

3. Variation in Test Results

An indication of the sampling, analytical, and operational errors associated with facility testing that might affect dioxin and furan emissions can be obtained by evaluating the variability in individual tests done at the same facility. Table 8 shows the variation in emissions (expressed in TEFs) for 11 facilities at which more than one test was performed. Emissions varied between tests at the same facility by as much as a factor of 3.8. The ratio of the standard deviation for all tests to the arithmetic mean for tests done at a given facility ranges from about 2% to nearly 60%.

4. Estimation of Toxic Equivalent Emissions

Because of the high cost of isomer-specific emissions testing, it would save time and resources if toxic equivalent emissions could be estimated from only a few congeners or

TABLE 8

Variation in Dioxin/Furan Emission Test Results[79]

(ng/Nm^3 @ 12% CO_2, dry)

Plant Tested (Reference No.)	Dioxin/Furan Test Results (U.S. EPA Toxic Equivalent)						High/ Low Ratio	Average	Standard Deviation	Ratio of Standard Deviation to Average
	#1	#2	#3	#4	#5	#6				
Pittsfield (53)	1.069	0.897	2.297	---	---	---	2.561	1.421	0.763	0.537
Saugus (48)	6.902	6.125	9.650	13.455	10.879	8.355	2.197	9.228	2.705	0.293
North Andover (47)	8.354	9.377	6.211	3.788	5.148		2.475	6.576	2.289	0.348
Westchester (31)	1.477	1.231	2.160	---	---	---	1.755	1.623	0.481	0.297
Stapelfeld (8)	1.652	2.360	2.068	---	---	---	1.429	2.027	0.356	0.176
Wurzburg (35)	0.671	0.277	0.178	---	---	---	3.770	0.375	0.261	0.695
Borsigstrasse (8)	6.083	6.311	---	---	---	---	1.037	6.197	0.161	0.026
Stellinger Moor (8)	5.479	5.911	---	---	---	---	1.079	5.695	0.305	0.[illegible]4
Marion County (40)	---	0.058	0.156	---	---	---	2.690	0.107	0.069	0.[illegible]48
Chicago, N.W. (21)	4.413	3.429	3.692	---	---	---	1.287	3.845	0.509	[illegible].133
Tulsa County (41)	0.760	0.646	0.699	---	---	---	1.176	0.702	0.057	0.081

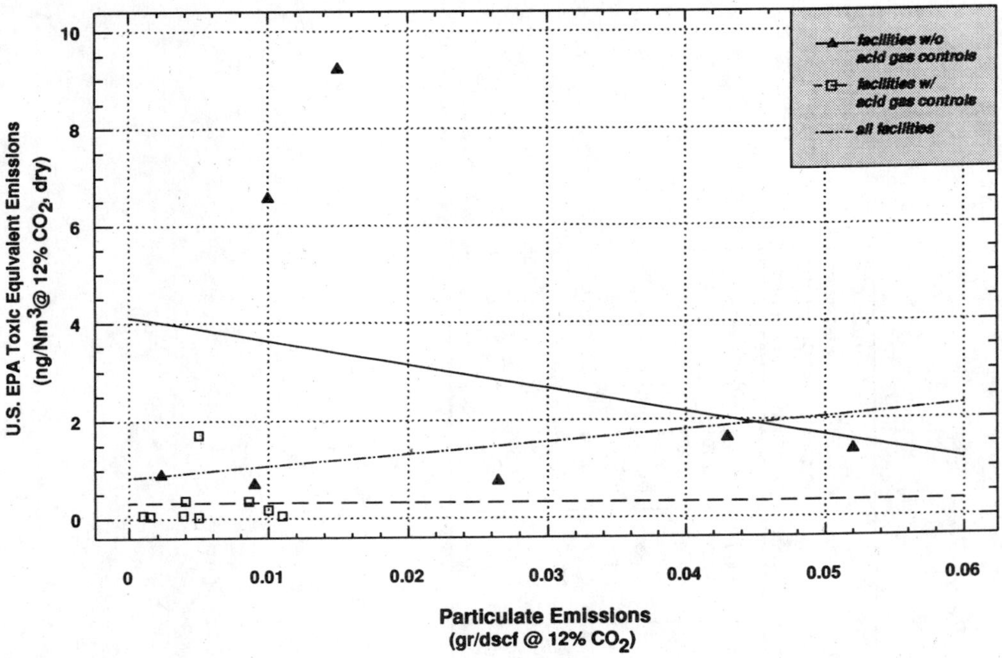

FIGURE 4. Relationship between particulate and dioxin/furan emissions (expressed in toxic equivalents).

homologs. Multiple regressions of the 1988 database were used to assess such approximations.[63] When all heat recovery facilities were considered (i.e., those data in column B of Table 1), preliminary results appeared encouraging. Regression coefficients of over 90% were obtained when total PCDD/PCDF emissions were estimated from as few as two (2,3,7,8) isomers of two homologs. When selected data for massburn excess air facilities were examined (i.e., those data in column E of Table 1), however, it became apparent that the use of such approximations is premature. Without emission data from other types of facilities, regression coefficients decreased to below 70% when emissions were estimated from four congeners or homologs. More significantly, the regressions often predicted decreasing toxic equivalent emissions with increasing emissions of a particular congener or homolog.

5. Correlation between Particulate and Dioxin/Furan Emissions

The correlation between dioxin and furan emissions and particulate emissions is presented in Figure 4. Although the data are too limited to draw any definitive conclusions, and the relationship does not appear to be linear (even with the logarithm of the TEF), a general trend toward a positive correlation between particulate and dioxin/furan emissions can be inferred.

Furthermore, the data presented in Table 6 do not support the argument that baghouses are always superior to ESPs in terms of particulate emissions. For example, consider the comparison between the Tulsa and Marion County facilities, which are both Ogden Martin facilities. Tulsa is controlled with an ESP and has an outlet particulate loading of 0.009 grains per dry standard cubic feet (gr/dscf) at 12% CO_2, where 7000 grains equal 1 lb, while Marion County, controlled with a dry scrubber/baghouse, has an outlet loading of 0.011 gr/dscf (Table 6).

6. Facility Age

Facility age can affect emissions, since improved combustion processes and pollution control technologies continue to be developed over time, and existing combustion techniques and control equipment can deteriorate with age. The WESTON database includes only those facilities that are less than 20 years old. Figure 5 and Table 6 reflect the effect of increasing plant

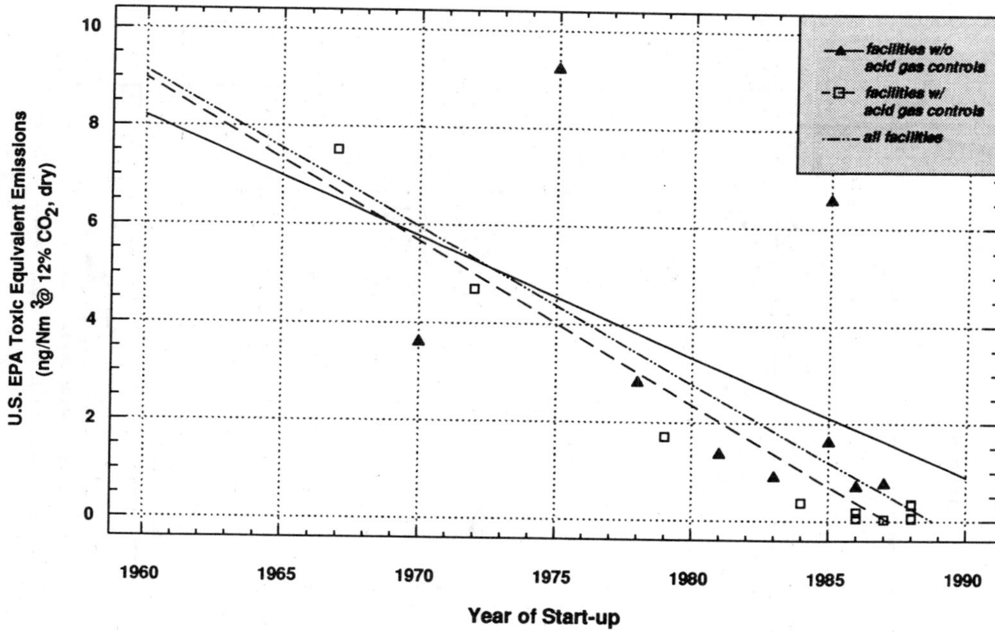

FIGURE 5. Relationship between plant age and dioxin/furan emissions (expressed in toxic equivalents).

age on PCDD/PCDF emissions (expressed in TEFs). While emissions from newer facilities are generally lower than emissions from older facilities, there are some noticeable exceptions to this trend. In particular, the North Andover facility, which became operational in 1985, exhibits the third highest emission rate, exceeding facilities that started operating in 1970. The potential increase in emissions due to facility aging must be considered in setting long-term permit limits. The use of lower-bound emissions values may not be prudent without considering some form of deterioration factor. However, because data measured at the same facility at several points in time are extremely limited, the effect of potential deterioration cannot be ascertained.

7. Multiple Regression

Since no single variable appears to correlate well with dioxin/furan emissions, a stepwise multiple linear regression analysis was conducted on the 1988 database to determine the likelihood that some combination of the control equipment and operating parameters listed in Table 6 influence the magnitude of PCDDs and PCDFs (expressed in TEFs) emitted from resource recovery facilities.[63] The various models tested are summarized in Table 9. If values had been available for all of the variables at all facilities tested, then only the first four models wold have been analyzed. Because data were not available for all variables at all facilities, other models were used that considered only variables for which all or most values were available. The step-wise regression routine selects the significant contributors to estimated dioxin and furan emissions, assuming one of the models listed in Table 9. For data of this nature, a regression coefficient (R^2) of 0.50 or greater is generally considered acceptable. A number of the models have regression coefficients (R^2) or 0.75 to 0.85

In general, models in which the logarithm of dioxin/furan emissions were used as the dependent variable yielded better fits than models that used untransformed emissions data, which was expected, since emissions data have been shown to be log-normally distributed (Figure 1). The parameter that accounted for most of the variability in PCDD/PCDF emissions was outlet temperature. Dioxin/furan emissions were directly proportional to outlet temperature (i.e., as outlet temperature decreased, emissions also decreased) (Figure 6). Facility age also significantly contributed to the variability in PCDD/PCDF emissions, with newer facilities

TABLE 9
Results of Stepwise Multiple Regression Analysis of Dioxin/Furan Emissions[69]

Model Number	No. of Facilities Considered [a]	Y U.S. EPA TEF	X_1 Start-up Year	X_2 No. of Units	X_3 Total Capacity	X_4 Particulate Loading [b]	X_5 Particulate Control (filter vs ESP)	X_6, X_{70} Acid Gas Controls (Yes/No)	X_{71} Wet Scrubber (Yes/No)	X_{72} Unspecified Dry Scrubber	X_{73} Spray Dryer (Yes/No)	X_{74} Dry Injection (Yes/No)	X_8 Outlet Temperature	Regression Coefficient (R^2)
1	9	LIN [c]	(LIN) [d]	(LIN)	LIN	LIN	LIN	LIN					[LIN] [e]	0.831
2	9	LOG [f]	LIN	LIN	LIN	LIN	LIN	LIN					[LIN]	0.791
3	9	LIN	(LIN)	(LIN)	LIN	LOG	LIN	LIN					[LIN]	0.831
4	9	LOG	LIN	LIN	LIN	(LOG)	LIN	LIN					[LIN]	0.865
5	15	LIN	LIN				[LIN]	LIN						0.347
6	15	LOG	(LIN)				[LIN]	(LIN)						0.782
7	12	LIN	LIN				LIN	LIN					[LIN]	0.541
8	12	LOG	LIN				LIN	(LIN)					[LIN]	0.842
9	15	LIN							LIN	LIN	(LIN)	(LIN)		N/A
10	15	LIN	LIN				[LIN]	LIN	LIN	LIN	LIN	LIN		0.340
11	15	LOG	(LIN)				LIN	LIN	LIN	LIN	[LIN]	(LIN)		0.830 [g]
12	11	LIN	[LIN]			LOG	LIN	LIN	LIN	LIN	LIN	LIN		0.420
13	11	LOG	[LIN]			LOG	LIN	(LIN)	LIN	LIN	LIN	LIN		0.738 [h]

[a] 15 total facilities. For any given model, only facilities with values for all variables considered are included.
[b] Front-half only.
[c] LIN = (Linear) Variable Considered
[d] ◯ = Significant Independent Variable
[e] ☐ = Independent Variable Responsible for Most of Prediction
[f] LOG = (Logarithm) Variable Considered
[g] R^2 was 0.837 when also included X_5, but X_5 removed as it did not meet criteria to be retained in model although was first variable entered.
[h] Same as footnote g, except R^2 was 0.818 with X_5 included.

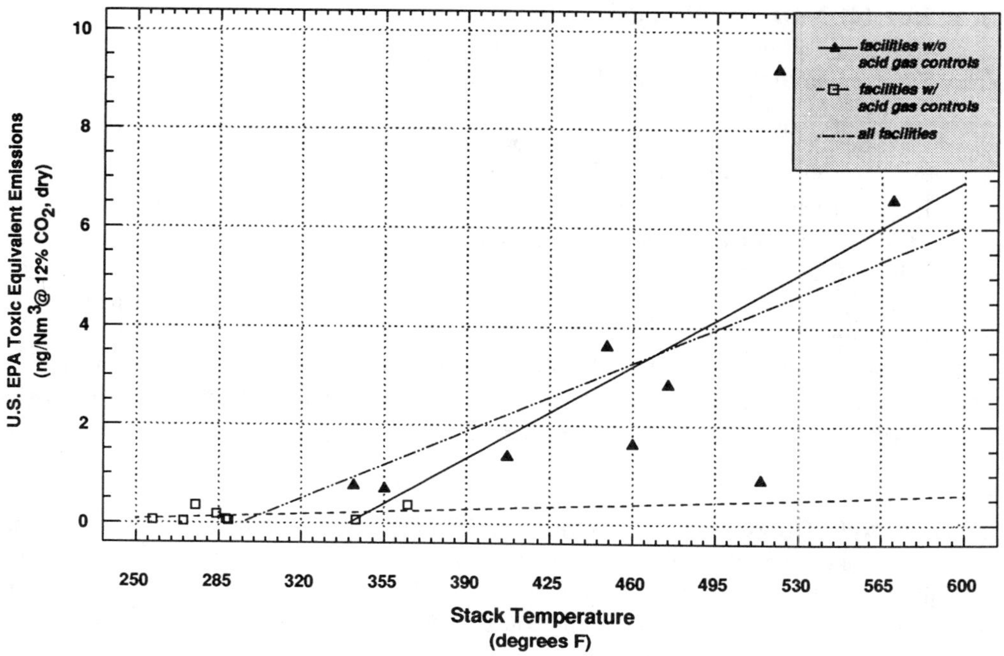

FIGURE 6. Relation between stack temperature and dioxin/furan emissions (expressed in toxic equivalents).

tending to have lower emissions than older facilities. In many cases, the type of pollution control equipment used substantially affected PCDD/PCDF emissions, with fabric filters and acid gas controls, particularly spray dryers and dry injection systems, correlating with lower emissions. The type of control equipment used was more often significant when temperature was not considered in a model. Because outlet temperature, type of particulate and acid gas controls, and, to a lesser degree, facility age, appear to be interdependent, it is impossible to clearly ascertain which combination of parameters is the crucial factor affecting reduced dioxin/furan emissions.

IV. ANALYSIS OF TRACE METALS EMISSIONS

Controlled emission factors (i.e., emission rates that reflect the use of some type of air pollution control system) for trace metals are presented in pounds of metal emitted per ton of MSW combusted. Because metals are present in MSW and are not destroyed during combustion, they are consistently emitted from resource recovery facilities. The amount emitted, however, is a function of the quantity of metal in the MSW, the physical and chemical properties of the metal, the combustion conditions within the facility, and the characteristics and effectiveness of the air pollution control equipment installed.

Most metals vaporize during combustion. As exhaust gases cool (during heat recovery, in ducting, and in the air pollution control devices), some metals condense. Temperatures within the facility in conjunction with the availability of reaction components determine what metallic compounds will be formed. Condensed (vaporized) metals usually adsorb to particle surfaces in the gas stream. Because adsorption rates are dependent on particle surface area (rather than volume or mass), and because smaller particles have more surface area per volume than larger particles, theoretically, more metals will adsorb onto smaller (less easily controlled) particles, a process called "fine particle enrichment". Only a few metals, however, clearly exhibit this tendency. Mercury is an exception, because it exists almost completely as a vapor at typical stack exit temperatures.

A. SELECTION OF FACILITIES

Five criteria were used to select facilities included in the final data base for all metals. These criteria are similar to the criteria used for estimating PCDD/PCDF emissions and include:

- Facilities that experienced abnormal conditions during testing were eliminated, leaving those in column A of Table 10.
- Facilities that do not recover heat were eliminated, leaving those in column B of Table 10.
- Facilities that burn RDF instead of mass burning MSW were eliminated, leaving those in column C of Table 10.
- Facilities that have a furnace capacity of less than 50 tons/d were eliminated, leaving those in column D of Table 10.
- Facilities that have no air pollution controls were eliminated and those with only particulate controls were segregated, leaving those in column E of Table 10. (Those facilities with a fabric filter rather than a high-efficiency electrostatic precipitator are noted in column F.)

Similar to PCDD/PCDF emissions, metals emissions are affected by the efficiency of the combustion and type and effectiveness of the pollution control equipment used. Since the available test data represent the use of diverse combustion processes and control equipment, only those facilities with nonmodular, massburn technology were considered. Acceptable test data were available from ten facilities that use massburn (nonmodular, excess air) technology with acid gas control with fabric filters of ESPs (Table 10). The ten facilities were Wurzburg, West Germany; Munich, West Germany, Marion County, OR; Stockholm, Sweden; Commerce, CA; Bristol, CT; Stanislaus, CA; Millbury, MA; Bridgeport, CT; and Jackson County, MI.[5,10,35,37,39,40,66-69]

Test data were also considered for six similar facilities without acid gas control (Table 10) to determine the effect of acid gas control on metals emissions. These six facilities include: Pinellas County, FL; Hillsborough County, FL; Baltimore, MD; Tulsa County, OK; North Andover, MA; and Dayton, OH.[11,13,20,38,41,71]

Individual test data for all 16 facilities were converted to common units (pounds of metal emitted per ton of MSW burned). Facility emission factors were calculated by averaging results for each individual test run at a given facility. The best estimate of controlled metals emissions for massburn facilities were developed by assuming a log-normal distribution and computing the geometric mean of the facility values both with acid gas controls in conjunction with high-efficiency particulate controls and with high-efficiency particulate controls only.

Hexavalent chromium (chromium VI) emissions were estimated from controlled total chromium emissions from massburn facilities similar in design assuming that hexavalent chromium is approximately 3.4% of total chromium emitted.[2] This assumption is based on emissions tests conducted by the U.S. EPA at the Marion County facility and is conservatively high, since this value of the hexavalent fraction was selected from the test run with the highest hexavalent to total chromium ratio.

An additional analysis was performed to assess the effect of acid gas controls and combustion temperature on metals emissions. Environment Canada[17] conducted a study on two pilot-scale air pollution control systems at the Quebec City resource recovery facility. One facility had a dry injection fabric filter system which was operated at various temperatures. WESTON performed a regression analysis of the emissions data from that facility, excluding values which were below the detection limit. These data showed that lower emissions tended to be correlated with reduced scrubber temperature for arsenic, cadmium, mercury, and lead but not for chromium, copper, and nickel (Figures 7 and 8). Correlation coefficients were significant for arsenic (50%), cadmium (97%), and mercury (82%) only. The Quebec City test results were not used for metals emissions factor development for three reasons: (1) the testing was done on a

TABLE 10
Facilities Investigated for Metals Emissions Data

	Facility	Ref. No.	A Best Data	B Plants w/ Heat Recovery	C Massburn	D Design Capacity Over 50 TPD/unit	E Acid Gas Controls
	Albany RDF, New York	30	X	X			
	Alexandria, Minnesota	22	X	X	X		
	Alexandria/Arlington, Virgina	36	X	X	X	X	
	Baltimore RESCO, Maryland	20	X	X	X	X	
	Braintree, Massachusetts	19	X				
	Bridgeport, Connecticut	69	X	X	X	X	X
	Bristol, Connecticut	37	X	X	X	X	X
	Cattaraugus, New York	27	X	X	X		
	Chicago, N.W., Illinois	21	X	X	X	X	
	Commerce Refuse-to-Energy, California	10	X	X	X	X	X
	Dayton, Ohio	71	X	X	X	X	X
	Hillsborough, Florida	38	X	X	X	X	
	Jackson Co., Michigan	66	X	X	X	X	X
a	Malmo, Sweden	54	X				
	Marion County, Oregon	40	X	X	X	X	X
	Millbury, Massachusetts	67	X	X	X	X	X
	Munich, West Germany	5	X	X	X	X	X
	Niagara RDF, New York	28	X	X			
	North Andover, Massachusetts	11	X	X	X	X	
b	Nuremberg, West Germany	46	X				
	Oneida, New York	29	X	X	X		
	Orrington, Maine	50	X	X			
	Oswego, New York	31	X	X	X		
	Philadelphia, N.W., Pennsylvania	51	X				
	Pigeon Point, Delaware	52	X	X			
	Pinellas County, Florida	13	X	X	X	X	
	Prince Edward Island, Canada	16	X	X	X		
c	Quebec City, Canada	17	X	X	X		
	Stanislaus, California	68	X	X	X	X	X
	Stockholm - Hogdalen, Sweden	39	X	X	X	X	X
	Tsushima, Japan	6	X				
	Tulsa County, Oklahoma	41	X	X	X	X	
	Wurzburg, West Germany	35	X	X	X	X	X

a The facility can burn refused derived fuel (RDF) and nonprocessed municipal solid waste. Tests were performed on both types of waste. Also the air pollution control technology used was a cyclone, scrubber, ESP and a fabric filter.

b Because of the lack of operating and facility information, the data could not be changed to lb/ton.

c The results are based on an average of six sets of tests (12 total). A pilot study of various control configurations conducted by Environment Canada and Flakt on a gas slipstream of about 6 to 9 percent of the flow from one of four 250 TPD units of a massburn facility (i.e., pilot unit equivalent to 23 TPD, maximum).

pilot-scale, modular unit controlling only a portion of the total gas flow, (2) the quantity of the MSW burned that resulted in the controlled portion of the flow could not be determined, and (3) inlet particulate loadings appeared atypically low (0.004 gr/dscf at 12% CO_2). Outlet loadings were not reported.

Ten facilities with and six facilities without acid gas control were considered separately. These facilities are basically the same ones considered in the U.S. EPA Report to Congress[56] except that new test results were added, and several facilities that the U.S. EPA considered were not included in our study. The Barron, Red Wing, and Tuscaloosa plants are modular facilities,[56] which are quite different from massburn, excess air facilities. These facilities were not included

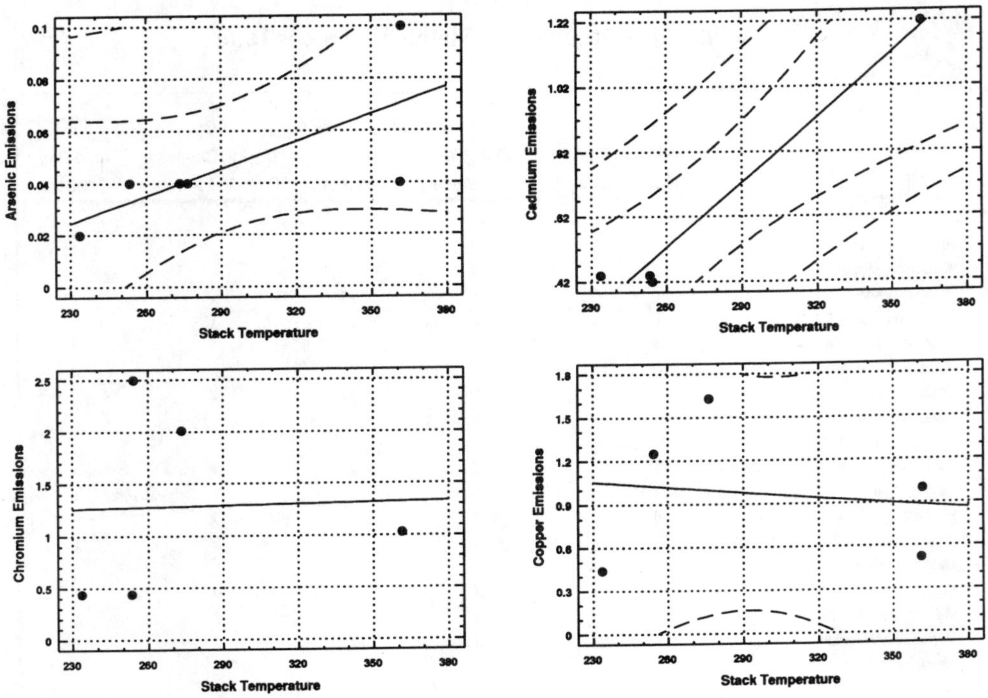

FIGURE 7. Correlation between Quebec City metals emissions (expressed in μg/Nm³, dry at 8% O_2) and stack temperature for arsenic, cadmium, chromium, and copper.

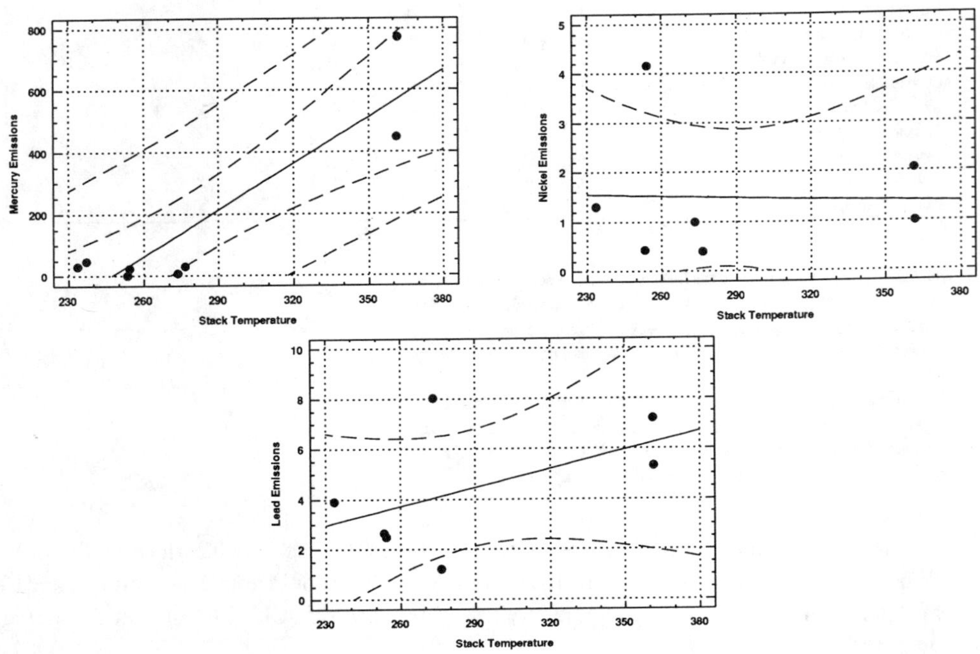

FIGURE 8. Correlation between Quebec City metals emissions (expressed in μg/Nm³, dry at 8% O_2) and stack temperature for lead, mercury, and nickel.[63]

in our analysis, because design differences probably alter the types and amounts of metallic compounds formed. The Tushima, Japan, facility did not have heat recovery[6] and was also excluded from our analysis.

Metals emissions factors for the remaining facilities with and without acid gas control are given in Table 11. For antimony, cobalt, manganese, selenium, tin, vanadium, and zinc, no data were available from facilities without acid gas control. For cadmium, copper, molybdenum, and nickel, data were available from only one facility without acid gas control. Thus, a comparison of emissions difference between facilities with and without acid gas control could be performed for arsenic, beryllium, chromium, lead, and mercury only (see also Figures 9 and 10).

In 1988, the available data seemed to indicate that significant differences between emissions from facilities with and without acid gas controls were evident for arsenic, beryllium, and lead.[63] The additional data that has become available, however, does not support a distinction between emissions from facilities with and without acid gas controls. Therefore, the geometric mean of test data from all nonmodular, heat recovery facilities, with and without acid gas control, was computed for all metals to obtain a single emission factor. Data from all facilities were used, since the selected facilities use high-efficiency particulate control equipment. The range of expected values was assumed to be the 95% confidence intervals (log-normal) for seven or more data points, the geometric mean plus or minus the geometric standard deviation for five or six data points, and the range of values for four or fewer data points.

Again, it should be noted that many facilities selected to develop metals emissions factors are located in other countries where the waste may be considerably different from that generated in the U.S. Any such differences would be expected to most dramatically affect metals emissions, because metals originate only in the waste and are not formed but only transformed during combustion. Nevertheless, since metals emissions data are so limited, the foreign data were considered.

B. STATISTICAL ANALYSES

A regression analysis was performed to determine if a relationship existed between particulate and trace metal emissions.[63] Although only facilities with heat recovery were considered (with the exception of the Japanese facility), both modular and nonmodular facilities as well as controlled and uncontrolled emissions were considered. In addition, much of the data are not from primary test reports, so that various assumptions had to be made to estimate emission factors. As a result, this analysis does not have the same level of accuracy as the former analyses presented, but it can be used to assess general trends.

Evaluation of the regression coefficients listed in Table 12 show that metals emissions tend to decrease as particulate emissions decrease for arsenic, cadmium, chromium, lead, and selenium. The relationship between particulate emissions and beryllium and nickel emissions are unclear. The near-zero regression coefficient for mercury shows that mercury emissions are essentially independent of particulate emissions as was expected for metals that are emitted primarily as a vapor.

V. EMISSIONS OF OTHER ORGANICS

In addition to PCDDs, PCDFs, and metals, other toxic organic pollutants known to be emitted from resource recovery facilities which are also routinely evaluated include polychlorinated biphenyls (PCBs), polycyclic aromatic hydrocarbons (PAHs), and aldehydes. Since very few facility tests included emissions data for these other organics, a statistical analysis similar to the one performed for PCDDs and PCDFs could not be done. We did assume, however, that emissions of other organic compounds emitted from resource recovery facilities were also log-normally distributed, and thus, the geometric mean was used to compute expected emission factors for organics. Sufficient data were not available to explore the effects of acid gas control

TABLE 11
Metal Emissions from Facilities with High Efficiency Particulate Controls Comparing with and without Acid Gas Controls
(lb/ton of refuse)

Technology Facility (Ref. No.)	Antimony	Arsenic	Beryllium	Cadmium	Chromium	Cobalt	Copper	Lead
Facilities With Acid Gas Control								
Wurzburg, West Germany (35)	3.97E-06	3.96E-08	------	3.96E-05	3.54E-06	------	------	7.84E-05
Munich, West Germany (5)	2.01E-05	3.56E-06	3.86E-09	6.58E-05	8.04E-03	1.28E-04	4.57E-05	6.85E-04
Marion County, Oregon (40)	------	------	2.36E-08	------	------	------	------	2.62E-04
Stockholm, Sweden (39)	------	------	1.96E-07	9.81E-07	------	------	------	1.09E-04
Commerce, California (10)	2.42E-06	1.45E-06	1.45E-06	1.55E-05	1.84E-05	2.91E-06	4.21E-04	1.55E-05
Bristol, Connecticut (37)	------	2.99E-05	6.72E-08	------	1.49E-05	------	------	2.24E-04
Stanislaus, California (68)	2.73E-05	1.25E-05	4.85E-09	1.97E-05	1.11E-04	------	1.14E-04	3.04E-04
Millbury, Massachusetts (67)	3.57E-05	3.63E-05	3.82E-06	1.74E-04	6.77E-04	------	3.70E-04	2.70E-03
Bridgeport, Connecticut (69)	------	1.23E-05	1.10E-05	------	2.42E-04	------	------	2.26E-05
Jackson Co., Michigan (66)	------	5.36E-06	6.46E-06	8.22E-04	3.29E-04	------	------	------
Geometric Mean	1.13E-05	4.55E-06	2.36E-07	3.79E-05	1.21E-04	1.93E-05	1.69E-04	1.67E-04
a Upper 95% Confidence Interval	NA	2.85E-05	2.51E-06	2.66E-04	9.51E-04	NA	NA	5.75E-04
b Plus Standard Deviation	3.82E-05	4.09E-05	5.12E-06	3.12E-04	1.42E-03	NA	NA	8.35E-04
Maximum	3.57E-05	3.63E-05	1.10E-05	8.22E-04	8.04E-03	1.28E-04	4.21E-04	2.70E-03
a Lower 95% Confidence Interval	NA	7.25E-07	2.22E-08	5.39E-06	1.55E-05	NA	NA	4.82E-05
b Minus Standard Deviation	3.37E-06	5.06E-07	1.09E-08	4.60E-06	1.03E-05	NA	NA	3.32E-05
Minimum	2.42E-06	3.96E-08	3.86E-09	9.81E-07	3.54E-06	2.91E-06	4.57E-05	1.55E-05
Facilities Without Acid Gas Control								
Pinellas County, Florida (13)	------	2.95E-05	1.42E-07	6.67E-05	3.82E-05	------	2.15E-04	1.34E-04
Hillsborough, Florida (38)	------	------	1.00E-06	------	------	------	------	2.60E-03
Baltimore, Maryland (20)	------	5.26E-05	------	------	2.67E-04	------	------	------
Tulsa County, Oklahoma (41)	------	------	2.56E-08	------	------	------	------	4.16E-03
North Andover, Massachusetts (11)	------	------	6.46E-06	------	------	------	------	6.12E-04
Dayton, Ohio (71)	------	------	------	------	------	------	------	1.23E-03
Geometric Mean	NA	3.94E-05	3.91E-07	6.67E-05	1.01E-04	NA	2.15E-04	1.02E-03
a Upper 95% Confidence Interval	NA	NA	NA	NA	NA	NA	NA	NA
b Plus Standard Deviation	NA	NA	NA	NA	NA	NA	NA	3.92E-03
Maximum	NA	5.26E-05	6.46E-06	NA	2.67E-04	NA	NA	4.16E-03
a Lower 95% Confidence Interval	NA	NA	NA	NA	NA	NA	NA	NA
b Minus Standard Deviation	NA	NA	NA	NA	NA	NA	NA	2.64E-04
Minimum	NA	2.95E-05	2.56E-08	NA	3.82E-05	NA	NA	1.34E-04
All Facilities								
Geometric Mean	1.13E-05	7.00E-06	2.76E-07	4.06E-05	1.17E-04	1.93E-05	1.77E-04	3.18E-04
a Upper 95% Confidence Interval	NA	3.25E-05	1.49E-06	2.10E-04	5.73E-04	NA	NA	8.60E-04
b Plus Standard Deviation	3.82E-05	5.98E-05	4.50E-06	2.89E-04	1.08E-03	NA	4.43E-04	1.78E-03
Maximum	3.57E-05	5.26E-05	1.10E-05	8.22E-04	8.04E-03	1.28E-04	4.21E-04	4.16E-03

	Manganese	Mercury	Molybdenum	Nickel	Selenium	Tin	Vanadium	Zinc
a Lower 95% Confidence Interval	NA	1.51E-06	5.10E-08	7.88E-06	2.39E-05	NA	NA	1.18E-04
b Minus Standard Deviation	3.37E-06	8.21E-07	1.69E-08	5.71E-06	1.27E-05	NA	7.09E-05	5.68E-05
Minimum	2.42E-06	3.96E-08	3.86E-09	9.81E-07	3.54E-06	2.91E-06	4.57E-05	1.55E-05
Ratio of Facilities with Acid Gas Controls to Facilities without Acid Gas Controls								
Geometric Mean	NA	0.12	0.60	0.57	1.20	NA	0.79	0.16
Upper Range	NA	0.54	0.39	NA	3.56	NA	NA	0.15
Lower Range	NA	0.02	0.87	NA	0.40	NA	NA	0.18
Upper Range with acid gas controls to Lower Range without acid gas controls	NA	0.97	98.12	NA	24.88	NA	NA	2.18

a The 95% confidence interval was calculated for those metals which have more than 6 data points.
b The standard deviation was calculated for those metals which have more than 4 data points.

Technology Facility (Ref. No.)	Manganese	Mercury	Molybdenum	Nickel	Selenium	Tin	Vanadium	Zinc
Facilities With Acid Gas Control								
Wurzburg, West Germany (35)	1.01E-05	-------	-------	1.60E-06	-------	-------	-------	-------
Munich, West Germany (5)	8.95E-04	-------	-------	3.73E-03	-------	2.70E-04	2.01E-05	3.06E-03
Marion County, Oregon (40)	-------	2.20E-03	-------	-------	-------	-------	-------	-------
Stockholm, Sweden (39)	-------	-------	-------	-------	-------	-------	-------	2.40E-04
Commerce, California (10)	7.75E-06	3.24E-04	9.68E-05	4.84E-05	2.13E-05	1.45E-05	4.84E-07	3.00E-04
Bristol, Connecticut (37)	-------	8.96E-04	-------	4.48E-05	-------	-------	-------	-------
Stanislaus, California (68)	9.74E-05	4.95E-03	-------	2.06E-04	3.34E-06	2.25E-04	2.74E-05	1.17E-03
Millbury, Massachusetts (67)	1.82E-04	1.17E-02	1.32E-04	1.76E-04	7.71E-06	1.99E-04	4.07E-05	7.51E-03
Bridgeport, Connecticut (69)	-------	3.01E-03	-------	2.69E-04	-------	-------	-------	-------
Jackson Co., Michigan (66)	-------	-------	-------	-------	-------	-------	-------	-------
Geometric Mean	6.59E-05	2.19E-03	1.13E-04	1.03E-04	8.19E-06	1.15E-04	1.02E-05	1.14E-03
a Upper 95% Confidence Interval	NA	NA	NA	9.10E-04	NA	NA	NA	NA
b Plus Standard Deviation	4.90E-04	7.79E-03	NA	1.09E-03	NA	NA	NA	5.00E-03
Maximum	8.95E-04	1.17E-02	1.32E-04	3.73E-03	2.13E-05	2.70E-04	4.07E-05	7.51E-03
a Lower 95% Confidence Interval	NA	NA	NA	1.18E-05	NA	NA	NA	NA
b Minus Standard Deviation	8.87E-06	6.18E-04	NA	9.85E-06	NA	NA	NA	2.60E-04
Minimum	7.75E-06	3.24E-04	9.68E-05	1.60E-06	3.34E-06	1.45E-05	4.84E-07	2.40E-04

TABLE 11 (continued)
Metal Emissions from Facilities with High Efficiency Particulate Controls Comparing with and without Acid Gas Controls (lb/ton of refuse)

Technology Facility (Ref. No.)	Manganese	Mercury	Molybdenum	Nickel	Selenium	Tin	Vanadium	Zinc
Facilities Without Acid Gas Control								
Pinellas County, Florida (13)	------	7.44E-03	2.92E-05	2.42E-05	------	------	------	------
Hillsborough, Florida (38)	------	------	------	------	------	------	------	------
Baltimore, Maryland (20)	------	------	------	------	------	------	------	------
Tulsa County, Oklahoma (41)	------	3.84E-03	------	------	------	------	------	------
North Andover, Massachusetts (11)	------	------	------	------	------	------	------	------
Dayton, Ohio (71)	------	------	------	------	------	------	------	------
Geometric Mean	NA	5.35E-03	2.92E-05	2.42E-05	NA	NA	NA	NA
a Upper 95% Confidence Interval	NA	NA	NA	NA	NA	NA	NA	NA
b Plus Standard Deviation	NA	NA	NA	NA	NA	NA	NA	NA
Maximum	NA	7.44E-03	NA	NA	NA	NA	NA	NA
a Lower 95% Confidence Interval	NA	NA	NA	NA	NA	NA	NA	NA
b Minus Standard Deviation	NA	NA	NA	NA	NA	NA	NA	NA
Minimum	NA	3.84E-03	NA	NA	NA	NA	NA	NA
All Facilities								
Geometric Mean	6.59E-05	2.74E-03	7.20E-05	8.62E-05	8.19E-06	1.15E-04	1.02E-05	1.14E-03
a Upper 95% Confidence Interval	NA	7.23E-03	NA	5.59E-04	NA	NA	NA	NA
b Plus Standard Deviation	4.90E-04	8.75E-03	NA	8.07E-04	NA	NA	NA	5.00E-03
Maximum	8.95E-04	1.17E-02	1.32E-04	3.73E-03	2.13E-05	2.70E-04	4.07E-05	7.51E-03
a Lower 95% Confidence Interval	NA	1.04E-03	NA	1.33E-05	NA	NA	NA	NA
b Minus Standard Deviation	8.87E-06	8.59E-04	NA	9.21E-06	NA	NA	NA	2.60E-04
Minimum	7.75E-06	3.24E-04	2.92E-05	1.60E-06	3.34E-06	1.45E-05	4.84E-07	2.40E-04
Ratio of Facilities with Acid Gas Controls to Facilities without Acid Gas Controls								
Geometric Mean	NA	0.41	3.87	4.27	NA	NA	NA	NA
Upper Range	NA	1.05	NA	NA	NA	NA	NA	NA
Lower Range	NA	0.16	NA	NA	NA	NA	NA	NA
Upper Range with acid gas controls to Lower Range without acid gas controls	NA	2.03	NA	NA	NA	NA	NA	NA

a The 95% confidence interval was calculated for those metals which have more than 6 data points.
b The standard deviation was calculated for those metals which have more than 4 data points.

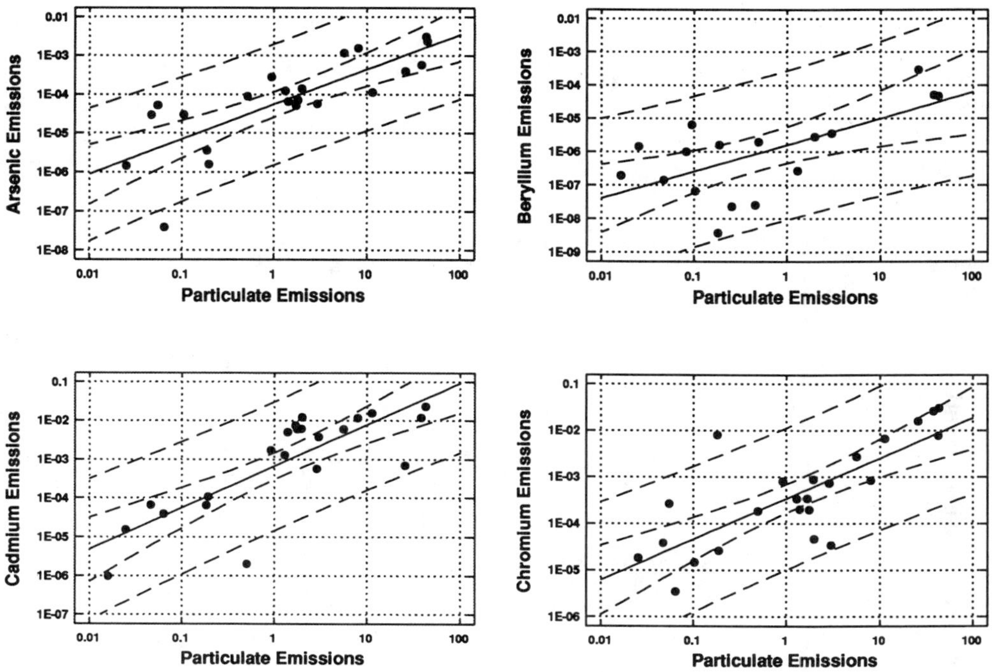

FIGURE 9. Regression analysis of metals vs. particulate emissions (expressed in lb/ton) for arsenic, beryllium, cadmium, and chromium.[63]

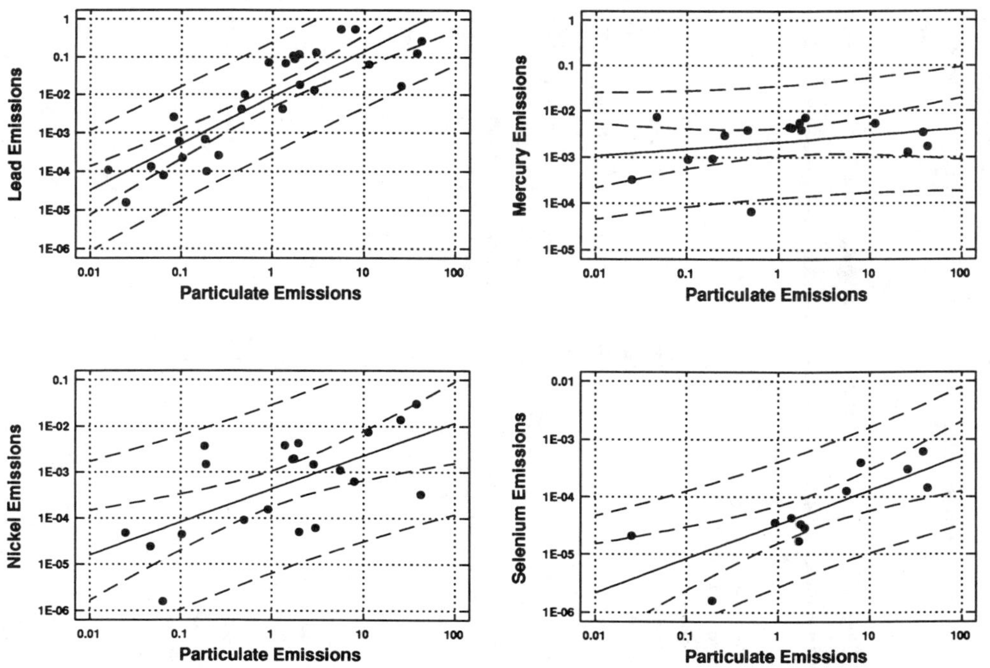

FIGURE 10. Regression analysis of metals vs. particulate emissions (expressed in lb/ton) for lead, mercury, nickel, and selenium.[63]

TABLE 12
Particulate and Metals Emissions from Various MSW Resource Recovery Facilities[63] (lb/ton)

Facility	Particulate	Arsenic	Beryllium	Cadmium	Chromium	Lead	Mercury	Nickel	Selenium
Hamilton, Ontario	11.300	1.13E-04	NA a	1.58E-02	6.78E-03	6.44E-02	5.39E-03	7.57E-03	NA
Hampton, VA	2.900	5.80E-05	NA	5.80E-04	7.30E-04	1.30E-02	NA	1.50E-03	NA
Little Rock, AK	3.030	NA	3.60E-06	3.86E-03	3.46E-05	1.34E-01	NA	6.20E-05	NA
Marysville, OH	2.010	NA	2.80E-06	1.23E-02	4.74E-05	1.79E-02	NA	5.10E-05	NA
Essen, West Germany	0.500	9.00E-05	2.00E-06	2.00E-06	1.85E-04	1.00E-02	6.70E-05	9.20E-05	NA
Nicosia, IN	8.000	1.60E-03	NA	1.20E-02	8.40E-04	5.52E-01	NA	6.32E-04	3.92E-04
SWRC #1, Canada	0.920	2.85E-04	NA	1.75E-03	8.00E-04	7.08E-02	NA	1.56E-04	3.59E-05
Alexander, VA	5.600	1.18E-03	NA	6.16E-03	2.74E-03	5.43E-01	NA	1.12E-03	1.29E-04
Munich, West Germany	0.183	3.56E-06	3.86E-09	6.58E-05	8.04E-03	6.85E-04	NA	3.73E-03	NA
Gallatin, TN	42.500	3.18E-03	4.80E-05	2.36E-02	7.86E-03	2.74E-01	1.71E-03	3.32E-04	1.44E-04
Japanese Plant (Uncontrolled)	25.700	4.00E-04	3.00E-04	7.00E-04	1.60E-02	1.70E-02	1.30E-03	1.40E-02	3.00E-04
Japanese Plant (Controlled)	0.190	1.60E-06	1.60E-06	1.10E-04	2.60E-05	1.00E-04	9.30E-04	1.50E-03	1.60E-06
PEI, Canada (Normal Operations)	1.690	5.07E-05	NA	7.58E-03	3.47E-04	1.10E-01	5.61E-03	1.95E-03	1.73E-05
PEI, Canada (Long Cycle)	1.767	7.20E-05	NA	6.06E-03	1.99E-04	8.89E-02	3.94E-03	2.00E-03	3.40E-05
PEI, Canada (High Temp.)	1.963	1.41E-04	NA	6.32E-03	8.88E-04	1.20E-01	7.20E-03	4.33E-03	2.93E-05
PEI, Canada (Low Temp.)	1.393	6.60E-05	NA	5.14E-03	2.03E-04	6.85E-02	4.32E-03	3.89E-03	4.40E-05
Pinellas Co., FL	0.047	2.95E-05	1.42E-07	6.67E-05	3.82E-05	1.34E-04	7.44E-03	2.42E-05	NA
Hillsborough, FL	0.083	NA	1.00E-06	NA	NA	2.60E-03	NA	NA	NA
Commerce, CA (Uncontrolled)	38.000	5.81E-04	5.16E-05	1.21E-02	2.66E-02	1.28E-01	3.44E-03	3.15E-02	6.15E-04
Commerce, CA (Controlled)	0.025	1.45E-06	1.45E-06	1.55E-05	1.84E-05	1.55E-05	3.24E-04	4.84E-05	2.13E-05
Baltimore, MD (Uncontrolled)	43.640	2.40E-03	NA	NA	3.05E-02	NA	NA	NA	NA
Baltimore, MD (Controlled)	0.054	5.26E-05	NA	NA	2.67E-04	NA	NA	NA	NA

Marion Co., OR	0.257	NA	2.36E-08	NA	NA	2.62E-04	2.97E-03	NA	NA
Tulsa Co., OK	0.458	NA	2.56E-08	NA	NA	4.16E-03	3.84E-03	NA	NA
Wurzburg, West Germany	0.064	3.96E-08	NA	3.96E-05	3.54E-06	7.84E-05	NA	1.60E-06	NA
Braintree, MA	1.300	1.25E-04	2.69E-07	1.30E-03	3.37E-04	4.22E-03	4.49E-03	NA	NA
North Andover, MA	0.095	NA	6.46E-06	NA	NA	6.12E-04	NA	NA	NA
Bristol, Connecticut	0.103	2.99E-05	6.72E-08	NA	1.49E-05	2.24E-04	8.96E-04	4.48E-05	NA
Stockholm, Sweden	0.016	NA	1.96E-07	9.81E-07	NA	1.09E-04	NA	NA	NA
b Regression of LN of Metal vs. LN of Particulate:									
Intercept		5.63E-05	1.58E-06	6.64E-04	3.39E-04	8.59E-03	2.10E-03	4.33E-04	3.31E-05
Slope		2.46	2.22	2.93	2.38	3.37	1.16	2.05	1.81
Regression Coefficient		62.62	42.87	65.94	59.49	74.54	7.08	40.23	61.67

a Data not available.

b Values of the regression coefficient greater than 50% indicate a significant relationship between metals and particulate emissions.

TABLE 13
Polychlorinated Biphenyls Emission Factors Based on Test Data from Selected Facilities[a] (μg/Nm^3 @12% CO_2, dry)[80]

Facility	Results for Individual Test Runs 1	2	3	4	5	6	Test Average
b Cattaraugus, New York (27)	0.603	2.017	1.347	----	----	----	1.322
Chicago (Northwest), Illinois (21)	0.033	0.017	0.124	0.023	0.015	0.107	0.053
Commerce, California (10)	< 0.113	< 0.657	----	----	----	----	0.385
b Oneida, New York (29)	0.210	1.271	1.174	----	----	----	0.885
Pittsfield, Massachusetts (53)	ND *c*	ND	----	----	----	----	NA
b Prince Edward Island, Canada (16)	ND	ND	2.620	----	----	----	2.620
Stanislaus, California (68)	< 0.1998	< 0.1849	< 0.2116	0.1697	< 0.1892	< 0.1793	0.189
All Facilities							
Geometric Mean							0.478
Plus 1 Standard Deviation (Geometric)							1.970
Minus 1 Standard Deviation (Geometric)							0.116

a Based on test results from massburn (excess air and modular) facilities
b Based on test results from modular unit (less than 50 tons/day/unit)
c Value was at or below the detection limit, but the detection limits were not supplied; therefore, it was not included in the calculations of the average.

on organics; therefore, the available data from facilities that did and did not have acid gas control were combined to estimate organics emissions.

A. POLYCHLORINATED BIPHENYLS

The six facilities used to estimate emission factors for polychlorinated biphenyls (PCBs) were Cattaraugus, NY; Chicago (Northwest), IL; Commerce, CA; Oneida, NY; Prince Edward Island, Canada; and Stanislaus, CA.[10,16,21,27,29,68] Test results from these massburn (excess air and modular) facilities were converted to standardized units. The results of individual tests and the averages for each facility are presented in Table 13. Expected PCB emissions for a massburn facility were estimated from the geometric mean of the distribution. The geometric mean and the ranges of values reported for each facility and the facility averages are also given in Table 13.

B. POLYCYCLIC AROMATIC HYDROCARBONS

The development of emission factors for PAHs was more difficult, since PAHs are a complex mixture of a various compounds with diverse toxicities. The International Agency for Research on Cancer (IARC)[57a] (Lyon, France) determined that 14 different PAHs are probable human carcinogens. Of those 14 compounds, benzo(a)pyrene (BaP) is the most studied and the most toxic PAH which is known to be emitted from resource recovery facilities. The U.S. EPA[58] assumes that all PAHs designated as probably human carcinogens emitted from resource recovery facilities are as toxic as BaP and that the toxicity of the other PAHs is negligible relative to the known toxicity of BaP.[58] The EPA assumption was also applied in this analysis.

Table 14 shows PAH data from massburn (excess air and modular) facilities after conversion to standardized units. Due to the lack of facility emissions test data, modular units were included in our analysis. Individual test results were averaged for each facility to determine the emissions of each carcinogenic PAH. Using these average emissions values, total carcinogenic PAH

TABLE 14
Polycyclic Aromatic Hydrocarbons Emission Factors Based on Test Data from Selected Facilities[a] ($\mu g/Nm^3$ @12% CO_2, dry)[80]

Facility / Possible Carcinogens	Results for Individual Test Runs 1	2	3	Test Average
b **Alexandria, Minnosota (22)**				
Benz(a)anthracene	0.720	-----	-----	0.720
Benzo(b)fluoranthene	0.513	-----	-----	0.513
Benzo(k)fluoranthene	0.568	-----	-----	0.568
Benzo(a)pyrene	0.643	-----	-----	0.643
Dibenz(a,h)anthracene	0.536	-----	-----	0.536
Indeno(1,2,3-cd)pyrene	0.739	-----	-----	0.739
Total Carcinogens				3.718
Total PAHs				12.032
b **Cattaraugus, New York (27)**				
Benzo(a)pyrene	0.928	0.928	0.879	0.912
Total Carcinogens				0.912
Total PAHs				0.912
Commerce, California (10)				
Benz(a)anthracene	0.0025	0.0015	-----	0.0020
Benzo(b)fluoranthene	0.0026	0.0032	-----	0.0029
Benzo(k)fluoranthene	0.0026	0.0032	-----	0.0029
Benzo(a)pyrene	0.0043	< 0.0016	-----	0.0030
Dibenz(a,h)anthracene	< 0.0017	< 0.0032	-----	0.0025
Indeno(1,2,3-cd)pyrene	0.0019	< 0.0028	-----	0.0024
Total Carcinogens				0.0156
Total PAHs				0.465
c **Hogdalen, Sweden (39)**				
Total Carcinogens	0.040	-----	-----	0.040
Total PAHs	0.040	-----	-----	0.040
Nuremberg, Germany (46)				
Benz(a)anthracene	< 0.0015	-----	-----	0.0015
Benzo(b)fluoranthene	< 0.0015	-----	-----	0.0015
Benzo(a)pyrene	< 0.0015	-----	-----	0.0015
Dibenz(a,h)anthracene	< 0.0031	-----	-----	0.0031
Total Carcinogens				0.0076
Total PAHs				0.0244
b **Oneida, New York (29)**				
Benzo(a)pyrene	11.921	11.597	9.991	11.17
Total Carcinogens				11.17
Total PAHs				11.17
c **Pittsfield, Massachusetts (53)**				
Total Carcinogens	0.204	0.095	-----	0.149
Total PAHs	0.204	0.095	-----	0.149
b **Prince Edward Island, Canada (16)**				
Benz(a)anthracene	0.248	0.269	0.065	0.194
Benzo(j,k)fluoranthene & Benzo(a)pyrene	0.139	0.492	0.133	0.255
Dibenz(a,h)anthracene	0.022	ND (d)	0.007	0.014
Indeno(1,2,3-cd)pyrene	0.059	ND (d)	0.009	0.034
Total Carcinogens				0.497
Total PAHs				7.731
Bristol, Connecticut (37)				
Benz(a)anthracene	< 0.007	< 0.006	< 0.003	0.005
Benzo(b)fluoranthene & Benzo(k)fluoranthene	< 0.01	< 0.01	< 0.007	0.009
Benzo(a)pyrene	< 0.007	< 0.007	< 0.003	0.006
Dibenz(a,h)anthracene	< 0.026	< 0.026	< 0.003	0.018
Indeno(1,2,3-cd)pyrene	< 0.015	< 0.014	< 0.007	0.012
Total Carcinogens				0.050
Total PAHs				0.579

TABLE 14 (continued)
Polycyclic Aromatic Hydrocarbons Emission Factors Based on Test Data from Selected Facilities[a] (mg/Nm³ @12% CO_2, dry)[80]

Facility Possible Carcinogens	Results for Individual Test Runs 1	2	3	Test Average
Stanislaus, California (68)				
Benz(a)anthracene	0.6676	0.4875	-----	0.578
Benzo(b)fluoranthene & Benzo(k)fluoranthene	0.9125	0.6507	-----	0.782
Benzo(a)pyrene	0.0993	0.0632	-----	0.081
Dibenz(a,h)anthracene	0.0109	0.0104	-----	0.011
Indeno(1,2,3-cd)pyrene	0.0795	0.0631	-----	0.071
Total Carcinogens				1.522
Total PAHs				3.260
Total Carcinogens				
Geometric Mean				0.252
Upper 95% Confidence Interval				1.434
Lower 95% Confidence Interval				0.044
Total PAHs				
Geometric Mean				1.234
Upper 95% Confidence Interval				6.201
Lower 95% Confidence Interval				0.246

a Based on test results from massburn (excess air and modular) facilities
b Based on test results from modular unit (less than 50 tons/day/unit)
c An analytical analysis was not performed or given in the test report. Therefore, WESTON made the most conservative assumption for risk assessment purposes, i.e, that all measured PAH is carcinogenic and, thus, using the EPA procedure, assumed to have the toxicity of benzo(a)pyrene.
d Value was at or below the detection limit, but the detection limits were not supplied; therefore, it was not included in the calculations of the average.

emissions were calculated for each facility. Expected emissions of carcinogenic PAHs for any given facility were calculated from the geometric mean of the log normal distribution.

C. ALDEHYDES

Formaldehyde (RCHO), which is one constituent of total aldehydes, has been identified as a potential human carcinogen.[59] Aldehyde test results were found for three massburn (excess air and modular) facilities (Table 15). Average test results were computed for each facility. The geometric mean of the facility results was then calculated as the best estimate of expected aldehyde emissions from massburn facilities assuming a log normal distribution.

VI. EMISSIONS OF CRITERIA POLLUTANTS AND ACID GASES

Criteria pollutants are those pollutants for which the U.S. EPA has issued air quality criteria and subsequent health- and welfare-based National Ambient Air Quality Standards (NAAQS). Criteria pollutants include particulate matter (PM) [originally total suspended particulate matter (TSP) and now particulate matter with aerodynamic diameters of 10 mm or less (PM-10)], sulfur oxides (SO_x), nitrogen dioxide (NO_2), carbon monoxide (CO), ozone, and lead. Acid gases emitted from resource recovery facilities include hydrogen chloride (HCl), hydrogen fluoride (HF), and sulfuric acid (H_2SO_4).

TABLE 15
Aldehydes Emission Factors Based on Test Data from Selected Facilities[a] ($\mu g/Nm^3$ @ 12% CO_2, dry)[80]

Facility	Results for Individual Test Runs 1	2	3	Test Average
b Cattaraugus, New York (27)	641.9	563.4	646.1	617.1
b Oneida, New York (29)	213.1	215.7	212.1	213.6
Westchester, New York (31)	745.6	397.3	512.6	551.8
All Facilities				
Geometric Mean				417.5
Maximum				617.1
Minimum				213.6

a Based on test results from massburn (excess air and modular) facilities
b Based on test results from modular unit (less than 50 tons/day/unit)

Emission levels of criteria pollutants and acid gases can be determined from: (1) the MSW stream composition; (2) state and federal regulations; (3) best available technology (BAT) policies; and (4) permit conditions. Emission estimates used in risk assessments must be the same or related to those used in air permit applications. If the MSW has been analyzed for a given facility, data on the average and range of individual components including carbon, hydrogen, oxygen, sulfur, chlorine, and fluorine may be available. The analysis should also provide information on the ash, moisture, and heating content of the waste. These data can be used to estimate the average and range of volumetric flow rates, as well as uncontrolled emissions for some criteria pollutants and acid gases. Again, emissions estimates derived from annual average data are more appropriate for estimating long-term impacts, such as those modeled in a risk assessment, while upper-bound values are more appropriate for permitting purposes. (Upper-bound values should be a better representation of short-term emission variations such as could be measured during permit compliance emissions testing.)

A. PARTICULATE MATTER

When MSW is burned, PM may originate from three sources: inorganic substances, organometallic substances, and unburned waste. Inorganic materials cannot be combusted, and consequently most of these material exit the combustion system as bottom ash with a small portion carried out of the stack on fly ash. Organometallic compounds appear as inorganic oxides or metal salts in the flue gas as a result of high-temperature oxidation. Unburned waste entrained in the combustion flue gas results from incomplete combustion and the agglomeration of small particles.

The size and quantity of PM emitted from resource recovery facilities depends on such factors as residence time within the combustion chamber, combustion temperature, oxidizing/reducing conditions during combustion, and the composition of the MSW. Longer residence times allow for more complete combustion of organic particles with a commensurate decrease in particulate matter. Longer residence times also lead to the increased likelihood that larger particles will drop out in the combustion chamber rather than becoming entrained as fly ash. Maintaining high temperatures in the combustion chamber promotes more complete burning and a greater proportion of small particles.

Examples of measured PM emissions (in terms of mass per unit volume emissions) are given in Tables 6 and 12. For permitting purposes, however, PM emission estimates are based on regulatory or policy limits. If permit limits are based on the lowest or even average value measured at one or all facilities, there is no guarantee that value will be attained at another facility. Equipment manufacturer guarantees and the experience with all similar facilities should be considered. The New Source Performance Standards (NSPS) for resource recovery facilities and other types of incinerators is 0.08 gr/dscf. For facilities with heat recovery, the NSPS for steam generating units with over 100 million Btu/heat input are required to limit PM emissions to 0.1 pounds per million Btu of heat input (roughly equivalent to 0.04 gr/dscf at 12% CO_2).

Neither of these federal standards, however, is a practical limitation on PM emissions from resource recovery facilities that use modern combustor technology and pollution control equipment, since many state and local regulations and BAT limits are more stringent than these federal standards. For example, in many jurisdictions, particulate emissions are typically limited to between 0.01 and 0.02 gr/dscf at either 12% CO_2 or 7% O_2. For consistency, proposed permit limits calculated from the waste composition in conjunction with stack flow rates can be used to calculate typical and upper-bound emissions. PM-10 emissions should be estimated from total PM emissions using the same size distribution assumed for deposition modeling.

B. SULFUR DIOXIDE AND SULFURIC ACID

Sulfur dioxide (SO_2) is formed by the oxidation of sulfur contained in the waste. Sulfur is found in items such as rubber, textiles, and plastics. Sulfuric acid (H_2SO_4) is formed by the reaction of sulfur trioxide (SO_3) and water. SO_3 results from the oxidation of SO_2; therefore, H_2SO_4 can be considered a by-product of SO_2 production.

Uncontrolled SO_2 emissions can be calculated by assuming that all sulfur present in the waste is oxidized. If state regulations of policies limit O_2 emissions, controlled emissions can be calculated from the amount of sulfur estimated to be present in the waste (given the waste composition). Many states require about a 70 to 80% reduction in SO_2 emissions unless emissions are reduced to a level below which acid gas removal becomes difficult (typically 30 to 50 ppm, dry volume at 12% CO_2 or 7% O_2). Other states specify only HCl reductions as an indication of good acid gas control design and operation. Acid gas emissions for facilities with and without acid gas control are presented in Table 16. It should also be noted that inaccurately high levels of sulfuric acid emissions may be measured and reported due to interference of hydrogen chloride in the analytical technique.[64]

C. NITROGEN OXIDES

Nitrogen oxides (NO_x) are products of all conventional combustion processes. Nitric oxide (NO) is the predominant form of NO_x produced with lesser amounts of nitrogen dioxide (NO_2). Once emitted, NO converts to NO_2 in the atmosphere. Hence, NO and NO_2 are referred to collectively as NO_x. The generation of NO_x from resource recovery facilities results from two mechanisms: fuel NO_x and thermal NO_x formation. Given typical operating temperatures maintained in the combustor, fuel NO_x is likely to be the dominant formation mechanism.[60] The majority (75 to 80%) of total NO_x emitted from burning MSW results from fuel nitrogen oxidation.[61] Since the nitrogen content of MSW is typically less than 1%, the potential for NO_x formation is relatively small. The NO_x produced by exposing the air present inside the combustion zone (ambient air contains 79% nitrogen by volume) to high temperatures of combustion is referred to as thermal NO_x formation. The main factors influencing thermal NO_x reactions are combustion temperature, residence time within the combustion zone, and oxygen levels present in the combustion zone.

Emissions data for NO_x from selected facilities are given in Table 16. These data are from newer facilities that do not have specific NO_x control devices but generally have some pollution control equipment and combustor design features that reduce NO_x emissions. In areas of the U.S. that do not attain NAAQS standards for NO_2 and in some areas that do not attain the NAAQS standards for ozone (because NO_2 participates in the photochemical reactions that form ozone),

TABLE 16
Emission Factors for Criteria Pollutants and Acid Gases by Facility
(ppm dv @12% CO_2)

Facility (Ref. No.)	Criteria Pollutants			Acid Gases			
	CO	NOx	THC	SO2	HCl	H2SO4	HF
Facilities With Acid Gas Control							
Commerce, California (10)	27.79	113.68	10.11	1.84	8.39	0.14	0.07
Stockholm, Sweden (39)	20.34	599.36	----	89.11	58.64	----	----
Marion County, Oregon (40)	20.88	352.11	2.30	48.09	11.95	----	----
Wurzburg, West Germany (35)	32.47	245.81	0.94	----	----	----	----
Stapelfeld, West Germany (8)	52.79	96.42	----	11.31	30.41	----	----
Bristol, Connecticut (37)	38.00	579.00	----	27.00	8.00	3.95	0.10
Stanislaus, California (68)	42.50	103.00	3.50	3.35	1.50	----	0.16
Millbury, Massachusetts (67)	12.35	221.45	5.38	55.60	14.25	3.18	1.40
Jackson Co., Michigan (66)	29.67	152.67	----	4.56	----	----	----
Number of Facilities in Sample	9	9	5	8	7	3	4
Log Normal Distribution							
Geometric Mean	28.39	218.74	3.33	14.58	11.73	1.21	0.20
b Upper 95% Confidence Interval	39.82	376.68	NA	49.05	34.18	NA	NA
c Plus Standard Deviation	44.09	443.63	8.16	62.22	37.29	NA	NA
Maximum	52.79	599.36	10.11	89.11	58.64	3.95	1.40
b Lower 95% Confidence Interval	20.24	127.02	NA	4.34	4.03	NA	NA
c Minus Standard Deviation	18.28	107.85	1.36	3.42	3.69	NA	NA
Minimum	12.35	96.42	0.94	1.84	1.50	0.14	0.07

TABLE 16 (continued)
Emission Factors for Criteria Pollutants and Acid Gases by Facility
(ppm dv @12% CO_2)

Normal Distribution							
Arithmetic Mean	30.75	273.72	4.45	30.11	19.02	2.42	0.43
b Upper 95% Confidence Interval	40.30	424.97	NA	56.52	37.18	NA	NA
c Plus Standard Deviation	43.17	470.49	8.01	61.70	38.66	NA	NA
b Lower 95% Confidence Interval	21.21	122.47	NA	3.70	0.86	NA	NA
c Minus Standard Deviation	18.34	76.95	0.88	-1.48	-0.62	NA	NA
Facilities Without Acid Gas Control							
a Alexandria, Minnesota (22)	25.5	281.54	----	120.77	293.8	----	----
a Cleburne, Texas (26)	56.84	162.19	26.53	125.18	505.33	----	----
Tulsa County, Oklahoma (42)	19.00	294.00	1.07	90.67	442.67	0.70	----
Pinellas, Florida (13)	3.80	275.00	----	166.00	677.00	----	7.60
a Pittsfield, Massachusetts (53)	29.13	102.96	----	----	----	----	----
Westchester, New York (31)	----	205.53	----	117.91	42.36	----	----
Hillsborough, Fl (38)	15.30	325.00	----	161.00	----	----	----
Chicago, Illinois (21)	263.14	----	3.08	----	----	----	----
North Andover, Massachusetts (47)	19.69	268.36	1.13	86.89	148.13	2.65	----
Saugus, Massachusetts (48)	36.74	----	0.29	----	----	----	----
Alexandria/ Arlington, Virginia (36)	20.00	207.00	----	36.00	163.00	----	----
Dayton, Ohio (71)	28.20	83.39	----	55.30	----	----	----

a The facility consist of modular furnaces.
b The 95% confidence interval was calculated for those metals which have more than 6 data points.
c The standard deviation was calculated for those metals which have more than 4 data points.

Facility (Ref. No.)	Criteria Pollutants			Acid Gases			
	CO	NOx	THC	SO2	HCl	H2SO4	HF
Number of Facilities in Sample	11	10	5	9	7	2	1
Log Normal Distribution							
Geometric Mean	26.74	202.79	1.96	96.81	239.73	1.36	7.60
b Upper 95% Confidence Interval	52.93	282.65	NA	142.34	578.16	NA	NA
c Plus Standard Deviation	73.88	322.57	10.51	159.85	621.04	NA	NA
Maximum	263.14	325.00	3.08	166.00	677.00	2.65	NA
b Lower 95% Confidence Interval	13.51	145.50	NA	65.85	99.40	NA	NA
c Minus Standard Deviation	9.68	127.49	0.36	58.64	92.54	NA	NA
Minimum	3.80	83.39	0.29	36.00	42.36	0.70	NA
Normal Distribution							
Arithmetic Mean	47.03	220.50	6.42	106.64	324.61	1.68	7.60
b Upper 95% Confidence Interval	96.02	279.74	NA	140.38	534.38	NA	NA
c Plus Standard Deviation	119.96	303.32	17.71	150.54	551.42	NA	NA
b Lower 95% Confidence Interval	-1.96	161.25	NA	72.89	114.85	NA	NA
c Minus Standard Deviation	-25.90	137.68	-4.87	62.73	97.80	NA	NA
All Facilities							
Number of Facilities in Sample	20	19	10	17	14	5	5
Log Normal Distribution							
Geometric Mean	27.47	210.20	2.55	39.73	53.03	1.27	0.41
b Upper 95% Confidence Interval	39.78	277.42	6.48	82.12	155.87	NA	NA
c Plus Standard Deviation	60.60	373.81	9.38	163.12	343.13	5.16	3.05

TABLE 16 (continued)
Emission Factors for Criteria Pollutants and Acid Gases by Facility
(ppm dv @12% CO_2)

Maximum	263.14	599.36	10.11	166.00	677.00	3.95	7.60
b Lower 95% Confidence Interval	18.97	159.26	1.01	19.22	18.04	NA	NA
c Minus Standard Deviation	12.45	118.19	0.70	9.67	8.20	0.31	0.06
Minimum	3.80	83.39	0.29	1.84	1.50	0.14	0.07
Normal Distribution							
Arithmetic Mean	39.71	245.71	5.43	70.62	171.82	2.12	1.87
b Upper 95% Confidence Interval	64.83	322.50	11.13	100.70	301.89	NA	NA
c Plus Standard Deviation	93.39	405.03	13.40	129.12	397.10	4.23	5.12
b Lower 95% Confidence Interval	14.58	168.92	-0.27	40.54	41.74	NA	NA
c Minus Standard Deviation	-13.98	86.39	-2.54	12.12	-53.46	0.02	-1.39

a The facility consist of modular furnaces.
b The 95% confidence interval was calculated for those metals which have more than 6 data points.
c The standard deviation was calculated for those metals which have more than 4 data points.

there is increasing pressure to reduce NO_x emissions through additional control devices. Thermal $DeNO_x$ devices installed at the Commerce, CA, facility achieved NO_x removal efficiencies of about 45%; however, the cost effectiveness of this and other selective noncatalytic reduction control techniques is debatable.

D. CARBON MONOXIDE

Carbon monoxide (CO) is the product of incomplete oxidation of carbon compounds. For MSW combustion, the quantity of CO remaining after combustion depends largely on the temperature, amount of available air, and the amount of turbulence within the combustion zone as well as exhaust gas residence time, which is determined by the design of the combustor/boiler. CO emissions data for selected facilities are presented in Table 16. A number of states limit CO levels to about 100 ppm (dry volume at 12% CO_2 or 7% O_2). The underlying purpose of these limits, however, is not to reduce adverse health effects per se but to ensure good combustion. While CO levels lower than 100 ppm do not necessarily mean better combustion, higher levels are indicative of poor combustion.

E. HYDROCARBONS

Hydrocarbons, or volatile organic compounds (VOCs), which are a precursor to ozone formation, are the result of incompletely oxidized organic matter. Good design and operating practices, which include providing sufficient oxygen, turbulence (for mixing), and residence time, reduce emissions of VOCs. Total hydrocarbon (THC) emissions data are given in Table 16.

F. HYDROGEN CHLORIDE AND HYDROGEN FLUORIDE

HCl and hydrogen fluoride (HF) are by-products of the combustion of chlorine and fluorine present in the waste stream. Chlorine is present in polyvinyl chloride (PVC), other chlorinated plastics, sodium chloride (salt), rubber, and paper. Fluorine is present primarily in fluorinated plastics and other fluorocarbon products.

Uncontrolled emissions of HCl and HF can be estimated by assuming that all of the chlorine and fluorine present in the waste is available for chemical reaction. Many state regulations limit HCl emissions generally to about 10% of uncontrolled emissions (i.e., 90% reduction by weight) unless emissions are below some cut-off level (typically 30 to 50 ppm, dry volume at 12% CO_2 or 7% O_2). Controlled emissions can be calculated from expected amounts of chlorine in the waste. The amount of hydrogen in the waste is not limiting because of the excess hydrogen available in the moisture in the combustion air. HCl emissions data for facilities with and without acid gas control are presented in Table 16.

HF emissions are usually an order of magnitude lower than HCl emissions. Although HF emissions are typically not regulated, HF is generally at least as reactive in an acid gas scrubber as HCl. Therefore, controlled HF emissions can be estimated from combustion calculations assuming the emission reduction projected for HCl (See Table 16).

VII. CONCLUSIONS

Acceptable emission test data for dioxins, furans, and trace metals are rapidly increasing. Emissions test data for other trace organics, criteria pollutants, and acid gases are also increasing but less rapidly. WESTON has developed a computerized system to facilitate the accurate and rapid incorporation of new test data into the development of pollutant-specific emission factors. Emission factors have been developed for the type of resource recovery facility most commonly built, massburn excess air. These emission factors are summarized in Table 17. The statistical variability of these emission factors was also examined. This variability is especially important to consider in regulatory development and compliance, as well as risk assessment. The effect of various process and air pollution control parameters upon emissions has been explored.

TABLE 17
MSW Incineration Emission Factors for Facilities With and Without Acid Gas Controls

	Acid Gas Controls			Without Acid Gas Controls		
PCDDs & PCDFs (ng/Nm^3 @12% CO2, dry)	**Geometric Mean**	**Lower 95% C.I**	**Upper 95% C.I**	**Geometric Mean**	**Lower 95% C.I**	**Upper 95% C.I**
2,3,7,8-T4CDD	0.04	0.02	0.12	0.33	0.14	0.78
T4CDD	0.89	0.23	3.40	3.84	1.53	9.69
PCDD	6.23	1.65	23.51	51.06	24.52	106.31
T4CDF	3.77	0.76	18.72	27.66	12.56	60.93
PCDF	10.97	2.66	45.23	97.43	43.58	217.79
U.S. EPA Toxic Equivalent	0.28	0.09	0.88	2.06	1.00	4.24
a **Metals (lb/ton)**	**Geometric Mean**	**Lower Range**	**Upper Range**	**Geometric Mean**	**Lower Range**	**Upper Range**
Antimony	1.13E-05 *b*	3.82E-05	3.37E-06	1.13E-05	3.82E-05	3.37E-06
Arsenic	7.00E-06	3.25E-05	1.51E-06	7.00E-06	3.25E-05	1.51E-06
Beryllium	2.76E-07	1.49E-06	5.10E-08	2.76E-07	1.49E-06	5.10E-08
Cadmium	4.06E-05	2.10E-04	7.88E-06	4.06E-05	2.10E-04	7.88E-06
Chromium	1.17E-04	5.73E-04	2.39E-05	1.17E-04	5.73E-04	2.39E-05
c Chromium (VI)	3.97E-06	1.95E-05	8.11E-07	3.97E-06	1.95E-05	8.11E-07
Cobalt	1.93E-05	1.28E-04	2.91E-06	1.93E-05	1.28E-04	2.91E-06
Copper	1.77E-04	4.43E-04	7.09E-05	1.77E-04	4.43E-04	7.09E-05
Lead	3.18E-04	8.60E-04	1.18E-04	3.18E-04	8.60E-04	1.18E-04
Manganese	6.59E-05	4.90E-04	8.87E-06	6.59E-05	4.90E-04	8.87E-06
Mercury	2.74E-03	7.23E-03	1.04E-03	2.74E-03	7.23E-03	1.04E-03
Molybdenum	7.20E-05	1.32E-04	2.92E-05	7.20E-05	1.32E-04	2.92E-05
Nickel	8.62E-05	5.59E-04	1.33E-05	8.62E-05	5.59E-04	1.33E-05
Selenium	8.19E-06	2.13E-05	3.34E-06	8.19E-06	2.13E-05	3.34E-06
Tin	1.15E-04	2.70E-04	1.45E-05	1.15E-04	2.70E-04	1.45E-05
Vanadium	1.02E-05	4.07E-05	4.84E-07	1.02E-05	4.07E-05	4.84E-07
Zinc	1.14E-03	5.00E-03	2.60E-04	1.14E-03	5.00E-03	2.60E-04
d **Other Organics (ug/Nm^3 @12% CO2, dry)**	**Geometric Mean**	**Lower Range**	**Upper Range**	**Geometric Mean**	**Lower Range**	**Upper Range**
Polychlorinated Biphenyls	0.478	1.97	0.116	0.478	1.97	0.116
Polycyclic Aromatic Hydrocarbons						
Carcinogenic	0.252	1.43	0.044	0.252	1.43	0.044
Total	1.234	6.20	0.246	1.234	6.20	0.246
Aldehydes	417	617	214	417	617	214
e **Criteria Pollutants/ and Acid Gases (ppmdv @ 7% O2)**	**Geometric Mean**	**Lower Range**	**Upper Range**	**Geometric Mean**	**Lower Range**	**Upper Range**
a Carbon Monoxide	27.5	19.0	39.8	27.5	19.0	39.8
a Nitrogen Oxides	210	159	277	210	159	277
d Total Hydrocarbons	2.55	1.01	6.48	2.55	1.01	6.48
Sulfur Dioxide	14.6	4.3	49.1	96.8	65.9	142.3
d Sulfuric Acid	1.27	0.31	5.16	1.27	0.31	5.16
Hydrogen Chloride	11.7	4.0	34.2	239.7	99.4	578.2
Hydrogen Fluoride	0.20	0.07	1.40	7.60	NA	NA

a The data showed that there is little to no difference in emissions between facilities with and without acid gas controls. Therefore, the emission factors shown are for all facilities.
b 1.13E-05 is an abbreviation for 1.13 x 10^-5, or 0.0000113
c 3.4 percent of the total chromium was assumed.
d There was not enough data to make any correlation between facilities with and without acid gas controls.
e The lower and upper ranges are based upon the normal distribution.

Lower dioxin/furan emissions seem to be related to reduced flue gas outlet temperature, particulate and acid gas controls, and the use of newer combustion and control technologies. Because of the scarcity of reliable test data, results presented herein can only be considered tentative. Although the specific mechanisms and parameters associated with lower pollutant emissions cannot yet be determined, particularly in light of their apparent interdependency, results of our analyses suggest that the newer facilities emit less PCDDs and PCDFs than older, existing facilities.

VIII. NOTE ADDED IN PROOF

Two significant developments have occurred since the preparation of the draft chapter. First, the U.S. EPA has adopted a new toxic equivalency weighting system for dioxins/furans. Second, the U.S. EPA has proposed new source performance standards (NSPS) for municipal waste combustors. The following sections outline these developments.

A. REVISED U.S. EPA TOXIC EQUIVALENCY FOR DIOXINS/FURANS

The U.S. EPA has adopted the international toxicity equivalency factors developed under the auspices of the North Atlantic Treaty Organization's Committee on Challenges of Modern Society with participation by the U.S. EPA.[77] This weighting system requires congener-specific emissions data in addition to the 2,3,7,8 and non-2,3,7,8 congeners of the tetra- through hepta-homologs of both PCDDs and PCDFs required for most previous weighting systems. Not only are the octa-homologs assigned toxicity equivalency values, but 1,2,3,7,8-P_5CDF and 2,3,4,7,8-P_5CDF are assigned different equivalency values. On the other hand, the new system assigns a zero toxicity equivalency value to all non-2,3,7,8 congeners. The overall effect of the revision is that the current (1989) U.S. EPA TEF is generally about twice the value of the previous (1987) U.S. EPA TEF.

B. PROPOSED NEW SOURCE PERFORMANCE STANDARDS

On December 20, 1989, the U.S. EPA proposed new source performance standards (NSPS Subpart Ea) for municipal waste combustors constructed, modified, or reconstructed after that date.[78] Standards were proposed for metals, organics, acid gases, and nitrogen oxides. Different standards were proposed for large and small facilities. For large facilities, defined as those with a total capacity greater than 250 tons/d, total dioxin/furan emissions are proposed to be limited to a value between 5 and 30 ng/Nm^3 at 7% O_2, dry; SO_2 emissions will be limited by an 85% reduction by weight or to 30 ppm, dry volume at 7% O_2, whichever is less stringent; HCl emissions will be limited by an 95% reduction by weight or to 25 ppm, dry volume at 7% O_2, whichever is less stringent; NO_x emissions are proposed to be limited to a value between 120 and 200 ppm, dry volume at 7% O_2. For small facilities, defined as those with a total capacity of 250 tons/d or less, total dioxin/furan emissions are proposed to be limited to 75 and 250 ng/Nm at 7% O_2, dry for massburn and RDF facilities, respectively; SO_2 emissions will be limited by an 50% reduction by weight or to 30 ppm, dry volume at 7% O_2, whichever is less stringent; HCl emissions will be limited by an 80% reduction by weight or to 25 ppm, dry volume at 7% O_2, whichever is less stringent. Metal emissions from all sizes of facilities would be controlled by limiting particulate emissions to 0.015 gr/dscf at 7% O_2, dry and by limiting opacity to 10% (on a 6-min average). As an operating practice to ensure good combustion for all sizes of facilities, CO emissions (dry volume at 7% O_2) are limited to 50 ppm for modular (starved air and excess air) facilities; 100 ppm for massburn waterwall and refractory, and bubbling or circulating fluidized bed combustors; and 150 ppm for massburn rotary waterwall, RDF spreader stokers, coal/RDF co-fired combustors, and other technologies. Somewhat less stringent standards were also proposed for existing municipal waste combustors. Final revised standards including specific numeric limitations for dioxins/furans and NO_x for both new and existing sources are expected to be promulgated in December 1990.

REFERENCES

1. **Trichon, M. and Feldman, J.,** The Formation of Trace Toxic Emissions Resulting from Incineration, presented at the 80th Annual Air Pollution Control Association Meeting, New York, June 21-25, 1987.
2. **Anderson, C. L., Wertz, K. L., and Gergen, W. P.,** Metals Analysis by NAA for the Marion County Emission Test Program Performed in September 1986. Revised October 20, 1987, Radian Corporation, personal memorandum.
3. **Cavallaro, A., et al.,** Summary of results of PCDDs analyses from incinerator effluents, *Chemosphere*, 2, 859, 1982.
4. **Clement, R., Tosine, H., et al.,** Levels of chlorinated organics in municipal incinerators, in *Chlorinated Dioxins and Dibenzo-Furans in the Total Environment II*, Keith, L., Rappe, C., and Choudhary, G., Eds., 1985, 489.
5. Cooper Engineers, Inc., Air Emissions Tests of a Deutsche Babcock Anlagen Dry Scrubber System at the Munich North Refuse-Fired Power Plant, unpublished data, 1985.
6. Cooper Engineers, Inc., West County Agency of Contra Costa County Waste Co-Disposal/Energy Recovery Project, unpublished data, 1985.
7. **DeFre, D.,** Dioxin levels in the emissions of Belgian municipal incinerators, *Chemosphere*, 15(9-12), 1255, 1986.
8. Dioxins and Furans Content in the Residues and Air Emissions from the Refuse Incineration Plant Ahrensburger Weg, unpublished data, 1984.
9. Energy Systems Associates, Air Emissions Tests at Commerce Refuse-to-Energy Facility, May 26-June 5, 1987, unpublished data, 1987.
10. Energy Systems Associates, Air Emissions Tests at Commerce Refuse-to-Energy Facility, unpublished data, 1988.
11. Entropy Environmentalists, Inc., Stationary Source Sampling Report — Massachusetts Refusetech, Inc., Resource Recovery Facility, North Andover, MA, unpublished data, 1985.
12. Entropy Environmentalists, Inc., Stationary Source Sampling Report — Ogden Martin Systems of Haverhill, Inc., Lawrence, Massachusetts Thermal Conversion Facility, unpublished data, 1987.
13. Entropy Environmentalists, Inc., Stationary Source Sampling Report — Signal RESCO Pinellas County Resource Recovery Facility, unpublished data, 1987.
14. Environcon Limited, Report on Combustion Testing Program at the SWARU Plant, Hamilton-Wentworth, Report ARB-45-84-ETRD, 1984.
15. Environmental Protection Service, Measurement of Emissions of Polychlorinated Dibenzo-*p*-Dioxin and of Polychlorinated Dibenzofuran from the Des Carrieres Incinerator in Montreal, Report EPS 5/UP/RQ/1, Environment Canada, Toronto, Ontario, Canada, 1984.
16. Environmental Protection Service, The National Incinerator Testing and Evaluation Program: Two-Stage Combustion-Prince Edward Island, Volumes 1-5. Report EPS 3/UP/1, Environment Canada, Toronto, Ontario, Canada, 1985.
17. Flakt Canada Ltd. and Environment Canada, The National Incinerator Testing and Evaluation Program: Air Pollution Control Technology, Volumes 1-5, Report EPS 3/UP/2, Environment Canada, Toronto, Ontario, Canada, 1986.
18. **Gizzi, F., et al.,** Polychlorinated dibenzo-*p*-dioxins and polychlorinated dibenzofurans in emissions from an urban incinerator — average and peak values, *Chemosphere*, 2, 577, 1982.
19. **Golenbiewski, M., Rnanth, K., et al.,** Environmental Assessment of a Waste-to-Energy Process: Braintree Municipal Incinerator, unpublished data, 1980.
20. **Haile, C., Blair, R., Lucas, T., et al.,** Midwest Research Institute, Assessment of Emissions of Specific Compounds from a Resource Recovery Municipal Refuse Incinerator, U.S. Environmental Protection Agency 560/5-84-002, 1984.
21. **Haile, C., Stanley, J., Nulton, C., et al.,** Midwest Research Institute, Comprehensive Assessment of the Specific Compounds Present in Combustion Processes, Volume I — Pilot Study of Combustion Emissions Variability, U.S. Environmental Protection Agency 560/5-83-004, 1983.
22. Interpoll Laboratories, Results of the July 1987 Emission Performance Tests of the Pope/Douglas Waste-to-Energy Facility MSW Incinerators in Alexandria, Minnesota, unpublished data, 1987.
23. Midwest Research Institute, Dioxins Analysis of Philadelphia Northwest Incinerator Summary Report, Volume 2, Appendices A-F, 1985.
24. Midwest Research Institute, Results of the Combustion and Emission Research Project at the Vicon Incinerator Facility in Pittsfield, Massachusetts, unpublished data, 1987.
25. Ministry of the Environment, Determination of Chlorinated Dibenzo-*p*-Dioxins, Chlorinated Dibenzofurans, Chlorinated Biphenyls, Chlorobenzenes, and Chlorophenols in Air Emissions and Other Process Streams at SWARU, Ontario, Canada, 1984.
26. Mullins Environmental Testing Company, Inc., Source Emission Survey of City of Cleburne Waste-Fired Incinerator Precipitator Stack-Cleburne, TX, unpublished data, 1986.

27. New York State Department of Environmental Conservation, Emission Source Test Report — Preliminary Report on Cattaraugus County, ERF, unpublished data, 1986.
28. New York State Department of Environmental Conservation, Emission Source Test Report — Preliminary Report on Occidental Chemical Corporation EFW, unpublished data, 1986.
29. New York State Department of Environmental Conservation, Emission Source Test Report — Preliminary Report on Oneida County ERF, unpublished data, 1986.
30. New York State Department of Environmental Conservation, Emission Source Test Report — Sheridan Avenue RDF Plant: "ANSWERS," unpublished data, 1985.
31. New York State Department of Environmental Conservation, Emission Source Test Report — Preliminary Report on Westchester RESCO, unpublished data, 1986.
32. New York State Department of Environmental Conservation, Phase I Resource Recovery Facility Emission Characterization Study Overview Report — Addendum, October 24, 1988.
33. New York State Department of Environmental Conservation, Phase I Resource Recovery Facility Emission Characterization Study — Overview Report, unpublished data, 1987.
34. Office of Environmental Protection, Bundesamt Fur Umweltsch, Environmental Pollution Through Dioxins and Furans from MSW Incineration Plants, unpublished data, 1982.
35. Ogden Martin Systems, Inc. and Martin GMBH, Air Emissions Testing at the Wurzburg, West Germany, Waste-to-Energy Facility, unpublished data, 1986.
36. Ogden Projects, Inc., Environmental Test Report Regarding the Alexandria/Arlington Resource Recovery Facility, unpublished data, 1987.
37. Ogden Projects, Inc., Environmental Test Report Regarding the Bristol Resource Recovery Facility, unpublished data, 1988.
38. Ogden Projects, Inc., Environmental Test Report Regarding the Hillsborough Resource Recovery Facility, unpublished data, 1988.
39. Ogden Projects, Inc., Environmental Test Report Regarding Hogdalen, unpublished data, 1986.
40. Ogden Projects, Inc., Environmental Test Report Regarding the Marion County Solid Waste-to-Energy Facility, unpublished data, 1986.
41. Ogden Projects, Inc., Environmental Test Report Regarding the Walter B. Hall Resource Recovery Facility, unpublished data, 1986.
42. Ogden Projects, Inc., Environmental Test Report Regarding the Walter B. Hall Resource Recovery Facility, unpublished data, 1988.
43. **Olie, K., Lustenhouwer, A., and Hutzinger, O.,** Polychlorinated dibenzo-*p*-dioxins and related compounds in incinerator effluent, in *Chlorinated Dioxins and Related Compounds Impact on the Environment,* Hutzinger, O., Frei, R. W., Merian, E., and Pocchiari, I., Eds., Pergamon Press, New York, 1982, 227.
44. **Palombella, R.,** Municipal Solid Waste Incineration; Air Emissions and Dioxins, GRCDA Presentation, August 28, 1985.
45. PEI Associates, Inc., Emissions Test Report — Method Development and Testing for Chromium: No. 2 Refuse-to-Energy Incinerator Baltimore RESCO, unpublished data, 1986.
46. **Pohle, R. and Asndt, R.,** *Mull and Abfall,* April 1986, "Environmental Test Report Regarding Primary and Secondary Measures for Emission Reduction in Thermal Waste Utilization in Nuremburg", Translated for Ogden Projects, Inc. by Information on Demand, June 8, 1987.
47. Radian Corporation, Final Emissions Test Report for Dioxin/Furans and Total Organic Chlorides Emission Testing — North Andover Resource Recovery Facility, unpublished data, 1986.
48. Radian Corporation, Final Emissions Test Report for Dioxin/Furans and Total Organic Chlorides Emission Testing — Saugus Resource Recovery Facility, unpublished data, 1986.
49. **Rappe, C., Marklund, S., et al.,** Polychlorinated dibenzo-*p*-dioxins, dibenzofurans, and other polynuclear aromatics formed during incineration and polychlorinated biphenyl fires, in *Chlorinated Dioxins and Dibenzofurans in the Total Environment,* Choudhary, G., et al., Eds., Butterworths, Reading, MA, 1983.
50. Roy F. Weston, Inc., Source Emissions Compliance Test Report — Orrington, Maine, unpublished data, 1988.
51. Roy F. Weston, Inc., Source Emissions Compliance Test Report — City of Philadelphia, Northwest Incinerator, unpublished data, 1988.
52. Roy F. Weston, Inc., Source Emissions Compliance Test Report — Energy Generating Facility, Pigeon Point, Delaware, unpublished data, 1988.
53. Roy F. Weston, Inc., Source Emissions Report for Dioxins and Dibenzofurans — Pittsfield (Vicon) Resource Recovery Facility, unpublished data, 1986.
54. Swedish Environmental Protection Agency, Operational Studies at the SYSAV Energy from Waste Plant Malmo, Sweden, Translated by Flakt Canada, Ltd., June 1983.
55. **Wurband, M. and Atkins, R.,** Interaction of Emission Factors, BACT Alternatives, and Predicted Impacts on Air Permit Applications for Resource Recovery Facilities, presented at the Spring 1986 Conference of the New England Section of the Air Pollution Control Association, Providence RI, April 22-26, 1986.
56. U.S. Environmental Protection Agency, Report to Congress, EPA/530-SW-87-021a, Office of Solid Waste and Emergency Response, Washington, D.C., 1987.

57. **Bellin, J. S. and Barnes, D. G.,** Interim Procedures for Estimating Risks Associated with Exposures to Mixtures of Chlorinated Dibenzo-*p*-Dioxins and -Dibenzofurans (CDDs and CDFs), EPA/625/3-87-012, Risk Assessment Forum, U.S. Environmental Protection Agency, Washington, D.C., 1987.

57a. *Polynuclear Aromatic Compounds, Part I: Chemical, Environmental and Experimental Data,* International Agency for Research on Cancer, Lyon, France, 1983.

58. *Methodology for the Assessment of Health Risks Associated with Multiple Pathway Exposure to Municipal Waste Combustor Emissions,* U.S. Environmental Protection Agency, Office of Air Quality Planning and Standards, Research Triangle Park, NC, and the Environmental Criteria and Assessment Office, Cincinnati, OH, 1986.

59. *Monograph on the Evaluation of Carcinogenic Risks to Humans, Supplement 7,* International Agency for Research on Cancer, Lyon, France, 1987.

60. **Russell, S. H. and Roberts, J. E.,** Oxides of nitrogen: Formation and control in resource recovery facilities, *Proceedings of the 1984 National Waste Processing Conference,* American Society for Mechanical Engineers, New York, 1984.

61. California Air Resources Board, Air Pollution Control at Resource Recovery Facilities, unpublished data, 1984.

62. **Siebert, P. C., Alston, D. R., Walsh, J. F., and Jones, K. H.,** Statistical Properties of Available Worldwide MSW Combustion Dioxin/Furan Emissions, presented at 80th Annual Air Pollution Control Association Meeting, New York, June 25, 1987.

63. **Siebert, P. C., Alston, D. R., Walsh, J. F., and Jones, K. H.,** Effect of Control Equipment and Operating Parameters on Municipal Solid Waste (MSW) Incinerator Trace Emissions, presented at the 81st Annual Air Pollution Control Association Meeting, Dallas, June 24, 1988.

64. Hydrogen Chloride Interference in U.S. EPA Method 8, Technical Bulletin No. 1, Ogden Projects, Inc., Emeryville, CA, June 17, 1988.

65. TRC Environmental Consultants, Air Emission Compliance Test Results and Related Data at the Mid-Connecticut Resource Recovery Facility, Hartford, CT, August 1988.

66. Clean Air Engineering, Inc., Report on Compliance Testing Performed at Jackson County Resource Recovery Facility, August 1988.

67. Entropy Environmentalists, Inc., Stationary Source Sampling Report Wheelabrator Resource Recovery Facility, Millbury, MA, August 1988.

68. State of New York, Official Compilation of Codes, Rules and Regulations of the State of New York, 6 NYCRR 219-1.1(b)(2), December 1988.

69. Emission Test Report for Bridgeport, Connecticut, Resource Recovery Facility, unpublished data, 1989.

70. Entropy Environmentalists, Inc., Stationary Source Sampling Report for Maine Energy Recovery Company, Biddeford, ME, unpublished data, 1987.

71. Pacific Environmental Services, Inc., Compliance Test Report on Air Emissions from Incinerators 1 and 3 at the Montgomery County North Reduction Plant, Dayton, OH, August 1988.

72. Scott Environmental Services, Sampling and Analysis of Chlorinated Organic Emissions from the Hampton Waste-to-Energy System, prepared for Bionetics Corporation, unpublished data, 1985.

73. **Eadon, G., Kaminsky, L., Silkworth, J., Aldous, K., Hilker, D., O'Keefe, P., Smith, R., Gierthy, J., Hawley, J., Kim, N., and Decaprio, A.,** Calculation of 2,3,7,8-TCDD equivalent concentrations of complex environmental contaminant mixtures, *Environ. Health Perspect.,* 70, 221, 1986.

74. California Air Resources Board, Staff Report: Public Hearing to Consider Adoption of a Regulatory Amendment Identifying Chlorinated Dioxins and Dibenzofurans as a Toxic Air Contaminant, Sacramento, CA, 1986.

75. Ontario Federal Government, Chlorinated Dioxins and Chlorinated Dibenzofurans: Ambient Air Guidelines, Human Health Services, Ministry of Labor, Ottawa, Canada, 1982.

76. Midwest Research Institute, Municipal Waste Combustion Study: Emissions Data Base for Municipal Waste Combustors, EPA/530-SW-87-021B, U.S. Environmental Protection Agency, Washington, D.C., 1987.

77. U.S. Environmental Protection Agency, Interim Procedures for Estimating Risks Associated with Exposures to Mixtures of Chlorinated Dibenzo-*p*-Dioxins and -Dibenzofurans (CDDs and CDFs) and 1989 Update, EPA/625/3-90/016, Risk Assessment Forum, U.S. Environmental Protection Agency, Washington, D.C., 1989.

78. U.S. Environmental Protection Agency, Standards of Performance for New Stationary Sources; Municipal Waste Combustors, 40 CFR Parts 51, 52 and 60, Federal Register, Vol. 54, No. 243, pp. 52251-52304, Washington, D.C., December 20, 1989.

79. **Jones, K. H., Siebert, P. C., Alston, D. G., and Walsh, J. F.,** Worldwide MSW Combustion Dioxin/Furan Emissions and Their Regulatory Implications, presented at the NITEP International Workshop on Municipal Solid Waste (MSW) Incineration, Toronto, Ontario, Canada, October 2, 1987.

80. **Siebert, P. C., Alston, D. R., and Jones, K. H.,** Toxic Trace Pollutants from Incineration, presented at the AIChE National Meeting, Philadelphia, August 23, 1989.

Chapter 2

REGULATORY ANALYSIS OF POLLUTANT EMISSIONS, INCLUDING POLYCHLORINATED DIBENZO-*p*-DIOXINS (CDDs) AND DIBENZOFURANS (CDFs), FROM MUNICIPAL WASTE COMBUSTORS

David H. Cleverly, Rayburn M. Morrison, Brenda L. Riddle, and Robert G. Kellam

TABLE OF CONTENTS

I. INTRODUCTION AND BACKGROUND

The combustion of municipal solid waste (MSW) represents an increasingly important element of the solid waste disposal problem in the U.S. At the time of the U.S. Environmental Protection Agency's (U.S. EPA) regulatory analysis, approximately 111 MWCs incinerated more than 6×10^6 metric tons of MSW annually.[1] This amount represents about 5 to 6% of the total annual mass of MSW generated in the U.S. The U.S. EPA estimates, however, that by the mid 1990s, combustion could account for as much as 25% of domestic MSW disposal in as many as 300 facilities.[2]

The anticipated growth in the municipal waste combustor (MWC) industry is largely the result of a consistent growth in the volume of MSW generated and increasing concern about the continuous availability and environmental impacts of MSW disposal in landfills. (U.S. EPA uses the term "municipal waste combustor" to refer to an incineration facility solely dedicated to the combustion of MSW or garbage.) While 80% of MSW is currently disposed of in about 10,000 landfills nationwide, many of these facilities are nearing capacity.[3] In addition, the siting of new landfills has become increasingly difficult, particularly in densely populated areas.

Although the combustion of MSW does not eliminate the need for landfilling, since residual ash must be disposed of, it does reduce the volume of waste by 70 to 90%, thus extending the operating lifetime of landfills.[3] Another benefit of incineration is the ability to recover energy from the heat of combustion in the form of steam or hot water, which can be used to partially offset the energy requirements of the facility or can be sold to industrial clients and local electric utilities.

Disposal of MSW by combustion, however, releases potentially harmful pollutants into the atmosphere. Currently, EPA only limits the emission of particulate matter (PM) to 180 mg per dry standard cubic meter (mg/dscm) (corrected to 12% CO_2) from MWCs built before 1986. In 1986 EPA promulgated a particulate matter (PM) standard equivalent to the emission of 70 mg/dscm for industrial boilers, including new energy recovery MWCs with capacities greater than approximately 180 metric tons/d. The present control of emissions is not intended to address health concerns specific to individual chemical constituents or other pollutants of MWC emissions. Because the current standard for the control of PM emissions may not sufficiently protect public health, an analysis to determine the need for additional federal regulation was undertaken in 1987. The following sections describe the methods and results of the U.S. EPAs regulatory analysis of MWC emissions, and U.S. EPAs recent determination that further federal regulation of these emissions is warranted.[3]

II. MUNICIPAL WASTE COMBUSTION TECHNOLOGY

MWCs operating in the U.S. can be classified into three general design categories: massburn, modular, and refuse-derived fuel (RDF). The first type is called massburn, because the waste is combusted without any preprocessing other than removal of items too large to go through the feed system. In a typical massburn combustor, refuse is placed on a grate that moves through the combustor. Combustion air in excess of stoichiometric amounts is supplied both below (under fire air) and above (overfire air) the grate. Massburn combustors are usually field-erected and can burn 50 to 1000 tons of MSW per day per incineration unit. Many massburn facilities have two or more combustor units comprising the incineration facility with a total refuse combustion capacity ranging from 45 to 2720 metric tons/d.[3]

Modular combustors also burn waste without preprocessing, but they are typically factory fabricated and shipped in preassembled sections to the intended site. Modular combustors generally incinerate 5 to 100 tons of refuse per day per incineration unit.[3] One of the most common types of modular combustors is the starved air or controlled air type, incorporating two combustion chambers. Air is supplied to the primary chamber at substoichiometric levels. The incomplete combustion products entrained in the combustion gases from the primary combustion chamber pass into the secondary combustion chamber where excess air is added, and combustion is completed. Another type of modular combustor, functionally similar to larger massburn units, uses excess air in the primary chamber; no additional air is added in the secondary chamber. Modular facilities are also comprised of one or more incineration units have a total combustion capacity of 4.5 to 270 metric tons of MSW per day.[3]

The third major type of MWC is designed to burn RDF. RDF is a general term describing MSW which has been processed into a physical form that theoretically enhances combustibility. RDF is a fuel that has been prepared by shredding, sorting, separating, and trommeling MSW to reduce the noncombustible content. It may vary in physical form from shredded waste having a uniform size to a finely divided fuel suitable for co-firing with pulverized coal. Some cases of RDF processing involves magnetic separation of ferrous metals and reduction of other noncombustible components of the refuse, such as aluminum and glass products. RDF facilities have a total combustion capacity of 227 to 2720 metric tons/d.[3]

Existing MWCs can be further classified as having or not having heat recovery boilers. All RDF MWCs have boilers to recover heat. While some RDF units ar designed to be co-fired with fossil fuels, only facilities burning RDF are defined as MWCs in this study. All facilities operating prior to 1986 were classified as existing MWCs. The U.S. EPA[1] has estimated the projected growth of MWCs from 1986 to 1995 in terms of facilities and total refuse incineration capacity. Projected facilities are defined as planned but not yet operating and were either under construction, had contracts for construction, or had formally been proposed at the time of the U.S. EPAs regulatory impact analysis. The number and total design capacity of existing and projected MWC facilities in the U.S. are summarized in Table 1. The geographical distribution of existing and projected MWC facilities is shown in Figure 1.

III. TOXIC EQUIVALENCY OF CDDs AND CDFs

In the past decade, significant advances have been made in understanding the biological mechanisms of CDD/CDF action on a macromolecular level. A mechanism of evoking the characteristic toxic response is postulated to occur by a halogenated dioxin or dibenzofuran compound that reversibly and noncovalently binds to a receptor protein in the cell cytosol and then translocates to the cell nucleus.[4,7,8,11] The parent compound, not a metabolite, is thought to trigger a series of intranuclear events, which include activation of messenger RNA(s) that lead directly to the induction of a group of enzymes within the endoplasmic reticulum.[4-6,9,10] Many halogenated aromatic compounds, such as chlorinated dioxins, are so fat soluble that without the action of an enzyme complex embedded within the endoplasmic reticulum, commonly referred to as the microsomal monooxygenase system, they would persist in the human body for long periods of time. This biochemical system is called an electron transport system, because the enzymes insert one atom of diatomic molecular oxygen into an aromatic compound to cleave the aromatic ring. Thus, a hydroxyl group is introduced into the xenobiotic compound during phase I metabolism followed by attack of the xenobiotic compound by conjugating enzymes in phase II metabolism.[5,6] The conjugating products are sufficiently hydrophilic that the detoxified products are more water soluble and readily excreted from the cell.

An example of phase I enzymes is a group of hemoproteins collectively referred to as cytochrome P-450 reductase.[5] Currently, at least seven species of cytochrome P-450 with distinct molecular weights, amino acid composition, and terminal amino acid sequences have been quantified in the cell cytosol of rat liver an chick embryos.[7] The most frequently observed enzyme mediated by the cytochrome P-450 when cells are challenged by chlorinated dioxin and dibenzofuran compounds *in vitro* is aryl hydrocarbon hydroxylase (AHH).[4-12] The induction of AHH appears to be mediated by a protein receptor within the cell cytoplasm. This receptor stereospecifically binds to the CDD/CDF compound. The strength of this CDD/CDF-receptor

TABLE 1
Number and Capacity of Existing and Projected MWCs in the U.S.

Design type	Existing MWCs: No. of MCWs	Existing MWCs: Capacity (metric tons/d)	Projected MWCs: No. of MCWs	Projected MWCs: Capacity (metric tons/d)
Massburn				
No heat recovery	21	13,000	0	0
With heat recovery	24	20,100	118	113,000
Modular				
No heat recovery	17	600	0	0
With heat recovery	39	3,900	24	5,000
RDF				
With heat recovery	10	11,400	31	39,000
Unknown	0	0	37	36,000
Total	111	49,000	210	193,000

bond appears to be directly proportional to the amount of AHH enzyme that is induced by the endoplasmic reticulum.[8-10] The most important finding of receptor binding research is that AHH enzyme induction has been strongly correlated with toxic manifestations, including shrinking of the thymus, thickening of skin cells, acute lethality, birth defects in offspring (such as cleft palate), tumor formation, and drastic weight loss, in animals exposed to 2,3,7,8-TCDD.[4,6,8]

The affinity of various CDD/CDF compounds for the protein receptor is largely a function of a molecule's geometry. Dioxin compounds with chlorine atoms substituted in the four lateral positions (i.e., in the 2, 3, 7, and 8 positions) have the greatest affinity for the receptor and maximum inducibility of AHH activity. Additional chlorine substitution at positions 1, 4, 5, or 9 decreases AHH activity, and removal of one or more of the laterally attached chlorine atoms diminishes activity even further.[4,13] By challenging *in vitro* bioassays with different chlorinated dioxin and dibenzofuran compounds, it is possible to rank the 135 chlorinated dioxins and 75 chlorinated dibenzofurans relative to the ability of 2,3,7,8-TCDD to bind to the receptor or express AHH activity in cell culture. Thus, if 2,3,7,8-TCDD is arbitrarily set to 1, representing maximum AHH induction *in vitro*, then other CDDs and CDFs will be measured as a fraction of 1 corresponding to their relative biological response. This comparative response is characterized as the toxic equivalency procedure.[13] In 1987, the U.S. EPA formally adopted an interim procedure for estimating the human health risks associated with exposures to mixtures of CDDs and CDFs.[14] The procedure utilizes a set of derived toxicity equivalency factors (TEFs) to convert the concentration of any CDD/CDF mixture present in an environmental sample into an equivalent concentration of 2,3,7,8-TCDD. This approach simplifies the assessment of both carcinogenic and noncarcinogenic risks involving exposures to mixtures of CDDs and CDFs, such as is emitted from MWCs. The concentration of CDDs and CDFs emitted from MWC stacks is reported in terms of toxic equivalents (TEQ), or in terms of the known toxicity of 2,3,7,8-TCDD. This exposure concentration is multiplied by the appropriate cancer potency estimate or compared to the risk specific dose associated with teratogenicity for 2,3,7,8-TCDD.

IV. EMISSIONS OF CDDs AND CDFs FROM MWCs

Emissions of polychlorinated dibenzo-*p*-dioxins (CDDs) and dibenzofurans (CDFs) from MWCs have been well documented by stack sampling,[15] but the precise mechanism of formation during combustion remains unknown and largely theoretical. Various theories proposed to explain the presence of CDDs and CDFs in municipal waste combustor emissions include: (1) they are present in bleached paper or other MSW constituents, and a portion of them survives

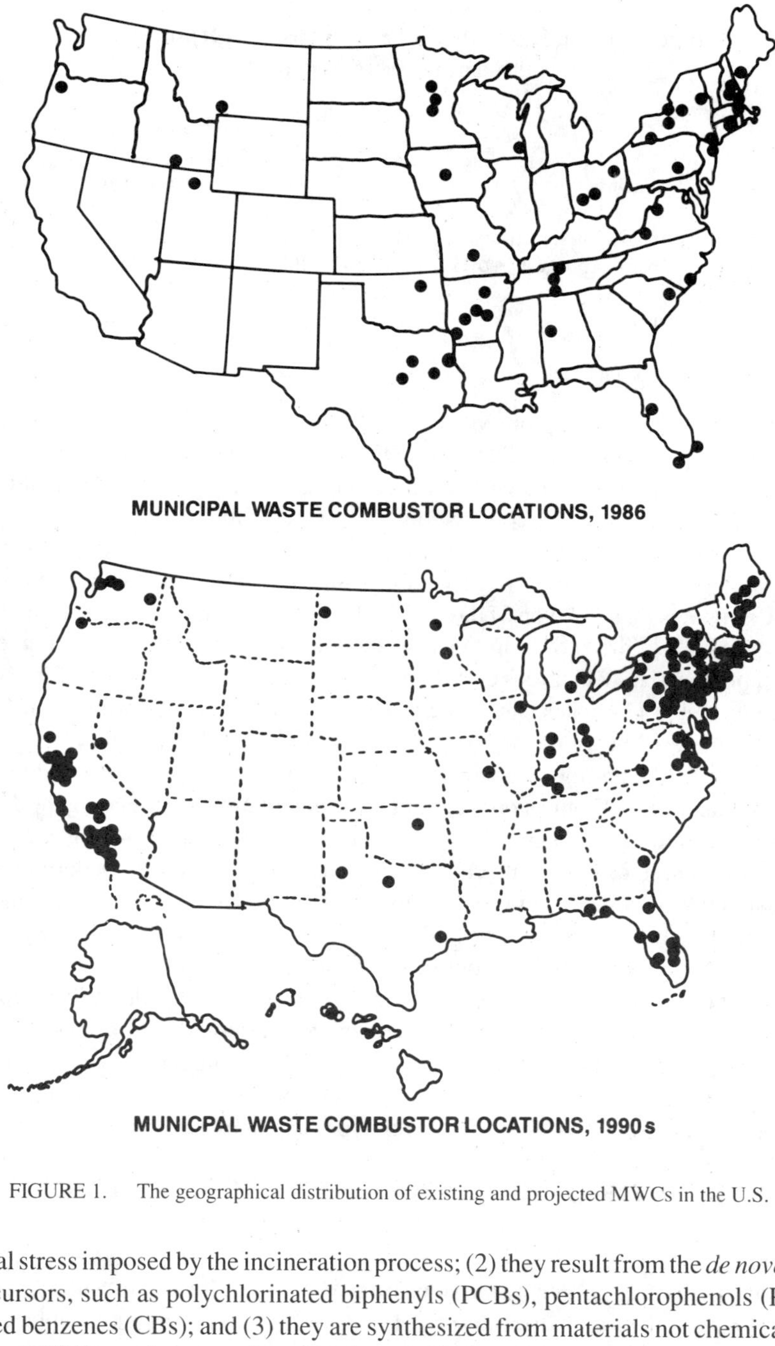

FIGURE 1. The geographical distribution of existing and projected MWCs in the U.S.

the thermal stress imposed by the incineration process; (2) they result from the *de novo* synthesis from precursors, such as polychlorinated biphenyls (PCBs), pentachlorophenols (PCPs), and chlorinated benzenes (CBs); and (3) they are synthesized from materials not chemically related to CDDs and CDFs, such as petroleum products, chlorinated hydrocarbons, inorganic chloride ions, and plastics.[16,17] CDDs and CDFs are suspected to be created on the surface of flyash downstream of the furnace in regions where combustion offgases have cooled to between 200° and 400°C.[18,19]

Vogg et al.[19] theorized that inorganic chloride ions present during combustion, such as copper chloride, may act as a catalyst for surface reactions on particulate matter to convert aromatic compounds to chlorinated dioxins and dibenzofurans. Direct evidence for the thermal conversion of polyvinyl chloride to CDDs and CDFs was demonstrated in laboratory pyrolysis experiments.[20] In these experiments, the isomeric distribution of CDDs and CDFs was similar

TABLE 2
Range of Toxic Equivalent (TEQ) Emissions from MWCs (milligrams per metric ton)

Facility type	Emissions factor	
	Existing	Projected
Massburn — heat recovery	0.14—3.2	0.008—0.2
Massburn — nonheat recovery	0.44—10.3	NA[a]
RDF	0.30—6.7	0.2—4.0
Modular	0.01—0.2	0.01—0.2

[a] Data not applicable.

to the pattern found in emissions from MWCs.[20,21] Choudhary et al.[22] proposed that lignin in the refuse may thermally degrade into CDDs and CDFs in the presence of chlorine. This theory was later confirmed by Liberti and Brocco[23] and Liberti et al.[24] who demonstrated that CDDs and CDFs can be produced from the combustion of pure vegetable extracts in the presence of chlorine. Since MSW is an extremely heterogenous waste, it is likely that most, if not all, of the theoretical formation mechanisms suggested may actually occur during combustion.

The EPA estimates that 24 kg of CDDs and CDFs (expressed in TEQ) are emitted annually from existing MWCs in the U.S.[3,25] In the absence of additional federal regulation, emissions from the 210 projected new facilities could add an additional 21 kg TEQ/year for a total of 45 kg TEQ/year by 1995. Estimates can only approximate, since actual emissions of CDDs and CDFs from MWCs are known to vary by as much as three orders of magnitude.[15,18]

A summary of TEQ emissions from each of the three major classes of MWCs are presented in Table 2. The range of TEQ emissions represents the uncertainty in estimating the distribution of toxic isomers within homolog groups as well as the uncertainty in measuring trace quantities of CDDs and CDFs in stack gases with conventional sampling protocols.[25] In deriving emission factors, a distinction was made between existing and projected MWCs. This distinction was based on the premise that new and proposed MWCs represent definite improvements in design, combustion efficiency, and pollution control compared to facilities built 1 decade ago. CDD/CDF emission factors were calculated by averaging emissions data for MWCs representing subcategories of technologies and dividing those average emissions by the rate of MSW fed into the incinerator and are expressed in units of grams of TEQ per megagram (metric ton) of MSW burned. Emissions factors for each MWC technology were determined by calculating the average emission rate for all facilities of a given type (e.g., massburn, modular, and RDF).

V. EMISSIONS OF OTHER POLLUTANTS OF CONCERN FROM MWCs

The evaluation of other pollutant emissions from MWCs was limited to pollutants for which emission test data and some indication of public health concern were available. Data were sufficient for analysis of 17 pollutants or classes of pollutants as summarized in Table 3. On a mass basis, the predominant pollutants emitted during the combustion of MSW are carbon monoxide, hydrogen chloride, nitrogen oxides, and sulfur oxides.[3,15]

In general, emission factors were derived by averaging available emissions data for a given pollutant and facility type. Different emissions factors were developed for each pollutant to represent emissions from existing and projected MWC populations. The existing MWC population was defined as all facilities operating as of 1986, while projected facilities were defined as facilities planned, constructed, or operational after that date.

Subjective judgments were made regarding the selection of certain facilities to represent

TABLE 3
Estimated Emission from Existing and Projected MWCs (metric tons per year)

Compound	Existing[a]	Projected[b]
Arsenic	2.7	13.0
Beryllium	0.1	0.6
Cadmium	10.4	19.9
Carbon monoxide	20000.0	84500.0
Chlorobenzenes	3.8	2.5
Chlorophenols	5.4	3.5
Chromium (+6)	0.6	2.6
CDDs/CDFs[c]	0.024	0.021
Formaldehyde	58.6	119.0
Hydrogen chloride	47700.0	195000.0
Lead	341.0	1260.0
Mercury	68.0	148.0
Nitrogen oxides	30500.0	134000.0
Particulate matter	10400.0	26700.0
Polychlorinated biphenyls	5.0	21.0
Polycyclic aromatic hydrocarbons	1.0	4.0
Sulfur oxides	13000.0	63000.0

[a] Based on existing air pollution control.
[b] Based on new MWCs achieving 99% control of particulate matter.
[c] Expressed in TEQs.

potential pollutant emissions from existing and projected MWCs. Emission factors for existing facilities were derived using emissions data from most tested facilities with the exception of those facilities that use specific state-of-the-art pollution control equipment. Only data from facilities judged to have combustion technologies representative of newer, more advanced control technologies were used to derive emission factors for projected facilities.

To determine if additional federal regulation of pollutant emissions from MWCs is warranted, emissions from existing MWCs that do not use modern, sophisticated pollution control equipment (i.e., baseline emissions) were compared to emissions from facilities that use a dry-scrubber combined with a fabric filter (i.e., controlled emissions). The following assumptions were made regarding the baseline scenario of control of pollutant emissions. For existing MWCs, the baseline scenario for organic emissions assumed uniform use of electrostatic precipitators (ESPs). Although a variety of particulate control devices are actually in place on existing units, available organic emissions data were collected from ESP-controlled units only. Baseline conditions assumed 20% control of organic pollutants and current levels of particulate matter control.[25] Controlled levels for metal emissions (except mercury) were assumed to be proportional to controlled levels for particulate emissions.[25]

For the projected population, baseline conditions assumed the uniform applications of ESPs, although some states are requiring more stringent control. Twenty and 99% control were assumed for organic and particulate matter emissions, respectively,[25] and metal emissions (with the exception of mercury) were assumed to be controlled in proportion to particulate matter emissions.[25] These controlled efficiencies reflect optimized containment of these pollutants using advanced control technologies. For baseline, no control of inorganic acid gases, such as sulfur dioxide or hydrogen chloride, was assumed for either existing or projected facilities. For both existing and projected facilities, controlled conditions assumed the uniform application of dry alkaline scrubbers in combination with fabric filters. The percent control of organic, metal (except for mercury), mercury, and HCl baseline and controlled emissions are summarized in

TABLE 4
Percent Emissions Controlled Under Different Scenarios Modeled for Existing and Projected MWCs

Control scenario	Technology	Organics	Metals (except Hg[a])	Hg	HCl[b]
Existing					
Baseline	ESP[c]	20	NA[d]	30	0
Controlled	Dry alkaline scrubber with fabric filter	95	99.5	50	90
Projected					
Baseline	ESP	20	99	30	0
Controlled	Dry alkaline scrubber with fabric filter	95	99.5	50	90

[a] Hg = mercury.
[b] HCl = hydrochloric acid.
[c] Electrostatic precipitator
[d] Proportional to actual particulate matter control levels.

Table 4. The development of the emission factors for the various pollutant classes are described in the following sections.

A. ORGANIC EMISSIONS FACTORS

Emission factors for organics, expressed in grams of pollutant per metric ton of MSW combusted, are presented in Table 5 for the different MWC technologies modeled. Organic emissions factors were developed by dividing the measured organic pollutant emission rate by the MSW combustion rate for each facility tested. Emission factors for each MWC technology modeled were determined by calculating the average emission rate for all facilities of a given type. The organic emission factors in Table 5 were used to estimate organic emissions from MWCs under the baseline scenario, assuming that 20% control of organics was achievable through the application of ESPs alone. Under the controlled scenario, 95% control of organic emissions was assumed to be achievable through uniform application of dry alkaline scrubbers combined with fabric filters, since full- and pilot-scale systems equipped with dry scrubbers in combination with a baghouse system (fabric filters) generally achieved better than 95% control of organic emissions when operated at temperatures below 200°C.[25-27] To estimate organic emissions after application of dry alkaline scrubbers in combination with fabric filters (i.e., controlled emissions factors), baseline emission factors were adjusted as follows:

$$EF_C = EF_B\,(1 - CE_C)/(1 - CE_B), \tag{1}$$

where: EF_C = emission factor for the controlled scenario; EF_B = emission factor for the baseline scenario; CE_C = controlled organic control efficiency for alkaline scrubber and fabric filters (0.95, Table 4); and CE_B = baseline organic control efficiency for ESP (0.20, Table 4).

B. METALS EMISSIONS FACTORS

Table 6 presents metal emissions factors estimated for the various MWC technologies. Metal emissions factors were derived by calculating the ratio of metals to particulate emissions in controlled particulate emission measurements for each tested facility. The average metal-to-particulate ratio was calculated for each combustor technology to determine baseline emission factors for metals. The metals emissions factors in Table 6 reflect the current control levels

TABLE 5
Organic Emissions Factors for MWCs
(g/mg of waste emitted)[a]

Organic	Facility type	Emission factor	
		Existing	Projected[b]
CDDs/CDFs[c,d]	Massburn/heat recovery	0.00096	0.00058
	Massburn/nonheat recovery	0.0031	NA[e]
	RDF	0.002	0.0012
	Modular	0.000069	0.000069
Benzo(a)pyrene	Massburn/heat recovery	0.048	0.048
	Massburn/nonheat recovery	0.048	NA
	RDF	0.13	0.13
	Modular	0.0038	0.0038
Polychlorinated biphenyls (PCBs)	Massburn/heat recovery	0.0021	0.00021
	Massburn/nonheat recovery	0.0021	NA
	RDF	1.4	1.4
	Modular	0.0045	0.0045
Formaldehyde	Massburn/heat recovery	5.2	2.5
	Massburn/nonheat recovery	5.2	NA
	RDF	1.1	1.1
	Modular	1.3	1.3
Chlorobenzenes	Massburn/heat recovery	0.32	0.0089
	Massburn/nonheat recovery	0.32	NA
	RDF	0.14	0.14
	Modular	0.018	0.018
Chlorophenols	Massburn/heat recovery	0.46	0.018
	Massburn/nonheat recovery	0.46	NA
	RDF	0.18	0.18
	Modular	0.018	0.018

[a] Emission factors were derived from an average of available emissions data for a given pollutant and facility type.

[b] Data used to calculate emission factors for projected facilities exclude test data for facilities not equipped with state-of-the-art combustion technology.

[c] CDDs/CDFs refers to the class of 75 chlorinated dibenzo-*p*-dioxins (CDDs) and 135 chlorinated dibenzofurans (CDFs).

[d] CDDs/CDFs emissions are reported as 2,3,7,8-TCDD toxic equivalent (TEQ) concentration based on homolog-specific data. Isomer-specific TEQ can be derived by dividing by 3 and 7 to produce a likely range of emissions factors.

[e] NA = not applicable. All projected MWC are assumed to be heat recovery facilities.

TABLE 6
Metals Emissions Factors for MWCs
(mg metal per gram of controlled particulate emissions)

Metal	Facility type	Emissions factor
Arsenic	MB water wall	335
MB refractory	11.9	
RDF	90	
Modular	353	
Beryllium	MB water wall	0.091
MB refractory	11.9	
RDF	65	
Modular	0.036	

TABLE 6 (continued)
Metals Emissions Factors for MWCs
(mg metal per gram of controlled particulate emissions)

Metal	Facility type	Emissions factor
Cadmium	MB water wall	686
MB refractory	1228	
RDF	106	
Modular	3310	
Chromium (+6)	MB water wall	9476
MB refractory	416	
RDF	10912	
Modular	818	
Lead	MB water wall	13425
MB refractory	61105	
RDF	10530	
Modular	72500	
Mercury	MB water wall	8163
MB refractory	6770	
RDF	868	
Modular	2340	
Nickel	MB water wall	6855
MB refractory	2812	
RDF	5770	
Modular	1610	

achieved by particulate matter control devices. In general, existing particulate controls may achieve up to 99% control of particulate and metals emissions, with the exception of mercury emissions, which are assumed to be controlled to about 30% by ESPs at temperatures above 200°C (Table 4).

To estimate metal emissions from MWCs under the baseline scenario, actual or estimated particulate control efficiencies were used existing MWC facilities. Particulate emissions were estimated for each facility or model plant using the particulate emission factors reported in Table 7 and either the actual or estimated particulate control efficiency show in Table 4. To generate estimates of emissions under the controlled scenario, fabric filters were assumed to be capable of achieving 99.5% control of metals emissions (except for mercury). Mercury emissions were assumed to be controlled by only 50%,[25,26] although recent tests of a pilot scale dry-scrubber fabric filter system demonstrated greater than 90% control of mercury when the combustion gases were cooled to 145°C preceding the fabric filter.[27] To estimate controlled emissions of all metals except mercury for existing MWCs, baseline emissions data were adjusted as follows:

$$EF_C = EF_B\,(1 - CE_C)/(1 - CE_B), \tag{2}$$

where: EF_C = emission factor for the controlled scenario; EF_B = emission factor for the baseline scenario; CE_C = metal control efficiency achieved with particulate controls under the controlled scenario (0.995, Table 4); and CE_B = metal control efficiency achieved with particulate controls for existing facilities under the baseline scenario (Table 4). To estimate metals emissions (except for mercury) from projected MWCs, emissions factors were adjusted as follows:

$$EF_C = EF_B\,(1 - 0.005)/(1 - CE_B), \tag{3}$$

where: EF_C = emission factor for the controlled scenario; EF_B = emission factor for the baseline scenario; and CE_B = metal control efficiency achieved with particulate controls for projected

TABLE 7
Uncontrolled Particulate Emissions Factors for MWCs
(k/metric ton of MSW burned)

Combustor type	Emissions factor
Massburn	18.0
Modular	1.0
RDF	40.0

TABLE 8
HCl[a] Emissions Factors for MWCs
(kg/metric ton of MSW burned)

	Emission factor	
Facility	Average	Maximum
Massburn/heat recovery	4.13	7.15
Massburn/nonheat recovery	4.13	7.15
RDF	2.57	NA[b]
Modular	3.58	4.43

[a] HCl = hydrochloric acid.
[b] Not available.

facilities under the baseline scenario (0.99, Table 4). The mercury emission factors for existing and projected facilities, based on estimated controlled particulate emissions, were adjusted as follows:

$$EF_C = EF_B (1 - CE_C)/(1 - CE_B), \quad (4)$$

where: EF_C = emission factor for the controlled scenario; EF_B = emission factor for the baseline scenario; CE_C = mercury control efficiency achieved with particulate controls under the controlled scenario (0.50, Table 4), and CE_B = mercury control efficiency achieved with particulate controls under the baseline scenario (0.30, Table 4).

C. HYDROCHLORIC ACID EMISSIONS FACTORS

Baseline HCl emission factors are presented in Table 8 for the types of MWCs modeled. HCl emission rates were divided by the waste combustion throughput for each tested facility are expressed in units of kilograms of HCl per metric ton of waste combusted. Average and maximum emission factors were estimated from actual test data from MWC emissions to model long- and short-term exposures.[15] Average emissions factors were determined by calculating the average of all HCl measurements for a given combustor type. Maximum emissions factors represent the highest measured value of HCl emissions per quantity of waste combusted for each combustor type.

Annual average and maximum baseline emissions factors for existing and projected facilities in Table 8 reflect 0% control (i.e., dry alkaline scrubbers were not used). Ninety percent control of HCl emissions can be achieved through the application of a dry alkaline scrubbers and fabric filters (Table 4).[25-27] Emissions data from existing and projected facilities with scrubbers were assumed to represent the controlled scenario. HCl emissions under the controlled scenario were estimated by adjusting the emission factors in Table 8 as follows:

$$EF_C = EF_B (1- CE_C), \quad (5)$$

where: EF_C = emission factor for the controlled scenario; EF_B = emission factor for the baseline scenario; and CE_C = HCl control efficiency achievable under the controlled scenario (0.9, Table 4).

VI. UNCERTAINTIES IN EMISSIONS ESTIMATES

Emissions estimates used in risk assessments are based on test data collected from a limited number of MWCs that vary widely in design and operational performance. Since the tests were conducted with different objectives, testing protocols and the level of detail reported were extremely variable. Direct comparison of the test results for different facilities is therefore difficult. In addition, specific sampling and analysis methods were frequently not described in detail and may not be consistent for all tests. As described above, average emission factors were developed from the test data to represent typical emissions from existing and projected MWCs.

VII. EXPOSURE ASSESSMENT

Estimation of potential human health risks associated with pollutants emitted from MWCs requires estimation of magnitude of population exposure. In the absence of monitored ambient air data in the vicinity of operating MWCs, the U.S. EPA used mathematical models to predict the atmospheric dispersion of emissions and subsequent potential for human exposure. Air quality modeling protocols are described in detail in another EPA report.[28]

Estimates of population exposure to ambient air concentrations of the emitted pollutants were developed using the U.S. EPAs Human Exposure Model (HEM).[29] The HEM accepts as inputs the location of a facility (geodetic map coordinates) and its emissions characteristics such as stack height, gas velocity, exit temperature, and the mass rate of pollutant emissions from actual or representative stationary combustion sources. Using site-specific source data and 5 years of average meteorological data from the geographically closest National Weather Service record station, the HEM estimates annual average ground-level ambient air concentrations in 16 wind directions at 10 radial distances from the modeled MWC facility. Through various interpolation techniques, these estimated annual average ambient air concentrations are paired with the population distribution obtained from the 1980 census data to yield an estimate of population inhalation exposure to predicted (modeled) ground-level ambient air concentrations for the 13 pollutants of concern.

An exposure analysis was conducted for each of the 111 existing MWCs that were operational in the U.S. as of 1986. Exposure to contaminants emitted from projected MWCs in the U.S. were analyzed using hypothetical facilities located in urban areas. The characteristics of the model facilities were developed for each technology subcategory from state permits describing facility size, capacity, and stack parameters of planned units.[25] The population exposure assessment derived from the model plant analyses for each technology subcategory was extrapolated to obtain a national population exposure estimate based on the national throughput of MSW expected to be incinerated in facilities projected to be built from 1987 to 1995.

VIII. NATIONAL ESTIMATES OF CANCER RISK

Population exposure for all facility-emitted pollutants of concern were combined with an evaluation of the carcinogenic potency for each chemical to predict the probability of developing cancer. For 2,3,7,8-TCDD, the data obtained by Kociba et al.[30] was applied to the modified multistage model developed by K. S. Crump, Co. and Watson[31,32] to estimate carcinogenic potency. The 95% upper-limit carcinogenic potency of 2,3,7,8-TCDD was determined to be 1.56×10^5 (mg/kg/d^{-1}).[32] Using this potency factor, the cancer risk associated with a continuous lifetime (70 year) exposure to an ambient air concentration of 1 pg of 2,3,7,8-TCDD per cubic meter of air was estimated to be 3.3×10^{-5}, assuming that the average adult weight 70 kg and breathes 20 m^3 of air per day, 75% of which is absorbed by the lungs.[32]

Two measures of excess cancer risk based on inhalation exposure are typically estimated: annual cancer incidence and maximum individual lifetime cancer risk (MIR). The annual cancer incidence is the aggregate population risk in units of average expected excess cancer cases per year in the entire population residing within 50 km (30 mi) of an MWC. It is calculated by multiplying the aggregate population exposure by the unit cancer risk factor, and dividing by the duration of exposure (70 years). The MIR is defined as the probability that the most exposed individual(s) in the population living near a MWC will contract cancer following a lifetime exposure. It is determined by multiplying the highest modeled ambient air concentration to which an individual is expected to be exposed by the unit cancer risk estimate. The MIR is a statement of probability that the exposed individual(s) may develop cancer and is expressed in scientific notation as a negative exponent of 10. For example, a risk (probability) of contracting cancer of 1 in 10,000 is expressed as 10^{-4}.

Estimated cancer risks resulting from inhalation exposure to predicted ambient concentrations of MWC emissions are depicted in Table 9. The ranges result from the quantitative uncertainties inherent in estimating carcinogenic risk for individual compounds or for classes of compounds.[25] Other uncertainties that may result in the over- or underestimation of risk include: (1) the assumption of additivity of risk across carcinogens; (2) the limiting of carcinogenic potency to discrete compounds in the emissions; (3) the use of models rather than actual measured data to quantify population exposure; and (4) questions concerning the mechanism of carcinogenesis from exposure to chlorinated dioxins and dibenzofurans. Concerning the mechanism of carcinogenesis, models used by the U.S. EPA implicitly assume that 2,3,7,8-TCDD acts as an initiator of cancer. If, however, 2,3,7,8-TCDD acts exclusively as a promoting agent to amplify the carcinogenic response of initiated cells, the present model may not be appropriate. Chronic exposure studies of 2,3,7,8-TCDD indicate initiating events of organs other than the liver and skin, whereas subchronic animals bioassays indicate 2,3,7,8-TCDD acts a promoter in liver and skin carcinogenesis.[13,32] In addition, equivocal epidemiologic evidence suggests an association between exposure to 2,3,7,8-TCDD and soft tissue sarcoma and non-Hodgkin's lymphoma.[13,32] Changes in the assumptions upon which cancer potency factors for CDDs and CDFs are based could lead to a reduction in the estimates of cancer risk associated with exposure to MWC emissions.

Our analysis suggests that most of the estimated cancer risk associated with direct inhalation of pollutants emitted from MWCs is attributable to the CDDs and CDFs. Specifically, 2 to 40 and 0.8 to 20 additional cancer cases per year may result from baseline CDD/CDF emissions from existing and projected MWCs, respectively (Table 10). The MIR ranges from 10^{-6} to 10^{-3} for existing MWCs and 10^{-4} to 10^{-6} for projected MWCs. Thus, in the absence of further control of emissions, CDDs and CDFs released from all 321 existing and projected MWCs expected to be operational in 1995 are estimated to cause about 4 to 60 excess cancer cases per year in the exposed U.S. population.

Table 11 summarizes estimated nationwide cancer risks from inhalation exposure to emitted pollutants if all MWCs were to use a dry scrubber combined with a fabric filter system. A comparison of the data presented in Table 10 representing the baseline scenario with the data presented in Table 11 representing the controlled scenario reveals that population exposure to carcinogenic compounds (and subsequent risk of developing cancer) can be reduced by more than tenfold with the application of dry scrubber-fabric filter technology. The U.S. EPA believes that this reduction is a signficant achievement in terms of protecting public health.

The EPA has also begun preliminary analysis of the potential for exposure from the deposition of emitted CDDs and CDFs on the earth surface and subsequent human exposure through indirect (food chain) pathways.[25,33] The preliminary analysis indicates that indirect

TABLE 9

Contribution of MWC Emission Constituents to Estimated Annual Cancer Incidence and Maximum Individual Lifetime Risk

Pollutant	Existing MWC annual cancer incidence[a,b]	Existing MWC max individual risk range[c,d]	Projected MWC annual cancer incidence[a,b]	Projected MWC max individual risk range[c,d]
Chlorinated dioxins and dibenzofurans (CDDs/CDFs)	2—40	10^{-6}—10^{-3}	0.8—20	10^{-6}—10^{-4}
Chlorophenols	0.0001—0.0003	10^{-9}—10^{-8}	0.0001—0.0003	10^{-10}—10^{9}
Chlorobenzenes	0.009—0.002	10^{-7}—10^{-6}	0.004—0.01	10^{-9}—10^{-7}
Formaldehyde	0.009	10^{-8}	0.02	10^{-8}—10^{-7}
Polycyclic aromatic hydrocarbons	0.01—0.6	10^{-7}—10^{-5}	0.05—3.0	10^{-7}—10^{-5}
Polychlorinated biphenyls	0.02	10^{-8}—10^{-5}	0.2	10^{-9}—10^{-6}
Arsenic	0.2	10^{-7}—10^{-4}	0.1	10^{-8}—10^{-7}
Beryllium	0.02	10^{-9}—10^{-6}	0.001	10^{-11}—10^{-8}
Cadmium	0.2	10^{-6}—10^{-4}	0.2	10^{-7}—10^{-6}
Chromium(6)	0.2	10^{-7}—10^{-4}	0.1	10^{-7}—10^{-6}
Rounded total:[e]	2—4	10^{-6}—10^{-3}	2—20	10^{-6}—10^{-4}

[a] The ranges in annual cancer incidence reflect assumptions made regarding the potential carcinogenicity of classes of organic compounds.

[b] Annual cancer incidence is defined as the average number of excess cancer cases expected annually in the exposed population.

[c] The ranges in maximum individual lifetime cancer risk reflect differences in emissions and the evaluation of emissions data from MWC technologies within the existing and proposed categories.

[d] Maximum individual risk is defined as the probability of developing cancer following a lifetime exposure to the maximum modeled long-term ambient air concentration. The probability is expressed as a negative exponent of 10. A risk of 1 chance in 10,000 is expressed as 10^{-4}.

[e] Apparent errors in totals are due to intentional rounding to one significant figure.

TABLE 10
Estimated Nationwide Cancer Risk from Inhalation Exposure to MWC Emissions (baseline scenario)

	Organics[a]		Metals[b]		Combined	
	Ann. incid.[c]	Max. indiv.[d]	Ann. incid.	Max. indiv.	Ann. incid.	Max. indiv.
Existing sources (1985)						
Massburn (nonheat recovery)	1—30[e]	10^{-4}—10^{-3}	0.2	10^{-5}	1—30	10^{-4}—10^{-3}
Massburn (heat recovery)	0.2—4	10^{-4}—10^{-3}	0.04	10^{-4}	0.2—4	10^{-4}—10^{-3}
RDF	0.1—3	10^{-5}—10^{-3}	0.2	10^{-5}	0.3—3	10^{-5}—10^{-3}
Modular	0.0008—0.01	10^{-6}—10^{-4}	0.01	10^{-4}	0.01—0.02	10^{-4}—10^{-4}
Existing total[f]	2—40	10^{-4}—10^{-3}	0.5	10^{-4}	2—40	10^{-4}—10^{-3}
Projected sources (1993)						
Massburn (heat recovery)	0.3—7	10^{-6}—10^{-5}	0.3	10^{-6}	0.6—7	10^{-6}—10^{-5}
RDF	0.8—10	10^{-5}—10^{-4}	0.1	10^{-6}	0.9—10	10^{-5}—10^{-4}
Modular	0.04—0.9	10^{-6}—10^{-5}	0.01	10^{-6}	0.05—0.9	10^{-6}—10^{-5}
Projected total[f]	1—20	10^{-5}—10^{-4}	0.4	10^{-6}	2—20	10^{-5}—10^{-4}
Combined total[f]	3—60	10^{-4}—10^{-3}	0.9	10^{-4}	4—60	10^{-4}—10^{-3}

[a] CDDs, chlorophenols, chlorobenzenes, formaldehyde, PCB, PAH. Risk ranges for organics result from assumptions about the carcinogencity of pollutant classes and the recovery efficiency for CDD/CDF in stack tests.

[b] Arsenic, beryllium, cadium, chromium (+6).

[c] Annual incidence is the modeled number of cancer cases per year in population within 50 km of all municipal waste combustors in the U.S.

[d] Maximum individual risk is the modeled portability that a person exposed to the highest modeled concentration of pollutants from a municipal waste combustor will develop cancer over his or her 70 year lifespan.

[e] rounded to one significant figure. See text for assumptions involved in producing these estimates.

[f] Totals do not add due to rounding.

TABLE 11
Estimated Nationwide Excess Cancer Risk from Inhalation Exposure to MWC Emissions if Dry Alkaline Scrubbers and Fabric Filters are Applied (controlled scenario)[a]

	Organics[b]		Metals[c]		Combined	
	Ann. incid.[d]	Max. indiv.[e]	Ann. incid.	Max. indiv.	Ann. incid.	Max. indiv.
Existing sources (1985)						
Massburn (nonheat recovery)	0.08[f]—2.0	10^{-5}—10^{-4}	0.05	10^{-6}	0.1—2.0	10^{-5}—10^{-4}
Massburn (heat recovery)	0.01—0.3	10^{-5}—10^{-4}	0.01	10^{-6}	0.02—0.3	10^{-5}—10^{-4}
RDF	0.01—0.2	10^{-6}—10^{-5}	0.03	10^{-6}	0.04—0.2	10^{-6}—10^{-5}
Modular	<0.0001	10^{-7}—10^{-6}	0.001	10^{-6}	0.001	10^{-6}
Existing total[g]	0.01—3.0	10^{-5}—10^{-4}	0.1	10^{-6}	0.2—3.0	10^{-5}—10^{-4}
Projected sources (1993)						
Massburn (heat recovery)	0.02—0.4	10^{-7}—10^{-6}	0.2	10^{-6}	0.2—0.6	10^{-6}
RDF	0.05—0.9	10^{-6}—10^{-5}	0.04	10^{-6}	0.09—0.9	10^{-6}—10^{-5}
Modular	0.001—0.04	10^{-7}—10^{-6}	0.001	10^{-6}	0.01—0.03	10^{-6}
Projected total[g]	0.07—1.0	10^{-6}—10^{-5}	0.2	10^{-6}	0.3—1.0	10^{-6}
Combined total[g]	0.2—4.0	10^{-5}—10^{-4}	0.3	10^{-6}	0.5—4.0	10^{-5}—10^{-4}

[a] Assumes 95% control of organics with DS/FF.
[b] CDDs, chlorophenols, chlorobenzenes, formaldehyde, PCB, PAH. Risk ranges for organics result from assumptions about the carcinogencity of pollutant classes and the recovery efficiency for CDD/CDF in stack tests.
[c] Arsenic, beryllium, cadium, chromium (+6).
[d] Annual incidence is the modeled number of cancer cases per year in population within 50 km of all municipal waste combustors in the U.S.
[e] Maximum individual risk is the modeled portability that a person exposed to the highest modeled concentration of pollutants from a municipal waste combustor will develop cancer over his or her 70 year lifespan.
[f] All numbers were rounded to one significant figure See text for assumptions involved in producing these estimates.
[g] Totals do not add due to rounding.

exposure via ingestion of dioxin contaminated foods resulting from MWC emissions deposited over 30 years are comparable to exposures through direct inhalation. Human exposure to facility-emitted contaminants via indirect pathways warrants further study.

IX. REGULATORY DETERMINATION

Based on the nationwide quantitative health risk assessment, the U.S. EPA Administrator determined that MWC emissions can reasonably be anticipated to contribute to the endangerment of public health and welfare[3] and warrant further federal regulation and control under the Clean Air Act (CAA). The U.S. EPA is currently revising the new source performance standard (NSPS) for control of MWC emissions under section 111(b) of the CAA. The NSPS will reflect the best system of continuous emission reduction that has been demonstrated for MWCs and will specifically address the control of dioxin and other carcinogenic compounds. The EPA also announced an intent to designate one or more currently unregulated pollutants in MWC emissions as candidates for monitoring to determine if adequate emission control is being achieved. This designation will require a U.S. EPA issuance of existing source guidelines under section 111(d) of the CAA for use by states in developing their emissions standards. The standards will include emissions limits and quantitative requirements for compliance monitoring. The proposed NSPS and draft existing source guidelines are currently scheduled for issuance in November 1989.[3] The NSPS under development will apply to MWCs for which construction has commenced on or after the date of proposal. Promulgation of the final NSPS requirement are scheduled for December 1990.[3]

In the interim, the EPA has issued operational guidance for use by state and local government regulatory agencies with permitting authority.[3] This operational guidance is expected to be followed by these regulatory agencies in reviewing best available control technology (BACT) determinations for new or substantially modified MWCs. Under this BACT guidance, dry alkaline scrubbers followed by high efficiency electrostatic precipitators or fabric filters are deemed effective in controlling toxic organic pollutants (such as CDDs and CDFs), metals, and acid gases.

X. CONCLUSION

The U.S. EPA evaluated pollutant emissions from MWCs currently operating as well as emissions from facilities expected to begin operation in the next several years. From these data a nationwide quantitative cancer risk assessment was done to determine the need for further federal control of MWC emissions. The analysis suggests that by the mid-1990s if no additional federal control were to occur, it is possible that 4 to 60 excess cancer cases per year could occur in the exposed U.S. population residing near MWCs. Most of the carcinogenic risk is attributable to emissions of the highly toxic CDDs and CDFs. The analysis was an important factor in the U.S. EPAs determination in July 1987 that further control of pollutant releases into air from MWCs is needed. Uniform application of dry alkaline scrubbers combined with either fabric filters or ESPs could potentially reduce the carcinogenic risk associated with the emission of CDDs, CDFs, and other pollutants by more than one order of magnitude.

REFERENCES

1. U.S. Environmental Protection Agency, Characterization of the Municipal Waste Combustion Industry, EPA/530-SW-87-021h, Office of Solid Waste and Emergency Response, Washington, D.C., June 1987.
2. U.S. Environmental Protection Agency, Report to Congress, EPA/530-SW-87-021a, Office of Solid Waste and Emergency Response, Washington, D.C., June 1987.
3. U.S. Environmental Protection Agency, Assessment of municipal waste combustor emissions under the Clean Air Act, *Fed. Reg.*, 52, 25399, 1987.
4. **Poland, A., Knutson, J., and Glover, E.,** A consideration of the mechanisms of action of 2,3,7,8-tetrachlorodibenzo-*p*-dioxin and related halogenated aromatic hydrocarbons, in *Human and Environmental Risks of Chlorinated Dioxins and Related Compounds,* Tucker, R. E., Young, A. C., and Gray, A. D., Eds., Plenum Press, New York, 1983, 539.
5. **Nebert, D. W. and Jearsen, N. M.,** The Ah locus: genetic regulation of the metabolism of carcinogens, drugs, and other environmental pollutants by cytochrome P-450 mediated monooxygenases, *CRC Crit. Rev. Biochem.,* 6, 401, 1979.
6. **Nebert, D. W., Negishi, N., and Eisen, H. J.,** Genetic differences in enzymes which metabolize drugs, chemical carcinogens, and other environmental pollutants, in *Human and Environmental Risks of Chlorinated Dioxins and Related Compounds,* Tucker, R. E., Young, A. C., and Gray, A. D., Eds., Plenum Press, New York, 1983, 441.
7. **Poland, A. P., Greenlee, W. F., and Kende, A. S.,** Studies on the mechanisms of action of chlorinated dibenzo-*p*-dioxins and related compounds, *Ann. N.Y. Acad. Sci.,* 320, 214, 1979.
8. **Poland A. P. and Knutson, J. C.,** 2,3,7,8-Tetrachlorodibenzo-*p*-dioxin and related halogenated aromatic hydrocarbons: examination of the mechanism of toxicity, *Ann. Rev. Pharm. Toxicol.,* 22, 517, 1982.
9. **Tukey, R. H., Hannah, R. R., Negishi, M., Nebert, D. W., and Eisen, H. J.,** The AH locus: correlation of intranuclear appearance of inducer receptor complex with induction of cytochrome P-450 messenger-RNA, *Cell,* 31, 275, 1982.
10. **Nebert, D. W., Eisen, H. J., Negishi, M., Lang, M.A., Kjelmela, C. M., and Okey, A. B.,** Genetic mechanisms of controlling the induction of polysubstrate mono-oxygenase (P-450) activities, *Ann. Rev. Pharm.,* 21, 431, 1981.
11. **Poland, A. P., Glover, E., and Kende, A. S.,** Stereospecific, high affinity binding of 2,3,7,8-tetrachlorodibenzo-*p*-dioxin by hepatic cytosol. Evidence that the binding species is the receptor for the induction of aryl hydrocarbon hydroxylase, *J. Biol. Chem.,* 251, 4936, 1976.
12. **Knutson, J. C. and Poland A.,** 2,3,7,8-Tetrachlorodibenzo-*p*-dioxin: failure to demonstrate toxicity in 23 cultured cell types, *Toxicol. Appl. Pharm.,* 54, 377, 1980.
13. **Mukerjee, D. and Cleverly, D.,** Risks from exposure to polychlorinated dibenzo-*p*-dioxins and dibenzofurans emitted from municipal incinerators, *Waste Manage. Res.,* 5, 269, 1987.
14. U. S. Environmental Protection Agency, Interim Procedures of Estimating Risks Associated with Exposures to Mixtures of Chlorinated Dibenzo-*p*-dioxins and Dibenzo-furans (CDDs and CDFs), EPA/625/3-87-012, Risk Assessment Forum, Washington, D.C., March 1987.
15. U.S. Environmental Protection Agency, Emissions Data Base for Municipal Waste Combustors, EPA/530-SW-87-021b, Office of Solid Waste and Emergency Response, Washington, D.C., June 1987.
16. **Hutzinger, O., Blumich, M. J., van den Berg, M., and Olie, K.,** Sources and fate of PCDDs and PCDFs: an overview, *Chemosphere,* 14, 581, 1985.
17. U.S. Environmental Protection Agency, Combustion Control of Organic Emissions, EPA/530-SW-87-021c, Office of Solid Waste and Emergency Response, Washington, D.C., June 1987.
18. **Commoner, B., Shapiro, K., and Webster, T.,** The origin and health risks of PCDD and PCDF, *Waste Manage. Res.,* 5, 327, 1987.
19. **Vogg, H., Metzger, M., and Stieglitz, L.,** Recent findings on the formation and decomposition of PCDD/PCDF in municipal solid waste incineration, *Waste Manage. Res.,* 5, 285, 1987.
20. **Markland, S., Kjeller, L.-O., Hansson, M., Tysklind, M., Rappe, C., Ryan, C., Collazo, H., and Dougherty, R.,** Determination of PCDDs and PCDFs in incineration samples and pyrolytic products, in *Chlorinated Dioxins and Dibenzofurans in Perspective,* Rappe, C., Choudhary, G., and Keith, L. H., Eds., Lewis Publishers, Chelsea, MI, 1986, 79.
21. **Rappe, C., Andersson, R., Bergqvist, R., Brohede, R. A., and Hansson, C.,** Sources and relative importance of PCDD and PCDF emissions, *Waste Manage. Res.,* 51, 225, 1987.
22. **Choudhary, C. G., Olie, K., and Hutzinger, O.,** Mechanism in the thermal formation of chlorinated compounds including polychlorinated dibenzo-*p*-dioxins, in *Chlorinated Dioxins and Related Compounds: Impact on the Environment,* Hutzinger, O., Frei, R. W., Merian, E., and Pocchiari, F., Eds., Pergamon Press, New York, 1982, 275.
23. **Liberti, A. and Brocco, D.,** Formation of polychlorinated dibenzo-dioxins and polychlorinated dibenzofurans in urban incinerator emissions, in *Chlorinated Dioxins and Related Compounds: Impact on the Environment,* Hutzinger, O., Frei, R., Merian, E., and Pocchiari, F., Eds., Pergamon Press, New York, 1982, 245.

24. **Liberti, A., Goretti, G., and Russo, M. V.,** PCDD and PCDF formation in the combustion of vegetable wastes, *Chemosphere,* 12, 661, 1983.
25. U.S. Environmental Protection Agency, Assessment of Health Risks Associated with Municipal Waste Combustion Emissions, EPA/530-SW-87-021g, Washington, D.C., June 1987.
26. Environment Canada, The National Incinerator Testing and Evaluation Program: Air Pollution Control Technology, Summary Report EPS 3/UP/2, Ottawa, Canada, 1986.
27. U. S. Environmental Protection Agency, Municipal Waste Combustion Multipollutant Study: Emission Test Report, Marion County Solid Waste-to-Energy Facility, Volumes 1 and 2: Summary of Results, EMB Report #86-MIN-03, Office of Air Quality Planning and Standards, Research Triangle Park, NC, 1987.
28. U.S. Environmental Protection Agency, Guidelines on Air Quality Models (Revised), EPA-450/2-78-027R, July 1986.
29. U.S. Environmental Protection Agency, User's Manual for the Human Exposure Model (HEM), EPA-450/5-86-001, Office of Air Quality Planning and Standards, Research Triangle Park, NC, June 1986.
30. **Kociba, R. J., Keyes, D. G., Beyer, J. E., Carreon, R. M., and Gehring, P. J.,** Long-term toxicological studies of 2,3,7,8-tetrachlorodibenzo-*p*-dioxin (TCDD) in laboratory animals, *Ann. N.Y. Acad. Sci.,* 320, 397, 1978.
31. **Howe, R. B.,** GLOBAL 83: An Experimental Program Developed for the U.S. Environmental Protection Agency as an Update to GLOBAL 82: A Computer Program to Extrapolate Quantal Animal Toxicity Data to Low Doses, K. S. Crump and Company, Inc., Ruston, LA, unpublished, 1983.
32. U.S. Environmental Protection Agency, Health Assessment Document for Polychlorinated Dibenzo-*p*-Dioxins, Final Report EPA/600/8-84/014F, Office of Research and Development, Washington, D.C., September 1985.
33. U.S. Environmental Protection Agency, Methodology for the Assessment of Health Risks Associated with Multiple Pathway Exposure to Municipal Waste Combustor Emissions, Office of Air Quality Planning and Standards, Research Triangle Park, NC, and Office of Environmental Criteria and Assessment Office, Cincinnati, OH, October 1986.

Chapter 3

MODELING THE DISPERSION OF TOXIC AIR POLLUTANTS EMITTED FROM MUNICIPAL SOLID WASTE INCINERATORS

Ray Kapahi

TABLE OF CONTENTS

I. INTRODUCTION

The use of air quality models in planning and decision making has experienced tremendous growth since the enactment of the Clean Air Act (CAA). Under the 1977 amendments to the Clean Air Act, the permitting of new sources requires the use of atmospheric dispersion models to estimate the possible consequences of discharging air contaminants into the atmosphere. While estimating air quality impacts from new sources has been and will continue to be an important application of air quality models, increasing focus has shifted to their applications to risk assessment. To conduct a risk assessment, estimates of air concentrations and deposition rates of various air contaminants (e.g., toxic air pollutants, carcinogens, etc.) are usually required. Air quality models provide a means of translating emission rates of various pollutants into air concentrations and deposition rates.

This chapter provides an overview and a description of the types of air quality models available and how they can be used to estimate air concentrations and deposition rates in the vicinity of an existing or a new source. The physical and chemical processes that affect air contaminants after they have been released into the atmosphere are first identified. (Air contaminant in the present context refers to any gaseous or solid material that is discharged into the atmosphere.) Then, the classes of air quality models currently in use and how they simulate those physical and chemical processes are reviewed. The final section is a survey of the sources of data that are needed to apply these models.

II. WHAT IS AN AIR QUALITY MODEL?

An air quality model is an abstraction that attempts to simulate the physical and chemical processes that occur in the atmosphere. It attempts to simulate the transport, dispersion, chemical transformations, and deposition of air contaminants as they travel from a source to a receptor. The function of an air quality model is schematically illustrated in Figure 1. In the context of estimating risk from airborne contaminants, an air quality model can be considered a black box that translates information about the physical properties of a specific air contaminant, its source, and the local and regional meteorological data into estimates of air concentrations and/or deposition rates at various locations. As noted by Angle,[1] a model allows the user to predict hypothetical future states of a system that may not exist. As such, a model can project the consequences of an action that would be expensive, difficult, or destructive to do in the real world. Models allow us to make cause and effect projections in a systematic and reproducible manner.

III. FACTORS AFFECTING THE ADVECTION, DIFFUSION, AND REMOVAL OF AIR CONTAMINANTS

Three principal processes control the fate of air contaminants as they move from a source to a receptor: (1) advection, (2) diffusion, and (3) removal. These factors are illustrated in Figure 2. A basic understanding of these processes is fundamental to understanding and applying air quality models.

A. ADVECTION

Advection refers to the physical transport or movement of air contaminants by the wind. Wind refers to the horizontal wind. The vertical component of wind is normally an order of magnitude lower than the horizontal component and in many cases is zero over averaging times of 1 h or greater. An exception to this rule occurs when significant topographical features such as mountains are present.

The ambient wind is characterized by its direction and speed. The former determines the

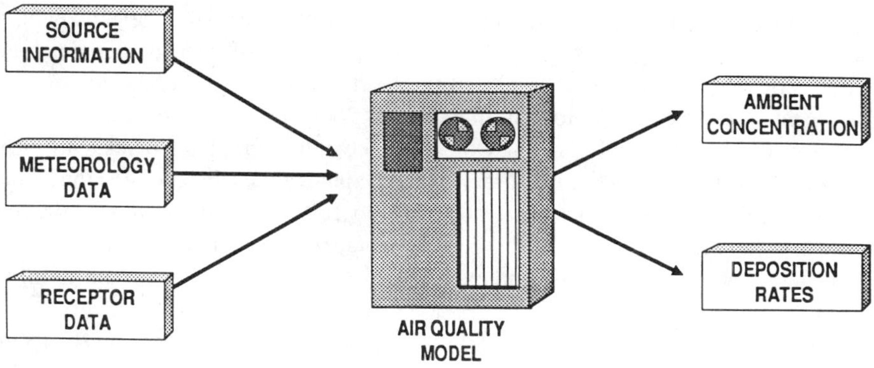

FIGURE 1. Overview of an air quality model.

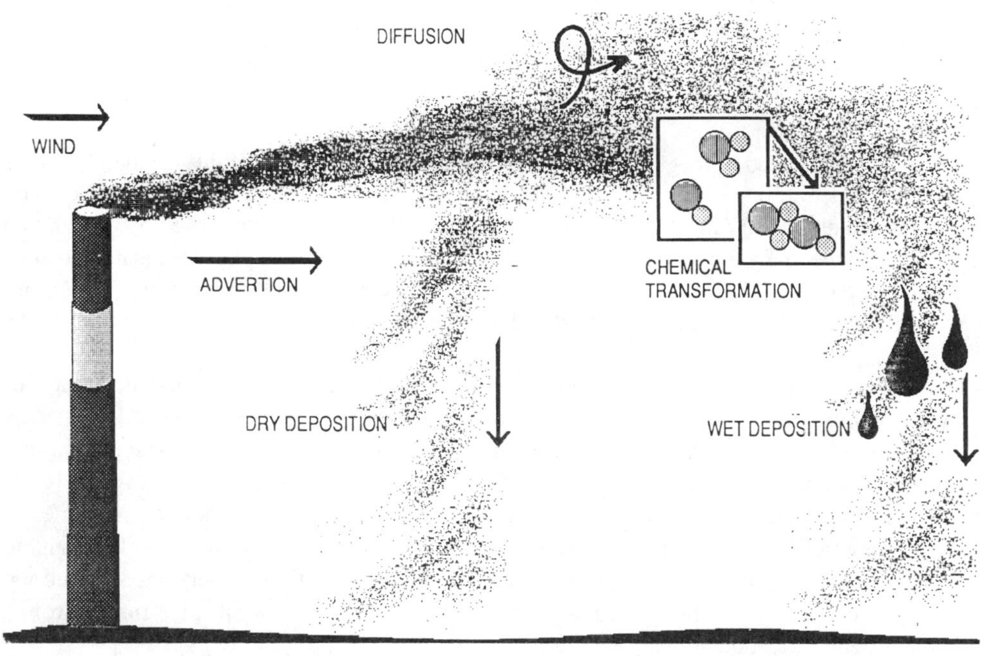

FIGURE 2. Factors affecting the advection, dispersion, and removal of air pollutants.

direction in which the air contaminant will travel, while the latter determines how far the material will be carried. Wind speed also affects the dilution of the contaminant. Other factors being equal, a higher wind speed allows a greater volume of ambient air to be mixed with the emission plume.

Generally, wind speed refers to a mean wind speed averaged over some time interval, typically 1 h. Closer inspection of wind speed data reveal that wind speed exhibits rapid and random motion about a mean value. For practical reasons, wind speed can be viewed as consisting of two components, a rapidly fluctuating component and a more slowly varying mean component. It is the mean component that advects pollutants and serves as the primary mechanism for the movement of air contaminants. The random component is the primary mechanism that diffuses or spreads the air contaminants.

One of the most important characteristics of the horizontal wind is its spatial and temporal variability. This variability is due to the various scales of motion present in the atmosphere. On a global scale, the uneven heating of the Earth and the Earth's rotation combine to form zonal winds. In North America, the westerlies represent an example of global scale wind patterns. On the scale of a few hundred to a thousand kilometers, wind is influenced by the presence of weather patterns and fronts. For example, in the Northern Hemisphere, the presence of low pressure systems result in wind blowing in a counterclockwise direction. On a scale of 50 to about 500 km, the winds are influenced by the presence of topographical barriers or proximity to large bodies of water. In this scale, examples include land and sea breezes and the channeling of wind by a mountain range. On a scale of a 500 to 1000 m wind speed and direction are influenced by the very presence of the Earth's surface. The increase in wind speed with increasing height above the ground is an example of how the surface of the Earth influences wind.

Temporally, wind exhibits seasonal and diurnal cycles. In addition, the wind shows marked variability in the presence of significant weather elements, such as fronts. These changes can occur over time scales of several hours or days.

Normally, wind is measured at a height of 10 m. The variation of wind speed with height depends on the underlying surface and atmospheric stability and can be estimated using the following power law

$$u_h = u_o\,(h/h_o)^a \qquad (1)$$

where: u_h = wind speed at height h; u_o = wind speed at the reference height; h = height above ground, h_o = the reference height; and a = a constant ranging from 0.2 to 0.5 that varies with atmospheric stability. Irwin[2] discusses the application of the power law to various urban and rural areas. Since many emission sources, such as roof vents or tall stacks, are located at some height above the ground, the power law serves as a convenient model for estimating winds aloft. Figure 3 illustrates the variation in wind speed with changes in height at several locations in the U.S.

In addition to its variability with height, wind speed and direction also fluctuate at various locations on the surface of the Earth. For distances of 1 to 50 km, surface wind is influenced by the presence of terrain, changes in surface roughness, and the proximity to bodies of water. Numerous wind models are available that provide spatial and temporal estimates of wind based on a limited set of observations. Wind models range from simple interpolation schemes to diagnostic models that incorporate changes in terrain and wind variation with height to sophisticated, three-dimensional dynamic models that incorporate time-dependent dynamic and thermodynamic forces and their influence on the wind. Figure 4 illustrates the output from a diagnostic wind model developed by Sherman.[3]

As one might expect, more sophisticated wind models require extensive input data before a realistic simulation can be conducted. In addition, detailed wind simulations are computationally intensive. As a result, such simulations are conducted only for short periods of time, typically ranging from a few hours to a few days.

Finally, the author wishes to note that because of the fundamental importance of wind in the transport of air pollutants, any uncertainty (spatial or temporal) in our knowledge of the wind translates into a corresponding uncertainty in our ability to estimate the movement of air pollutants.

B. TURBULENT DIFFUSION

Turbulent diffusion is primarily responsible for the spreading of pollutants about the mean wind. This spreading is caused by rapidly fluctuating, random variations in wind speed and direction about a slowly varying mean. These random variations are also referred to as turbulence or turbulent eddies. The concept of plume dilution by turbulent eddies is shown in Figure 5.

Turbulence is produced in the atmosphere either by mechanical and/or thermal means. For

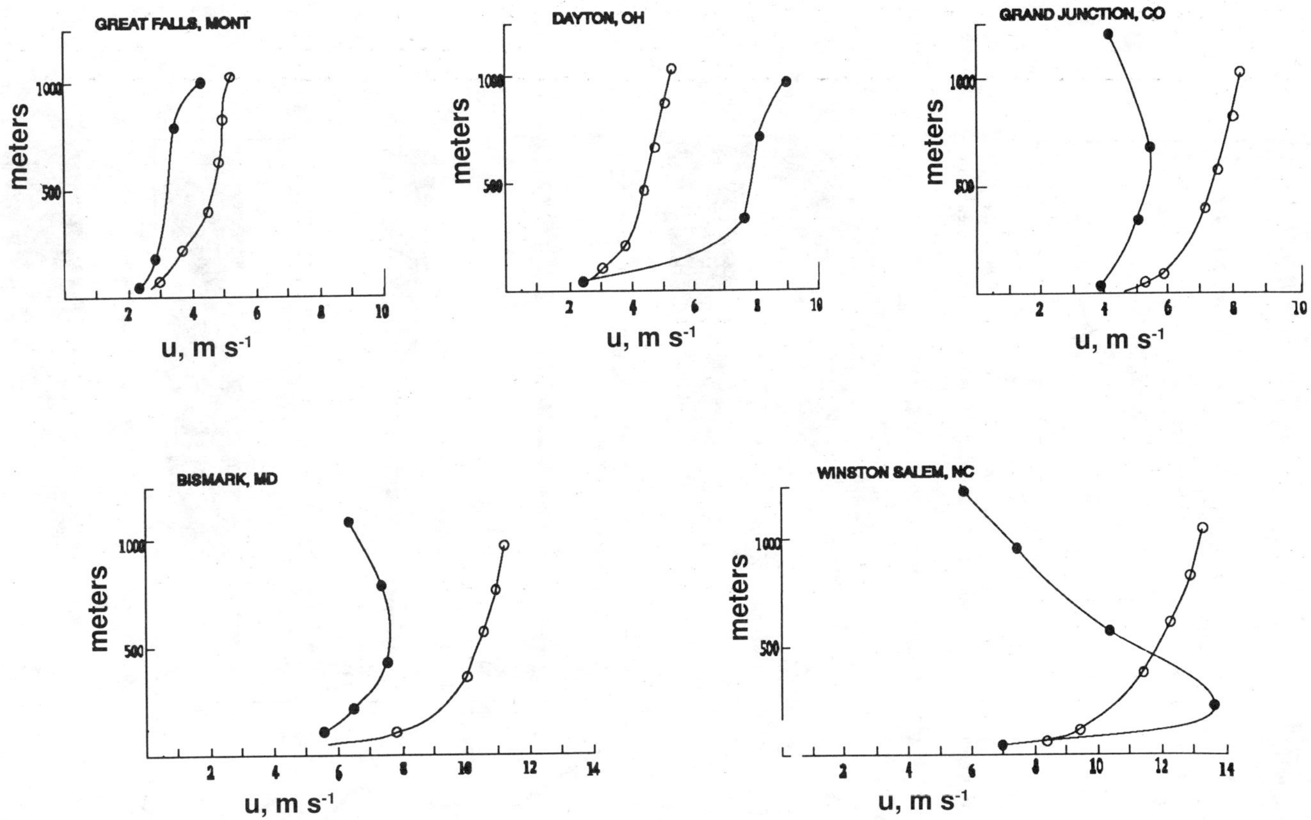

FIGURE 3. Wind variation with measured height (solid lines) and 1/7 power law (dashed lines).[15]

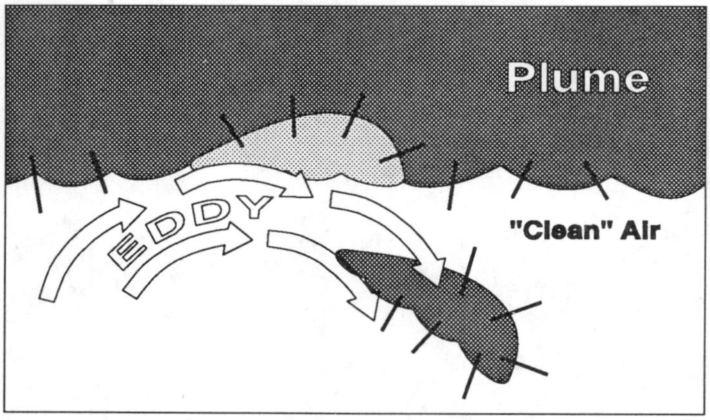

FIGURE 4. Grossly simplified view of how a turbulent eddy breaks up a plume and causes a much more rapid mixing of pollutants with ambient or "clean" air.[8]

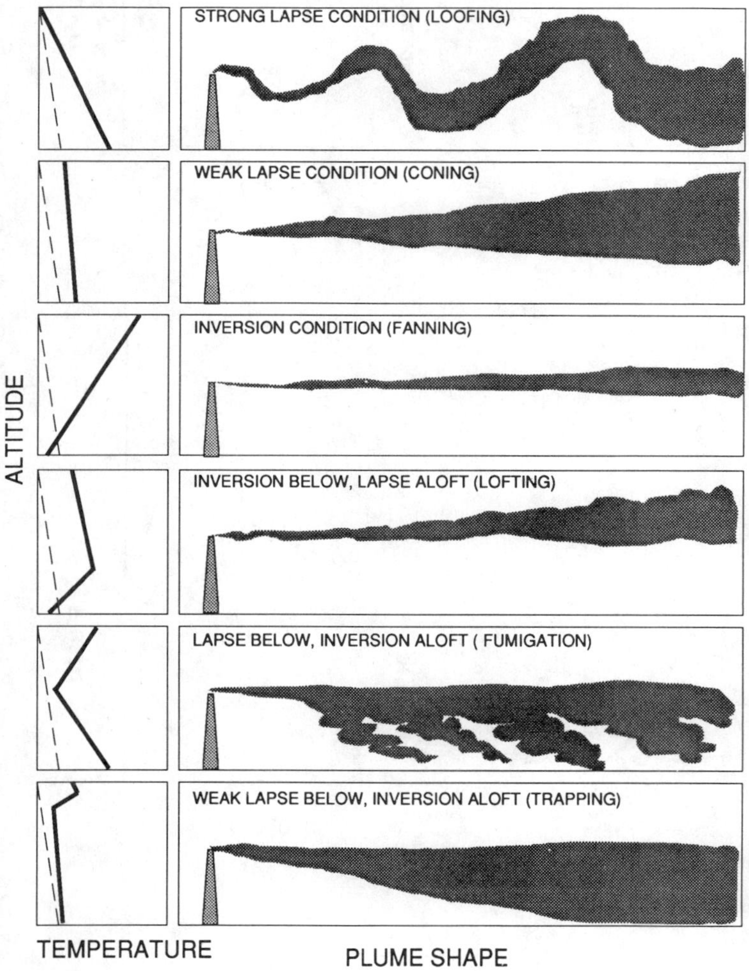

FIGURE 5. Particle expansion of a continuous plume related to vertical temperature structure. Dashed lines correspond to the dry adiabatic lapse rate.

TABLE 1
Commonly Used Methods of Estimating Atmospheric Turbulence

Turner's method	Relies on surface wind speed, amount of sky cover, and time of day
Sigma theta method	Measures the standard deviation of wind direction fluctuations in the horizontal and vertical planes
Lapse rate	Based on variations in temperature with height above ground level
Bulk Richardson number	Uses the ratio of thermal and mechanical indices of turbulence

example, in the lower 500 to 1000 m of atmosphere, friction between the wind and the underlying ground generates mechanical turbulence. Similarly, the heating of the ground by the sun leads to thermally generated turbulence. The relative amount of each component varies with time and space. During the night, turbulence is largely mechanical, whereas on a hot, sunny day with light winds, turbulence is largely thermal in origin.

The level of turbulence can also vary depending on human activity. In cities the so-called "urban heat island effect" can cause thermal turbulence. In addition, the presence of cities changes the surface roughness, which also enhances turbulence. For a discussion of the unique microclimate of cities and how it affects dispersion patterns, consult Landsburg.[4]

Several qualitative and quantitative measures of turbulence have been devised and are reported in the literature.[5-7] Table 1 summarizes some commonly used measures of atmospheric turbulence. The terms unstable and stable are frequently used to describe the atmosphere that contains greater or smaller amounts of turbulence, respectively. One practical method for estimating stability is the Pasquill-Gifford (PG) classification scheme that uses the time of day, surface wind speed, and the amount of cloud cover to provide a discrete measure of stability. Table 2 defines the PG stability classification scheme.

Another commonly used method is to measure the change of temperature with height. For a neutral atmosphere, the temperature cools by 1°C for every 100 m rise.[8] That is,

$$dt/dz = -1.0°C/100 \text{ m or } -0.01°C/m \tag{2}$$

where t is temperature and z is height. The negative sign indicates that temperature decreases with height. The change in temperature with height is referred to as the lapse rate. For a neutral atmosphere, the lapse rate of 0.01°C/m is referred to as the dry adiabatic lapse rate.[8] The effect of stability on the dispersal of air contaminants is illustrated in Figure 6. This figure shows that atmospheric stability has a profound effect on air contaminant dispersal.

C. REMOVAL PROCESSES

Removal of air contaminants discharged into the atmosphere refers to the reduction of mass by deposition or chemical transformation. Deposition refers to dry or wet processes that lead to the removal of gases or particulates from the air by gravitational settling, turbulent transfer to the ground, or by scavenging by rain or snow. Chemical transformation refers to the change in the physical or chemical properties of a discharged substance. Chemical reactions involving gas-to-gas, gas-to-particle, particle-to-gas, and particle-to-particle transformations can occur leading to the formation of new materials known as secondary pollutants.

A major challenge facing the modeling community is the identification and the parameterization of relevant removal processes. Frequently, detailed information about the behavior of

TABLE 2
Stability Categories for Six Surface Wind Speed Classes

Surfacc wind speed at 10 m (m/s)	Daytime: incoming solar radiation			Nighttime[a]: thinly overcast or	
	Strong[b]	Moderate	Slight[c]	≥4/8 clouds	≤3/8 clouds[d]
< 2	A[e]	A-B	B	U[f]	U
2-3	A-B	B	C	E	F
3-5	B	B-C	C	D[g]	E
5-6	C	C-D	D	D	D
>6	C	D	D	D	D

[a] The period 1 h before sunset and 1 h after sunrise.
[b] Strong incoming solar radiation corresponds to a solar altitude greater than 60° with clear skies.
[c] Slight incoming solar radiation corresponds to a solar altitude of 15° to 35° with clear skies.
[d] Incoming solar radiation (expected to be strong with clear skies) is likely to decreased to moderate levels with broken (5/8 to 7/8 cloud cover), medium-height clouds, and to slight with broken, low-level clouds.
[e] Class A is the most stable; class F the least stable.
[f] U equals unstable.
[g] Class D, the neutral class, should be used for overcast conditions during the day or night regardless of wind speed.

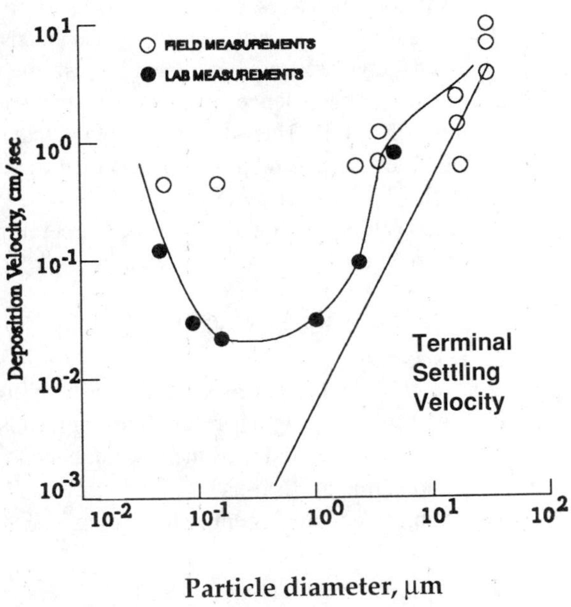

FIGURE 6. Laboratory and field measurements of deposition speeds of particles to grass.[10]

specific materials in the atmosphere is only partially or not known. For example, dioxins may be formed during the combustion of certain types of fuels. After their formation, they may be emitted in the gaseous phase, in the solid phase sorbed to flyash, or a mixture of both. The relative partitioning between the gaseous and particulate phases is not known for most emitted contaminants. Once emitted into the atmosphere, further transformation can occur as a result of cooling and/or due to the presence of other materials. At present, detailed information about many of these processes is not known. To account for these and other complexities, a prioritized list of relevant removal processes is prepared. Then a series of systematic modeling simulations are performed that address those removal processes. To estimate the importance of removal processes, an initial simulation is conducted without including any removal processes. Then the removal processes are introduced one by one, so that their relative impact on air concentrations can be assessed.

The two removal processes most commonly considered in risk assessments, chemical transformations and dry deposition, are described below along with commonly used parameterization schemes used to simulate them.

1. Chemical Transformations

The effect of chemical transformations in the atmosphere is twofold. First, materials released into the atmosphere (primary pollutants) may react to form new substances (secondary pollutants), both of which will behave differently in the atmosphere. A second result of a chemical transformation is the loss of the primary pollutant as a result of the chemical reaction.

In the context of performing risk assessments, it is the second of the two effects that concern us the most. Failure to take chemical transformations into account will result in an overestimate of the ambient concentration of the primary pollutant. For a simple, first-order chemical reaction such as

$$A \xrightarrow{k} B \tag{3}$$

the loss of the primary pollutant A can be parameterized in terms of its half-life according to the following equation.

$$\alpha = \frac{1}{k\, C_A} \tag{4}$$

where k equals the elimination half-life, and C_A is the concentration of pollutant A. Reactions involving several species can be modeled using a set of ordinary differential equations. For example, if

$$A + B \xrightarrow{k_1} D \tag{5}$$

and

$$D + E \xrightarrow{k_2} A \tag{6}$$

then the rate of change in compound D is given by:

$$dC_D/dt = k_1\, C_A\, C_B - k_2\, C_D\, C_E \tag{7}$$

Many air quality models have first-order (Equation 3) chemical reactions built into them. In such cases, one need only specify the chemical's half-life. If the half-life of a contaminant is not known, an exponential decay is assumed using the following equation:

$$C(t)/C(o) = e^{-t/Tc} \tag{8}$$

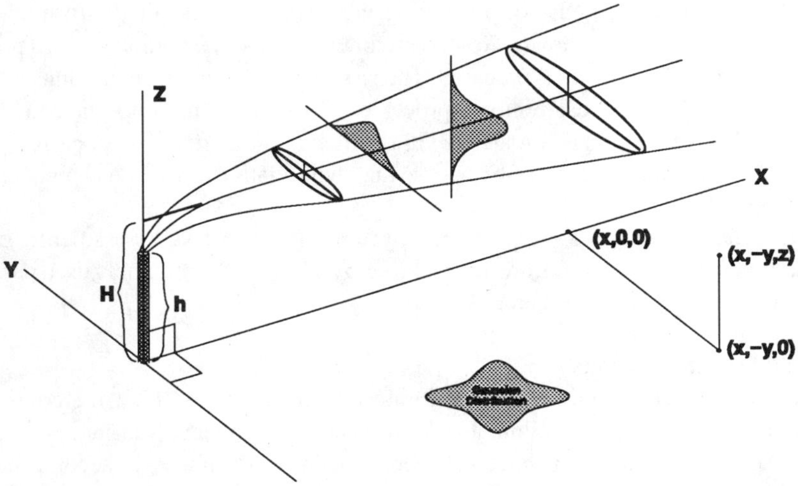

FIGURE 7. Coordinate system showing Gaussian distributions in the horizontal and vertical planes.[17]

where C(t) = concentration of the pollutant at time t, C(o) = concentration of the pollutant at time zero, and Tc = decay time constant.[9] Specific chemical reaction mechanisms that are applicable to various carcinogens and toxic air contaminants emitted from municipal solid waste (MSW) incinerators have yet to be developed for use in air quality models. Nevertheless, if these removal processes can be modeled as simple, first-order reactions, then they can be accounted for in the model using empirical chemical-specific half-life values.

2. Dry Deposition

Dry deposition refers to the transfer of air contaminants from the atmosphere to the surface of the Earth. Both gases and particulate matter (PM) are subject to dry deposition; however, in risk assessments dry deposition of gases is often ignored. The rate at which airborne material is deposited is called deposition flux, Fd. Fd is related to the local concentration of a chemical, Co. The relationship between Fd and Co can be defined as:

$$F_d = V_d C_o \tag{9}$$

where V_d is the constant of proportionality commonly referred as deposition velocity.

While gravity tends to deposit larger particles (e.g., those larger than 30 µm), gravity has only a small influence on particles smaller than 30 µm. Figure 7 presents laboratory measurements of deposition velocities for various size particles and compares the measured deposition velocity with the terminal velocity due to gravity. Figure 7 clearly illustrates that mechanism(s) other than gravity are responsible for dry deposition of smaller particles.[10] Research over the last 3 decades has shown atmospheric turbulence, rather than gravity, is the dominant mechanism responsible for dry deposition of particulates smaller than 30 µm.[10-12] Hence, the deposition velocity is a measure of the rate of deposition and is related to particle size, particle density, level of atmospheric turbulence, type of underlying surface, and atmospheric pollutant concentration.

Deposition velocities can be estimated for various sizes of particles under varying conditions of atmospheric turbulence using a series of monograms prepared by Sehmel.[13] For a given particle size, density, and level of atmospheric turbulence, these monograms give a value for the deposition velocity.

An alternative method for computing deposition rates is to first estimate the ground-level concentration of the air contaminant and then multiply this concentration (worst-case) deposition velocity, typically 1 to 2 cm/s. For small particulates, this method tends to provide a conservative estimate of the deposition rate. For example, assume that the ground-level concentration of a particular contaminant is 10 micrograms/cubic meter ($\mu g/m^3$). By assuming a deposition velocity of 2 cm/s, the dry deposition rate would be 0.2 $\mu g/m^2$-s (10 $\mu g/m^3 \times 0.02$ m/s).

IV. CLASSES OF AIR QUALITY MODELS

Currently, two main approaches are followed for estimating the dilution and spatial distribution of air contaminants in the atmosphere. These approaches have led to statistical models and gradient transfer or "K" theory models as the classes or categories of dispersion models.

The first approach is based on the fact that atmospheric turbulence is random and that the path a particle takes in the atmosphere can be described by a statistical function. In a homogeneous field of turbulence, the diffusive spread of particles from a continuous point source can be viewed as a random process. Therefore, the distribution of particles in planes perpendicular to the wind may be assumed to be approximately Gaussian according to the Central Limit Theorem of statistics. These models are frequently referred to as Gaussian plume models. By far, these are the most commonly used models today and are recommended by the U.S. Environmental Protection Agency (EPA).[14]

There are several reasons why statistical models are widely used. First, they are simple and easy to use. Input data and computer requirements are modest. Second, they can be used to simulate dispersion over a wide range of time scales. For example, hourly, seasonal, or annual average ambient air concentrations can be easily estimated using routinely available meteorological data. Finally, when predicted values are compared with field measurements, Gaussian models, for the most part, perform as well as other models.[15] The models are consistent with the random nature of atmospheric turbulence.

The second approach is based on the gradient transfer hypothesis, or "K" theory, that leads to the classical diffusion equation. This approach assumes that the diffusive flux of a substance is proportional to the gradient of its concentration. This concept is borrowed from early studies in heat conduction. It has been observed that when values of molecular diffusivity in air are used, the classical diffusion equation underestimates the rate of atmospheric diffusion by several orders of magnitude. The reason is that pollutants are diffused not only by molecular diffusion, but also to a much greater degree by turbulent eddies. This approach parameterizes turbulent eddies in terms of turbulent eddy diffusivity, or "K".

A. STATISTICAL MODELS

Mathematically and conceptually, the simplest air quality models are derived assuming that the distribution of pollutant concentrations in the crosswind (horizontal) and vertical directions assume a bivariate Gaussian distribution. In these models total pollutant mass across any cross-sectional area perpendicular to the plume centerline is conserved. The equation which forms the basis for these models is

$$C = \frac{Q}{2\pi\sigma_y\sigma_z} \bullet \exp\left(\frac{-y^2}{2\sigma y^2}\right) \bullet \left(\exp\left[\frac{-(z-H)^2}{2\sigma_z{}^2}\right] + \exp\left[\frac{-(z+H)^2}{2\sigma_z{}^2}\right]\right) \qquad (10)$$

where:

C = the concentration of a given chemical in air (g/m^3)
Q = the emission rate of that chemical from a point source (g/s)
z = the receptor height above ground level (m)
x = distance downwind from the pollutant source to the receptor (m)

y = lateral distance from the pollutant source to the receptor (m)
H = the effective source height (m)

The parameters σ_y and σ_z are the dispersion coefficients that describe how the plume spreads downwind from the source. Figure 8 illustrates the various parameters given in Equation 10. The dispersion coefficients, σ_y and σ_z, are related to the distance downwind from the point source, atmospheric stability, and the amount of anthropogenic activity. It is possible to estimate σ_y and σ_z from observations of turbulence at plume height. Such detailed measurements are not often available, and many models use empirical stability classification schemes and monograms that give values of σ_y and σ_z for various downwind distances.

Values of σ_y and σ_z have been tabulated by various researchers for use in rural and urban areas.[16] A method reported by Turner[17] uses six stability classes and gives the values of σ_y and σ_z for various downwind distances (Figure 9). The values of σ_x and σ_y are based on field experiments involving the diffusion of tracer materials. Various analytical formulae have been developed to estimate the variation in σ_y and σ_z with distance.[18] The relationships developed by Briggs are used in many Gaussian models.[19] They are listed in Table 3 for rural and urban areas. Qualitatively, Equation 10 predicts that:

- The air concentration of a given chemical at a fixed receptor is directly proportional to the emission rate.
- The air concentration of a given chemical is inversely proportional to the wind speed for a given effective release height.
- The predicted air concentration approaches zero at large distances from the source.

Since many MSW incinerators emit contaminants from tall stacks rather than at ground level, the effective height, H, in Equation 10 is the sum total of the source height plus any plume rise. Plume rise is related to the total volume of the flue gas emitted, the exit velocity, and the temperature of the flue gas. Numerous formulae have been developed to estimate plume rise. One commonly used model has been proposed by Briggs:[20,21]

$$H = h_s + DH \quad (11)$$

where H equals effective height above the source, h_s equals source height, and H is the plume rise estimated using the following equations for unstable or neutral conditions:

$$\Delta H = \frac{21.4}{u} F^{3/4} \text{ if } F \leq 55 \quad (12)$$

and

$$\Delta H = \frac{38.7}{u} F^{3/5} \text{ if } F \geq 55 \quad (13)$$

where u equals wind speed (m/s), and F, the buoyancy flux parameter (m^3/s^3), equals:

$$F = g/4 \bullet V_s \bullet d^2 \bullet [(T_s - T_a)/T_a] \quad (14)$$

where g = acceleration due to gravity (9.8 m/s), V_s = stack gas exit velocity (m/s), d = inside stack diameter (m), T_s = stack gas temperature (°K), and T_a = ambient air temperature (°K). The following equations can be used to estimate plume rise under stable conditions:

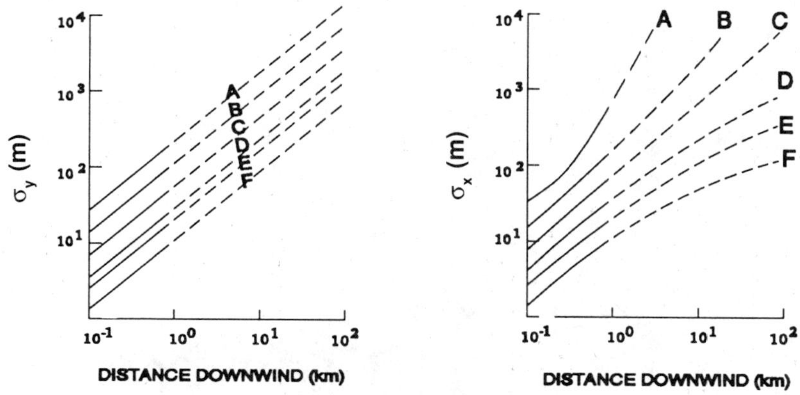

FIGURE 8. Pasquill-Gifford downwind (σ_y, left) and lateral (σ_x, right) dispersion coefficients.[6]

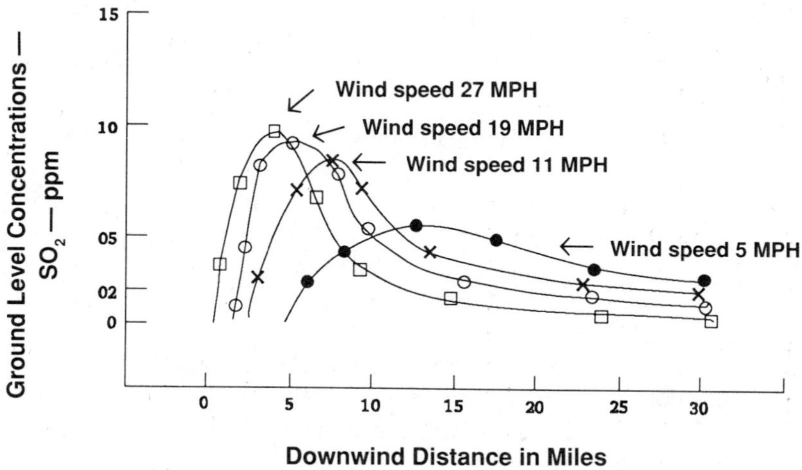

FIGURE 9. Maximum ground level (Z=O) SO_2 concentrations (1/2 h) 2000 NW 2-650 ft stacks, 2.5% sulfur oil stability "C" Briggs plume rise.

TABLE 3
Urban and Rural Dispersion Parameters for Distances between 100 and 10,000 m[6]

Pasquill type	σ_y (m)	σ_z (m)
Urban parameters		
A-B	$0.32x\ (1 + 0.0004x)^{-0.5}$	$0.24x\ (1 + 0.0001x)^{0.5}$
C	$0.22x\ (1 + 0.0004x)^{-0.5}$	$0.20x$
D	$0.16x\ (1 + 0.0004x)^{-0.5}$	$0.14x\ (1 + 0.0003x)^{-0.5}$
E-F	$0.11x\ (1 + 0.0004x)^{-0.5}$	$0.08x\ (1 + 0.0015x)^{-0.5}$
Rural parameters		
A	$0.22x\ (1 + 0.0001x)^{-0.5}$	$0.20x$
B	$0.16x\ (1 + 0.0001x)^{-0.5}$	$0.12x$
C	$0.11x\ (1 + 0.0001x)^{-0.5}$	$0.08x\ (1 + 0.0002x)^{-0.5}$
D	$0.08x\ (1 + 0.0001x)^{-0.5}$	$0.06x\ (1 + 0.0015x)^{-0.5}$
E	$0.06x\ (1 + 0.0001x)^{-0.5}$	$0.03x\ (1 + 0.0003x)^{-1}$
F	$0.04x\ (1 + 0.0001x)^{-0.5}$	$0.016x\ (1 + 0.0003x)^{-1}$

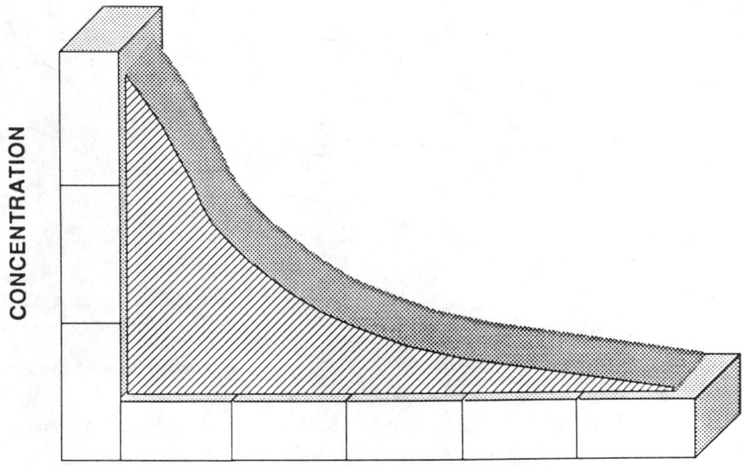

FIGURE 10. Variation of ambient concentration with distance for a ground-based emission source.

$$H = 4\,F^{1/4}\,S^{-3/8} \tag{15}$$

where

$$S = 9.8\,(T/z)/T \tag{16}$$

where T is the ambient temperature (°K) and z is height. For an elevated emission source, the general concentration pattern under various wind speeds is illustrated in Figure 10. The maximum concentration shifts closer to the source at higher wind speeds because of lower plume rise. For ground-level sources (z = 0), Equation 10 reduces to:

$$C = \frac{Q}{2\pi\sigma_y\sigma_z}\exp\left(\frac{-y^2}{2\sigma_y^{\,2}}\right)\left[2\exp-\frac{1}{2}\left(\frac{z^2}{\sigma_z^{\,2}}\right)\right] \tag{17}$$

which predicts that the concentration would be highest near the source and would decrease with increasing distance from the source.

Gaussian models employ several simplifying assumptions, the most important of which are

1. The pollutant is nonreactive and has no appreciable settling velocity.
2. The ground does not absorb any portion of the pollutant mass.
3. Wind speed, direction, and stability are assumed to be constant over time and space for a given period, usually 1 h.
4. The model is steady-state. As a result, these models cannot simulate situations where the concentration is strongly influenced by the past history of the plume.

In spite of these limitations, the basic Gaussian model has been adapted for use in simulations involving chemical reactions, deposition, and the presence of topographical features.

B. GRADIENT TRANSFER OR "K" THEORY MODELS

As noted earlier, these models utilize a modified diffusion equation to estimate the concentration of a particular pollutant. The basic equation governing the transport and diffusion of emitted pollutants is

$$\frac{\partial \bar{q}}{\partial i} = -\bar{u}\frac{\partial \bar{q}}{\partial x} + \frac{\partial}{\partial x}\left(K_x \frac{\partial \bar{q}}{\partial x}\right) + \frac{\partial}{\partial y}\left(K_y \frac{\partial \bar{q}}{\partial y}\right) + \frac{\partial}{\partial z}\left(K_z \frac{\partial \bar{q}}{\partial z}\right) \tag{18}$$

The first expression on the right side of Equation 18 quantifies the advection in the x direction caused by the mean wind (u). The terms inside the brackets represent the diffusive fluctuations in the x, y, and z directions, which are proportional to the gradients of concentration. The "K"s inside the brackets are also referred to as diffusivity coefficients. If a chemical undergoes deposition or chemical reactions, additional terms are added to the right side of Equation 18 to account for these removal mechanisms. Except in cases when the mean wind speed is very light, the second term on the right side of the Equation 18 can be disregarded.

The advantage of this method is that it allows for the specification of the eddy diffusivities ("K" values) at various heights. Thus, in principle, vertical variation in wind shear can be simulated. Generally, K is related to height, for example,

$$K = A(z/z_1)^B \tag{19}$$

where A and B are constants that are a function of stability, z is the height above ground in meters, and z_1 is a reference height.

Models based on "K" theory typically use a three-dimensional wind model that estimates wind speed, direction, and pollutant concentration within each grid cell. Typically, the cell dimensions are a few hundred to a few kilometers in size. This results in a more accurate regional simulation in that the variation in wind and turbulence can be explicitly taken into account. Since the results are averaged within a cell, it is generally not possible to estimate impacts associated with an individual plume, if the plume dimensions are less that the dimensions of the cell.

Typically, the computational domain used in a three-dimensional simulation with"K" theory models consists of an array of three-dimensional grid cells. Pollutants are advected from cell to cell by the wind. Topography is simulated by creating cell boundaries that prevent the air from penetrating that cell. Model results consist of concentrations of the pollutant in each cell. The predicted concentration is averaged within each cell.

The difficulty in using "K" theory models lies in the solution of Equation 17. Unfortunately, this equation cannot be solved analytically; therefore, numerical techniques must be employed. The methods of numerical solution are discussed in Long and Pepper.[22] Regardless of the technique used to solve the basic diffusion equation, extensive computations are needed. For this reason, these models are preferred for short-term (1 to 48 h) simulations involving many sources on a regional scale.

V. AVAILABILITY OF AIR QUALITY MODELS

Many states as well as the federal government serve as repositories of air quality models and meteorological data. For example, in California, the Air Resources Board located in Sacramento makes computer codes available to the public. Similarly, other states and local air pollution control districts often makes computer models available to the general public. At the federal level, the EPA has developed a set of air quality models which are available through the National Technical Information Service (NTIS) in Springfield, VA. Meteorological input data for use in dispersion models are available from the National Climatic Data Center in Asheville, NC.

REFERENCES

1. **Angle, R. P.,** *Air Quality Modeling and User's Needs,* Alberta Environment, Edmonton, Alberta, Canada, 1979.
2. **Irwin, J. S.,** A theoretical variation of the wind profile power law exponent as a function of surface roughness and stability, *Atmos. Environ.,* 13, 191, 1979.
3. **Sherman, C. E.,** A mass consistent model for wind fields over complex terrain, *J. Appl. Meteorol.,* 17, 312, 1978.
4. **Landsburg, H. E.,** *The Urban Climate,* Academic Press, New York, 1981.
5. **Pasquill, F.,** *Atmospheric Diffusion,* 2nd ed., Halstead Press, New York, 1974.
6. **Gifford, F. A.,** Turbulent diffusion typing schemes, *Nucl. Safety,* 17, 72, 1976.
7. **Cramer, H. E.,** A practical method for estimating the dispersal of atmospheric contaminants, Proc. 1 Natl. Conf. on Applied Meteorology, American Meteorological Society, Boston, 1957.
8. **Williamson, S. J.,** *Fundamentals of Air Pollution,* Addison Wesley, Reading, MA, 1973.
9. **Hanna, S. R., Briggs, G. A., and Hosker, R. P.,** *Handbook of Air Pollution,* Technical Information Center, U.S. Department of Energy, Washington, D. C., 1982.
10. **McMahon, T. A. and Denison, P. J.,** Empirical atmospheric deposition parameters: a survey, *Atmos. Environ., 13, 572, 1979.*
11. **Wesley, M. L. and Hicks, B. B.,** Some factors that affect deposition rates of SO_2 and similar gases on vegetation, *J. Air Poll. Control Assoc.,* 27, 1110, 1977.
12. **Hicks, B. B.,** Dry deposition processes in acidic deposition phenomena and its effects, Altshuller, A. P. and Linthurst, R. A., Eds., U.S. Environmental Protection Agency, Research Triangle Park, NC, 1983.
13. **Sehmel, G. A.,** Particle and gas dry deposition: a review, *Atmos. Environ.,* 14, 983, 1980.
14. Guideline on Air Quality Models (Revised), EPA-450/2-78-027R, U.S. Environmental Protection Agency, Research Triangle Park, NC, 1986.
15. **Stern, A. C., Boubel, R. W., Turner, D. B., and Fox, D. L.,** *Fundamentals of Air Pollution,* 2nd ed., Academic Press, New York, 1984.
16. **Briggs, G. A.,** Diffusion estimates for small emissions, Atmospheric and Turbulent Diffusion Laboratory, Contribution File No. 79, National Oceanic and Air Administration, Oak Ridge, TN.
17. **Turner, D. B.,** Workbook of Atmospheric Dispersion Estimates, 999-AP-26, U.S. Department of Health, Education and Welfare, Washington D.C., 1979.
18. **Busse, A. D. and Zimmerman, J. R.,** User's Guide for the Climatological Dispersion Model, U.S. Environmental Protection Agency, Research Triangle Park, NC, 1973.
19. **Hanna, S. R.,** Air pollution modeling over short distances, in *Handbook of Meteorology,* McGraw-Hill, New York, 1978.
20. **Briggs, G. A.,** Plume Rise, U.S. Atomic Energy Commission Critical Review Series, TID-25075, National Technical Service, Springfield, VA, 1969.
21. **Briggs, G. A.,** Plume rise predictions, in *Lectures on Air Pollution and Environmental Impact Analysis,* Haugen, D. A., Ed., American Meteorological Society, Boston, 1975.
22. **Long, P. E. and Pepper, D. W.,** A comparison of six numerical schemes for calculating the advection of atmospheric pollutants, Preprints of 3rd Symposium on Atmospheric Turbulence, Diffusion, and Air Pollution, American Meteorological Society, Raleigh, NC, 1976.

Chapter 4

ASSESSING THE EXTENT OF HUMAN EXPOSURE THROUGH THE FOOD CHAIN TO POLLUTANTS EMITTED FROM MUNICIPAL SOLID WASTE INCINERATORS*

Holly A. Hattemer-Frey and Curtis C. Travis

TABLE OF CONTENTS

* Research was sponsored by the U.S. Environmental Protection Agency under an Interagency agreement applicable under Martin Marietta Energy Systems, Inc. Contract No. DE-AC05-840R21400 with the U.S. Department of Energy.

I. INTRODUCTION

Human exposure to pollutants emitted from a municipal solid waste (MSW) incinerator can occur via inhalation, ingestion of contaminated food items, infant consumption of mother's milk, and dermal absorption. Of particular concern are potential exposures from ingesting contaminated food items, since the food chain has been shown to be a primary source of human exposure to a large class of organics, including DDT, dioxin, pentachlorophenol, benzo(a)pyrene, most pesticides,[1-5] and metals.[6] The purpose of this chapter is to quantify the extent of human exposure to a representative organic (2,3,7,8-tetrachlorinated dibenzo-*p*-dioxin, commonly referred to as TCDD) and a representative metal (cadmium) through the terrestrial food chain from an arbitrary MSW incinerator.

II. THE TERRESTRIAL FOOD CHAIN MODEL

The Terrestrial Food Chain (TFC) model, developed by Oak Ridge National Laboratory's Office of Risk Analysis, is a multimedia transport model designed to assess the magnitude of human exposure to compounds emitted from a point source (in this case, an MSW incinerator). The TFC model offers improvements over existing food chain models, such as those generated by the U.S. Environmental Protection Agency[7] and Travis et al.[8] Methodologies described in these earlier models were modified to account for the effects of direct deposition of pollutants onto plant, water, and soil surfaces and for the effects of air-to-leaf transfer of organic vapors. The TFC model uses annual atmospheric deposition rates (μg/m^2/year) and annual average air concentrations (μg/m^3) to estimate the amount of pollutant entering the terrestrial food chain and the average daily intake of pollutants by individuals living near an arbitrary MSW incinerator.

III. ACCUMULATION OF POLLUTANTS IN THE TERRESTRIAL FOOD CHAIN

Assessing the magnitude of exposure to pollutants emitted from an MSW incinerator depends largely on being able to predict their bioaccumulation in the food chain. The octanol-water partition coefficient (K_{ow}) is used extensively to estimate the bioconcentration potential of pollutants in biological systems. Chemicals with large K_{ow} values, or lipophilic compounds, such as TCDD, DDT, and polychlorinated biphenyls (PCBs), are soluble in organic matter. These chemicals tend to sorb strongly to air particulates, soil, and sediment, bioaccumulate in living organisms, and transfer to humans through the food chain. Conversely, chemicals with small K_{ow} values, such as trichloroethylene tetrachloroethylene, and benzene, tend to partition mostly into air or water, and inhalation is the primary pathway of human exposure.

The TFC model was used to assess the amount of pollutant entering the terrestrial food chain and the extent of human exposure from inhalation and from ingestion of contaminated agricultural produce (fruits, grains, and vegetables), meats, dairy products, eggs, and soil.

A. ACCUMULATION IN VEGETATION

The accumulation of facility-emitted compounds in vegetation is a complex process that involves root uptake, foliar uptake (i.e., air-to-leaf transfer), and deposition onto outer plant surfaces.

1. Root Uptake

Root uptake of organics has been correlated with K_{ow}[9,10] and can be estimated from B_v, the soil-to-root bioconcentration factor (BCF). B_v is defined as the equilibrium concentration of pollutant in plant tissue (μg/g dry weight, DW) divided by the concentration of pollutant in soil (μg/g DW) and can be estimated from the following geometric mean regression equation developed by Travis and Arms[11] in the absence of measured data:

$$B_v = 38.73\ K_{ow}^{-0.578} \tag{1}$$

(Halfon[12] recommended that a geometric mean regression be used to determine the relationship between K_{ow} values and various measures of bioaccumulation, since both of these parameters are usually measured with error. The traditional linear regression method assumes that one variable is measured with error and the second is known precisely.)

The organic matter content of soil is the single most important factor affecting root uptake of pollutants.[13] Because lipophilic compounds are expected to sorb strongly to soil, root uptake is not expected to be a major pathway of vegetative contamination for this group of compounds. Conversely, root uptake is the principal means of metals incorporation into vegetation.[14]

The concentration of pollutant in vegetation on a DW basis due to root uptake (CVR) in micrograms of pollutant per gram of plant dry weight basis (μg/g DW) can be estimated from the following equation:

$$\text{CVR}\ (\mu\text{g/g DW}) = B_v * C_s \tag{2}$$

where B_v equals the plant-specific, chemical-specific soil-to-root BCF, and C_s is the concentration of pollutant in soil. The TFC model calculates the maximum cumulative pollutant concentration in soil according to the following equation:

$$C_s\ (\mu\text{g/g})\ \text{DW}) = \frac{Dy^*\ [1.0 - \exp(-K_s * Tc)]}{Z * D_s * K_s * K_1} \tag{3}$$

where:

D_y = annual atmospheric deposition rate (μg/m²/year)
K_s = pollutant-specific soil loss constant 1/year or year^{-1}
T_c = length of long-term pollutant build-up in the soil (65 years)
Z = soil mixing depth (20 cm)
D_s = bulk density of soil (1.4 g/cm³)
K_1 = 100^2 to convert cm² to m²

The TFC model assumes that deposition of facility-emitted pollutants onto the soil occurs continuously throughout the operating lifetime of the facility. The parameter, T_c, reflects the total time period of pollutant deposition, which corresponds to the operating lifetime of the incinerator. The U.S. EPA[7] estimated that the lifetime of an MSW incinerator ranges from 30 to 100 years. An average value of 65 years was used in this analysis. The soil loss constant, K_s, reflects the loss of organics from the soil due to leaching, degradation, and volatilization. Since metals do not decay or degrade in the environment, K_s values for metals equal zero. The TFC model also assumes that crops are grown in tilled soil and that pollutants are distributed evenly within this tilled soil to a depth of 20 cm. To model direct soil ingestion, soil is considered to be untilled. In this case, effective depth to which pollutants are distributed was assumed to be 1 cm.

2. Air-to-Leaf Transfer

While some organics are taken up by plants directly from contaminated soil and translocated to upper plant parts, other compounds volatilize from the soil, and their vapors are absorbed by aerial plants parts.[15-19] Volatilization of organics from polluted soil may be as or more important than root uptake as a source of plant contamination. Buckley,[20] Nash and Beall,[21] and Mosbaek et al.[22] showed that PCB, DDT, and mercury contamination of plant foliage was mainly due to volatilization of these compounds from the soil.

The accumulation of volatilized organics by upper plant parts can be estimated from B_{va}, the air-to-plant BCF, which is defined as the equilibrium concentration of organic in upper plant parts (μg/g DW) divided by the equilibrium concentration of organic in air *as a vapor* (μg/g). B_{va} has been shown to be positively correlated with K_{ow} divided by Henry's Law Constant (H) in atm-m^3/mol.[23] Given this relationship, lipophilic compounds must have a low H value to be accumulated via this pathway.

B_{va} can be estimated from the following geometric mean regression equation derived from the experimental data collected by Bacci et al.[18] for ten chemicals with varying physicochemical properties:

$$B_{va} = 5.0 \times 10^{-6}\, K_{ow} \div H \tag{4}$$

To calculate the concentration in vegetation due to air-to-leaf transfer, the model assumes that organics deposited onto the soil volatilize. The concentration of pollutant in air due to volatilization (C_{av}) is determined by assuming that a uniform amount of pollutant is emitted from the soil per unit area. A Gaussian plume line model[24] was used to estimate the loss of pollutant as it is transported away from the air-soil-water interface to plant height. Assuming that the area over which the pollutant volatilizes is essentially circular and that the soil emits a uniform amount of pollutant per unit area, atmospheric pollutant concentration is estimated by:

$$C_{av} = \sum_{i=1}^{6} F_i \int_0^R \frac{2q}{\sqrt{2\pi}\sigma_{zi}\, u} \exp\left[-\frac{1}{2}\left\{\frac{H}{\sigma_{zi}}\right\}^2 \right] dx \tag{5}$$

where:

C_{av} = the partial concentration of pollutant in air due to volatilization from soil (μg pollutant/m^3 air)
F_i = fraction of time of the ith atmospheric stability (unitless)
R = radius of the contaminated area (m)
q = volatilization rate of pollutant from soil (μg/m^2/s)
H = height of plant above soil (m)
u = average wind speed above contaminated area (m/s)
σ_{zi} = Gaussian plume dispersion parameter for the vertical direction (m) for the ith Pasquill atmospheric stability category

The Gaussian plume dispersion parameter for the vertical direction (m) for the ith Pasquill atmospheric stability category, σ_{zi}, is calculated according to the formula:

$$\sigma_{zi} = ai\,(x/K_3)^{bi} \tag{6}$$

where ai and bi are empirically determined parameters for each Pasquill atmospheric stability category,[25] x is the downward distance from source (m), and K_3 is a units conversion factor of 1000.

Thus, the concentration in vegetation due to the air-to-leaf transfer (CVA) can be estimated from the following equation:

$$\text{CVA } (\mu g/g \text{ DW}) = \frac{[C_{av} + (C_a * F_v)] * B_{va}}{D_a} \tag{7}$$

where:

C_{av} = the pollutant concentration in air due to revolatilization (µg/m^3 air)
C_a = the pollutant concentration in air due to facility emissions (µg/m^3)
F_v = the fraction of pollutant in the vapor phase
B_{va} = the plant-specific, chemical-specific air-to-leaf BCF (µg/g plant DW ÷ µg/g in air as a vapor DW)
D_a = density of air (1190 g/m^3)

It was assumed that protected produce, including potatoes and other root vegetables, legumes, and garden fruits, whose edible portions either grow underground or are protected by pods, shells, or nonedible peels, were not contaminated via this pathway. If accumulation via foliar uptake is a surface phenomenon only (i.e., organics are not being incorporated into plant tissue), then this assumption is valid. If, on the other hand, organic vapors are being incorporated into plant tissue through the leaves and other aerial plant parts, then this assumption may not be valid. With the exception of mercury,[22,26] air-to-leaf contamination for most metals is equal to zero, since metals do not volatilize from the soil and do not exist in air as a vapor.

3. Deposition

Deposition of pollutants onto outer plant surfaces can also contribute substantially to vegetative contamination, especially for lipophilic compounds that are expected to sorb strongly to particulates. The concentration of pollutant on vegetation due to atmospheric deposition at the time of harvest (CVD) can be estimated using the following equation:

$$\text{CVD } (\mu g/g \text{ DW}) = \frac{D_y * r * [1.0 - \exp(-K_e * T)]}{Y * K_e} \tag{8}$$

where:

D_y = annual pollutant deposition rate (µg/m^2/year)
r = intercept fraction for the edible plant portion (unitless)
K_e = weathering half-life 1/year or year^{-1}
T = length of the growing season (year)
Y = yield or standing crop biomass of the edible plant portion (g DW/m^2)

The TFC model assumes that deposition occurs only during the growing season and that protected produce are not contaminated via this pathway. The length of the growing season for food crops consumed by humans (T_h) and for grains (T_g) was set at 58 d (0.16 year), while the length of the growing season for forage crops (T_f) equals 51 d (0.12 year).[26] Thus, all food crops consumed by humans are assumed to be subject to deposition for the same time period. Thus, the value of 58 d represents the longest time required for any food crop to mature.

During the time period of deposition, pollutants may be subjected to a variety of removal processes. At a minimum, the model assumes that all pollutants are subject to weathering. Plant concentrations are corrected for pollutant loss due to weathering using the elimination rate constant K_e. K_e can be estimated from the following equation:

$$K_e \text{ (years)}^{-1} = \ln 2/T_{1/2} \text{ (years)} \tag{9}$$

where ln 2 is the natural log of two, and $T_{1/2}$ is the half-life of the compound on the plant surface

TABLE 1
Average Intercept Fractions (r) for Vegetation Consumed by Animals and Humans[26,39]

Plant group	Mean	Range
Forage (r_f)	0.47	0.02—0.82
Grains (r_g)	0.00	—
Potatoes (r_p)	0.00	—
Leafy vegetables (r_{lv})	0.16	0.08—0.38
Legumes (r_{le})	0.008	0.005—0.01
Root vegetables (r_r)	0.00	—
Garden fruits (r_{gf})	0.05	0.004—0.08

TABLE 2
Yield (Y) or Standing Crop Biomass of Vegetation Consumed by Humans and Animals (G DW)[26,28,39]

Vegetation type	Mean	Range
Forage	310	20—750
Grain	300	140—450
Potatoes	480	410—560
Leafy vegetables	180	90—350
Legumes	100	80—130
Root vegetables	340	90—440
Garden fruits	110	10—250

in years. In the absence of measured data, it is generally assumed that organics deposited onto outer plant surfaces have a 14-d weathering half-life.[26]

Intercept fraction (r) and yield (Y) values account for the fact that not all airborne pollutants deposited within a given area actually settle on edible plant parts. Vegetation-specific differences in the intercept fraction and productivity values for plants consumed by animals and humans (Tables 1 and 2) require that different estimates of CVD be made for forage, grains, and food crops consumed by humans.[26] The total concentration of pollutant on vegetation consumed by animals and humans is determined by summing the contribution from each of the three pathways of vegetative contamination (CVR + CVD + CVA).

B. ACCUMULATION IN ANIMAL TISSUES

Assessing the environmental fate of chemicals depends largely on the ability to predict the extent to which they will bioaccumulate in living organisms, including animals and humans. A traditional measure of a chemical's potential to accumulate in biota is the BCF, which is defined as the equilibrium concentration of pollutant in an organism or tissue (µg/kg) divided by the equilibrium concentration of pollutant (µg/kg) in food for terrestrial organisms. This concept, however, is not readily applied to humans and other terrestrial organisms because the amount of chemical in various food items can vary markedly. Thus, for risk assessment purposes, it is more efficacious to use biotransfer factors (BTFs), which relate the equilibrium concentration of pollutant in an organism or tissue (µg/kg) to total daily intake of pollutant (µg/d).[11]

In the absence of measured data, BTFs for organics in beef (B_b) and cow milk (B_m) (whole tissue basis) can be estimated from the following geometric mean regressions developed by Travis and Arms:[11]

$$B_b \text{ (d/kg)} = 2.51 \times 10^{-8} * K_{ow} \text{ (n= 36; r = 0.81)} \quad (10)$$

$$B_m \text{ (d/kg)} = 7.94 \times 10^{-9} * K_{ow} \text{ (n = 28; r = 0.74)} \quad (11)$$

Measured concentrations of organics in beef or milk fat can be converted to a fresh meat or whole milk basis assuming that beef and milk are 25 and 3.7% fat, respectively. Daily intake was computed assuming an average feed ingestion rate of 16 kg/d for lactating cows and 8 kg/d for nonlactating cows.[11]

Farm animals are exposed to facility-emitted pollutants via ingestion of contaminated feed (forage and grains) and soil. Ingestion of contaminated feed and soil can be a significant exposure pathway for terrestrial organisms, because lipophilic organics, such as DDT, TCDD, and PCBs tend to accumulate in soil and plants. Travis and Hattemer-Frey[27] showed that ingestion of contaminated soil accounted for 20% of the total daily intake of TCDD by cows.

The predicted daily intake of pollutants by agricultural animals is estimated by multiplying the concentration of pollutant in forage, grains, and soil times the quantity of forage, grains, and soil consumed daily that originated from the study area (i.e., that area around the facility affected by facility emissions) and the appropriate chemical-specific biotransfer factor. Pollutant concentrations in a given animal tissue are calculated according to the following equation:

$$C_t = [(Q_f * F_f * C_f) + (Q_g * F_g * C_g) + (Q_s * C_s)] * B_a \tag{12}$$

where:

C_t = pollutant concentration in an animal tissue (μg/g)
Q_f = quantity of forage crops consumed daily by the animal (kg DW/d)
F_f = fraction of forage crops that are locally grown (unitless)
C_f = pollutant concentration in forage crops (μg/g DW)
Q_g = quantity of grains consumed daily by the animal (kg DW/d)
F_g = fraction of grains that are locally grown (unitless)
C_g = pollutant concentration in grains (μg/g DW)
Q_s = quantity of contaminated soil consumed by the animal (kg DW/d)
C_s = pollutant concentration in soil (μg/g DW)
B_a = biotransfer factor for the animal tissue (kg/d)

Beef cattle consume about 2.66 kg DW of forage and 5.0 kg DW of grains per day, while dairy cattle consume 11.0 and 7.1 kg DW of forage and grains per day, respectively.[28] It was assumed that 100% of the forage and 60% of the grains consumed by agricultural animals are locally grown (i.e., Q_f equals 1.0 and Q_g equals 0.6).

Fries[29] concluded that since lactating dairy cattle are rarely pastured, it is unlikely that soil ingestion would exceed 1% of dry matter intake. Although beef cattle do spend some time on pasture, they tend to pass most of their time confined to feed lots.[29] Hence, we assumed that soil ingestion accounts for 1% of dry matter intake for both dairy cattle (0.18 kg DW/d) and beef cattle (0.08 kg DW/d).

Sheep ingest 1.6 kg DW of forage per day (no grains),[28] while chickens and pigs consume 0.07 and 3.5 kg DW of grains per day (no forage).[30] Fries[29] reported that sheep are likely to ingest slightly more soil as represented by the percent of dry matter intake, while chickens and pigs, which are rarely pastured, would ingest similar amounts of soil as beef cattle. Hence, we assumed that soil ingestion equals 2% of dry matter intake for sheep (0.03 kg DW/d) and 1% of dry matter intake for chickens and pigs (0.001 and 0.035 kg DW/d, respectively).

IV. ESTIMATING THE DAILY INTAKE OF POLLUTANTS THROUGH THE FOOD CHAIN

For risk assessment purposes, an important objective in evaluating the environmental fate of facility-emitted pollutants is predicting the pathways of human exposure. Three routes exist by

TABLE 3
Average Daily U.S. Human Consumption Values (DW basis)

Food group	Intake[a] (g/d)
Milk and dairy products	40.17
Beef	33.73
Beef liver	6.30
Poultry	12.05
Lamb	1.64
Pork	10.86
Eggs	6.99
Potatoes	30.06
Leafy vegetables	2.59
Legumes (dried)	3.11
Legumes (nondried)	0.78
Root vegetables	3.34
Garden fruits	10.84
Grains and cereals	177.60

[a] Adapted from wet weight consumption rates reported in Yang and Nelson[49] and wet-to-dry conversion factors from Baes et al.[26]

TABLE 4
Average Percent of Food Items that are Locally Produced[50]

			Households	
Food group	All	Urban	Rural, nonfarm	Rural, farm
All meat except poultry	5.25	0.78	6.14	44.2
Poultry	7.38	2.85	11.9	34.3
Dairy products	4.00	0.00	3.99	39.9
Eggs	7.51	0.62	9.14	47.9
Potatoes	9.30	1.21	14.6	44.8
Vegetables	21.2	7.55	35.7	70.0
Legumes (dried)	4.44	2.63	6.90	16.7
Legumes (nondried)	21.2	7.55	35.7	70.0
Root vegetables	21.2	7.55	35.7	70.0
Garden fruits	8.11	3.24	13.2	31.3
Grains and cereals	0.43	0.13	0.82	1.56

which humans can be exposed to pollutants emitted from an MSW incinerator through the terrestrial food chain: (1) direct ingestion of contaminated soil, (2) consumption of contaminated plants, and (3) consumption of contaminated animal tissues (i.e., meats and dairy products).

Daily intake from the ingestion of contaminated food items is estimated by multiplying the concentration of pollutant in meats, dairy products, and plants by the average daily human consumption value (g DW/d) (Table 3) and the fraction of food consumed that is locally grown (Table 4). Hence, consumption of each food group was weighted by the fraction of that food group that originates from the study area.

A. DIRECT INGESTION OF CONTAMINATED SOIL

Humans can be exposed to pollutants by intentionally eating contaminated soil or by inadvertent hand-to-mouth transfer of soil. The TFC model calculates human daily pollutant intake from ingesting contaminated soil by:

$$I_s = C_s * Q_s * E \quad (13)$$

where:

I_s = pollutant intake due to direct ingestion of contaminated soil (µg/d)
C_s = concentration of pollutant in untilled soil after the total time period of deposition (µg/g DW)
Q_s = quantity of soil ingested by adults (g DW/d)
E = exposure duration adjustment (unitless)

The intentional ingestion of nonfood items, or pica behavior, occurs in children from ages 1 to 6 years. The exposure duration adjustment factor is used to more effectively distribute this exposure over the entire lifetime of the individual. We recommend the use of an exposure duration adjustment factor when modeling carcinogenic pollutants. E should be set equal to 1.0 for all noncarcinogens and for all chemicals when determining the total daily intake of pollutants by adults. When estimating daily pollutant intake by children from ingestion of contaminated soil, E is calculated by dividing the duration of exposure (e.g., 5 years) by the assumed lifetime (e.g., 70 years).

B. CONSUMPTION OF CONTAMINATED PLANTS

Humans can also be exposed to pollutants emitted from MSW incinerators by consuming contaminated plants. The total amount of pollutant ingested daily as a result of this pathway is calculated using the following equation:

$$I_p = C_p * F_p * E_p \quad (14)$$

where:

I_p = daily intake due to the consumption of a contaminated plant group (µg/d)
C_p = concentration of pollutant in the plant group (µg/g DW)
F_p = fraction of the plant group that originates from the contaminated area (unitless) (Table 4)
E_p = dult daily dietary consumption of the plant group (g DW/d)

Homegrown fruits and vegetables are assumed to be contaminated by facility-emitted pollutants. F_p represents the portion of each plant group that is locally grown. Plants imported from outside of the affected area are assumed to be uncontaminated. Total daily intake of pollutant via ingestion of contaminated plants is calculated by summing the contributions of each individual plant group modeled.

C. CONSUMPTION OF CONTAMINATED ANIMAL TISSUES

The total amount of pollutant intake by humans due to the consumption of a single, contaminated animal tissue is given by:

$$I_a = C_t * F_a * C_h \quad (15)$$

where:

I_a = pollutant intake from the consumption of a contaminated animal tissue (μg/d)
C_t = concentration of pollutant in the animal tissue (μg/g DW)
F_a = fraction of the animal tissue that is locally produced (Table 4)
C_h = human consumption of the animal tissue (g DW/d)

Total daily intake of pollutant via consumption of animal tissues is calculated by summing the contribution of each individual tissue group modeled.

D. TOTAL DAILY INTAKE OF POLLUTANT

Total daily intake of pollutant by humans is calculated by:

$$I_t = I_i + I_s + \sum_{i=1}^{np} I_{pi} + \sum_{i=1}^{na} I_{oi} \qquad (16)$$

where:

I_t = total daily intake of pollutant by humans (μg/d)
I_i = total daily intake of pollutant from inhalation, [concentration of pollutant in air (μg/m^3) times an average adult inhalation rate 20 m^3/d)] (μg/d)
I_s = total daily intake of pollutant from ingesting contaminated soil (μg/d)
I_{pi} = total daily intake of pollutant from the consumption of the ith contaminated plant group (μg/d)
np = total number of plant groups
I_{ai} = total daily intake of pollutant from the consumption of the ith contaminated animal tissue (μg/d)
na = total number of animal tissues modeled

The TFC further assumes that the concentration of pollutant on vegetation consumed by humans is not decreased by food preparation and washing activities and that 100% of all pollutants ingested are absorbed through the gut. Although gastrointestinal (GI) absorption factors are widely used in health risk assessments, we believe that is not correct to use GI factors when estimating exposure because risk estimates are obtained by multiplying exposure times potency. Potency values are derived from laboratory experiments in which laboratory animals are given a specific dose of a carcinogen by one of several routes. In these experiments, potency values are based upon *administered* dose, not effective dose to the target tissue. Risk estimates should also reflect administered dose. Hence, reducing exposure estimates by applying GI factors is not appropriate.

V. APPLICATION OF THE TERRESTRIAL FOOD CHAIN MODEL TO TCDD

Although there are actually more than 75 different molecular forms of dioxin, 2,3,7,8-tetrachlorodibenzo-*p*-dioxin (TCDD) is often referred to as dioxin. TCDD is a highly lipophilic, extremely persistent compound that sorbs strongly to soil and sediment and bioaccumulates in the food chain. Because TCDD is the most potent chemical ever evaluated by the U.S. EPA,[31] there is widespread fear that exposure to even small amounts of dioxin could lead to serious adverse human health effects. Because combustion is widely recognized to be a source of TCDD,[32-36] many people who live near MSW incinerators are concerned that they will be exposed to high levels of dioxin and subsequently develop cancer.

A. MODEL INPUTS AND RESULTS

The TFC model was used to quantify the magnitude of exposure to TCDD emitted from an arbitrary MSW incinerator in the U.S. based on the following inputs. The deposition rate and maximum air concentration predicted for an arbitrary MSW incinerator, 1.43×10^{-3} μg/m²/year and 1.1×10^{-8} μg/m³, respectively[37] (see Chapter 10) were used in this assessment. Since the purpose of this exercise was to estimate the magnitude of human exposure to TCDD emitted from an arbitrary, modern U.S. MSW incinerator, we entered generic mean values for all site-specific input parameters. When the TFC is applied to specific cases, site-specific vs. generic input data should be used, and the user should anticipate that the major pathways and extent of human exposure may vary.

1. Uptake of TCDD by Plants

Since TCDD is extremely lipophilic, and root uptake is inversely correlated with lipophilicity,[10] root uptake was not expected to be a major source of vegetative contamination for TCDD. Using a soil concentration of 8.0×10^{-8} μg/g, a soil-to-plant BCF 4.25×10^{-3} for vegetation consumed by animals and humans,[11] a soil loss constant of 0.06/year,[40] a soil concentration of 8.0×10^{-8} μg/g, and Equations 2 and 3, the concentration of TCDD on all plants due to root uptake is estimated to be 0.34 fg/g DW.

Using an air-to-leaf BCF of 9838 for vegetation consumed by animals and humans (K_{ow} and H for TCDD equal 7.1×10^{6} and 3.6×10^{-3})[5,41] and Equation 7, the concentration of TCDD on forage due to foliar uptake is estimated to be 43 fg/g DW. The predicted concentration due to air-to-leaf transfer on legumes (dried and fresh) and leafy vegetables was 43 fg/g DW, while the predicted concentration of TCDD on garden fruits was 41 fg/g DW. Again, it was assumed that grains were not contaminated via this pathway.

Deposition of TCDD onto outer plant surfaces was expected to contribute substantially to vegetative contamination, since it was assumed that all of the TCDD emitted from an incinerator is sorbed onto particulates. Isensee and Jones[38] demonstrated that 94 and 63% of the TCDD applied to soybeans and oats remained on the leaves for 21 d, indicating that most of the dioxin deposited from the atmosphere onto outer plant surfaces is likely to persist until the plant is harvested or consumed by animals.

Using an atmospheric deposition rate of a 1.4×10^{-3} μg/m²/year, a weathering constant (K_e) of 14 d (18.2/year),[26] an intercept fraction (r) of 0.47[39] (Table 1), 0.12 years as the length of the growing season,[26] a crop yield of 310 g/m² (Table 2),[38] and Equation 8, the concentration of TCDD on forage due to deposition was estimated to be 106.5 fg/g DW. Because grains have a protective outer husk, it was assumed that grains were not contaminated via deposition of facility-emitted pollutants.

Using the plant-specific intercept fractions and crop yields for fruits and vegetables listed in Tables 1 and 2, respectively, and 0.16 years as the length of the growing season,[26] the predicted concentration of TCDD on plants consumed by humans due to deposition ranged from 43 fg/g DW for legumes and leafy vegetables to 41 fg/g DW for garden fruits (Table 5).

Thus, the total concentration of TCDD on forage equals 150 fg/g DW, 71% of which was due to deposition, 29% to foliar uptake, and 0.2% to root uptake (Table 5). Forage was about 500 times more contaminated with TCDD due to facility emissions than were grains (150 fg/g vs. 0.3 fg/g). The total concentration of TCDD on fruits and vegetables consumed by humans ranged from 0.34 fg/g DW for potatoes, root vegetables, and grains, 100% of which was attributable to root uptake, to 110 fg/g DW for leafy vegetables, 60% of which was due to deposition, 39% to foliar uptake, and 0.3% to root uptake (Table 5).

2. Uptake of TCDD by Animals

Because of its low water solubility and high lipophilicity, TCDD readily accumulates in

TABLE 5
Percentage of Vegatative Contamination due to Deposition, Foliar Uptake, and Root Uptake of TCDD

	TCDD conc. (fg/g)[a]	Percentage		
		Due to deposition	Due to foliar uptake	Due to root uptake
Human crops				
Potatoes	0.3	0	0	100
Leafy vegetables	110.0	60	39	0.3
Legumes	49.1	12	88	0.7
Root vegetables	0.3	0	0	100
Garden fruits	76.8	45	54	0.4
Grains	0.3	0	0	100
Weighted average	6.1	42	53	6
Animal crops				
Forage	150.0	71	29	0.2
Grains	0.3	0	0	100

[a] Femtograms per gram dry weight.

TABLE 6
Predicted TCDD Concentration in Animal Tissues from Ingesting Contaminated Soil, Forage, and Grains

		Percentage		
Tissue group	Total conc. (fg/g DW)[a]	Due to forage ingestion	Due to grains ingestion	Due to soil ingestion
Beef and beef liver	1090	76	0.2	24
Dairy products	449	85	0.1	15
Pork	32	0	1	99
Chicken	193	0	1	99
Lamb	6	82	0	18
Eggs	146	0	1	99

[a] Femtograms per gram dry weight basis.

living organisms. Ingestion of contaminated plants and soil represents primary exposure pathways for terrestrial organisms. Washburn[42] (see Chapter 10) used the elimination rate constant to calculate steady-state BTFs for cow milk and beef. These steady-state BTFs were converted to a dry weight basis using wet-to-dry weight conversion factors of 0.385 and 0.13 for beef and milk, respectively.[26] Thus, BTFs of 2.08 and 0.23 kg/d were used to predict the accumulation of TCDD in cow beef and milk.

Data on the bioaccumulation of pollutants in other agricultural livestock, including pigs, chickens, and sheep, are scarce. As a result, we backcalculated the BTFs for chickens, sheep, and pigs that were required to predict the measured concentration of TCDD in poultry, eggs, lamb, and pork reported by Beck et al.[1] This approach yielded BTFs (kg/d DW, whole tissue basis) of 0.02, 0.56, 119, and 90 for sheep, pigs, chickens, and eggs, respectively. Because there are no measured data on the concentration of TCDD in beef liver, the BTF used for beef was also applied to beef liver.

The predicted concentration of TCDD in animal tissues from the ingestion of contaminated soil, grains, and forage is given in Table 6. These data show that ingestion of contaminated soil

TABLE 7
Predicted Average Daily Intake of TCDD through the Food Chain

Food group	Daily intake (pg/d)[a]	Daily intake (%)
Fruits and vegetables	0.14	3.9
Potatoes	0.001	0.03
Leafy vegetables	0.061	1.7
Nondried legumes	0.008	0.2
Dried legumes	0.007	0.2
Root vegetables	0.0002	0.01
Garden fruits	0.067	1.8
Grains and cereals	0.0003	0.01
Milk and dairy products	0.72	19.7
Meat (total)	2.43	65.9
Beef	1.95	53.0
Beef liver	0.37	10.0
Pork	0.018	0.5
Poultry	0.086	2.4
Lamb	0.001	0.01
Eggs	0.08	2.1
Soil ingestion[b]	0.06	1.6
Inhalation[c]	0.25	6.8
Total	3.68	100.0

[a] Picograms per day.
[b] Calculated using a soil concentration of 1.6×10^{-6} μg/g DW.
[c] Calculated using an average inhalation rate of 20 m^3/d.

was the major source of contamination for pork, chicken, and eggs, while ingestion of contaminated forage was the primary route of contamination for beef, beef liver, dairy products, and lamb. The data confirm that ignoring soil ingestion as a source of exposure by agricultural animals may lead to an underestimation of the concentration of pollutant in animal tissues. Beef and dairy cattle accumulated the highest concentration of TCDD due to facility emissions.

3. Total Daily Intake of TCDD by Humans

Table 7 gives the predicted average daily intake of TCDD by individuals living near a typical, modern MSW incinerator from ingesting contaminated food items, incidental soil ingestion, and inhalation. These data show that the food chain (including incidental soil ingestion) accounts for about 93% of human exposure to TCDD. Ingestion of beef and dairy products together accounted for 86% of total daily intake. Consumption of contaminated produce and inhalation were comparable (each pathway accounts for about 5% of total intake), while consumption of lamb, pork, eggs, and soil were not major pathways of human exposure. The average daily intake of TCDD from both direct and indirect pathways of human exposure is estimated to be 3.7 pg/d.

VI. VALIDATION OF THE TERRESTRIAL FOOD CHAIN MODEL

To validate the TFC model, input parameters for TCDD were modified to simulate background environmental conditions, and model predictions of the background daily intake of TCDD by humans and concentrations of TCDD in various media were compared to actual measured data. To estimate background levels of TCDD using the TFC model, the measured background concentration of TCDD in urban air (2.0×10^{-8} g/m^3)[43] and the fraction of TCDD present in the vapor phase (0.3)[44] were entered into the TFC to derive the background annual

TABLE 8
Comparison of Background Concentrations of TCDD Predicted by the Terrestrial Food Chain Model with Measured Background Environmental Concentrations

Media	Conc. predicted by TFC (pg/g WW)[a]	Background conc. (pg/g WW)	Ref. for background data
Beef	0.33	0.15	1
Beef liver	0.26	NA[b]	—
Dairy products	0.05	0.008	51
		0.015	52
		0.009	53
Pork	0.008	0.008	1
Poultry	0.05	0.045	1
Lamb	0.002	0.003	1
Eggs	0.02	0.02	1
Exposed produce	0.007	0.005	1
Protected produce	0.02	0.006	45
Forage	0.11	0.08	45
Grains	0.0002	0.006	45
Soil	0.97	0.96	54
		1.4	43

[a] Picograms per gram wet weight or whole tissue basis. Model predictions were converted to a dry weight basis using conversion factors from Baes et al.[26]
[b] Measurement not available.

average atmospheric air concentration. Second, the measured background air concentration for TCDD (2.0×10^{-8} µg/m^3) was multiplied by the fraction that exists in the particulate phase (0.7) and a deposition velocity of 6.3×10^4 m/year^2 to derive a "background" deposition rate of 8.8 $\times 10^{-4}$ µg/m^2/year.

Using the modified deposition rate and air concentration inputs and assuming that all of the food items (both plant and animal) consumed by individuals living near a facility originated from the study area, results show that TFC model predictions simulating background concentrations agree well with measured values reported in the literature (Table 8). The total daily intake of TCDD obtained from the TFC model (53 pg/d) agrees reasonably well with estimates of 39 and 16 pg/d reported by Travis and Hattemer-Frey[45] and Beck et al.[1] These results demonstrate that the TFC is a valid model for quantifying total daily intake of pollutants emitted from an MSW incinerator.

VII. APPLICATION OF THE TERRESTRIAL FOOD CHAIN MODEL TO CADMIUM

Ingestion is the primary route by which humans are exposed to metals present in the environment.[6] Metals are most notably different from organic compounds in that they do not decay or degrade in the environment. Unlike organics, metals can enter plant tissues only through root uptake and direct deposition.

A. MODEL INPUTS AND RESULTS

The TFC model was used to quantify the magnitude of exposure to cadmium emitted from an arbitrary MSW incinerator in the U.S. based on the following inputs. A deposition rate and maximum air concentration of 7.3 µg/m^2/year and 4.0×10^{-6} µg/m^3, respectively, were used in this assessment.[46] Again, since the purpose of this exercise was to estimate the magnitude of human exposure to cadmium emitted from an arbitrary, modern U.S. MSW incinerator, we entered generic mean values for all site-specific input parameters.

TABLE 9
Percentage of Vegetative Contamination Due to Deposition, Foliar Uptake, and Root Uptake for Cadmium

		Percentage		
	Total conc. (ng/g)[a]	Due to deposition	Due to foliar uptake	Due to root uptake
Human crops				
Potatoes	0.09	0	0	100
Leafy vegetables	1.4	14	0	86
Legumes (nondried)	0.26	7	0	93
Legumes (dried)	0.26	7	0	93
Root vegetables	2.0	0	0	100
Garden fruits	1.3	8	0	92
Grains	0.05	0	0	100
Weighted average	0.16	5	0	95
Animal crops				
Forage	0.73	47	0	53
Grains	0.05	0	0	100

[a] Nanograms per gram dry weight basis.

1. Uptake of Cadmium by Plants

Using a soil-to-plant BCF of 0.39 for forage, 0.05 for grains, 0.09 for potatoes, 1.18 for leafy vegetables, 0.24 for fresh and dried legumes, 1.98 for root vegetables, and 1.16 for garden fruits,[26,47,48] a soil concentration of 1.0×10^{-3} μg/g, and assuming that cadmium does not degrade in soil (i.e., K_s equals zero), the concentration of cadmium on forage, grains, and food crops consumed by humans (a weighted average) is 0.39, 0.05, and 0.15 ng/g, respectively. The concentration of cadmium in individual plants due to root uptake is listed in Table 9.

Using an atmospheric dry deposition rate of 7.3 μg/m^2/year,[46] a weathering constant (K_e) of 31.88/year,[26,48] an intercept fraction (r) of 0.47 (Table 1),[39] 0.12 years as the length of the growing season,[26] and a crop yield of 310 g/m^2[6,39] (Table 2), the concentration of cadmium on forage due to deposition was estimated to be 0.34 ng/g DW. Because grains have a protective outer husk, it was assumed that grains were not contaminated via deposition of facility-emitted pollutants.

Using the plant-specific intercept fractions and crop yields for fruits and vegetables listed in Tables 1 and 2, respectively, and 0.16 years as the length of the growing season,[26] the weighted-average concentration of TCDD on exposed vegetables and fruits consumed by humans due to deposition equals 0.01 ng/g DW. The concentration of TCDD due to deposition on individual food crops is listed in Table 9.

Thus, the total concentration of cadmium on forage equals 0.73 ng/g DW, 53% of which was due to root uptake and 47% to deposition (Table 9). In this case, forage was only about 15 times more contaminated with cadmium due to facility emissions than were grains (0.7 vs. 0.05 ng/g). The total concentration of TCDD on fruits and vegetables consumed by humans (weighted-average) was 0.16 ng/g DW, 95% of which is due to root uptake (Table 9). Root vegetables, garden fruits, and leafy vegetables accumulated the highest levels of cadmium of plant groups consumed by humans.

2. Uptake of Cadmium by Animals

Using BTFs (dry weight basis) 0.003, 0.135, 0.03, 0.03, 12.45, 0.006, and 0.65 were entered into the model to predict the bioaccumulation of cadmium in beef, beef liver, lamb, pork, poultry, dairy products, and eggs, respectively.[46] The predicted concentration of cadmium by livestock

TABLE 10
Predicted Concentration of Cadmium in Animal Tissues from Ingesting Contaminated Soil, Forage, and Grains

Tissue	Total conc. (pg/g)[a]	Percentage: Due to forage ingestion	Due to grains ingestion	Due to soil ingestion
Beef	10.9	53	4	43
Beef liver	489.0	53	4	43
Dairy products	65.0	68	2	30
Pork	22.5	0	13	87
Lamb	48.7	65	0	35
Poultry	274.0	0	7	93
Eggs	14.3	0	10	90

[a] Picograms per gram dry weight basis.

from ingestion of contaminated soil, grains, and forage as well as the total concentration of TCDD in beef, beef liver, dairy products, pork, chicken, lamb, and eggs due to facility emissions are given in Table 10. These data show that ingestion of contaminated soil was the major pathway of exposure for pigs and chickens, while ingestion of contaminated forage was the primary route of exposure for beef cattle, dairy cattle, and sheep. Chickens accumulated the highest concentration of cadmium due to facility emissions.

3. Total Daily Intake of Cadmium by Humans

Table 11 gives the predicted average daily intake of cadmium by individuals living near an arbitrary MSW incinerator from ingesting contaminated food items, incidental soil ingestion, and inhalation. These data show that the food chain accounts for 98% of human exposure to cadmium. Unlike TCDD, the primary pathway of human exposure was consumption of contaminated plants, which accounted for 73% of total daily intake of cadmium. Consumption of contaminated meat and dairy products accounted for only 11% of total daily intake, and inhalation was a negligible source of human exposure. The average daily intake of cadmium from both direct and indirect pathways of human exposure is estimated to be 5.0 ng/d.

VIII. CONCLUSIONS

Since organic compounds end up in the media in which they are most soluble, assessing the magnitude of exposure to pollutants emitted from an MSW incinerator largely depends on being able to predict their bioaccumulation in the food chain. Lipophilic compounds, such as TCDD, are most soluble in organic matter. These chemicals tend to sorb strongly to soil, sediment, and vegetation, bioaccumulate in the living organisms, and transfer to humans through the food chain. Thus, for many pollutants, the food chain may be an important contributor to total human exposure.

The TFC model, a multimedia transport model, was used to assess the magnitude of human exposure to compounds resulting from the combustion of MSW. The model quantifies the amount of pollutant entering the terrestrial food chain and the extent of human exposure from ingestion of agricultural produce, meat, dairy products, eggs, and soil. Results show that the food chain, especially beef and dairy products, accounts for about 93 and 98% of human exposure to TCDD and cadmium emitted from an arbitrary MSW incinerator. Inhalation was not a major pathway of human exposure for either compound. The average daily intake of TCDD due to facility emissions is estimated to be 3.7 pg/d. Similarly, the food chain accounted for about 98% of total daily intake of cadmium, which was estimated to be 4.9 ng/d. Unlike TCDD, however, consumption of contaminated produce was the major pathway of human exposure to facility-emitted cadmium.

TABLE 11
Predicted Average Daily Intake of Cadmium by Individuals Living Near an Arbitrary MWC

	Daily intake (ng/d)[a]	Daily intake (%)
Fruits and vegetables (total)	3.6	72.6
Potatoes	0.24	5.0
Leafy vegetables	0.75	15.0
Dried legumes	0.04	0.9
Nondried legumes	0.03	0.7
Root vegetables	1.4	28.1
Garden fruits	1.1	22.2
Grains and cereals	0.035	0.7
Milk and dairy products	0.11	2.1
Meat (total)	0.42	8.7
Beef	0.02	0.4
Beef liver	0.16	3.2
Pork	0.01	0.3
Poultry	0.23	4.7
Lamb	0.004	0.1
Eggs	0.01	0.2
Soil ingestion[b]	0.74	15.0
Inhalation[c]	0.08	1.6
Total	4.9	100

[a] Nanograms per day.
[b] Calculated using a soil concentration of 2.0×10^{-2} μg/g.
[c] Calculated using an average adult inhalation rate of 20 m^3/d.

REFERENCES

1. **Beck, H., Eckart, K., Mathar, W., and Wittowski, R.,** PCDD and PCDF body burden from food intake in the Federal Republic of Germany, *Chemosphere,* 18, 587, 1989.
2. **Hattemer-Frey, H. A. and Travis, C. C.,** Pentachlorophenol: environmental partitioning and human exposure, *Arch. Environ. Contam. Toxicol.,* 18, 482, 1989.
3. **Lioy, P. L., Harkov, R., Waldman, J. M., Pietarinen, C., and Greenberg, A.,** The total human environmental exposure study (THEES) to benzo(a)pyrene: comparison of the inhalation and food pathways, *Arch. Environ. Health,* 43, 304, 1988.
4. **Travis, C. C. and Arms, A. D.,** The food chain as a source of toxics exposure, in *Toxic Chemicals, Health, and the Environment,* Lave, L. B. and Upton, A. C., Eds., Plenum Press, New York, 1987.
5. **Travis, C. C. and Hattemer-Frey, H. A.,** Human exposure to 2,3,7,8-TCDD, *Chemosphere,* 16, 2331, 1987.
6. **Belcher, G. D. and Travis, C. C.,** The Food Chain as a Source of Human Exposure from Municipal Waste Combustion, presented at the International Conference on Municipal Waste Combustion, Hollywood, FL, April 11-14, 1989.
7. *Methodology for the Assessment of Health Risks Associated with Multiple Pathway Exposure to Municipal Waste Combustor Emissions,* U.S. Environmental Protection Agency, Office of Air Quality Planning and Standards, Research Triangle Park, NC, and the Environmental Criteria and Assessment Office, Cincinnati, 1986.

8. **Travis, C. C., Holton, G. A., Etnier, E. L., Cook, C., O'Donnell, F. R., Hetrick, D. M., and Dixon, E.,** Assessment of inhalation and ingestion population exposures from incinerated hazardous wastes, *Environ. Int.*, 12, 553, 1986.
9. **Baes, C. F., III,** Prediction of radionuclide K_d values from soil-plant concentration ratios, *Trans. Am. Nuclear Soc.*, 41, 53, 1982.
10. **Briggs, G. G., Bromilow, R. H., and Evans, A. A.,** Relationships between lipophilicity and root uptake and translocation of non-ionized chemicals by barley, *Toxicol. Environ. Chem.*, 7, 173, 1982.
11. **Travis, C. C. and Arms, A. D.,** Bioconcentration of organics in beef, milk, and vegetation, *Environ. Sci. Technol.*, 22, 271, 1988.
12. **Halfon, E.,** Regression method in ecotoxicology: a better formulation using the geometric mean functional regression, *Environ. Sci. Technol.*, 19, 747, 1985.
13. **Nash, R. G.,** Plant uptake of insecticides, fungicides, and fumigants from soil, in *Pesticides in Soil and Water*, Guenzi, W. D., Ed., Soil Science Society of America, Madison, WI, 1974.
14. **Dowdy, R. H. and Larson, W. E.,** The availability of sludge-borne metals to various vegetable crops, *J. Environ. Quality*, 4, 278, 1975.
15. **Bacci, E. and Gaggi, C.,** Chlorinated hydrocarbon vapours and plant foliage: kinetics and applications, *Chemosphere*, 16, 2515, 1987.
16. **Bacci, E. and Gaggi, C.,** Chlorinated pesticides and plant foliage: translocation experiments, *Bull. Environ. Contam. Toxicol.*, 37, 850, 1986.
17. **Bacci, E. and Gaggi, C.,** Polychlorinated biphenyls in plant foliage: translocation or volatilization from contaminated soils?, *Bull. Environ. Contam. Toxicol.*, 35, 673, 1985.
18. **Bacci, E., Calamari, D., Gaggi, C., and Vighi, M.,** Bioconcentration of organic chemical vapours in plant leaves: experimental measurements and prediction, *Chemosphere*, in press.
19. **Beall, M. L., Jr. and Nash, R. G.,** Organochlorine insecticide residues in soybean plant tops: root versus vapor sorption, *Agron. J.*, 63, 460, 1971.
20. **Buckley, E. H.,** Accumulation of airborne polychlorinated biphenyls in foliage, *Science*, 216, 520, 1982.
21. **Nash, R. G. and Beall, M. L., Jr.,** Distribution of silvex, 2,4-D, and TCDD applied to turf in chambers and field plots, *J. Agric. Food Chem.*, 28, 614, 1980.
22. **Mosbaek, H., Tgell, J., and Sevel, T.,** Plant uptake of airborne mercury in background areas, *Chemosphere*, 17, 1227, 1988.
23. **Travis, C. C. and Hattemer-Frey, H. A.,** Uptake of organics by aerial plant parts: a call for research, *Chemosphere*, 17, 277, 1988.
24. *Industrial Source Complex (ISC) Dispersion Model Users' Guide*, Vol. 1, EPA-450/4-79-030, U.S. Environmental Protection Agency, Washington, D.C., 1979.
25. **de Temmerman, L., Vandeputte, R., and Guns, M.,** Biological monitoring and accumulation of airborne mercury in vegetables, *Environ. Pollut.*, 41, 139, 1986.
26. **Baes, C. F., III, Sharp, R. D., Sjoreen, A., and Shor, R.,** *A Review and Analysis of Parameters for Assessing Transport of Environmentally Released Radionuclides Through Agriculture*, ORNL-5786, U. S. Department of Energy, Oak Ridge National Laboratory, Oak Ridge, TN, 1984.
27. **Travis, C. C. and Hattemer-Frey, H. A.,** Human exposure to 2,3,7,8-TCDD, *Chemosphere*, 16, 2331, 1987.
28. **Shor, R. W., Baes, C. F., III, and Sharp, R. D.,** *Agricultural Production in the United States by County: A compilation of Information from the 1975 Census of Agriculture for Use in Terrestrial Food Chain Transport and Assessment Models*, ORNL-5786, U. S. Department of Energy, Oak Ridge National Laboratory, Oak Ridge, TN, 1982.
29. **Fries, G. F.,** Assessment of potential residues in food derived from animals exposed to TCDD contaminated soil, *Chemosphere*, 16, 2123, 1987.
30. **Ng, Y., Colsher, C., and Thompson, S.,** *Transfer Coefficients for Assessing the Dose from Radionuclides in Meat and Eggs*, NUREG/CR-2967, U. S. Government Printing Office, Washington, D. C., 1982.
31. *Health Assessment Document for Polychlorinated Dibenzo-*p*-Dioxins*, EPA-600/8-84-014F, U. S. Environmental Protection Agency, Washington, D. C., 1985.
32. **Hutzinger, O., Blumich, M. J., van den Berg, M., and Olie, K.,** Sources and fate of PCDDs and PCDFs: an overview, *Chemosphere*, 14, 581, 1985.
33. **Lao, R. C., Thomas, R. S., Chiu, C., Li, K., and Lockwood, J.,** Analysis of PCDD-PCDF in environmental samples, in *Chlorinated Dioxins and Dibenzofurans in the Total Environment II*, Keith, L. H., Rappe, C., and Choudhary, G., Eds., Butterworths, Boston, 1983.
34. **Marklund, S., Kjeller, L.-O., Hansson, M., Tysklind, M., Rappe, C., Ryan, C., Collazo, H., and Dougherty, R.,** Determination of PCDDs and PCDFs in incineration samples and pyrolytic products, in *Chlorinated Dioxins and Dibenzofurans in Perspective*, Rappe, C., Choudhary, G., and Keith, L. H., Eds., Lewis Publishers, Chelsea, MI, 1986.
35. **Olie, K., Lustenhouwer, J. W. A., and Hutzinger, O.,** Polychlorinated dibenzo-*p*-dioxins and related compounds in incinerator effluents, in *Chlorinated Dioxins and Related Compounds: Impact on the Environment* Hutzinger, O., Frei, R. W., Merian, E., and Pocchiari, F., Eds., Pergamon Press, Oxford, 1982.

36. **Weerasinghe, N. C. A. and Gross, M. L.,** Origins of polychlorodibenzo-*p*-dioxins (PCDD) and polychlorodibenzofurans (PCDF) in the environment, in *Dioxins in the Environment,* Kamrin, M. A. and Rodgers, P. W., Eds., Hemisphere Publishing, Washington, D. C., 1985.
37. **Hattemer-Frey, H. A. and Travis, C. C.,** Characterizing the extent of human exposure to PCDDs and PCDFs emitted from municipal waste combustors, in this volume.
38. **Isensee, A. R. and Jones, G. E.,** Absorption and translocation of root and foliage applied 2,4-dichlorophenol, 2,7-dichlorodibenzo-*p*-dioxin, and 2,3,7,8-tetrachlorobenzo-*p*-dioxin, *J. Agric. Food Chem.,* 19, 1210, 1971.
39. **Hoffman, F. O. and Baes, C. F.,** *A Statistical Analysis of Selected Parameters for Predicting Food Chain Transport and Internal Dose of Radionuclides,* ORNL/NUREG/TM-882, Oak Ridge National Laboratory, Oak Ridge, TN, 1979.
40. **Miller, C. and Hoffman, F.,** An examination of the environmental half-time for radionuclides deposited onto vegetation, *Health Phys.,* 45, 731, 1983.
41. *Superfund Public Health Evaluation Manual,* EPA 540/1-86-060, U. S. Environmental Protection Agency Office of Emergency Response and Remedial Action, Washington, D. C., 1986.
42. **Washburn, S. T.,** The accumulation of chlorinated dibenzo-*p*-dioxins and dibenzofurans in beef and milk, in this volume.
43. **Rappe, C. and Kjeller, L.-O.,** PCDDs and PCDFs in environmental samples air, particulates, sediments and soil, *Chemosphere,* 16, 1775, 1987.
44. **Bidleman, T. F.,** Atmospheric processes, *Environ. Sci. Technol.,* 22, 361, 1988.
45. **Travis, C. C. and Hattemer-Frey, H. A.,** Dioxin: research needs for risk assessment, *Chemosphere,* in press.
46. **Belcher, G. D. and Travis, C. C.,** in this volume.
47. Environmental Profiles and Hazard Indices for Constituents of Municipal Sludge: Cadmium, Office of Water Regulations and Standards, U. S. Environmental Protection Agency, Washington, D. C., 1985.
48. **Heffron, C. L., Reid, J. T., Elfving, D. C., Stoewsanson, G. S., Haschek, W. M., Telford, J. N., Furr, A. K., Parkinson, T. F., Bach, C. A., and Gutenmann, W. H.,** Cadmium and zinc in growing sheep fed silage corn grown on municipal sludge amended soil, *J. Agric. Food Chem.,* 28, 58, 1980.
49. **Yang, Y. and Nelson, C. B.,** An estimation of daily food usage factors for assessing radionuclide intakes in the U. S. population, *Health Phys.,* 50, 245, 1986.
50. Household Food Consumption Survey, 1965-1966, Report No. 12, Food Consumption of Households in the United States, Seasons and Year, 1965-1966, U. S. Department of Agriculture, Washington, D. C., 1966.
51. **Beck, H., Eckert, K., Kellert, M., Mathar, W., Ruhl, Ch.-S, and Wittowski, R.,** Levels of PCDDs and PCDFs in samples of human origin and food in the Federal Republic of Germany, *Chemosphere,* 16, 1977, 1987.
52. **Fürst, P., Fürst, C., and Groebel, W.,** Levels of PCDDs and PCDFs in Food-Stuffs from the Federal Republic of Germany, presented at the 9th Int. Symp. on Chlorinated Dioxins and Related Compounds, September 20-23, Toronto, Canada.
53. **Startin, J. R., Rose, M., Wright, C., and Gilbert, J.,** Surveillance of British Foods for PCDDs and PCDFs, presented at the 9th Int. Symp. on Chlorinated Dioxins and Related Compounds, September 20-23, Toronto, Canada.
54. **Nestrick, T. J., Lamparski, L. L., Frawley, N. N., Hummel, R. A., Kocher, C. W., Mahle, N. H., McCoy, J. W., Miller, D. L., Peters, T. L., Pillepich, J. L., Smith, W. E., and Tobey, S. W.,** Perspectives of a large scale environmental survey for chlorinated dioxins: overview and soil data, *Chemosphere,* 15, 1453, 1986.

Chapter 5

2,3,7,8-TETRACHLORODIBENZO-*p*-DIOXIN IN THE AGRICULTURAL FOOD CHAIN: POTENTIAL IMPACT OF MSW INCINERATION ON HUMAN HEALTH

Jeffrey B. Stevens

TABLE OF CONTENTS

I. INTRODUCTION

Most municipalities currently dispose of their municipal solid waste (MSW) in landfills. However, the known potential for groundwater contamination and limited landfill space have forced many cities to pursue other options, e.g., incineration. One fundamental issue associated with incineration, however, is its potential impact on human health and the environment. Emissions from MSW incinerators have been shown to contain numerous toxic substances, including polychlorinated dibenzo-*p*-dioxins (PCDDs), polychlorinated dibenzofurans (PCDFs), polynuclear aromatic hydrocarbons (PAHs), and toxic metals.[1] Of these emissions, 2,3,7,8-tetrachlorodibenzo-*p*-dioxin (TCDD) is viewed as the most hazardous.[2]

TCDD is a highly lipophilic, persistent chemical.[3] Consequently, TCDD is expected to bioaccumulate in the food chain. Since many MSW incinerators have already been or will be built in rural agricultural areas, it is important to address the relative magnitude of human exposure to facility emissions via all exposure routes, including inhalation, ingestion of contaminated soil, agricultural produce, and beef and milk products, and dermal absorption. The purpose of this study was to identify the major agricultural food chain sources of human exposure to TCDD emitted from a rural MSW incinerator. An environmental transport approach was used to model both the environmental fate of TCDD and its disposition in common farm livestock.

II. AGRICULTURAL FOOD CHAIN MODEL

The overall approach of this model is presented in Figure 1. For this assessment TCDD was assumed to be entirely particle-bound when released from the facility.[1] This assumption enabled a conservative calculation of TCDD deposition onto soil, surface waters, and plants as a function of time. It was additionally assumed that the MSW incinerator emits TCDD at a fixed rate of 1 ng per dry standard cubic meter (ng/dscm) continuously for a 70-year period.

Various routes of exposure to TCDD were then modeled for common farm livestock. Typical exposure routes included inhalation; ingestion of contaminated soil, water, and foodstuffs; and skin contact with contaminated soil/dust. The bioconcentration of TCDD in these animals and its distribution to human food items (milk, beef, pork, chicken, and eggs) were modeled subsequently. Finally, a total food chain TCDD dose to humans was calculated by multiplying daily food consumption values for a rural, adult family member by the concentration of TCDD in various food items.

A. ENVIRONMENTAL FATE OF TCDD

1. TCDD Level in the Air Around an MSW Incinerator

The first step is to determine ambient air concentrations of TCDD around the MSW incinerator. The ambient air concentration of TCDD at ground level around a point source such as an MSW incinerator can be calculated using the Industrial Source Complex-Long Term (ISCLT) model.[4] The accuracy of this model requires input data that is site-specific (i.e., topographical measurements in the vicinity of an incinerator as well as average annual meteorological data for that region.) For this study, an incinerator located approximately 100 m west of the Minneapolis/St. Paul metropolitan area was chosen. Five years of meteorological data were used in the ISCLT model to predict the ambient air concentration of TCDD around this facility. Once released into the atmosphere, TCDD may exist in the vapor phase or is sorbed to particulates. It was conservatively assumed (due to a lack of appropriate data in the literature) that all of the TCDD released from the MSW incinerator was either bound to particulates 1 μm in diameter or in the gaseous state, and that TCDD does not photodegrade while airborne.

TCDD dose measurements were then calculated, assuming that all of the agricultural products ingested by the maximally exposed individual originate from the area in which the highest ambient air concentration of TCDD is predicted (0.0086 pg/m^3). Thus, these TCDD dose

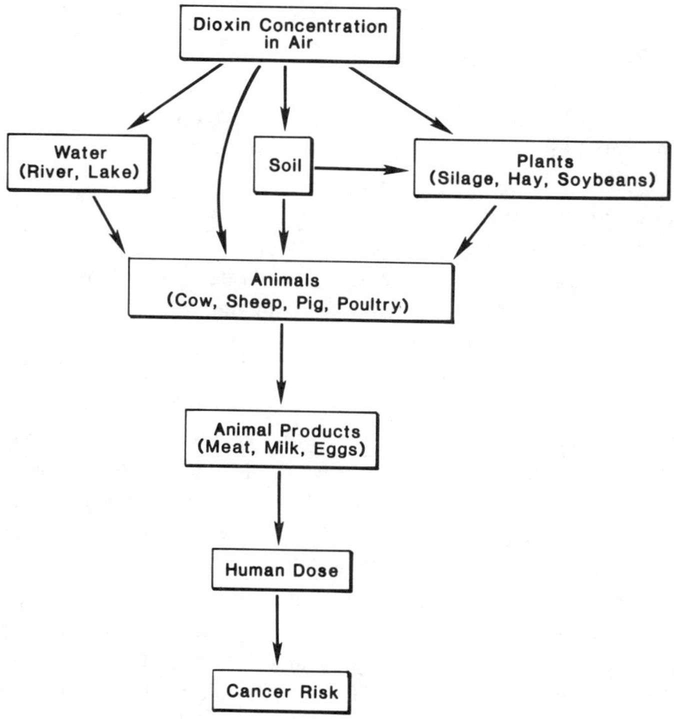

FIGURE 1. Schematic of agricultural food chain model.

estimates should be regarded as upper-bound (worst case) values. An average particle deposition rate of 0.3 cm/s (259 m/d) was used to calculate TCDD accumulation over time in the various environmental compartments (e.g. soil, plants).[5]

2. TCDD Accumulation in Soil

The persistence of TCDD in soil is an important factor in determining potential exposure to this substance. For example, long-term observations in Seveso, Italy and in other locations indicate the half-life of TCDD in soil is on the order of 10-12 years.[3,6-8] A soil half-life value of 12 years was used in this model. The accumulation of TCDD in soil around the MSW incinerator was calculated assuming continuous and constant deposition of TCDD onto the soil for a 70-year period. The deposited material was estimated to remain in the top 1 cm for soil ingested by dairy cows, beef cows, and humans.[3] It was assumed that agricultural practices (plowing, etc.) for crop production mix topsoil to a depth of 15 cm.[9] Annual year-end TCDD levels in soil for each of the 70 years of assumed facility operation were calculated using Equation 1:

$$C_t = \frac{I}{d*sd*k_{el}} (1 - e^{-kel*t}), \qquad (1)$$

where: C_t = TCDD concentration in soil at end of year "t" (pg/g)
I = total TCDD input to soil
= ambient air concentration of TCDD (0.0086 pg/m^3) X particle deposition rate (9.45×10^4 m/year)
k_{el} = elimination rate constant for TCDD in soil (1n $2/T^{1/2}$) (year^{-1})
t = number of years (year)

sd = soil density (assumed to be 1.6×10^6 g/m^3)
d = mixing depth of dioxin (m).

Year-end values were then used to calculate a 70-year average soil level of TCDD. Average soil TCDD levels calculated in this manner are presented in Table 1.

3. TCDD Accumulation in Plants

Plants may be contaminated with TCDD via aerial deposition onto their outer surfaces and/or via root uptake from contaminated soil. Various researchers have shown, however, that plants do not absorb significant levels of TCDD from contaminated soil.[10-13] Therefore, it was assumed that root uptake of TCDD by plants was negligible.

The fractional deposition factors of airborne particulates containing TCDD onto plant surfaces were derived from Baes et al.[14] These investigators measured deposition of radioactive particulates onto plants and developed equations that correlate particle deposition (mass/m^2) with crop yield (Table 2). These equations are specific for plant type and take into account the change in plant surface area due to plant growth. Annual average crop yields and harvest times for each plant species were obtained from the Minnesota State Agricultural Office for the two county area surrounding the incinerator (Table 3).[15]

It was assumed that TCDD on plant surfaces does not degrade. However, weathering of particulates from plant surfaces does occur. An average particle half-life on outdoor plant surfaces has been estimated to be 14 d.[16-18] We assumed this half-life for the TCDD-containing particles in this model. The nonsteady-state TCDD accumulation by plants was estimated using the following equation:

$$C_t = \frac{I}{k_{el} * Y} (1 - e^{-kel*t}), \qquad (2)$$

where:
C_t = plant concentration of TCDD at harvest time (pg/kg)
I = total TCDD input onto plant surfaces
= ambient air concentration (0.0086 pg/m^3) $\times$ particle deposition rate (259 m/d) $\times$ the fraction of surface area comprised of plant surface (f values calculated as in Table 2)
k_{el} = rate constant of weathering of particulates on plants (d^{-1}) ($\ln 2/T^{1/2} = 0.05$ d^{-1})
t = time to harvest (days) (Table 3)
Y = crop yield (kg/m^2) (Table 3)

Table 2 provides the TCDD levels in the various plant species modeled for this assessment.

B. BIOCONCENTRATION OF TCDD IN FARM ANIMALS

1. Dairy Cow

a. Daily TCDD Dose to the Dairy Cow

Because TCDD is extremely lipophilic, it readily bioaccumulates in living organisms, including agricultural animals and humans. Agricultural animals can be exposed to TCDD by ingesting contaminated forage, soil and drinking water, and via inhalation. The daily dose of TCDD to beef and dairy cows and young cows was calculated according to the following approach.

An average dairy cow inhales about 150 m^3 of air per day.[9] It was estimated that only 25% of the TCDD-containing particles inhaled actually deposit in the alveolar region of the cow lung,[19] and that the inhalation bioavailability of particulate-bound TCDD is 0.40.[20,21] Hence, the TCDD exposure via inhalation was estimated to be 0.1 pg per animal per day.

TABLE 1
Average TCCD Levels in Soil

Soil conditions	70-Year average TCDD level[a] (pg/g)
Undisturbed soil[b]	0.675
Tilled soil[c]	0.045

[a] Equation 1 in the text was used to calculate the yearly average soil values, with the following data: ambient air concentration of TCDD (0.0086 pg/m^3; ISCLT dispersion modeling results of the facility); particle deposition rate (259 m/d); soil density (1.6×10^6 g/m^3); mixing depth (m); k_{el} (0.058 $year^{-1}$); 365 d/year.

[b] Soil level used to model soil ingestion by dairy cows, beef cows, and humans.

[c] Soil level used to model TCDD uptake into plants.

TABLE 2
TCDD Build-Up in Plants

Crop	Fractional deposition equation[a]	TCDD levels[b] (pg/kg)
Corn silage	f = 1 - e–0.768x	4.4
Hay/grasses	f = 1 - e–2.88x	81.6
Lettuce	f = 1 - e–0.15x	3.3
Tomatoes	f = 1 - e–0.068x	2.0
Beans	f = 1 - e–0.073x	8.6
Potatoes	—	0.0
Carrots, beets	—	0.0

[a] TCDD levels were assumed to originate from aerial deposition of TCDD-containing particulates only. TCDD absorption from soil was assumed to be negligible for all plants.

[b] Equations taken from Baes et al.[14] "f" represents the fraction of total surface area impacted by aerial deposition of particulates that is plant surface and is a unitless value. Plants in which the edible portions are below ground (potatoes, carrots, and beets) were assumed to have no aerial deposition. "x" represents crop yield in kg/m^2 (see Table 3).

TABLE 3
Annual Crop Yields and Harvest Times for Various Plant Species[15]

Plant	Crop yield (kg/m^2)	Harvest time (d)
Corn	2.69	120
Hay/grasses	0.23	40
Lettuce	8.6	45
Tomatoes	12.0	70
Beans	2.7	53

The dose of TCDD from ingestion of contaminated plant material was estimated using information reported by Gullikson[22] and from data provided to us by the University of Minnesota Agricultural Extension Office.[23] A 600-kg dairy cow in Minnesota ingests, on the average, 4.5 kg grain, 16.3 kg corn silage, and 6.8 kg grasses per day for a total of 27.6 kg of dry matter per day. To calculate the food ingestion dose, these daily consumption data were multiplied by the TCDD levels in each plant type (Table 2) and then summed. Grain was assumed to contain negligible quantities of TCDD.[10] It was additionally estimated that a lactating dairy cow ingests 0.5% of its daily dry matter (or 0.14 kg) of soil per day.[24] The oral bioavailability of particulate-bound TCDD was estimated at 0.40.[20,21] Hence, the combined food and soil ingestion dose was determined to be 353.1 pg per animal per day.

In Minnesota, water is provided to dairy cattle from troughs filled daily from groundwater supplies. Therefore, it was assumed that ingestion of water contaminated with TCDD was negligible. Dermal contact with soil contaminated with TCDD was also assumed to be a negligible source of exposure.[5] Summation of the inhalation dose, food dose, and soil dose yields a total daily TCDD dose from all routes of exposure of 353.2 pg per animal per day.

b. Fractional Distribution of TCDD to Milk

Jensen and Hummel[25] fed TCDD (as a component in 2,4,5-T) to adult dairy cows to determine TCDD translocation to milk. After being exposed for 21 d at the highest dose (500 ppm), these investigators found that approximately 12% of the daily TCDD intake dose was translocated to milk each day. Since the TCDD half-life in dairy cattle was reported to be 41 d by Jensen and Hummel,[25] it is apparent that steady-state conditions were not in effect during this experiment. A steady-state translocation factor can be determined, however, from the above data. The fraction of steady-state transfer achieved after only 21 d of exposure to TCDD (C_{21}) (the duration of Jensen and Hummels,[25] experiment) can be determined from the following equation:

$$\frac{C_{21}}{C_{SS}} = 1 - e^{-kel*t}, \tag{3}$$

where:

C_{21} = the concentration of TCDD in cow milk after a 21-d exposure
C_{SS} = the maximum concentration of TCDD in cow milk
k_{el} = the half-life of TCDD in cow adipose tissue
= $\ln 2/41 \text{ d} = 0.017 \text{ d}^{-1}$
t = duration of exposure to TCDD = 21 d

Hence, after a 21-d exposure, only 30% of the steady-state concentration was achieved. Since Jensen and Hummel[25] found that 12% of the daily intake of TCDD transferred to the milk, the amount of TCDD translocated to the milk under steady-state conditions can be estimated by dividing the fractional distribution of TCDD to milk at 21 d (12%) by the fraction of steady-state value (0.30) to yield a value of 40%. Therefore, under steady-state conditions, 40% of the cow's daily intake of TCDD would be expected to transfer to milk. The TCDD translocated at steady state (353.2 pg/d × 0.40 = 141.3 pg/d) would distribute in 15.6 l of milk each day.[51,54] Thus, under steady-state conditions, a TCDD level in cow's milk would be 9.1 pg/l.

2. Beef Cattle

a. TCDD Bioconcentration in Cows

Jensen et al.[26] fed beef cattle a diet containing 24 ppt TCDD for 28 d and found that the level of TCDD in adipose tissue was 3.75 times higher than the TCDD level in the diet. Using a pharmacokinetic model and a half-life of 115 d,[26] a steady-state bioconcentration factor (BCF_{ss}; concentration in adipose tissue divided by concentration in feed) of 24.8 was determined. This BCF_{ss} was used in this assessment.

b. Daily TCDD Dose/Distribution to Fat

An inhalation volume of 100 m^3/d was used for beef cattle in this study.[9] All other variables for calculating the inhalation dose were the same as those used for the dairy cow. The inhalation dose of TCDD by beef cattle was estimated to be 0.1 pg per animal per day.

The daily dose of TCDD from plant ingestion was based on the three feeding phases in the life of a beef cow: (1) an initial nursing plus pasture ingestion phase lasting approximately 180 d, (2) an additional pasture ingestion phase which lasts 55 d, and (3) a final grain ingestion phase which lasts 130 d.

The nursing calf primarily receives TCDD through milk and via inhalation. Assuming that TCDD partitions in the female beef cow similarly to its partitioning in the female dairy cow, a nursing calf dose was modeled identically to the calculations for estimating TCDD levels in cow milk using the following parameters: (1) an adult female cow ingests 4.4 kg corn silage and 17.6 kg grasses per day; (2) the TCDD bioavailability factor for the adult female cow is 0.40; (3) an adult female cow gives birth to only one calf per year;[10] (4) the bioavailability factor of TCDD for the nursing calf from ingestion of mother's milk was assumed to be 0.60;[20] and (5) the nursing calf inhales 30 m^3 air per day.[9] A daily nursing dose of 144 pg TCDD per calf per day was calculated from these data.

For calculation of the TCDD dose from pasture ingestion, it was estimated that a young beef cow ingests 13.6 kg food per day (3.4 kg corn silage and 10.2 kg grasses) and 0.1 kg soil per day (0.5% of dry matter intake).[9] Assuming a TCDD bioavailability factor of 0.40, a daily pasture ingestion dose of 371 pg TCDD per calf per day was estimated.

An average daily TCDD dose for the first 235 d of the animal's life was then calculated using a time-weighted averaging equation and the appropriate daily route-specific doses, (i.e., the nursing dose, the pasture dose, and the inhalation dose). This average daily TCDD dose was estimated to be 481 pg TCDD per animal per day. Equations 4 and 5 were then used to calculate accumulation of TCDD in young cow adipose tissue.

$$BCF_{235} = BCF_{ss}\,(1 - e^{-kel*t}), \tag{4}$$

where BCF_{235} = bioconcentration factor at day 235
BCF_{ss} = steady-state bioconcentration factor = 24.8[26]
k_{el} = elimination rate constant = 0.006 d^{-1} [26]
t = number of days = 235

This equation yields a BCF_{235} value of 18.8.

$$C_{235} = D * BCF_{235} \div M, \tag{5}$$

where C_{235} = TCDD concentration in adipose tissue at day 235 (pg/kg)
D = TCDD dose = 481 pg/d
BCF_{235} = bioconcentration factor at day 235 = 18.8
M = mass of daily ingested foodstuffs (plants plus milk) = 13.6 kg/d[9]

These data yield an adipose tissue TCDD level of 665 pg/kg.

During the last 130 d of a beef cow's life, it is placed on a grain only diet. It was assumed that grain was not contaminated with TCDD.[10] Therefore, during the grain-fed time period the animal would experience a loss of TCDD from its body. Using a half-life of 115 d,[26] the TCDD level in adipose tissue at the time of sacrifice was estimated according to the following equation:

$$C_{365} = C_{235}\, e^{-kel*t}, \tag{6}$$

where C_{365} = TCDD concentration in adipose tissue at day 365 (end of the animal's life)
t = 130 d (365 to 235)

which yields a TCDD adipose tissue concentration of 305 pg/kg (or 0.31 pg/g).

c. TCDD Level in Table Beef

It was estimated that the average cut of edible beef contains 18% adipose tissue,[27] and that the lean portion does not contain appreciable levels of TCDD. Hence, the TCDD level in edible beef was assumed to be 0.056 pg/g (0.31 pg/g × 0.18).

3. Other Livestock

For any agricultural foodchain assessment to be complete, other livestock animals such as pigs, sheep, and chickens (including broilers and layers) also need to be evaluated. Unfortunately, no data exist which address TCDD behavior in these animals. Therefore, an assumption was made that TCDD disposition in these animals mimics that in dairy and beef cows. TCDD levels in these other agricultural foodstuffs were thus calculated based on this assumption.

Earlier work[5] using the above assumption showed that compared to cows, these other animals do not contribute substantial amounts of TCDD to the human diet. As shown in Table 4, only the lamb is expected to contain levels of TCDD that are similar to levels estimated in cows.

C. HUMAN TCDD DOSE FROM THE AGRICULTURAL FOOD CHAIN

It was conservatively assumed that an adult farmer in the area surrounding the facility receives all of his/her daily milk, milk products, and agricultural foodstuffs from the local region. The Minnesota Department of Health[27] estimates that an average 70-kg adult ingests agricultural foodstuffs in the amounts shown in Table 5. It was assumed that the TCDD present in each of these foodstuffs was 87% bioavailable when the chemical was present in a nonparticulate form (i.e., in milk and meat) and 58% bioavailable when bound to particulates (i.e. on plant surfaces).[21,28] It was also assumed that produce was not washed prior to being ingested. Thus, TCDD doses to humans from the various agricultural food chain components can be calculated using the following equation:

$$\begin{matrix}\text{Total human} \\ \text{food chain} \\ \text{dose} \\ \text{(pg/d)}\end{matrix} = \sum_{i=1}^{7} \begin{bmatrix}\text{Amount of} \\ \text{foodstuff} \\ \text{ingested} \\ \text{(g/d)}\end{bmatrix} \times \begin{bmatrix}\text{TCDD} \\ \text{conc} \\ \text{in foodstuff} \\ \text{(pg/g)}\end{bmatrix} \times \begin{bmatrix}\text{TCD} \\ \text{bioavailability} \\ \text{factor}\end{bmatrix} \quad (7)$$

An overall food chain dose was calculated to be 10.8 pg TCDD/d (Table 5). A comparable inhalation TCDD dose can be calculated using the following equation:

$$\begin{matrix}\text{Human} \\ \text{inhalation} \\ \text{dose} \\ \text{(pg/d)}\end{matrix} = \begin{bmatrix}\text{Ambient} \\ \text{air TCDD} \\ \text{conc} \\ \text{(pg/m)}\end{bmatrix} \times \begin{bmatrix}\text{Ventilation} \\ \text{rate} \\ \text{(m}^3\text{/d)}\end{bmatrix} \times \begin{bmatrix}\text{Fraction of} \\ \text{particulate} \\ \text{transferred to} \\ \text{deep lung}\end{bmatrix} \times \begin{bmatrix}\text{TCDD} \\ \text{bioavailability} \\ \text{factor}\end{bmatrix} \quad (8)$$

The ambient air concentration of TCDD was determined to be 0.0086 pg/m^3, an average adult ventilation rate is 20 m^3/d,[29] the fraction of particulates depositing into the alveolar region of the tissue is 0.25, and the TCDD bioavailability via the inhalation route of exposure is 0.58.[28] These input data yield an inhalation dose of 0.03 pg/d. Thus, the agricultural food chain contributed approximately 360 times the inhalation dose of TCDD to an adult individual.

It seems clear from these results that exposure dose modeling of emission substances from incinerators through pathways other than inhalation alone are needed if risk managers are to understand the full impact of these facilities on public health. Calculation of the inhalation dose only for the above facility could have underestimated the receptor dose by two orders of magnitude.

TABLE 4
TCDD Level in Edible Foodstuffs

Animal	TCDD conc in human foodstuffs
Milk cow	9.100 pg/l milk
Beef cow	0.056 pg/g beef
Pig	0.012 pg/g pork
Sheep	0.065 pg/g lamb
Broiler chicken	0.000 pg/g chicken
Laying hen	0.000 pg/g egg

TABLE 5
Total Adult Daily Dose of TCDD Emitted From a Rural MSW Incinerator

Foodstuff	Daily consumption	Daily TCDD dose[a] (pg/d)	Percentage of total dose
Inhalation	20 m^3	0.03	<1
Vegetables/fruits[b]	582 g	0.58	5
Milk and milk products	1 l	5.46	50
Beef	125 g	4.20	39
Pork	46 g	0.33	3
Lamb	7 g	0.27	2.5
Chicken	65 g	0.00	0
Eggs	1	0.00	0
Total		10.9	

[a] Daily TCDD dose obtained by multiplying the amount of foodstuff ingested per day by the calculated TCDD concentration in that foodstuff, and then multiplying the dose calculated by the appropriate bioavailability factor (see Equation 8).

[b] Overall value presented in this table. Individual consumption values are potatoes, 142 g; leafy vegetables, 100 g; legumes, 98 g; root vegetables, 100 g; and fruit, 132 g.[15]

This type of analysis is also useful for two other purposes: site selection of proposed facilities and facility permit requirements for environmental sampling. For example, the overall health risk calculated for the facility modeled in this study but located in two different land-use areas could vary as much as two orders of magnitude, if one area contains dairy and beef cattle for local milk and beef consumption, while the other does not. Obviously, then, known behavioral patterns at specific sites can play an enormous role in calculating the magnitude of health risk for a facility. With respect to environmental sampling requirements in facility permits, these types of analyses suggest that if the facility is affecting the environment, excess levels of emission substances, such as TCDD, should first "show up" in bovine milk and/or beef fat. Thus, the requirement for periodic sampling of agricultural products originating within an impact zone of a facility might be a good approach by which the "real" environmental impact of a particular facility can be ascertained on a continuing basis.

III. DISCUSSION

The fate of TCDD was modeled in various agricultural animals and crops to estimate the magnitude of human exposure to TCDD emitted from a rural MSW incinerator. Ingestion of beef and dairy products accounted for 89% of the estimated total agricultural dose of TCDD (Table 5). The total dose from ingestion of contaminated food items was 360 times higher than the estimated inhalation dose of TCDD. Although beef and lamb contain the highest TCDD concentrations of any of the food chain components, ingestion of beef and milk products represented the greatest source of human exposure. Therefore, these foodstuffs might represent valuable sentinels to evaluate the impact of MSW facilities on surrounding rural areas.

Fat-soluble, persistent chemicals such as TCDD are of most concern with respect to environmental biomagnification. Therefore, it is not surprising that this chemical concentrates in agricultural crops and animals, and that human exposure to MSW incinerator emissions by ingestion of contaminated agricultural foodstuffs is highly significant. Using an emission rate of 1 ng TCDD/dscm yielded an agricultural food chain dose of 10.8 pg/d, which is 360 times the TCDD dose to an adult via inhalation. It should be stressed, however, that many conservative assumptions have gone into the modeling of both the food chain dose and the inhalation dose, so the magnitude of this ratio should not be considered accurate. For example, if TCDD is assumed to be 100% bioavailable to the human via inhalation and if 100% of the inhaled particles are assumed to be deposited into the deep lung, then a food chain-to-inhalation ratio of 52 would result.

A fugacity-food chain model for TCDD distribution in the environment has recently been published.[30] Using this model the agricultural food chain dose contributed 44 times more TCDD than the inhalation dose to an adult individual. Since these authors assumed 100% particle deposition in the lungs and 100% TCDD bioavailabity, their ratio (44) is comparable to our ratio of 52. Both studies, therefore, underscore the significance of the food chain in assessing TCDD exposure.

It should also be noted that in this model the calculated dose from the agricultural food chain is highly dependent upon the human and animal diets and human behavioral assumptions used. For example, it was assumed in this study that all of an adult individual's agricultural diet originated in the deposition zone of this facility. Therefore, this food chain dose estimate must be considered highly conservative.

A number of other, general assumptions have been used in this model which are important to clarify. First, it was assumed that all of the TCDD emissions from this facility behave as submicron particles in the atmosphere. Although this assumption may not be accurate with respect to the physical nature of TCDD, the ISCLT model that was used in this assessment does not distinguish between submicron particles and a chemical in the gaseous state. Thus, the uncertainty associated with the form of TCDD emitted from the facility (i.e., particulate vs. gaseous) should not significantly affect the results reported.

Secondly, pharmacokinetic modeling data for TCDD in farm animals was found only for the beef cow and the dairy cow. Thus, an assumption had to be made with respect to TCDD disposition in the other farm livestock. Uncertainty in this model as a result of this animal-to-animal extrapolation cannot be determined, however. Until pharmacokinetic data are generated in these other animals, a more accurate assessment of the contribution of these other foodstuffs to the total agricultural food chain dose cannot be made.

REFERENCES

1. U.S. Environmental Protection Agency, *Methodology for the Assessment of Health Risks Associated with Multiple Pathway Exposure to Municipal Waste Combustor Emissions,* (draft submitted for review to the Science Advisory Board), Office of Air Quality Planning and Standards, North Carolina, and the Environmental Criteria and Assessment Office, Cincinnati, 1986.
2. U.S. Environmental Protection Agency, *Interim Procedures for Estimating Risks Associated with Exposures to Mixtures of Chlorinated Dibenzo-p-dioxins and -dibenzofurans (CDDs and CDFs),* EPA 625/3-87/012, Office of Solid Waste and Emergency Response and Office of Pesticides and Toxic Substances, Washington, D.C., 1987.
3. **Hay, A.,** Chlorinated dioxins and the environment, *Nature,* 289, 351, 1981.
4. U.S. Environmental Protection Agency, *Industrial Source Complex (ISC) Dispersion Model User's Guide,* EPA-450/4-79-030 (Volume I) and EPA-450/4-79-031 (Volume II), U.S. Environmental Protection Agency, Research Triangle Park, NC, 1979.

5. **Stevens, J. B. and Gerbec, E. N.,** *Dioxin in the Food Chain: A Health Risk Assessment Model for RDF Incinerators,* Prepared for Northern States Power Co., Minnesota Pollution Control Agency and Minnesota Department of Health, Minneapolis, 1986.
6. **Di Domenico, A., Viviano, G., and Zapponi, G. A.,** Environmental persistence of 2,3,7,8-TCDD at Seveso, in *Chlorinated Dioxins and Related Compounds — Impact on the Environment,* Hutzinger, O., Frei, R. W., Merian, E., and Pocchiari, F., Eds., Pergamon Press, Oxford, 1982, 105.
7. **Young, A. L.,** Long-term studies on the persistence and movement of TCDD in a natural ecosystem, in *Human and Environmental Risks of Chlorinated Dioxins and Related Compounds,* Tucker, R. E., Ed., Plenum Press, New York, 1983, 173.
8. **Kimbrough, R., Falk, H., Stehr, P., and Fries, G.,** Health implications of 2,3,7,8-tetrachlorodibenzo-*p*-dioxin contamination of residential soil, *J. Toxicol. Environ. Health,* 14, 47, 1984.
9. University of Minnesota, personal communication from Department of Veterinary Medicine, St. Paul Campus, 1986.
10. **Wipf, H. K., Homberger, E., Neimer, N., Ranalder, V. B., Vetter, W., and Vuilleumeir, J. P.,** TCDD levels in soil and plant samples from the Seveso area, in *Chlorinated Dioxins and Related Compounds: Impact on the Environment,* Hutzinger, O., Frei, R. W., Merian, E., and Pocchiari, F., Eds., Pergamon Press, New York, 1982, 115.
11. **Cocucci, S., DiGerolamo, F., Verderio, A., Cavallaro, A., Colli, G., Gorni, A., Invernizzi, G., and Luciani, L.,** Absorption and translocation of tetrachlorodibenzo-*p*-dioxin by plants from polluted soil, *Experientia,* 35, 482, 1979.
12. **Isensee, A. R. and Jones, J. R.,** Absorption and translocation of root and foliage applied 2,4-dichlorophenol, 2,7-dichlorodibenzo-*p*-dioxin, and 2,3,7,8-tetrachlorodibenzo-*p*-dioxin, *J. Agric. Food Chem.,* 19, 1210, 1971.
13. **Jensen, D. J., Getzendaner, M., Hummel, R., and Turley, J.,** Residue studies for (2,4,5-trichlorophenoxy) acetic acid and 2,3,7,8-tetrachlorodibenzo-*p*-dioxin in grass and rice, *J. Agric. Food Chem.,* 31, 118, 1983.
14. **Baes, C. F., Sharp, R., Sjoreen, A., and Shor, R.,** *A Review and Analysis of Parameters for Assessing Transport of Environmentally Released Radionuclides through Agriculture,* Rep. No. 5786, Oak Ridge National Laboratory, Oak Ridge, TN, 1984.
15. Minnesota State Agricultural Office, St. Paul, personal communication, 1986.
16. **Miller, C. W. and Hoffman, F. O.,** An examination of the environmental half-life for radionuclides deposited on vegetation, *Health Phys.,* 45, 731, 1983.
17. **Witherspoon, J. P. and Taylor, F. G., Jr.,** Retention of 1-44Cm simulated fallout particles by soybean and sorghum plants, *Health Phys.,* 21, 673, 1971.
18. **Highland, J.,** *Generic Risk Assessment Vol. 2,* for Ontario Waste Management Corporation, ENVIRON Corporation, Washington, D.C., 1986.
19. **Hinds, W. C.,** *Aerosol Technology Properties, Behavior, and Measurement of Airborne Particles,* John Wiley & Sons, New York, 1982, 211.
20. **Fries, G. and Marrow, G.,** Retention and excretion of 2,3,7,8-tetrachloro-dibenzo-*p*-dioxin by rats, *J. Agric. Food Chem.,* 23, 265, 1975.
21. **Van den Berg, M., Olie, K., and Hutzinger, O.,** Uptake and selective retention in rats of orally administered chlorinated dioxins and dibenzofurans from fly-ash and fly-ash extract, *Chemosphere,* 12, 537, 1983.
22. **Gullickson, T.,** *Feeding Dairy Cattle,* Webb Book Publishing, St. Paul, MN, 1943.
23. University of Minnesota Agricultural Extension Office, personal communication, 1986.
24. **Fries, G., Marrow, G., and Snow, P.,** Soil ingestion by swine as a route of contaminant exposure, *Environ. Toxicol. Chem.,* 1, 201, 1982.
25. **Jensen, D. and Hummel, R.,** Secretion of TCDD in milk and cream following the feeding of TCDD to lactating dairy cows, *Bull. Environ. Contam. Toxicol.,* 29, 440, 1982.
26. **Jensen, D. J., Hummel, R., Mahle, N., Kocher, C., and Higgins, H.,** Residue study on beef cattle consuming 2,3,7,8-tetrachlorodibenzo-*p*-dioxin, *J. Agric. Food Chem.,* 29, 2665, 1981.
27. Minnesota Department of Health, personal communication, 1986.
28. **Poiger, H. and Schlatter, C.,** Pharmacokinetics of 2,3,7,8-TCDD in man, *Chemosphere,* 15, 1489, 1986.
29. U.S. Environmental Protection Agency, *Development of Statistical Distributions or Ranges of Standard Factors Used in Exposure Assessments,* OHEA-E-161, Office of Health and Environmental Assessment, Washington, D.C., 1985.
30. **Travis, C. C. and Hattemer-Frey, H. A.,** Human exposure to 2,3,7,8-TCDD, *Chemosphere,* 16, 2331, 1987.

Chapter 6

ASSESSING THE RISKS OF INCINERATING MUNICIPAL SOLID WASTE: THE DEVELOPMENT AND APPLICATION OF A METHODOLOGY*

Larry Fradkin, Randall J. F. Bruins, David H. Cleverly, Norman C. Possiel, Robert G. Kellam, Rayburn Morrison, Patricia M. McGinnis, Gaynor Dawson, and Annie M. Jarabek

TABLE OF CONTENTS

* The views in this paper are those of the authors and do no necessarily reflect the views or policies of the U.S. Environmental Protection Agency.

I. INTRODUCTION

Most analyses of human health risks associated with atmospheric emissions from combustion sources have focused only on exposures occurring by direct inhalation. Recent studies, however, have linked elevated levels of pollutants in soils, lake sediments, and cow milk to atmospheric transport and deposition of pollutants from combustion sources.[1-4] These studies indicate that deposition of atmospherically emitted pollutants could result in additional, indirect avenues of exposure for humans and other organisms. Therefore, a methodology was developed to examine both direct and indirect exposures and risks from atmospheric emission sources. The methodology has recently been used to study potential risks to human health and the environment from municipal waste combustors (MWCs).

At least 4% of the over 155 million tons of municipal solid waste (MSW) generated each year is incinerated at 111 facilities in the U.S. (Table 1, Figure 1). Faced with the dwindling capacities of existing landfills and the difficulties of siting and operating new landfills, communities are increasingly turning to incineration for waste disposal. Incineration can be an attractive alternative to landfilling because it reduces the volume of waste by 70 to 90%, thereby extending the time until landfill capacities are exhausted. In addition, most existing solid waste incinerators and all of the planned facilities have energy recovery systems.

The U.S. Environmental Protection Agency (EPA) projects that by the year 2000, more than 300 facilities nationwide will incinerate over 900 million tons of MSW.[5] At present, there are 210 facilities in the planning stage or under construction that are projected to be completed by 1993. This represents a nearly threefold increase in the number of facilities in a relatively short period of time. There is a limited database regarding the type and quantity of pollutants that may be emitted during the combustion of MSW. While many MWCs have been tested for compliance with particulate emission standards, a limited number of tests for specific pollutants have been conducted. The U.S. EPAs database on MWC pollutant emissions consists of test results from three types of MWCs located in the U.S. and Canada (Table 2).

II. MUNICIPAL WASTE COMBUSTOR TECHNOLOGY

Three major types of municipal waste combustion technologies are commonly used today: massburn, small modular, and refuse-derived fuel incinerators.

A. MASSBURN INCINERATORS

Massburn incinerators burn refuse without any preprocessing of the waste other than removal of oversized objects. The waste is loaded by an overhead crane into a feed chute where it is carried over three grates in the combustion chamber. The first grate functions as a drying section for the waste before it enters the second combustion grate section. The third grate is the burnout grate where final combustion takes place and the ash is collected by disposal. Some massburn incinerators have a waterwall for energy recovery.

B. SMALL MODULAR INCINERATORS

These incinerators consist of modular units that are factory made. Single modules can burn up to 100 tons of refuse per day. Additional capacity can be achieved by using multiple modules. The units are designed with two chambers. The first chamber is fed by a hopper and a ram feed system. The material is ignited by a gas or oil burner in the primary chamber under conditions of low air velocities. The secondary chamber is where complete combustion takes place under excess air conditions. This chamber has an auxiliary burner to maintain desired temperatures. Heat can be recovered by waste heat boilers.

C. REFUSE-DERIVED FUEL INCINERATORS

Refuse can be processed into what is referred to as refuse-derived fuel (RDF) and then burned

TABLE 1
Number and Capacity of Existing and Planned MWCs by Design Type[22]

Design type	Installed or projected design capacity[a] (tons/d)	No. of facilities
Existing facilities		
Massburn with heat recovery	20,100	24
Massburn without heat recovery	13,000	21
Modular with heat recovery	3,900	39
Modular without heat recovery	600	17
RDF[b] with heat recovery	11,400	10
RDF without heat recovery	0	0
Total	49,000	111
Projected facilities		
Massburn[c]	113,00	118
Modular[d]	5,000	24
RDF	39,000	31
Undetermined[e]	36,000	37
Total	193,000	210

[a] Installed design capacity applies to existing facilities. Projected applies to planned facilities.
[b] RDF = refuse-derived fuel.
[c] Includes both starved air and excess air designs.
[d] Includes both overfeed stoker and rotary combustion designs.
[e] Design type has either not been specified or data on design type were not provided in the reference.

in refuse incinerators. Processing the waste produces a more homogenous fuel with a greater heat value per pound. This fuel requires less costly grate systems. The RDF can be co-fired with coal or other fossil fuels. Refuse processing techniques include shredding, production of compressed pellets, pulverization into powder, and wet pulping. Waste heat boilers are used with these incinerators for heat recovery.

III. OVERVIEW OF MODELING APPROACHES

Emissions from MWCs are known to contain heavy metals and toxic organic compounds that may be either adsorbed onto the surface of emitted particulates or remain as gas or vapor. These pollutants are transported through the atmosphere, fall to the earth's surface by forces of gravity (dry deposition), and are washed out of the atmosphere by precipitation (wet deposition). Once deposited on the earth's surface, pollutants may further translocate to surface or groundwater or to plant and animal tissues. Human exposure results directly from the inhalation of pollutants in the atmosphere; indirectly from ingestion of contaminated soil, groundwater, and food; and from skin contact with pollutants present in soil.

This methodology estimates exposure to pollutants via six computer models that simulate the fate, transport, and bioaccumulation of pollutants through both direct and indirect exposure pathways:

- Industrial Source Complex Short-Term Air Dispersion Model (ISCST)
- Human Exposure Model (HEM)
- Terrestrial Food Chain Model (TFC)
- Surface Runoff Model (SRM)
- Groundwater Infiltration Model (GIM)
- Dermal Exposure Model (DEM)

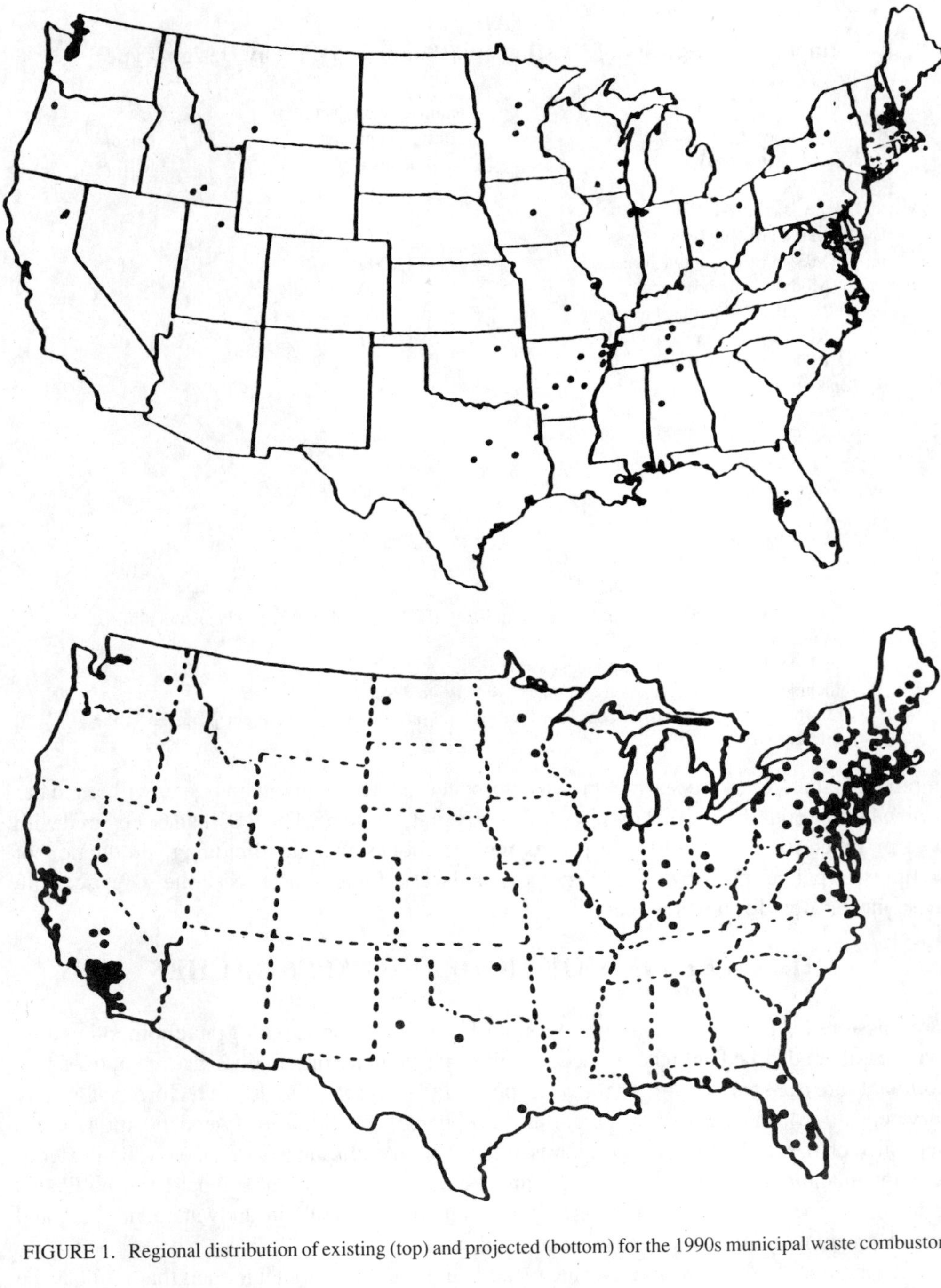

FIGURE 1. Regional distribution of existing (top) and projected (bottom) for the 1990s municipal waste combustors.

TABLE 2
Number of Municipal Waste Combustors with Emissions Tests Data[23]

	Massburn incinerators	RDF[a] incinerators	Modular incinerators	All facilities
Metals				
Arsenic	6	2	3	11
Beryllium	5	1	1	7
Cadmium	9	1	2	12

TABLE 2 (continued)
Number of Municipal Waste Combustors with Emissions Tests Data[23]

	Massburn incinerators	RDF[a] incinerators	Modular incinerators	All facilities
Chromium	9	2	2	13
Mercury	5	2	2	9
Nickel	7	2	2	11
Lead	10	2	2	14
Organics				
CDDs/CDFs[b]	8	5	6	19
Polycyclic aromatic hydrocarbons (PAHs)	1	1	1	3
PCBs	2	2	2	6
Formaldehyde	2	3	2	7
Chlorobenzenes	2	2	1	5
Chlorophenols	2	2	1	5

[a] Refused-derived fuel incinerators.
[b] Polychlorinated dibenzo-*p*-dioxins and dibenzofurans.

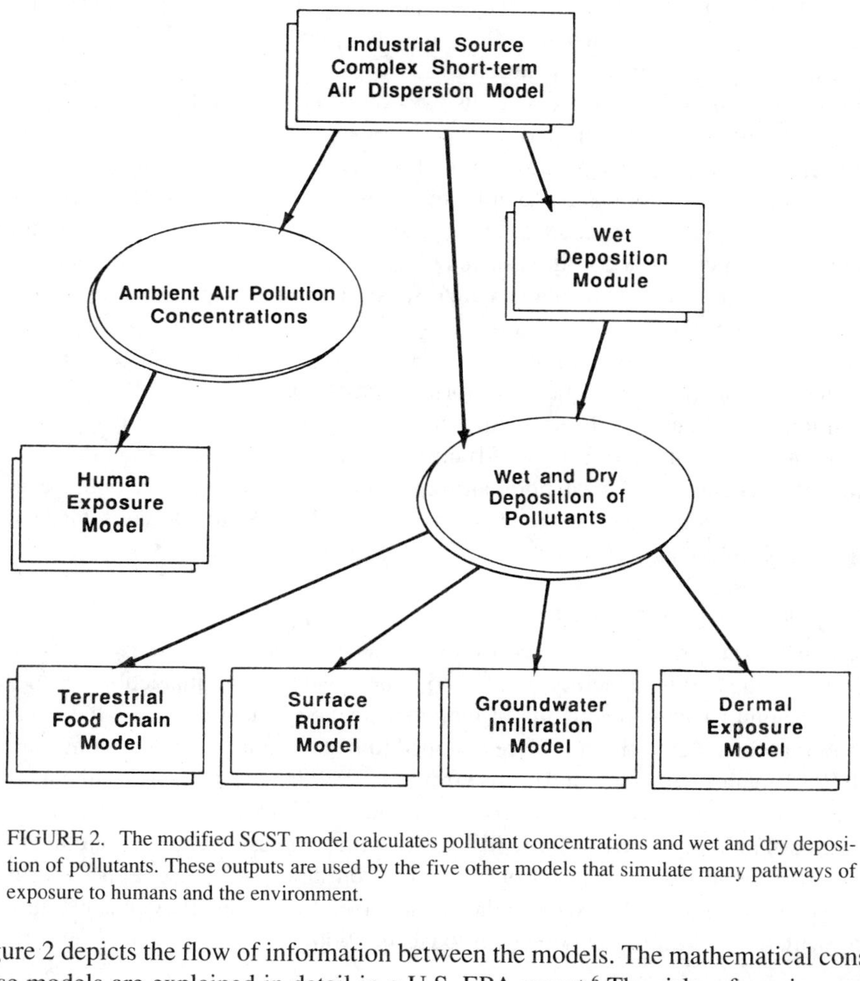

FIGURE 2. The modified SCST model calculates pollutant concentrations and wet and dry deposition of pollutants. These outputs are used by the five other models that simulate many pathways of exposure to humans and the environment.

Figure 2 depicts the flow of information between the models. The mathematical constructs of these models are explained in detail in a U.S. EPA report.[6] The risks of carcinogenic and noncarcinogenic (or systemic) effects resulting from exposure to each chemical are calculated separately for each pathway. Risks may be expressed in terms of aggregate risk or as risk to the most exposed individual (MEI). Aggregate risk, which is calculated only for carcinogens, is the sum of the estimated number of disease cases among the entire population of individuals exposed to each emitted pollutant. Since the risk for each pollutant is based on an upper-limit estimate of its carcinogenic potency,[7] the calculation of aggregate risk as a sum of the upper limits of risk is inherently conservative. The MEI risk refers to the risk of disease for an individual in a scenario constructed to represent a maximum level of exposure based on worst-case exposure assumptions rather than statistical measures of error in the exposure assessment. The MEI risk is computed for both carcinogenic and systemic toxic effects.

IV. METHODOLOGY COMPONENTS

A. INDUSTRIAL SOURCE COMPLEX SHORT-TERM AIR DISPERSION MODEL

The ISCST was developed to simulate the behavior of emission plumes from a wide variety of industrial sources.[8-11] It predicts the concentration of pollutants emitted into the air and the rate of pollutant deposition in the vicinity of the MWC. The rate of emission from the MWC stack in grams per second is needed as an input to the model for each chemical to be studied. These rates are estimated based on available data describing MWC emissions (see Table 2) and using assumptions regarding the capture efficiency of various types of air pollution control devices.[6]

Particle-mass fractionalization data indicate that the concentrations of inorganic pollutants generally increase with a decrease in particle size. Based on available test data for inorganics, assumptions are made regarding the size distribution of emitted particles. Although recent studies indicate that a significant proportion of toxic organics, such as 2,3,7,8-TCDD, may exist as vapor in ambient air,[12,13] the deposition calculations assume that all organic pollutants are in particulate form and, therefore, are available for deposition. Dispersion calculations consider particulate size; wind velocities at particular stack heights; and the aerodynamics resulting from the contours of the incinerator building, surrounding buildings, and the terrain. Many of these assumptions are based on limited data and are discussed as limitations of the model in a later section. The output of the ISCST model is a set of air pollutant concentrations at each of 160 radially arranged points at distances of 0.2 to 50 km from the emissions source (Figure 3).

The estimated ambient air concentrations form the basis from which exposure due to wet and dry deposition of pollutants is determined. The ISCST model predicts the dry gravitational settling of contaminants at each receptor grid point. A module was added to the model to simulate the process of wet deposition.[6] Total pollutant deposition rates resulting from both wet and dry processes are determined for each of the 160 grid points and are used to estimate total deposition over time periods of 30 and 100 years.

B. HUMAN EXPOSURE MODEL

The Human Exposure Model (HEM) estimates the extent of exposure by the population within 50 km of the MWC to facility-emitted pollutants due to direct inhalation.[14] Exposure is calculated assuming a static population of individuals who reside at the same location for 70 years. Indoor pollutant concentrations are assumed to be the same as outdoor concentrations.

The HEM contains a population database from the 1980 U.S. Census. Thus, the number of exposed individuals and the ambient air concentration of pollutants are available for each grid point in a 50-km radius around a specific MWC. Aggregate exposure to each pollutant through inhalation is expressed as the product of the population times the ambient air concentration summed over all grid points. The MEI for the inhalation pathway is an individual residing at the grid point with the highest mean ambient air concentration.

C. TERRESTRIAL FOOD CHAIN MODEL

MWC emissions settle onto the soil as a result of dry and wet deposition. The Terrestrial Food

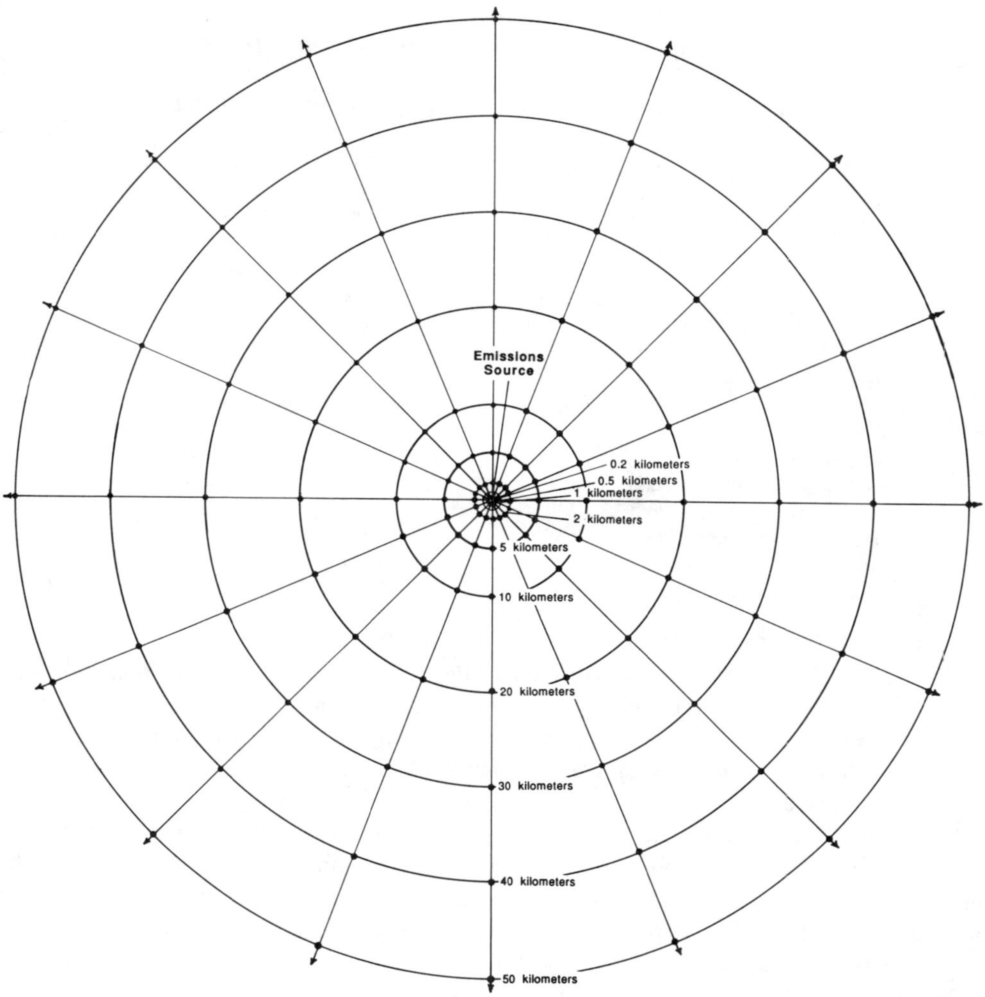

FIGURE 3. The SCST model calculates the concentration of each pollutant at each of the 160 grid points formed by the intersection of 16 wind direction vectors (radial lines) and 10 distance rings (concentric circles) surrounding the emissions source.

Chain Model (TFC) simulates the eventual transport of these deposited pollutants into the human and the ecological food chain. The model focuses on four pathways of human exposure. These pathways consider plant uptake of contaminants with subsequent introduction to the food chain, as well as direct ingestion of contaminated soil by children and animals. Table 3 describes the pathways addressed and the scenarios that identify the MEI for each exposure pathway. Exposure to the various pollutants is calculated separately for each pathway. Pathways of exposure for the ecological food chain are discussed in the Ecological Models section of this chapter.

The total level of pollutants deposited on the soil during a specified time period is obtained from the results of the modified ISCST model. The TFC model accounts for pollutant losses from soil due to leaching, volatilization, and chemical and biological degradation. Pollutant uptake rates into plant and animal tissues were established based on empirical data in the current literature. The exposure calculations for each pathway are based on assumptions about the behavior of the MEI, such as the volume and variety of home-grown foods consumed by an individual assumed to reside in the area of maximum pollutant deposition near the MWC, or the volume of soil consumed by a child through direct ingestion or by putting dirty objects in their mouth.

TABLE 3
Pathways of Exposure and MEI Definitions Used in the Models

Pathway of exposure	Definition of MEI
1. Deposition of pollutants onto soil, crop uptake of pollutants from contaminated soil, and human consumption of contaminated crops.	1. Rural individual residing in the area near the MWC with the highest rate of deposition. This individual's diet consists mainly of food grown in a backyard garden.
2. Deposition of pollutants onto soil, and human consumption of contaminated soil.	2. Child 1 to 6 years old who ingests soil incidentally or intentionally. The child lives at the site of maximum pollutant deposition.
3. Deposition of pollutants onto soil, uptake of pollutants into feed crops, ingestion of feed crops and soil by agricultural animals, and human consumption of meat and dairy products.	3. Rural individual who produces a substantial portion of his own meat and dairy products using animal feed grown in the area of maximum pollutant deposition.

D. SURFACE RUNOFF MODEL

The Surface Runoff Model (SRM) simulates the transport of soil-departed pollutants into rivers or other surface water bodies by precipitation. Humans are exposed to these pollutants through consumption of contaminated water or fish. Aquatic life is also impacted by these exposures. The risk to aquatic organisms and their wildlife predators is described in the section on ecological risks.

The SRM uses the Universal Soil Loss Equation[15] to estimate contaminant loading to a medium-sized stream of a hypothetical watershed area radiating 1 km^2 from the MWC. (The dimensions of the actual watershed can be entered if available.) The calculations are based on the frequency and effect of storms on water flows and erosion in the watershed, the dissolution of pollutants deposited onto the soil, the flow rate of the stream, and the degradation of pollutants. The MEI for this pathway is a person who obtains drinking water or eats fish exclusively from this stream. Exposure calculations are based on the concentration of contaminants that have accumulated in the water and in fish multiplied by the amount of water or fish consumed by the MEI.

E. GROUNDWATER INFILTRATION MODEL

The Groundwater Infiltration Model (GIM) characterizes the dissolution and movement of pollutants deposited onto the soil into underlying aquifers. Individuals will be exposed if the contaminated groundwater is used as a drinking water supply. Aquatic life in lakes and streams fed by contaminated aquifers could become contaminated, also resulting in human exposure.

The model focuses on the impacts to a small well fed by an aquifer with a recharge area defined by a circle with a radius of 200 m surrounding the MWC. Exposure is calculated using factors for annual rainfall, pollutant leaching rates, pollutant degradation, aquifer recharge rate, and drinking water consumption. The MEI is an individual using this aquifer as his or her sole drinking water source or consuming fish from a surface water body fed by the contaminated groundwater.

F. DERMAL EXPOSURE MODEL

Skin contact with contaminated soil may adversely affect human health. The Dermal Exposure Model (DEM) estimates the level of exposure due to absorption of pollutants through the skin. The model examines exposure to soil through gardening activities for adults and through outdoor play for preschool children. There is a paucity of data, however, regarding dermal absorption rates for various pollutants, particularly when found in a soil matrix. This model is proposed as a preliminary method to quantify such exposure, but reliable data with which to execute the methodology are presently lacking for most pollutants.

V. OVERVIEW OF THE EXPOSURE ASSESSMENT METHODOLOGY

A. DETERMINING HUMAN HEALTH RISKS

The estimates of exposure obtained from the HEM, TFC, SRM, GIM, and DEM models are used to determine the cancer and noncancer risks to humans. The risks from direct inhalation of pollutants are analyzed separately from the risks of indirect exposures. The HEM in combination with the ISCST is capable of estimating both the aggregate and MEI risks associated with direct inhalation of pollutants. Estimating the aggregate risks associated with the indirect pathways, however, cannot be performed with a meaningful degree of accuracy at this time. For the indirect pathways, therefore, only the risks to the MEI are examined.

B. ASSESSING THE RISK OF CANCER

For the direct inhalation pathway, cancer risks are assessed by application of unit risk estimates of cancer potency for the various pollutants. Unit risk estimates represent the lifetime cancer risk resulting from continuous inhalation of a unit concentration of the pollutant (e.g., 1 $\mu g/m^3$). These values were derived by the U.S. EPAs Office of Health and Environmental Assessment based on dose-response data available in the scientific literature and EPA Guidelines for Carcinogen Risk Assessment.[7] When adequate human data were available, the unit risk estimates of cancer potency were based on data describing human cancer incidence. For most pollutants, however, only the results of animal bioassays using relatively high dose rates were available. In these cases, conservative extrapolations were made to apply the results to the analysis of human risks at lower doses. The unit risk estimates derived in these cases are considered upper-bound estimates. It is not likely that the true risks would exceed these estimates, and it is likely that the true risks are lower.

The risks to the total exposed population and to the MEI from direct inhalation are estimated using the calculated ambient air concentration for each pollutant as estimated from the HEM and unit risk values. Aggregate risk, or the aggregate annual cancer incidence, is expressed in terms of the expected increase in the total number of cancer cases each year within the exposed population. MEI risk is expressed as the increase in an individual's risk of cancer over a lifetime of exposure.

Cancer risks for the indirect pathways are estimated for the MEI only. In place of inhalation exposure, estimates of daily exposure through ingestion of contaminated crops, animal products (e.g., beef, poultry, pork, eggs, and dairy products), soil, water, and aquatic organisms derived from the TFC, SRM, or GIM models are used. These exposures estimates are used along with cancer potency values to estimate the lifetime risk to the MEI.

Differences in disposition of toxic agents can result from exposure route differences. For example, several metals known to be carcinogens when inhaled (e.g., cadmium and chromium) are not carcinogens when ingested. Procedures for estimating effective target tissue concentrations after the dermal route of exposure have not been established. Potency by the dermal route is assumed equivalent to that for the oral route after adjustment for differential rates of absorption between the two routes has been made.

C. ASSESSING THE RISKS OF NONCANCER DISEASE

Systemic toxicity is a term often used to describe toxic effects other than cancer, such as diseases of the liver, kidney, nervous, or circulatory systems. The reference dose (RfD) for a given pollutant is one basis for determining the human risk of systemic toxicity resulting from exposure to a specific pollutant. RfDs, established by the U.S. EPA, are the daily dosages that during a lifetime of exposure, are projected to be without appreciable risk of producing an adverse effect. They are expressed as a dose rate in terms of milligrams of the pollutant per kilogram of body weight per day. (It should be noted that RfDs have an associated uncertainties of approximately one order of magnitude.[16])

For the direct inhalation pathway, the potential for systemic health effects is determined by comparing exposure levels derived from the HEM with existing federal air standards. If exposure is greater than the reference air concentration for a given pollutant, the possibility of experiencing systemic effects is suggested. It cannot be concluded, however, that systemic toxicity will necessarily occur. These comparisons are made for aggregate and MEI exposures.

For indirect pathways, the risk of systemic impairment is assessed in a similar manner. Exposure levels associated with ingestion of contaminated crops, animal products, and soil from the TFC model are compared to adjusted reference intake values (RIVs). RIVs are based on the RfD for each pollutant and take into account body weight and background dietary intake of the pollutant. Water concentrations established in the SRM and GIM models are compared with reference water concentrations (RWCs), which are based on RfDs and consider the amount of contaminated water and aquatic organisms consumed. Each of the reference values (i.e., RIVs and RWCs) takes into account the background (i.e., preexisting) pollutant exposures and determines whether the added exposure by a given pathway is likely to result in a total exposure that exceeds an RfD.

D. DETERMINING ECOLOGICAL RISKS

The effects of MWC emissions on components of the ecosystem other than humans are also addressed in this methodology. The TFC model is used to analyze the risk to soil biota, their predators, plants, and herbivorous animals. Soil organisms, such as microbial flora and fauna, earthworms, and insects, may accumulate soil-deposited contaminants. Small mammals and birds that prey on soil organisms are thereby exposed. Plants also assimilate and accumulate pollutants from the soil. Herbivorous animals are exposed to pollutants through ingestion of contaminated soil and plants. The SRM and GIM were used to predict exposure of aquatic organisms and of wildlife species that feed on aquatic organisms. Elevated contaminant concentrations in surface waters, resulting from either runoff or contaminated groundwater, result in acute or chronic exposure of aquatic organisms. Bioconcentration of pollutants by these organisms may result in exposures of predatory mammals or birds.

The effects of exposures in aquatic organisms are determined by comparison with ambient water quality criteria established by the U.S. EPA.[17] Because such criteria are not available for terrestrial organisms, the effects of exposure in these organisms are determined using toxicity thresholds established in the scientific literature. The literature sources on this topic are diverse and incomplete and, therefore, careful judgments must be made in their use. Indicators of toxicity may include reductions in growth rate, fecundity, and life span. Toxicity thresholds thus determined are compared to the estimated terrestrial or aquatic exposure levels to assess the ecological risks associated with pollutants from MWCs.

VI. LIMITATIONS OF THE METHODOLOGY

The methodology, and the reliability of results derived, are limited by uncertainties associated with the models themselves, and the quality and quantity of the data available to serve as inputs to the models. As to the limitations of the models themselves, some important issues should be noted. There has been no validation of modeled pollutant wet deposition rates predicted by the modified ISCST model. Atmospheric dispersion models are currently being modified to allow estimation of pollutant deposition in complex terrain. In addition, the HEM model does not account for the true activity patterns of the individuals living in the exposed area. The model assumes that the populations are exposed to the same ambient air concentrations whether they are indoors or outdoors and that exposed individuals inhale pollutants at a constant rate. Similar models, used to analyze the effects of ozone and carbon monoxide exposures have integrated person activity patterns into the model that account for the time spent indoors as opposed to outdoors where the microenvironment pollutant concentration and respiration rate may be different.[18] Furthermore, the TFC model currently considers only impacts from pollutant deposition onto soil and subsequent transport to other elements of the food chain. It is being revised to incorporate exposure from direct deposition of contaminants onto foliar surfaces and the uptake of pollutant vapors by aerial plant plants (also known as air-to-leaf transfer). Determining ranges of exposure (e.g., average to worse-case scenarios) for the MEI and constructing a probability distribution of exposure are also being evaluated.

The data supporting some aspects of the air dispersion model are insufficient in quality and quantity. The emissions database used in the analysis was not compiled from systematic tests of MWCs. A variety of data sources do not reflect the newer, more efficient MWCs in operation. Data regarding control efficiencies for MWCs are also incomplete. A few studies have measured metal removal efficiencies resulting from the installation of various air pollution control devices, including electrostatic precipitators, fabric filters, and dry scrubbers in combination with fabric filters. There are limited data, however, on control efficiencies for organic compounds. The present methodology assumes that a 20% reduction in organics can be achieved with the addition of an electrostatic precipitator and a 95% reduction with the addition of a dry scrubber in combination with a fabric filter. Field validation of assumed removal efficiencies is needed.

In addition, there are limited data documenting the distribution of various-sized particles in MWC emissions, and it is not known how organic pollutants behave in the emission plume (i.e., what portion exist in the particulate rather than vapor phase and to what size particles these pollutants are most attracted). This application of the ISCST model assumes that all pollutants, including organic compounds, remain in particulate form and, therefore, are available for deposition. There is some evidence that organics previously assumed to be largely in particulate form can be transported as a vapor.[13] There are more, but still limited, data regarding the particle-size affinities of inorganic compounds (i.e., heavy metals) in MWC emissions.

Data describing the subsequent fate and transport of deposited pollutants are incomplete for many pathways and pollutants. For example, many of the plant uptake rates used as inputs to the model are taken from studies of pollutants in soils amended with sewage sludge. There are no studies that address plant uptake rates of pollutants deposited by incinerator emissions, although a few examine impacts from other combustion sources, such as smelter emissions.[19,20] The bioconcentration factors used to estimate pollutant levels in aquatic organisms are based on water-column concentrations and do not account for the fact that these pollutants are actually sediment-associated as they enter the receiving water. Lack of sampling and analysis techniques to measure both the fate and transport of pollutants and the effects of pollutants on human health also represent limitations to the methodology and epitomize areas in need of further research. Most of the existing sampling, and analytical, and quality assurance protocols were developed for application to other pollution sources, such as hazardous waste incinerators and fuel combustion power plants.

The quality and scope of the data describing human and animal responses to individual pollutants vary widely and as already mentioned, there are limited data regarding the absorption and health effects of pollutants contacting the skin. In addition, the effects of possible synergism and antagonism between exposure pathways and/or pollutants has not been addressed.

Notwithstanding the data and model limitations, the proposed composite risk assessment methodology is important for two reasons. First, it is unique in its comprehensiveness by addressing potential health effects resulting from exposure through secondary routes following pollutant deposition, as well through direct inhalation. In addition to human health effects, general effects on the ecosystem are addressed. As the supporting data are developed, the accuracy of the model results will be improved. Where uncertainties exist, the tendency is to err on the conservative side (i.e., the risks predicted tend to be higher than what might actually be experienced). This is not necessarily a disadvantage if the models are used for screening pollutants to determine situations in which a more in-depth assessment may be appropriate.

Second, this methodology can be applied to studies assessing the risks of other pollution sources that involve contamination deposition onto soil and reductions in air quality. Once the risks associated with various types of MWCs, and the industry as a whole, are well established, these risks can be compared with those of alternative disposal options (e.g., Reference 21). In addition, useful comparisons of exposure and risk can be made with other combustion technologies, such as hazardous waste and sewage sludge incinerators, pulp and paper mills, and coal-fired power plants.

VII. RESULTS OF THE EXPOSURE ASSESSMENT

This methodology was used as shown in Figure 4 to evaluate potential impacts from the growth of MWC in the U.S.[6,22] Two emission control scenarios were compared: a baseline scenario representing MWCs that use a prevalent emission control technology (electrostatic precipitators, or ESPs), and a controlled scenario representing MWCs that use a sophisticated, state-of-the-art technology (an alkaline scrubber followed by a fabric filter). The baseline scenario assumed 20% control of organic pollutants and 30% control of mercury. For existing facilities, control of metals besides mercury was assumed proportional to actual particle control efficiencies at individual facilities. For projected facilities, a 99% control efficiency for metals was assumed. The efficiency for organic pollutants by particulate controls was derived from preliminary results of tests on ESPs showing ranges of 0 to 50%.[6,23]

The controlled scenario assumed 95% control for organics, 99.5% for metals (except mercury), and 50% for mercury. Risks from existing facilities (those operational as of 1985) and projected facilities (those designated for completion by 1993) were also compared under each control scenario. Impacts for the direct inhalation pathway were assessed using actual facility design, meteorology, and demographic data for all existing and planned facilities. Pollutant emission rates per ton of waste burned were assumed to be the same for all facilities within a given scenario. In the indirect pathways, existing facilities were represented by a plant in Virginia that in many ways reflects worst-case parameters for MWCs currently in operation. A hypothetical facility in Florida was used to represent the projected facilities. The higher volume burned daily by the latter plant (3000 tons compared to 120 tons at the former facility) occasionally resulted in higher total emissions than the Virginia facility for some pollutants.

A. RESULTS FOR THE DIRECT INHALATION PATHWAY

Analyses of cancer risks resulting from the direct inhalation pathway were conducted for both the existing and projected MWCs under both the baseline and controlled scenarios.[22] Risks from inhaling each of the chemicals or chemical classes listed in Table 2 were assessed for the population living within 50 km of MWCs in the U.S. These are not the only contaminants of concern in MWC emissions but are contaminants known to be present in MWC emissions for which some data on health effects are available. Maximum predicted annual-average air concentrations were compared with existing ambient air quality standards. Lead concentrations did not exceed the ambient air quality standard for lead,[24] nor did mercury concentrations surpass the National Emission Standard for Hazardous Air Pollutants (NESHAP) level for mercury.[25] Therefore, no adverse noncarcinogenic health effects are expected to be associated with exposure to those chemicals via inhalation. Results for nickel were not determined because it

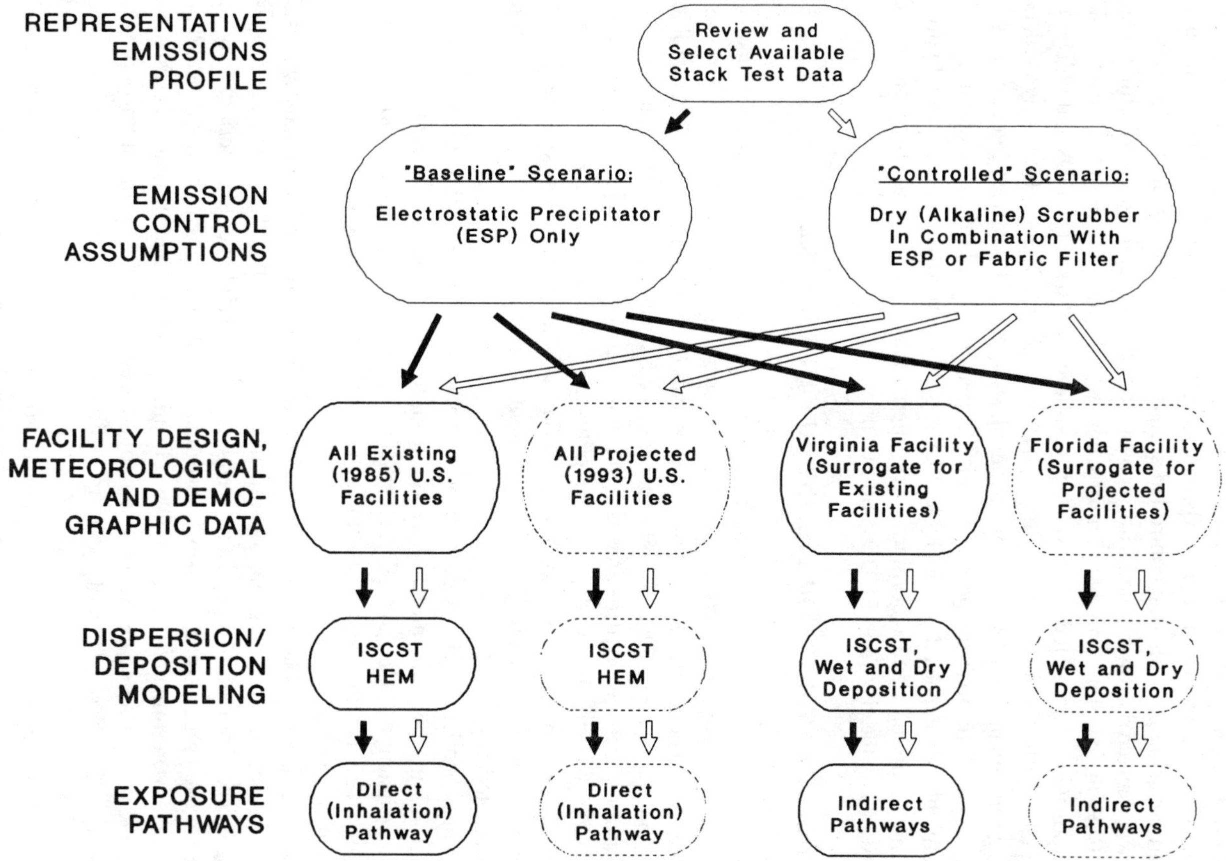

FIGURE 4. Procedures used to evaluate potential impacts from the increase in municipal waste combustors in the U.S.

was uncertain whether the forms of nickel likely to be emitted by MWCs include the nickel species known to be carcinogenic (nickel subsulfide).[26]

For inhalation of the remaining carcinogenic pollutants, the probability of developing cancer was calculated as both aggregate population risk (expressed as an annual cancer incidence) and as the maximum individual lifetime risk (expressed as lifetime probability of cancer for the MEI). A range of probabilities were used to reflect uncertainties inherent in estimating cancer risks. Both aggregate and MEI risks estimated for existing facilities were greater than the risks estimated for projected facilities for both the baseline and controlled emission scenarios (see Figure 4). The risks associated with human exposure to pollutants emitted from MWCs that use state-of-the-art emission controls (the controlled scenario) were approximately one order of magnitude lower than the risks associated with exposure to pollutants emitted from facilities that used only an ESP (the baseline emission control scenario). The majority of the estimated risk was due to exposure to chlorinated dibenzodioxins and dibenzofurans (PCDDs and PCDFs). The risk associated with exposure to facility-emitted PCDDs and PCDFs was four to five orders of magnitude greater than the risks from chlorophenols, which had the lowest calculated risk, for aggregate risk and three to five orders of magnitude greater for MEI risk. The aggregate and MEI risks from chlorobenzenes, formaldehyde, PCBs, beryllium, polycyclic aromatic hydrocarbons, arsenic, cadmium, and chromium were intermediate (i.e., one to three orders of magnitude lower than the risk from PCDDs and PCDFs).

B. PRELIMINARY RESULTS FOR THE INDIRECT EXPOSURE PATHWAYS

Preliminary analyses were conducted for indirect exposure to deposited contaminants for existing and projected facilities. Risk was calculated separately for each pathway. To date, assessments have been conducted for 9 of the 13 chemicals or chemical classes listed in Table 2. Since many pollutants accumulate in soil over time, the risks associated with contaminated soil will vary according to the length of time an MWC is assumed to operate. Exposure resulting from 30 and 100 years of continuous deposition of pollutants was examined at both facilities for both the baseline and controlled emission scenarios. Thirty years was chosen to represent the probable lifetime of an MWC facility; 100 years assumed replacement of the facility several times at the same location.

The following chemicals were evaluated for the indirect exposure pathways: beryllium, hexavalent chromium, inorganic mercury, nickel, lead, formaldehyde, chlorinated benzenes, polychlorinated biphenyls (PCBs), PCDDs, and PCDFs. Classes of chemical compounds were evaluated using an equivalency approach similar to that described in the interim methods for toxicity equivalency factors.[37]

Subject to the assumptions and uncertainties in the methodology, the preliminary analysis suggests that for some persistent organic constituents (PCDDs, PCDFs, PCBs, and chlorobenzenes), the MEI risks from indirect exposure to emissions deposited over long periods may be comparable to or even greater than the risks from direct inhalation. A possible human health or ecological risk was determined to exist if a given chemical or chemical class had an estimated indirect exposure that exceeded the RfD for compounds considered to be noncarcinogenic, or exceeded a lifetime cancer risk level of 10^{-6} (1 in 1 million) for chemicals considered possible carcinogens.

For all exposure scenarios (both the existing and projected facilities, both emission control technologies, and 30 and 100 years of continual deposition) PCDDs and PCDFs, and chlorinated benzenes pose a possible human health risk for the food ingestion pathways considered, while PCBs pose a risk only when the baseline control technology is used. A possible human risk also exists for PCDDs and PCDFs for the soil ingestion pathway. Mercury appears to pose a possible human health risk at the existing facility after 100 years of continual deposition and at both time intervals at the projected facility for both the baseline and controlled emission technology. A child's exposure to deposited lead and mercury through soil ingestion also appears to warrant further study for potential health effects.

Risk from ingestion contaminated water was not evaluated for the existing plant in Virginia since local waters are tidal and therefore not potable. A possible risk exists for PCDD/PCDF exposure at the projected Florida facility for both control technologies and deposition intervals. Inorganic mercury, PCDDs, and PCDFs pose a concern when the fish ingestion pathways of the SRM are considered for all exposure scenarios. Chlorinated benzenes and PCBs also demonstrate possible human health risk in all scenarios for the fish ingestion pathway, except when alkaline scrubbers and fabric filters (controlled technology scenario) are used at the hypothetical Florida site. Results of the GIM did not indicate a possible risk for any of the chemicals.

Data on dermal absorption from contaminated soil were available only for 2,3,7,8-TCDD,[28] which showed a possible health risk for all exposure scenarios at the existing facility and under baseline conditions after 100 years at the projected facility.

The potential for ecological risks from lead, mercury, PCDDs, PCDFs, chlorinated benzenes, and PCBs also appear to warrant further study, as possible risks were found for several of the exposure scenarios. Indirect exposures to nickel, hexavalent chromium, beryllium, and formaldehyde are not indicated as areas of concern by the analyses.

VIII. CONCLUSIONS

A preliminary methodology for assessing both direct and indirect pathways of exposure to MWC emissions has been presented. The model is dynamic in that it is undergoing revisions to include new pathways and is being modified as additional information becomes available. Revisions to the methodology will serve to reduce its uncertainties and limitations. This methodology has been applied to the preliminary assessment of human health and ecological risks for several exposure scenarios and the preliminary results have been presented. This approach, with further development, will enhance the ability of the scientific community to comprehensively evaluate the human health and environmental impact posed by these facilities.

REFERENCES

1. East Helena Source Apportionment Study, Particulate Source Apportionment Using the Chemical Mass Balance Receptor Model, Vol. 1, prepared for the Department of Health and Environmental Sciences, State of Montana, Helena, 1982.
2. **Swackhamer, D. L.,** Estimation of the atmospheric and nonatmospheric contributions and losses of polychlorinated biphenyls for Lake Michigan on the basis of sediment records of remote lakes, *Environ. Sci. Technol.*, 20, 879, 1986.
3. **Czuczwa, J. M. and Hites, R.,** Environmental fate of combustion-generated polychlorinated dioxins and furans, *Environ. Sci. Technol.*, 18, 444, 1984.
4. **Rappe, C., Nygren, M., Lindstrom, G., Buser, H. R., Blaser, O., and Wuthrich, C.,** Polychlorinated dibenzofurans and dibenzo-*p*-dioxins and other chlorinated contaminants in cow milk from various locations in Switzerland, *Environ. Sci. Technol.*, 21, 964, 1987.
5. Municipal Waste Combustion Study: Characterization of the Municipal Waste Combustion Industry, EPA/530-SW-8702h, Office of Solid Waste and Emergency Response, Office of Air and Radiation, Office of Research and Development, U.S. Environmental Protection Agency, Washington, D.C., 1987.
6. Municipal Waste Combustion Study: Assessment of Health Risks Associated with Municipal Waste Combustion Emissions, Office of Solid Waste and Emergency Response, Office of Air and Radiation, Office of Research and Development, U.S. Environmental Protection Agency, Washington, D.C., 1987.
7. *Fed. Reg.*, 51, 33992, 1985.
8. Industrial Source Complex Dispersion Model User's Guide, 2nd ed., EPA-450/4-86-005a, Office of Air Quality Planning and Standards, U.S. Environmental Protection Agency, Research Triangle Park, NC, 1986.

9. An Evaluation for the Industrial Source Complex (ISC) Dispersion Model, EPA-450/4-81-002, Office of Air Quality Planning and Standards, U.S. Environmental Protection Agency, Research Triangle Park, NC, 1981.
10. Tests of the Industrial Source Complex (ISC) Dispersion Model at the Armco Middletown, Ohio, Steel Mill, EPA-450/4-82-006, Office of Air Quality Planning and Standards, U.S. Environmental Protection Agency, Research Triangle Park, NC, 1982.
11. **Scire, J. S. and Schulman, L. L.,** Evaluation of the BLP and ISC Models with SF6 Tracer Data and SO2 Measurements at Aluminum Reduction Plants, presented at Air Pollution Control Assoc. Specialty Conf. on Dispersion Modeling for Complex Sources, St. Louis, 1981.
12. **Czuczwa, J. M. and Hites, R. A.,** Airborne dioxins and dibenzofurans: sources and fate, *Environ. Sci. Technol.,* 20, 195, 1986.
13. **Eitzer, B. D. and Hites, R. A.,** Concentrations of dioxins and dibenzofurans in the atmosphere, *Int. J. Environ. Anal. Chem.,* 27, 215, 1986.
14. The User's Model for the Human Exposure Model (HEM), EPA-450/5-86-001, Office of Air Quality Planning and Standards, U.S. Environmental Protection Agency, Research Triangle Park, NC, 1986.
15. **Wischmeier, W. H. and Smith, D. D.,** Predicting Rainfall Erosion Losses — A Guide to Conservation Planning, Hand. No. 537, U.S. Department of Agriculture, U.S. Government Printing Office, Washington, D.C., 1978.
16. Reference Dose (RfD): Description and Use in Health Risk Assessments, Integrated Risk Information System (IRIS), Online, Intra-Agency Reference Dose (RfD) Work Group, Office of Health and Environmental Assessment, Environmental Criteria and Assessment Office, U.S. Environmental Protection Agency, Cincinnati, 1988.
17. Quality Criteria for Water 1986, EPA/440-5-86-001, Office of Water Regulations and Standards, U.S. Environmental Protection Agency, Washington, D.C., 1987.
18. The NAAQS Exposure Model (NEM) Applied to Carbon Monoxide, EPA-450/5-83-003, Office of Air Quality Planning and Standards, U.S. Environmental Protection Agency, Research Triangle Park, NC, 1983.
19. **Roberts, T. M., Gizyn, W., and Hutchinson, T. C.,** Lead contamination of air, soil, vegetation, and people in the vicinity of secondary lead smelters, *Conf. Trace Subst. Environ. Health,* 8, 155, 1974.
20. **Munshower, F.,** Cadmium accumulation in plants and animals of polluted and nonpolluted grasslands, *J. Environ. Qual.,* 6, 411, 1977.
21. **Hwang, S.,** *Measurement of Toxic and Related Air Pollutants,* EPA 600/9-88-015, *Proceedings of the 1988 EPA/APCA International Symposium,* Air Pollution Control Association, Pittsburgh, 1988.
22. *Fed. Reg.,* 52, 25399, 1987.
23. Municipal Waste Combustion Study: Emission Data for Municipal Waste Combustors, EPA/530-SW-87-021b, Office of Solid Waste and Emergency Response, Office of Air and Radiation, Office of Research and Development, U.S. Environmental Protection Agency, Washington, D.C., 1987.
24. *Fed. Reg.,* 43, 46246, 1978.
25. Review of National Emissions Standards for Mercury, EPA-450/3-84-014, Office of Air Quality Planning and Standards, U.S. Environmental Protection Agency, Research Triangle Park, NC, 1984.
26. Integrated Risk Information System (IRIS). Risk Estimate for Carcinogenicity for Nickel Subsulfide, Online, Office of Health and Environmental Assessment, Environmental Criteria and Assessment Office, U.S. Environmental Protection Agency, Cincinnati, 1987.
27. Interim Procedures for Estimating Risks Associated with Exposures to Mixtures of Chlorinated Dibenzo-*p*-Dioxins and -Dibenzofurans (CDDs and CDFs), EPA 625/3-87-012, Risk Assessment Forum, U.S. Environmental Protection Agency, Washington, D.C., 1987.
28. **Poiger, H. and Schlatter, C.,** Influence of solvents and adsorbents on dermal and intestinal absorption of TCDD, *Food Cosmet. Toxicol.,* 18, 477, 1980.

Chapter 7

ESTIMATING THE CANCER RISKS OF EXPOSURE TO 2,3,7,8-TETRACHLORODIBENZO-*p*-DIOXIN

Michael Gough and Duncan Turnbull

TABLE OF CONTENTS

I. INTRODUCTION

"Not in my backyard!" greets proposals for the siting of many facilities including halfway houses, mental health institutions, waste dumps, and municipal solid waste (MSW) incinerators. Citizens understandably oppose new installations that may alter their neighborhoods or lifestyles. In the case of MSW incinerators, many people are concerned about possible health effects from airborne emissions and ash. Whether those objections reflect only health concerns or wether they are, in fact, surrogates for other concerns about aesthetics, property values, equity of social contracts, and other factors, has little importance. For whatever reasons, health effects stand at the forefront of any discussion on MSW incinerators.

Most urban dwellers know that some visible stack emissions are nothing but water vapors, while other emissions are scrubbed and cleaned to acceptable regulatory levels. Emissions from MSW incinerators, however, elicit greater concern, since they have been shown to contain the highly toxic chlorinated dibenzo-*p*-dioxins (PCDDs) and dibenzofurans (PCDFs).[1] Frequently, disputes about MSW incinerators focus on a single issue: How many additional cancers are expected to occur from the emitted PCDDs and PCDFs?

Despite the fact that the U.S. Environmental Protection Agency (EPA) states that its estimates of the potency of chemical carcinogens almost certainly exaggerate the risk and that lower-bound risk estimates may actually be zero,[2] risk estimates generated in health risk assessments are often taken as precise and accurate. Furthermore, estimated risks are often compared to an arbitrary value that marks the dividing line between "acceptable" and "unacceptable" risks. In practice, many federal agencies that regulate chemical carcinogens have arbitrarily defined that dividing line as a risk level of 1 in 1 million (10^{-6}),[3,4] despite that fact that the background risk of developing cancer in the U.S. is 0.2 (2×10^{-1}). The concept of a virtually safe dose (VSD) is usually equated with an exposure corresponding to a 10^{-6} risk level of developing cancer. When risk estimates are above that level, the argument that health risks are too high is accepted de facto by many people.

Often overlooked in this debate about acceptable levels of exposure to chemical carcinogens are questions about how to evaluate the carcinogenic potency of these substances. All U.S. regulatory agencies routinely use linear-at-low-dose/nonthreshold models (hereafter nonthreshold models) to estimate cancer risks. In particular, the Centers for Disease Control (CDC), the Food and Drug Administration (FDA), and the EPA use slightly different nonthreshold models, all of which are acceptable under current U.S. science policy guidelines,[5] to estimate the potency of various chemical carcinogens. These models postulate that cancer risk varies directly with dose at all exposure levels and that risk does not reach zero until exposure is zero (i.e., there is no threshold dose below which health risks are not expected to occur).

In contrast, several foreign governments[6,7] use threshold models to estimate the potency of some chemical carcinogens. Threshold models are based on the premise that some chemical carcinogens may not cause adverse human health effects until a certain dose, the threshold dose, is exceeded. Because threshold models equate zero risk with low doses of some chemical carcinogens, potency estimates derived from threshold models are generally lower than estimates obtained from nonthreshold models.

In late 1987, the EPA proposed a revision in its cancer potency estimate for 2,3,7,8-tetrachlorodibenzo-*p*-dioxin (TCDD). The proposal was the first time that a U.S. regulatory agency treated nonthreshold and threshold models equally to estimate the carcinogenic potency of a substance.[2,8] Adoption of the new potency estimate for TCDD would dramatically reduce health risk estimates for MSW incinerators. This chapter discusses the EPAs recent reevaluation of the cancer potency estimate for TCDD, the justifications for using a threshold or a nonthreshold model to estimate the carcinogenic potency of TCDD, and the policy implications associated with choosing between the models. Since the EPAs justifications are based on knowledge of the mechanism of TCDD carcinogenicity, this chapter begins with a review of current knowledge on that subject.

II. STAGES IN CARCINOGENESIS

Experimental and epidemiological evidence exists to suggest that cancer is at least a two-stage process.[9-11] The first step, initiation, is widely believed to involve a heritable alteration in the DNA structure (i.e., a mutation-like event) of a cell.[12] Many chemical carcinogens display genotoxic activity, or cause heritable changes in DNA or other effects indicative of interaction with and alteration of the normal structure and function of cellular DNA.[13] Thus, genotoxic carcinogens, or initiators, enhance the cancer process by altering cellular DNA.[11] Not all mutation-like events lead to cancer, however.

For instance, an average of 1000 oxidized thymidine residues (chemicals that are released from damaged DNA during the process of DNA repair) are produced by each of the human body's 6×10^{13} cells daily.[14] There are, in addition, other types of damaged DNA products, which have so far not been measured. It is possible, therefore, that the sum of thymidine and other kinds of damage is much greater than 1000 per cell per day. Clearly each case of DNA damage, as indicated by the appearance of repair products, does not lead to cancer. Some damage may occur in a stretch of DNA that is not involved in growth processes and cannot contribute to cancer. Most damage is probably repaired, and it is also possible that another event (or series of events) is necessary to push an initiated cell along the path to cancer.

A later step, called promotion, increases tumor incidence without directly interacting with DNA. Although the mechanisms of promotion are not precisely known, enhanced cell proliferation appears to play a crucial role.[15,16] Such proliferation can enhance the probability that a second mutation will occur at an already affected site, transforming an initiated cell into a cancerous one. Promoters, or nongenotoxic chemical carcinogens, can enhance the activity of naturally occurring genotoxic carcinogens or enhance the effects of spontaneous DNA damage. The two-stage theory of carcinogenesis does not exclude the possibility that more than two changes at the cellular level are involved in the progression of a normal to a transformed cell, but it does imply that two sequential processes must occur.[17]

The theory that cancer involves at least two distinct phases, initiation and promotion, was first proposed in the 1940s.[18-21] Experiments done on mouse skin cells showed that a single application of an initiator followed by repeated treatment with a promoter caused cancer, while application of either chemical by itself did not cause cancer. Initiation appeared to cause irreversible changes in affected cells, since cancer could be induced even if treatment with the promoter was delayed for several months.[21] Conversely, the promotional phase was reversible. Tumors did not develop if application of the promoter was terminated short of some critical value, or if the promoter was administered before the initiator.[21] Since the original demonstration of the combined action of initiators and promoters in mouse skin, similar effects have been demonstrated in other organs such as rodent livers.[11]

This hypothesis is consistent with observations that all known initiators are genotoxic. That is, they are capable of causing mutation-like changes in DNA. Once a mutation is replicated, it is essentially irreversible[11,12,22] since it is extremely unlikely that the original mutation and all its replicated progeny can be repaired. A mutation, theoretically, can result from a single interaction between an initiator and DNA. Therefore, some risk is associated with any exposure to an initiator, and risk increases as exposures increase. This correlation is supported by regulatory agencies in both the U.S., where risks from all chemical carcinogens, whether they are initiators or promoters, are estimated using nonthreshold models,[5] and in other countries where risks from genotoxic carcinogens are estimated with nonthreshold models.[23] About half of all chemicals that are carcinogenic in tests conducted by the National Toxicology Program are not mutagenic.[24] Risks from all of those chemicals, however, would be estimated using nonthreshold models in the U.S., whereas the risks from all nonmutagenic chemicals would be assessed with threshold models in European countries.

Thus, chemical carcinogens can be divided into two groups. Chemicals that alter DNA are called initiators, while those that function to increase cell turnover rates are called promoters. One way to reduce the incidence of cancer in human populations is to identify initiating and promoting agents and reduce or eliminate human exposure to them. Reduction or elimination of exposure to initiators would decrease the possibility of mutation, which would limit the number of initiated cells and reduce the likelihood of tumor formation. Although promoters are not carcinogenic or mutagenic in themselves, their absence would slow the rate at which initiated cells progressed to the tumor phase. The following sections review the scientific evidence concerning the carcinogenic effects of TCDD.

III. EVIDENCE THAT TCDD IS NOT GENOTOXIC

Two sources provide evidence that TCDD is not genotoxic: mutagenicity assays and other tests for genotoxicity (e.g., DNA repair assays), and *in vivo* and *in vitro* tumor initiation/promotion assays.

A. ASSAYS WITH BACTERIA

Of the seven studies that evaluated the mutagenicity of TCDD in 13 different strains of *Salmonella typhimurium* bacteria,[25-31] only two tests[25,26] indicated that TCDD is mutagenic. Results obtained by Seiler[26] are probably an artifact, since when McCann et al.[13] repeated the tests, no indication of a positive result was observed. Similarly, other studies using the same strain of bacteria have not confirmed the findings of Hussain et al.,[25] and there is no confirmation of their conclusions that TCDD is mutagenic.

B. ASSAYS WITH MAMMALIAN CELLS

1. Unscheduled DNA Synthesis

Two studies examined the ability of TCDD to damage DNA as measured by unscheduled DNA synthesis (UDS), a sensitive measure of the ability of a chemical to cause DNA damage. Loprieno et al.[32] observed no increase in UDS in human cells treated in culture with TCDD. Furthermore, Althaus et al.[33] observed no increase in UDS in cultured primary rat hepatocytes, which are capable of metabolizing TCDD and which represent target cells for TCDDs carcinogenicity. These results indicate that TCDD does not interact with DNA and that its promotional effects do not involve UDS.

2. Cell Transformation

Abernethy et al.[34] found no increase in the number of colonies with altered growth patterns (i.e., foci) associated with transformation when mouse cells were treated with TCDD alone. When TCDD was administered after a known mutagen, *N*-methyl-*N′*-nitro-*N*-nitrosoguanine (MNNG), however, an increase in cell transformation was observed. These findings support the hypothesis that TCDD is a promoter.

3. Sister Chromatid Exchange

Toth et al.[35] tested TCDD for its ability to induce sister chromatid exchanges in cultured Chinese hamster cells and observed negative results.

C. ASSAYS IN ANIMALS

1. Mutagenicity Assays

TCDD has been tested for mutagenicity in fruit flies and in rats. No positive results were observed. Zimmering et al.[36] injected 50, 250, or 500 ppm TCDD into Drosophila. Although these treatments induced up to 27% lethality, no increase in recessive lethal mutations ensued. Similarly, Khera and Ruddick[37] administered 4, 8, or 12 mg/kg/d TCDD orally to male rats for

7 d. Although signs of toxicity (and some deaths) did occur in treated animals, no increase in dominant lethal mutations was observed.

2. Cytogenetic Studies

Several studies have examined the ability of TCDD to induce chromosomal aberrations in bone marrow cells of rats and mice.[32,38-40] Two of these studies reported an increase in aberration frequency. Green et al.[38] examined the frequency of chromosomal aberrations in bone marrow cells of Osborne Mendel rats that had received 0.25, 0.5, 1.0, 2.0, or 4.0 mg TCDD per kilogram by gavage two times a week for 13 weeks. The frequency of cells with nonspecified aberrations in females at the highest dose level and in males at the two highest dose levels was significantly greater than the frequency at the lowest dose level. However, the fact that no control group was included, that the apparent increase in aberrations occurred only at toxic dose levels (the LD_{50} for TCDD in rats is between 25 and 60 mg/kg),[41] and that the frequency of aberrant cells in the low-dose males was unusually low (0.47%), suggest that these findings may not be reliable.

Loprieno et al.[32] reported an increase in the frequency of chromatid gaps and aberrations in bone marrow cells of mice given a single dose of 10 mg TCDD per kilogram of body weight. No increases in chromosomal aberrations, however, were observed in female rats given TCDD at 0.01, 0.1, or 1.0 mg TCDD per kilogram once a week for 45 weeks. Furthermore, Meyne et al.[40] found no increase in chromosome damage in bone marrow cells of mice given high doses of TCDD (up to 150 mg/kg), nor did Green and Moreland[39] observe an increase in chromosome damage in (1) rats gavaged with 10 mg/kg/d of TCDD for 5 d or with 20 mg/kg as a single dose or (2) in rats given a single intraperitoneal injection of 5, 10, or 15 mg TCDD per kilogram. Overall, there is no convincing evidence that TCDD causes structural chromosome damage in rats or mice.

D. CYTOGENETIC STUDIES IN HUMANS

Several studies have examined the frequency of chromosomes aberrations in humans exposed to TCDD. Two studies that evaluated the effects of occupational exposure to TCDD on chemical workers produced conflicting results. An increase in aberrations was reported by Czeizel and Kiraly,[42] while Hay,[43] who examined 800 workers with chloracne from TCDD exposure, found no significant increase in the frequency of chromosomal aberrations. The exact cause of the increase reported by Czeizel and Kiraly[42] is unclear, because the workers in that study were exposed to other chemicals in addition to TCDD.

Studies that examined individuals exposed to TCDD as a result of an accident in Seveso, Italy, which released a large quantity of TCDD into the surrounding countryside, revealed no significant increase in chromosome aberrations despite high exposures to TCDD.[30,43-53] These data provide no convincing evidence that TCDD causes genetic damage in humans. The few apparently positive studies are not reproducible or were confounded by other exposures.

IV. EVIDENCE THAT TCDD IS A TUMOR PROMOTER

Poland et al.[54] showed that TCDD acts as a skin tumor promoter in hairless mice. A single application of 5 mmol of the initiator MNNG or 0.2 mmol of the initiator dimethylbenz(a)anthracene (DMBA) followed by twice weekly application of TCDD (3.75 to 50 ng per mouse) resulted in an increased incidence of skin tumors compared to animals treated only with an initiator (MNNG or DMBA) or only with TCDD.[54]

TCDD has also been shown to act as a promoter of liver tumors in rats (Table 1). Pitot et al.[55] subjected female Sprague-Dawley rats to a 70% partial hepatectomy (PH) followed by a single 10 mg/kg dose of the potent initiator diethylnitrosamine (DEN) 24 h later. Control groups were not given DEN. Starting 1 week later, initiated and control rats received biweekly subcutaneous injections of 0.14 or 1.4 mg/kg TCDD. Rats subjected to a PH and DEN only developed many

TABLE 1
Promoting Effects of TCDD on Hepatocarcinogenesis by a Single Dose of Diethylnitrosamine (DEN) and Partial Hepatectomy[54]

Treatment	N[a]	No. of foci/cm³ of liver	Mean vol of foci (mm³)	Vol occupied by foci (%)	No. of rats with carcinoma
PH only	?	0	0	0	0
PH + DEN	4	309 ± 98	0.02	0.7	0
PH + TCDD (low dose)	4	34 ± 17	0.05	0.2	0
PH + TCDD (high dose)	5	25 ± 7	0.04	0.1	0
PH + DEN + TCDD (low dose)	5	1068 ± 166	0.08	9.0	0[b]
PH + DEN + TCDD (high dose)	7	871 ± 66	0.49	43.0	5[c]

[a] n is the number of animals per group. The number in the PH-only group was not reported.
[b] Three rats had neoplastic nodules.
[c] One other rat had a neoplastic nodule.

small enzyme-altered foci but no tumors. Rats given a PH plus TCDD alone developed only a few foci and no tumors. No foci were observed in rats subjected to only a PH. TCDD, when administered after DEN, enhanced the number and size of the enzyme-altered foci and led to the development of tumors. These results indicate that TCDD promotes tumor formation in cells previously initiated with DEN.

V. MECHANISMS OF TUMOR INDUCTION BY NONGENOTOXIC CARCINOGENS

Much emphasis has been placed on the initiation stage and its effect on cancer incidence. Nevertheless, the fact that cancer develops spontaneously in control animals and humans not known to be exposed to carcinogens and that the incidence of cancer in cells already initiated can be greatly enhanced by application of a promoter indicate that both stages of cancer (initiation and promotion) affect cancer incidence.

Although the exact mechanism of promotion is not precisely known, enhanced cell proliferation appears to play a crucial role.[15,16] Farber and Cameron[22] noted that:

> ...the essence of promotion is the selective stimulation of initiated cells to grow relative to surrounding cells; i.e., it is the creation of a differential growth environment, leading to amplification of the initiated cell population.

Reitz et al.[56] suggested that tissue damage leading to regenerative hyperplasia (i.e., increase in cell number) influences the activity of nongenotoxic carcinogens. During normal DNA replication, spontaneous genetic alterations occur at a very low rate. Increasing DNA replication, as would occur during hyperplasia, increases the likelihood of replication errors. Some of the resulting aberrations would be similar to the DNA lesions produced by a genotoxic carcinogen. Additionally, increased DNA replication reduces opportunities for DNA repair before lesions are replicated.[12]

Support for this general hypothesis was offered by Berman et al.[57] and Mirsalis et al.[58] Berman et al.[57] found that increasing the rate of normal replication increased the susceptibility of cultured cells to mutagens. Mirsalis et al.[58] examined seven liver carcinogens for their capacity to cause UDS and increased cellular proliferation. Four carcinogens caused no increase in UDS but did cause a significant increase (>100-fold) in cell proliferation.

Other hypotheses have been proposed to explain the action of promoters. A number of structurally unrelated, nongenotoxic chemicals that cause increased proliferation in cellular organelles called peroxisomes have been shown to induce liver cancer in rats.[58] Reddy et al.[60] hypothesized that prolonged proliferation of peroxisomes activated certain peroxisomal enzymes and elevated levels of hydrogen peroxide, which were responsible for the carcinogenic effect. Butterworth,[61] on the other hand, suggested that these agents may also increase cell proliferation in the liver and that it is the cell proliferation rather than the peroxisome replication that actually leads to cancer.

Gap junctions are areas for intercellular communication. Some promoters are suspected to affect those junctions or other aspects of cellular communication. It is possible that interference with such communication interferes with the body's ability to detect and inactivate or otherwise control abnormally growing (cancer) cells.[12]

Thus, mutation is necessary but not sufficient for carcinogenicity. Farber[62,63] argued that cell proliferation is also required to induce carcinogenicity, and chemicals that directly increase cell proliferation rates or cause cell damage with resultant regeneration can function as promoters. Nongenotoxic carcinogens may act by enhancing the activity of naturally occurring genotoxic carcinogens or by enhancing the effects of spontaneous genetic damage. This enhancement may occur by increasing the proportion of initiated cells, by interfering with DNA repair, or by some other mechanism that would increase the expression of induced or spontaneous genetic damage.

A. ROLE OF THE Ah RECEPTOR

The capacity of TCDD to activate a number of enzymes involved in the metabolism of exogenous chemicals is probably related to its carcinogenicity. Of particular interest is its ability to activate cytochrome P-450-dependent enzymes. These enzymes, originally discovered in the liver, bind to membranes within cells[64] and are "known to metabolize almost everything in the Merck Index from small molecules such as ethanol . . . to large molecules such as TCDD, benzo(a)pyrene and synthetic steroids. Virtually all drugs are substrates."[65]

Poland and Glover[66] and Poland et al.[67] investigated the ability of PCDDs to induce aryl hydrocarbon hydroxylase (AHH), one of the cytochrome P-450-dependent enzymes, activity in chick embryos and found a well-defined structure-activity relationship. TCDD most effectively induced AHH activity, while other PCDD congeners lacking chlorines in any one of the 2, 3, 7, and 8 positions or having chlorines in the 1, 4, 6, or 9 positions were less active or totally inactive. Hence, the number and location of chlorines determined binding activity. They also demonstrated that all halogenated dioxins and furans competed with TCDD for protein binding sites, which suggested that all PCDDs and PCDFs bind to the same receptor.

Concurrently, Nebert and Gielen[68] found that different strains of mice varied in their ability to respond to TCDD (i.e., the ability of TCDD to include AHH was genetically controlled). These two findings (i.e, the structure-activity relationship and the genetic segregation of AHH inducibility) led Poland and Glover[66] to postulate that enzyme induction was mediated by a receptor, which is designated the Ah receptor. The dominant allele for responsiveness is indicated by Ahb, while the recessive allele for nonresponsiveness is indicated by Ahd.[69] The receptor, when bound by TCDD or another ligand, is hypothesized to attach to a particular region(s) of DNA and to activate the genes for the cytochrome P-450-dependent enzymes. The unbound receptor is much less active or inactive.

TCDD is the most potent known inducer of AHH. It is only very slowly metabolized by cytochrome P-450-dependent enzymes[41] and, therefore, remains unaltered in the cell for long periods of time. Its resistance to metabolism probably contributes to its potency as an inducer.

Safe[70,71] studied the correlation between the effectiveness of PCDDs and PCDFs as inducers of cytochrome P-450-dependent enzymes and their toxicity in rats. PCDDs and PCDFs that were most active as AHH inducers were also the most toxic, which implied that a correlation between enzyme induction activity, acute toxicity, and carcinogenicity must exist. Established parallels between receptor binding affinity, enzyme induction, acute toxicity, and teratogenicity also support the conclusion that all of those effects are, at least in part, modulated by the Ah receptor. Thus, if the carcinogenic mechanism of TCDD involves a threshold, then the carcinogenic mechanism for all PCDDs and PCDFs also involves a threshold.

The induction of cytochrome P-450-dependent enzymes may explain the interactions between TCDD and some other carcinogens. Berry et al.[72] observed that treatment of CD-1 mice with TCDD before application of the initiator DMBA inhibited the development of skin tumors. The effect of TCDD was greatest when applied 3 to 5 d before DMBA, which caused 94 to 96% inhibition of tumor appearance. TCDD was ineffective, however, when administered only 5 min before initiation with DMBA[72] or when administered concurrently with DMBA.[73] The time required for effective inhibition of tumor development paralleled the time course for induction of the AHH enzyme. This temporal similarity suggest that the inhibitory effect of TCDD is due to its enzyme-inducing ability. If those data are correct, then TCDD-induced enzymes must inactivate the initiating chemical before it can reach its target.

The role of the Ah receptor in carcinogenicity is not clear. However, the proliferative effects of TCDD appear to be mediated by the Ah receptor, and carcinogenicity appears to depend on cell proliferation and terminal differentiation. Although the binding of TCDD to the Ah receptor is linear, the biological response of cells to chemicals that act through receptors is generally curvilinear. That generalization, if it characterizes the biological effects of TCDD, argues against the application of linear-at-low-dose/nonthreshold models for assessing the potency of TCDD.

VI. IMPLICATIONS FOR RISK ASSESSMENT

Assessing the risk associated with exposure to chemical carcinogens, such as TCDD, whose toxicity in humans is not known directly, necessitates the use of information obtained from other species. Because of limitations in the sensitivity (biological and statistical) of experimental procedures, major assumptions must be made to extrapolate experimental results across species (from laboratory animals to humans) and from high-dose exposures (to which laboratory animals are exposed) to low-dose exposures (levels to which humans are likely to be exposed in the environment).

A. INTERSPECIES EXTRAPOLATION

The toxic effects produced by TCDD differ substantially from species to species. In humans, TCDD has been shown to cause skin lesions (chloracne) and induce cytochrome P-450 dependent liver enzymes.[74] Despite the parallels between those documented effects in humans and similar effects observed in other species, there is little information available concerning the susceptibility of humans to the carcinogenic effects of TCDD. If the mechanism of TCDD carcinogenesis involves the Ah receptor, humans may show greater variability in response, and, perhaps, lower susceptibility than the mice and rats tested in long-term animal bioassays because of the variable and generally lower levels of Ah receptor in human cells.[75,76]

TCDD, as many other carcinogens, primarily affects the liver in rats.[77,78] Furthermore, rats have a very high number of spontaneously occurring preneoplastic lesions that can be stimulated to give rise to tumors through the application of various promoting agents,[79-81] and a high background (spontaneous) incidence of liver cancer. Conversely, liver cancer is rare in humans in the Western World. These differences in the occurrence of liver tumors in rats and humans have led many scientists to question the relevance of using results from long-term bioassays done on laboratory animals to estimate humans risks.[82-87]

B. HIGH-TO-LOW DOSE EXTRAPOLATION

Because doses administered in experimental studies are generally substantially higher than environmental exposures, a mathematical model must be applied to the experimental data to determine the dose-response relationship at low doses. Thus, information about cancer incidence observed in laboratory animals exposed to high dose levels is used to estimate the risk to humans exposed to the generally low levels of chemical carcinogens that exist in the environment. U.S. regulatory agencies typically assume that thresholds do not exist for carcinogens, which means that a finite risk is associated with all dose levels and that risk increases linearly as dose increases. These conservative assumptions are designed to guard against underestimating the true risk, and, as a corollary, they tend to overestimate the risk to humans.

The assumption that the dose-response curve for carcinogens is linear at low doses stems from the idea that carcinogens are genotoxic. As demonstrated earlier, however, there is no convincing evidence to confirm that TCDD is genotoxic. Hence, applying a linear-at-low-dose/nonthreshold model such as is typically used by the EPA, CDC, and FDA to assess human risk from exposure to TCDD is inappropriate.[87,88] In other countries, regulatory agencies evaluate genotoxic and nongenotoxic carcinogens differently. Nonthreshold models are used for genotoxic carcinogens, and threshold models are used for nongenotoxic carcinogens.[23]

Another procedure for extrapolation across species and dose is to use a tolerance distribution model of risk, such as the log-probit model. This model assumes that thresholds exist for toxic effects and that those thresholds are distributed log-normally within a population.[89] The major disadvantage to using this procedure is the same as that for all models used in cancer risk assessment: the assumption of a log-normal distribution of thresholds cannot be confirmed experimentally at low dose levels. Even a small deviation from assumed log-normality at the extremes of the distribution can have a large effect on estimates of risk at low dose levels. Uncertainty about the shape of the dose-response curve because of the small number of experimental dose levels in carcinogenicity bioassays also contributes to the uncertainty of risk estimates for low exposure levels.

Thorslund[90] applied the Moolgavkar-Venzon-Knudson (M-V-K) extrapolation model to liver tumor data obtained from rats and estimated the number of human liver cancers that would occur assuming that TCDD acts as a promoter (i.e., that TCDD has a threshold). When two different versions of the M-V-K model were used,[91,92] the estimated risks differed by a factor of 10,000. The higher risk estimate, however, was still about two orders of magnitude *lower* than the risk number derived by EPA using its standard nonthreshold model. The EPA[93] dismissed the use of the M-V-K model because of "concerns about it biological bases and assumptions" and the "lack of use and 'experience' with (the model for) other chemicals".

VII. DETERMINING AN ACCEPTABLE DOSE FOR TCDD

Both threshold and nonthreshold models have been used to estimate the cancer risk from TCDD. Threshold models, which embody the assumption that there is no risk associated with doses below a certain critical level, are used by U.S. regulatory agencies to set allowable exposure levels for all toxic substances *except* carcinogens. The FDA sets levels called "acceptable daily intakes" (ADIs), while the EPA calls allowable doses "reference doses" (RfDs).

The basis for setting an acceptable exposure level is the observation of a no-observed-adverse-effect level (NOAEL) in exposed human or animal populations. If a NOAEL is based on human data, the acceptable daily dose is calculated by dividing the NOAEL dose by a safety factor of 10 to account for variations in sensitivities among individual humans. If it is based on animal data, then the NOAEL dose is divided by a safety factor of 100. The additional safety factor of 10 accounts for variations between animals and humans.[94]

Based on a study of tumor development in rats, a NOAEL of 1 ng of TCDD per kilogram body weight per day (ng/kg/d) has been established for TCDD.[95] Using that NOAEL, an ADI can be calculated as follows:

TABLE 2
Acceptable Daily Intake (ADI) Levels and Virtually Safe Doses for TCDD

Government	NOAELa (pg/kg/d^b^)	Safety factor	ADI (pg/kg/d)
Ontario[6]	1000	100	10
Netherlands[7]	1000	250	4
Denmark[96]	1000	200	5
Germany[98]	1000	100—1000	1—10
			VSD (μg/hg/d)
U.S. EPA[99]	—	—	0.006
CDC[88]	—	—	0.028
FDA[101]	—	—	0.057
U.S. EPA[102]	—	—	0.100

[a]No-observed-adverse-effect-level in laboratory animals.
[b]Picograms (10^{-12} g) per kilogram body weight per day.

$$ADI = NOAEL/Safety\ Factor = 10\ pg/kg/d. \qquad (1)$$

where the NOAEL equals 1.0 ng/kg/d,[95] and the safety factor equals 100.[94]

Canada, Denmark, West Germany, and Holland[6,7,96-98] have all adopted a NOAEL/safety factor approach for determining an acceptable level of exposure to TCDD (Table 2), which incorporates the idea of a threshold for TCDDs cancer-causing activity. The ADIs listed in Table 2 are based on the NOAEL determined by Kociba et al.[95] and safety factors that range from 100 to 1000. The use of higher safety factors yields lower ADIs, and it can be surmised that higher safety factors were chosen because of greater concern for the toxicity of TCDD compared to other chemicals.

Nonthreshold models predict VSDs rather than ADIs or RfDs. Generally, VSDs correspond to an additional lifetime cancer risk of 10^{-6}. Use of slightly different nonthreshold models by the EPA, CDC, and FDA produced three different VSD estimates for TCDD. The EPA[99] estimated the highest VSD for TCDD, 0.006 pg/kg/d TCDD per kilogram body weight per day (pg/kg/d), which is equivalent to a 10^{-6} risk level. The CDC published a range of potencies based on cancer rates observed in different organs of rats and mice exposed to TCDD.[88] The CDC VSD that is most comparable to the EPAs, 0.028 pg/kg/d, is based on the frequency of liver cancer in dioxin-exposed female rats and is about five times higher than the EPA estimate. Conversely, the FDAs VSD, 0.06 pg/kg/d, yields a cancer potency estimate which is about ten times lower than the EPAs estimate. (See Reference 100 for a discussion of the estimates from the three agencies and Scheuplein[101] for an application of the FDA method.)

Table 2 demonstrates that choosing between a nonthreshold model to calculate a VSD and a threshold model to derive an ADI for TCDD has a profound effect on the level of exposure that is considered acceptable. With nonthreshold models, some finite risk is associated with all exposures greater than zero. In the case of TCDD, doses corresponding to a 10^{-6} cancer risk calculated using a threshold model are typically about two or three orders of magnitude greater than the VSDs estimated from nonthreshold models. U.S. regulatory agencies estimate the acceptable dose of TCDD ranges from 0.006 to 0.06 pg/kg/d, while estimates of safe doses obtained from threshold models by other countries cluster at 1.0 to 10.0 pg/kg/d.

VIII. CANCER POTENCY OF TCDD AND SCIENCE POLICY IMPLICATIONS

In late 1987, the EPA circulated a draft report that considered both nonthreshold and threshold models for estimating the carcinogenic potency of TCDD[9,102,103] and recommended 0.1 pg/kg/d as the VSD for TCDD (i.e., the dose which corresponds to an additional lifetime cancer risk of 1 in 1 million). The revised VSD is approximately midway between the VSDs estimated from nonthreshold and threshold models and is about 17 times higher than the EPAs earlier VSD.

Should the revised VSD be adopted, it would dramatically change potential health risk estimates for TCDD. For example, use of the old EPAs potency estimate would probably render all fish in U.S. bodies of water unsafe to eat, assuming about 10^{-6} is the standard for acceptable risk. Although none of the seven fish taken from Maine waterways[104] had TCDD levels that exceeded the level of contamination established by the FDA as acceptable (25 parts per trillion),[105] the five fish with detectable levels of TCDD had concentrations that were 20 to 160 times higher than the 0.07 parts per trillion level that the EPA equates with a 10^{-6} risk.[104] Similarly, using its old potency factor, the EPA estimated that TCDD emissions from all existing and projected MSW incinerators would result in about 10 to 120 additional cancer cases per year.[106] If the revised potency estimate is applied to the risks associated with MSW incinerators, cancer incidence drops to between less than 1 and 17 cases annually.

Consideration of the revised cancer potency estimate for TCDD is a step in the right direction, because it shows that the EPA can and will consider mechanistic information. Finkel[107] discussed possible consequences of the revision. If the revision is adopted, the EPA may receive many requests for reevaluation of potency estimates for other chemical carcinogens, especially in light of the fact that about half of the chemicals tested by the National Toxicology Program were not mutagenic.[24] It seems reasonable that regulated industries may request reconsideration of nongenotoxic compounds.

The EPA draft report elicited controversy,[8] and the EPA responded with a revised draft in June 1988.[93] The revised draft was reviewed by a subpanel of the EPAs Science Advisory Board (SAB) in November 1988, and the SAB Subpanel concluded that there was insufficient reason to change the cancer risk estimate, but that the old estimate was not well justified. The SAB recommended that EPA staff continue working on the development of a new potency estimate.

IX. CONCLUSIONS

The debate about how to model cancer risks from TCDD is a single example of a more general problem. Chemicals can influence the carcinogenic process in two distinct ways: (1) by direct interaction with genetic material (i.e., initiation) and (2) by increasing the growth rate of normal or initiated cells (i.e., promotion). Classification of chemical carcinogens as either initiators or promoters is not universally accepted.

Currently, U.S. science policy does not separate carcinogens into initiators and promoters; all chemical carcinogens are treated as initiators and risks are estimated using nonthreshold models. U.S. regulatory agencies will almost certainly favor the use of the more conservative nonthreshold models until and unless there is a general change in U.S. science policy that governs risk assessments for chemical carcinogens.

Nonthreshold models were developed for estimating cancer risks from genotoxic carcinogens. Despite the fact that TCDD is not genotoxic, U.S. regulatory agencies persisted in using nonthreshold models to estimate the potency of TCDD until 1987, when the EPA proposed a new potency number that appeared to be a compromise between estimates generated with threshold and nonthreshold models. The new risk estimate was not endorsed by the EPAs SAB, and it remains unclear if the Agency will try again to generate a new potency estimate for TCDD.

It is ironic that TCDD represents one of the first serious efforts to limit the use of nonthreshold models for all carcinogens, largely because of its notoriety as the most potent chemical carcinogen ever evaluated by the EPA. Had TCDD been less toxic and less well known, reevaluation of its potency estimate might have attracted little attention. In retrospect, there are many arguments against a reevaluation of the most potent carcinogen. Certainly if the potency of the most toxic chemical carcinogen can be lowered, all potency estimates, especially those for nongenotoxic carcinogens, would be fair game for reevaluation.

Manufacturers, faced with reducing exposures to carcinogens to comply with regulatory limits, would favor the use of threshold models for assessing the risks from nongenotoxic compounds. Similarly, people faced with cleaning up waste sites would welcome the lower risk levels predicted from threshold models. Alternately, people who are convinced of the risk of environmental carcinogens are likely to oppose any relaxation of use of the nonthreshold models. Those opposing views will surely come into conflict again, but for the time being, the U.S. science policy that all carcinogens are the same has been upheld.

REFERENCES

1. **Rappe, C., Andersson, R., Bergqvist, P.-A., Brohede, C., Hansson, M., Kjeller, L.-O., Lindstrom, G., Marklund, S., Nygren, M., Swanson, S. E., Tysklind, M., and Wiberg, K.,** Overview on environmental fate of chlorinated dioxins and dibenzofurans. Sources, levels and isomeric pattern in various matrices, *Chemosphere,* 16(8/9), 1603, 1987.
2. Health Assessment Document for Acetaldehyde, EPA-600/8-86-015a, U.S. Environmental Protection Agency, Washington, D.C., 1987.
3. Environmental Protection Agency, National emissions standards for hazardous air pollutants: benzene, *Fed. Reg.,* 49, 23558, 1979.
4. Food and Drug Administration, Chemical compounds in food producing animals, *Fed. Reg.,* 44, 17070, 1979.
5. Office of Science and Technology Policy, Chemical carcinogens: a review of the science and its associated principles, *Fed. Reg.,* 50, 10372, 1985.
6. Ontario Ministry of the Environment, Scientific Criteria Document for Standard Development No. 4-84: Polychlorinated Dibenzo-*p*-Dioxins (PCDDs) and Polychlorinated Dibenzofurans (PCDFs), Ministry of the Environment, Toronto, Canada, 1985.
7. **van der Heijden, C. A., Knaap, A. G. A. C., Kramers, P. G. N., and van Logtem, M. J.,** Evaluation of the Carcinogenicity and Mutagenicity of 2,3,7,8-Tetrachlorodibenzo-1,4-Dioxin (TCDD): Classification and No-Effect Level, State Institute of National Health, The Hague, 1982.
8. **Shabecoff, P.,** Estimate of risk of dioxin is cut in cancer study, *N. Y. Times,* December 9, p. #A1, 1987.
9. **Cohen, S. M., Greenfield, R. E., and Ellwein, L. B.,** Multistage carcinogenesis in the urinary bladder, *Environ. Health Perspect.,* 49, 202, 1983.
10. **Moolgavkar, S. H., Day, N. E., and Stevens, R. G.,** Two-stage model for carcinogenesis: epidemiology of breast cancer in females, *J. Natl. Cancer Inst.,* 65, 3, 559, 1980.
11. **Slaga, T. J.,** Overview of tumor promotion in animals, *Environ. Health Perspect.,* 50, 3, 1983.
12. **Zeise, L., Wilson, R., and Crouch, E. A. C.,** Dose-response relationships for carcinogens: a review, *Environ. Health Perspect.,* 73, 259, 1987.
13. **McCann, J., Choi, E., Yamasaki, E., and Ames, B. N.,** The detection of carcinogens as mutagens in the Salmonella microsome test: assay of 300 chemicals, *Proc. Natl. Acad. Sci. U.S.A.,* 72, 5135, 1975.
14. **Ames, B. N.,** Measuring Oxidiative Damage in Humans: Relation to Cancer and Aging, paper presented at seminar Detection Methods for DNA-Damaging Agents in Man, Espoo, Finland, September 2 to 4, 1987.
15. **Goldfarb, S., Pugh, T. D., Koen, H., and He, Y.-Z.,** Preneoplastic and neoplastic progression during hepatocarcinogenesis in mice injected with diethylnitrosamine in infancy, *Environ. Health Perspect.,* 50, 149, 1983.
16. **Cohen, S. M.,** Promotion in urinary bladder carcinogenesis, *Environ. Health Perspect.,* 50, 51, 1983.

17. **Greenfield, R. E., Ellwein, L. B., and Cohen, S. M.,** A general probabilistic model of carcinogenesis: analysis of experimental urinary bladder cancer, *Carcinogenesis,* 5(4), 437, 1984.
18. **Rous, P. and Kidd, J.,** Conditional neoplasms and subthreshold neoplastic states: a study of the tar tumors of rabbits, *J. Exp. Med.,* 73, 369, 1941.
19. **Berenblum, I.,** The cocarcinogenic action of croton resin, *Cancer Res.,* 1, 44, 1941.
20. **Mottram, J. C. A.,** A developing factor in experimental blastogenesis, *J. Pathol. Bacteriol.,* 56, 181, 1944.
21. **Berenblum, I. and Shubik, P.,** A new quantitative approach to the study of the stages of chemical carcinogenesis in the mouses' skin, *Br. J. Cancer,* 1, 383, 1949.
22. **Farber, E. and Cameron, R.,** The sequential analysis of cancer development, *Adv. Cancer Res.,* 31, 125, 1980.
23. **Kroes, R.,** Contribution of toxicology toward risk assessment of carcinogens, *Arch. Toxicol.,* 60, 224, 1987.
24. **Tennant, R. W., Margolin, B. H., Shelby, M. D., Zeiger, E., Haseman, J. K., Spalding, J., Caspary, W., Resnick, M., Stasiewicz, S., Anderson, B., and Minor, R.,** Prediction of chemical carcinogenesis in rodents from *in vitro* genetic toxicology assays, *Science,* 236, 933, 1987.
25. **Hussain, S., Ehrenberg, L., Lofroth, G., and Gejvall, T.,** Mutagenic effects of 2,3,7,8-TCDD on bacterial systems, *Ambio,* 1, 32, 1972.
26. **Seiler, J. P.,** A survey on the mutagenicity of various pesticides, *Experientia,* 29, 622, 1973.
27. **Nebert, D. W., Thorgeirsson, S. S., and Felton, J. S.,** Genetic differences in mutagenesis carcinogenesis and drug toxicity, in *In Vitro Metabolic Activation in Mutagenesis Testing,* de Serres, F. J., et al., Eds., Elsevier/North Holland, Amsterdam, 1976.
28. **Geiger, L. E. and Neal, R. A.,** Mutagenicity testing of 2,3,7,8-tetrachlorodibenzo-*p*-dioxin in histidine autotrophs of *Salmonella typhimurium, Toxicol. Appl. Pharmacol.,* 59, 125, 1981.
29. **Gilbert, P. G., Saint-Ruf, F., Poncelet, F., and Mercier, M.,** Genetic effects of chlorinated anilines and azobenzenes on *Salmonella typhimurium, Arch. Environ. Contam. Toxicol.,* 9, 533, 1981.
30. **Wassom, J. S.,** A review of the genetic toxicology of chlorinated dibenzo-*p*-dioxins, *Mutat. Res.,* 47, 141, 1977.
31. **Mortelmans, I., Haworth, S., Speck, W., and Zeiger, E.,** Mutagenicity testing of Agent Orange components and related chemicals, *Toxicol. Appl. Pharmacol.,* 75, 137, 1984.
32. **Loprieno, N. Sbrana, I., Rusciano, R., Lascialfari, D., and Lari, T.,** *In vivo* cytogenetic studies on mice and rats exposed to 2,3,7,8-tetrachlorodibenzo-*p*-dioxin, in *Chlorinated Dioxins and Related Compounds: Impact on the Environment,* Hutzinger, O., Frei, R. W., Merrian, E., and Pocchiari, P., Eds., Pergamon Press, New York, 1981, 419.
33. **Althaus, F. R., Lawrence, S. D., Sattler, G. L., Longfellow, D. G., and Pitot, H. C.,** Chemical quantification of unscheduled DNA synthesis in cultured hepatocytes as an assay for the rapid screening of potential chemical carcinogens, *Cancer Res.,* 42, 3010, 1982.
34. **Abernethy, D. J., Greenlee, W. F., Huband, J. C., and Boreiko, C. J.,** 2,3,7,8-Tetrachlorodibenzo-*p*-dioxin (2,3,7,8-TCDD) promotes the transformation of C3H/10T1/2 cells, *Carcinogenesis,* 6(4), 651, 1985.
35. **Toth, K., Olah, E., Somfai-Relle, S., and Sugar, J.,** Effect of herbicide 2,4,5-TCPE containing dioxin on mutation and induction of sister chromatid exchange, *Carcinogenesis,* 5, 1725, 1984.
36. **Zimmering, S., Mason, J. M., Valencia, R., and Woodruff, R. C.,** Chemical mutagenesis testing in Drosophila. II. Results of 20 coded compounds tested for the National Toxicology Program, *Environ. Mutagen.,* 7, 87, 1985.
37. **Khera, K. S. and Ruddick, J. A.,** Polychlorodibenzo-*p*-dioxin: perinatal effects and the dominant lethal test in Wistar rats, in *Chlorodioxins—Origin and Fate,* Blair, E. H., Ed., American Chemical Society, Washington, D.C., 1973.
38. **Green, S., Moreland, F., and Shu, C.,** Cytogenetic effect of 2,3,7,8-2,3,7,8-TCDD on rat bone marrow cells, *FDA By-Lines,* 7, 292, 1977.
39. **Green, S. and Moreland, F.,** Cytogenetic study of several dioxins in the rat, *Toxicol. Appl. Pharmacol.,* 33, 161, 1975.
40. **Meyne, J., Allison, D. C., Bose, K., Jordan, S. W., Ridolpho, P. F., and Smith, J.,** Hepatotoxic doses of dioxin do not damage mouse bone marrow chromosomes, *Mutat. Res.,* 157, 63, 1985.
41. **Neal, R. A.,** Biological effects of 2,3,7,8-tetrachlorodibenzo-*p*-dioxin in experimental animals, in *Public Health Risks of the Dioxins,* Lowrance, W. W., Ed., William Kaufman, Los Altos, CA, 1984, 15.
42. **Czeizel, E. and Kiraly, J.,** Chromosome examinations in workers producing klorinol and buvinol, in *The Development of a Pesticide as a Complex Scientific Task,* Banki, L., Ed., Medicina, Budapest, 1976, 139.
43. **Hay, A.,** Identifying carcinogens, *Nature,* 269, 468, 1977.
44. **Hay, A.,** Toxicology of dioxins, in *The Chemical Scythe: Lessons of 2,4,5-T and Dioxin,* Plenum Press, New York, 1982, 41.
45. **Reggiani, G.,** Estimation of the 2,3,7,8-TCDD toxic potential in the light of the Seveso accident, *Arch. Toxicol.,* Suppl. 2, 291, 1979.
46. **Reggiani, G.,** Acute human exposure to 2,3,7,8-TCDD in Seveso, Italy, *J. Toxicol. Environ. Health,* 6, 27, 1980.
47. **Pocchiari, F., Silano, V., and Zampieri, V.,** Human health effects from accidental release of tetrachlorodibenzo-*p*-dioxin (2,3,7,8-TCDD) at Seveso, Italy, *Ann. N.Y. Acad. Sci.,* 320, 311, 1979.

48. **Tenchini, M. L., Crimaudo, C., Pacchetti, G., Mottura, A., Agosti, S., and DeCarli, L.,** A comparative cytogenetic study on cases of induced abortions in 2,3,7,8-TCDD-exposed and non-exposed women, *Environ. Mutagen.*, 5, 73, 1983.
49. **Tenchini, M. L., Crimaudo, C., Simoni, G., DeCarli, L., Giorgi, R., and Nuzzo, F.,** Approaches to the evaluation of genetic damages after a major hazard in chemical industry: preliminary cytogenetic findings in 2,3,7,8-TCDD-exposed subjects after the Seveso accident, in *Genetic Damage in Man Caused by Environmental Agents*, Berg, K., Ed., Academic Press, New York, 1979, 301.
50. **Homberger, E., Reggiani, G., Sambeth, J., and Wipf, H. K.,** The Seveso accident: its nature, extent, and consequences, *Ann. Occup. Hyg.*, 22, 327, 1979.
51. **Mottura, A., Zei, G., Nuzzo, F., Crimaudo, C., Giorgi, R., Paggini, P., Mocarelli, P., Fraccaro, M., Nicoletti, B., and DeCarli, L.,** Evaluation of results of chromosome analysis on lymphocytes of 2,3,7,8-TCDD exposed subjects after the Seveso accident, *Mutat. Res.*, 85, 238, 1981.
52. **DiLernia, R., Crimaudo, C., and Pacchetti, G.,** The study of X-rays and 2,3,7,8-TCDD effects on satellite associations may suggest a single model for application of environmental mutagenesis, *Hum. Genet.*, 61(1), 42, 1982.
53. **Mocarelli, P., Pocchiari, F., and Nelson, N.,** Preliminary report: 2,3,7,8-tetrachlorodibenzo-*p*-dioxin exposure to humans — Seveso, Italy, *Morbid. Mortal. Wkly. Rep.*, 37, 733, 1988.
54. **Poland, A., Palen, D., and Glover, E.,** Tumor promotion by 2,3,7,8-TCDD in skin of HRS/J mice, *Nature*, 300, 271, 1982.
55. **Pitot, H. C., Goldsworthy, T., Campbell, H. A., and Poland, A.,** Quantitative evaluation of the promotion by 2,3,7,8-tetrachlorodibenzo-*p*-dioxin of hepatocarcinogenesis from diethylnitrosamine, *Cancer Res.*, 40, 3616, 1980.
56. **Reitz, R. H., Quast, J. F., Schumann, A. M., Watanabe, P. G., and Gehring, P. J.,** Non-linear pharmacokinetic parameters need to be considered in high dose/low dose extrapolation, *Arch. Toxicol.*, suppl. 3, 79, 1980.
57. **Berman, J. J., Tong, C., and Williams, G. M.,** Enhancement of mutagenesis during cell replication of cultured liver epithelial cells, *Cancer Lett.*, 4, 277, 1978.
58. **Mirsalis, J. C., Tyson, C. K., Loh, E. N., Steinmetz, K. L., Bakke, J. P., Hamilton, C. M., Spak, D. K., and Spalding, J. W.,** Induction of hepatic cell proliferation and unscheduled DNA synthesis in mouse hepatocytes following *in vivo* treatment, *Carcinogenesis*, 6, 1521, 1985.
59. **Reddy, J. K. and Lalwani, N. D.,** Carcinogenesis by hepatic peroxisome proliferators: evaluation of the risk of hypolipidemic drugs and industrial plasticizers to humans, *Crit. Rev. Toxicol.*, 12, 1, 1983.
60. **Reddy, J. K., Azarnoff, D. L., and Hignite, C. E.,** Hypolipidemic hepatic peroxisome proliferators form a novel class of chemical carcinogens, *Nature*, 283, 297, 1980.
61. **Butterworth, B.,** The genetic toxicology of di(2-ethylhexyl)phthalate (DEHP), *CIIT Activities*, 4, 1, 1984.
62. **Farber, E.,** The multistep nature of cancer development, *Cancer Res.*, 44, 4217, 1984.
63. **Farber, E.,** Cellular biochemistry of the stepwise development of cancer with chemicals, *Cancer Res.*, 44, 1984.
64. **Whitlock, J. P., Jr.,** the regulation of cytochrome P-450 gene expression, *Ann. Rev. Pharmacol. Toxicol.*, 26, 333, 1986.
65. **Ebert, D. W., Negiski, M., and Eisen, H. J.,** Genetic differences in enzymes which metabolize drugs, chemical carcinogens, and other environmental pollutants, in *Human and Environmental Risks of Chlorinated Dioxins and Related Compounds*, Tucker, R. E., Young, A. L.,and Grey, A. P., Eds., Plenum Press, New York, 1983, 441.
66. **Poland, A. and Glover, E.,** Chlorinated dibenzo-*p*-dioxins: potent inducers of delta-aminolevulinic acid synthetase and aryl hydrocarbon hydroxylase II. A study of the structure-activity relationship, *Mol. Pharmacol.*, 9, 736, 1973.
67. **Poland, A. P., Glover, E., and Kende, A. S.,** Stereospecific, high affinity binding of 2,3,7,8-tetrachlorodibenzo-*p*-dioxin by hepatic cytosol. Evidence that the binding species is the receptor for the induction of aryl hydrocarbon hydroxylase, *J. Biol. Chem.*, 251, 4936, 1976.
68. **Nebert, D. W. and Gielen, J. E.,** Genetic regulation of aryl hydrocarbon hydroxylase induction in the mouse, *Fed. Proc.*, 31, 1315, 1972.
69. **Eisen, H. J., Hannah, R. R., Legraverend, C., Okey, A. B., and Nebert, D. W.,** The Ah receptor: controlling factor in the induction of drug-metabolizing enzymes by certain chemical carcinogens and other environmental pollutants, in *Biochemical Actions of Hormones*, Litwack, G., Ed., Academic Press, New York, 1983, 227.
70. **Safe, S.,** Determination of 2,3,7,8-TCDD toxic equivalent factors (TEFs): support for the use of the *in-vitro* AHH induction assay, *Chemosphere*, 16, 791, 1987.
71. **Safe, S. H.,** Comparative toxicology and mechanism of action of polychlorinated dibenzo-*p*-dioxins and dibenzofurans, *Ann. Rev. Pharmacol. Toxicol.*, 26, 371, 1986.
72. **Berry, D. L., Slaga, T. J., DiGiovanni, J., and Juchau, M. R.,** Studies with chlorinated dibenzo-*p*-dioxins, polybrominated biphenyls, and polychlorinated biphenyls in a two-stage system of mouse skin tumorigenesis: potent anticarcinogenic effects, *Ann. N.Y. Acad. Sci.*, 320, 405, 1979.

73. **DiGiovanni, J., Viaje, A., Berry, D. L., Slaga, T., and Juchau, M. R.,** Tumor initiating ability of 2,3,7,8-tetrachlorodibenzo-*p*-dioxin (2,3,7,8-TCDD) and Arochlor 1254 in the two stage system of mouse skin carcinogenesis, *Bull. Environ. Contam. Toxicol.*, 18(5), 552, 1977.
74. **Poland, A. and Knutson, J. C.,** 2,3,7,8-Tetrachloro-dibenzo-*p*-dioxin and related halogenated aromatic hydrocarbons: examination of the mechanism of toxicity, *Ann. Rev. Pharmacol. Toxicol.*, 22, 517, 1982.
75. **Cook, J. C.,** personal communication, Chemical Industry Institute of Toxicology, Research Triangle Park, NC, 1986.
76. **Okey, A. B., Roberts, E. A., Golas, C. L., and Manchester, D. K.,** Ah Receptor for 2,3,7,8-Tetrachlorodibenzo-*p*-Dioxin (2,3,7,8-TCDD): Identification and Characterization in Human Tissue, presented at 5th Int. Symp. on Chlorinated Dioxins and Related Compounds, Bayreuth, West Germany, September 16 to 19, 1985.
77. **Greisemer, R. A. and Cueto, C.,** Toward a Classification Scheme for Degrees of Experimental Evidence for the Carcinogenicity of Chemicals for Animals, IARC Sci. Publ., No. 27, International Agency for Research on Cancer, Lyon, 1980, 259.
78. **Haseman, J. K., Crawford, D. D., Huff, J. E., Boorman, G. A., and McConnell, E. E.,** Results from 86 two-year carcinogenicity studies conducted by the National Toxicology Program, *J. Toxicol. Environ. Health*, 14, 621, 1984.
79. **Ogawa, K., Onoe, T., and Takeuchi, M.,** Spontaneous occurrence of gama-glutamyl transpeptidase-positive hepatocytic foci in 105-week-old Wistar and 72-week old Fischer 344 male rats, *J. Natl. Cancer Inst.*, 67, 407, 1981.
80. **Schulte-Hermann, R., Timmermann-Trosiener, I., and Schuppler, J.,** Promotion of spontaneous preneoplastic cells in rat liver as a possible explanation of tumor production by nonmutagenic compounds, *Cancer Res.*, 43, 839, 1983.
81. **Ward, J. M.,** Increased susceptibility of livers of aged F344/NCr rats to the effects of phenobarbital on the incidence, morphology, and histochemistry of hepatocellular foci and neoplasms, *J. Natl. Cancer Inst.*, 71, 815, 1983.
82. U.S. Environmental Protection Agency, Proposed guidelines for carcinogen risk assessment, *Fed. Reg.*, 49, 46294, 1984.
83. **Butler, W. H. and Newberne, P. M.,** Eds., *Mouse Hepatic Neoplasia*, Elsevier, Amsterdam, 1975.
84. Nutrition Foundation, The Relevance of Mouse Liver Hepatoma to Human Carcinogenic Risk. A report of the International Expert Advisory Committee to the Nutrition Foundation, Washington, D.C., September 1983.
85. Occupational Safety and Health Administration, Identification classification and regulation of potential occupational carcinogens, *Fed. Reg.*, 45, 5001, 1980.
86. **Tomatis, L., Partensky, C., and Montesano, R.,** The predictive value of mouse liver tumor induction in carcinogenicity testing: a literature survey, *Int. J. Cancer*, 12, 1, 1973.
87. *Report of the Dioxin Update Committee*, Office of Pesticides and Toxic Substances, U.S. Environmental Protection Agency, Washington, D.C., August 28, 1986.
88. **Kimbrough, R. D., Falk, H., Stehr, P., and Fries, G.,** Health implications of 2,3,7,8-tetrachlorodibenzo-*p*-dioxin (2,3,7,8-TCDD) contamination of residential soil, in *Public Health Risks of the Dioxins*, Lowrence, W. W., Ed., William Kaufmann, Los Altos, CA, 1984.
89. **Brown, C. C.,** High to low-dose extrapolation in animals, in *Assessment and Management of Chemical Risks*, ACS Symp. Ser. No. 239, Rodricks, J. V. and Tardiff, R. G., Eds., American Chemical Society, Washington, D.C., 1984, 57.
90. **Thorslund, T. W.,** *Quantitative Dose-Response Model for the Tumor Promoting Activity of TCDD*, Carcinogen Assessment Group, Washington, D.C., 1987.
91. **Moolgavkar, S. H. and Venzon, D. J.,** Two-event models for carcinogenesis: incidence curves for childhood and adult tumors, *Math. Biosci.*, 47, 55, 1979.
92. **Moolgavkar, S. H. and Knudson, A. G., Jr.,** Mutation and cancer: a model for human carcinogenicity, *J. Natl. Cancer Inst.*, 66, 1037, 1981.
93. *A Cancer Risk-Specific Dose Estimate for 2,3,7,8-TCDD External Review Draft*, EPA-600/6-88/007Aa, Office of Research and Development, U.S. Environmental Protection Agency, U.S. Government Printing Office, Washington, D.C., 1988.
94. National Research Council, Safe Drinking Water Committee, *Drinking Water and Health*, Vol. 1, National Academy of Sciences, Washington, D.C., 1977.
95. **Kociba, R. J., Keyes, D. G., Beyer, J. E., Carreson, R. M., Wade, C. E., Dittenber, D. A., Kalmins, P. O., Franson, L. F., Park, P. N., Barnard, S. D., Hummel, P. A., and Humiston, C. G.,** Results of a two-year chronic toxicity and oncogenicity study of 2,3,7,8-TCDD in rats, *Toxicol. Appl. Pharmacol.*, 46, 279, 1978.
96. Formation and Emission of Dioxins Especially in Connection with Waste Incineration, Miljostyrelsen, Strandgrade 29, 1401, Copenhagen, 1984.
97. **Cook, B. T. and Page, N. P.,** Comparisons of Carcinogenic Risk Assessments of 2,3,7,8-Tetrachlorodibenzo-*p*-Dioxin (2,3,7,8-TCDD), prepared by the Dynamac Corporation for the Carcinogen Assessment Group, Office of Research and Development, Environmental Protection Agency, Washington, D.C., 1986.

98. Federal Environmental Agency and Federal Health Office of West Germany, *Report on Dioxins — Update to November 1984*.
99. Health Assessment Document for Chlorodibenzo-*p*-Dioxins, Office of Health and Environmental Assessment of the Office of Air Quality Planning and Standards, U.S. Environmental Protection Agency, Washington, D.C., 1985.
100. Ambient Water Quality Criteria for 2,3,7,8-Tetrachlorodibenzo-*p*-Dioxin, Office of Water Regulation and Standards, U.S. Environmental Protection Agency, Washington, D.C., 1984.
101. **Scheuplein, R.,** Proposed Food and Drug Administration approach to tolerance-setting for dioxins in foods, in *Public Health Risks of the Dioxins,* Lowrance, W. W., Ed., William Kaufmann, Los Altos, CA, 1984.
102. A Cancer Risk-Specific Dose Estimate for 2,3,7,8-TCDD (Draft Report), Office of Research and Development, U.S. Environmental Protection Agency, Washington, D.C., 1987.
103. **Gough, M.,** Science policy choices and the estimation of cancer risk associated with exposure to TCDD, *Risk Anal.,* 8, 337, 1988.
104. National Dioxin Study: Report to Congress, EPA-530/SW-87-025, Office of Solid Waste and Emergency Response, U.S. Environmental Protection Agency, Washington, D.C., 1987.
105. National Dioxin Study: Tiers 3,4,5, and 7, EPA-440/4-87-003, Office of Water Regulation and Standards, Monitoring and Data Support Division, U.S. Environmental Protection Agency, Washington, D.C., 1987.
106. Municipal Waste Combustion Study: Report to Congress, EPA-530/SW-87-021a, Office of Solid Waste and Emergency Response, U.S. Environmental Protection Agency, Washington, D.C., 1987.
107. **Finkel, A.,** Dioxin: are we safer now than before?, *Risk Anal.,* 8, 168, 1988.

Chapter 8

THE ACCUMULATION OF CHLORINATED DIBENZO-*p*-DIOXINS AND DIBENZOFURANS IN BEEF AND MILK

Stephen T. Washburn

TABLE OF CONTENTS

I. INTRODUCTION

The problem of municipal solid waste (MSW) disposal has reached crisis proportions in many parts of the world. The factors contributing to our MSW disposal problem are numerous, including dwindling landfill capacity, rising landfill costs, and the scarcity of effective recycling and waste minimization programs. In an attempt to reduce the amount of waste requiring land disposal, many communities have begun to turn to MSW incineration as a possible solution. At the same time, the potential human health risks posed by MSW incinerators have fueled widespread public opposition to the construction and operation of these facilities.

Many of the concerns regarding MSW incineration have focused on emissions of a family of structurally similar compounds known as polychlorodibenzo-*p*-dioxins (PCDDs) and polychlorodibenzofurans (PCDFs). A variety of PCDDs and PCDFs can be produced inadvertently during the combustion of MSW, and many forms are suspected to be highly toxic to both animals and humans.

Chronic exposure to certain forms of PCDDs and PCDFs has been linked to a variety of adverse human health effects, including birth defects and liver damage[1,2] and has been shown to produce cancer in laboratory animals.[3] The most potent of the PCDDs and PCDFs is widely believed to be 2,3,7,8-tetrachlorodibenzo-*p*-dioxin (2,3,7,8-TCDD), which is one of the most carcinogenic compounds to be identified in laboratory experiments. Some other PCDDs and PCDFs are also believed to be potentially carcinogenic.[4]

The hazards of PCDDs and PCDFs are compounded by their environmental persistence and their tendency to accumulate in living organisms. Although PCDDs and PCDFs can photodegrade under some conditions,[5] most natural degradation mechanisms are very slow. PCDDs and PCDFs have been detected in soil,[6] surface water sediments,[7] terrestrial animals,[8] aquatic organisms,[9,10] and plants.[11,12]

Human exposure to PCDDs and PCDFs emitted from MSW incinerators may occur through a variety of pathways. Although inhalation is the most direct exposure pathway, PCDDs and PCDFs may settle onto soil, plants, and surface water, resulting in the potential for more complex, indirect pathways of human exposure. For example, vegetables and grains may become contaminated by the direct deposition of PCDDs and PCDFs onto leaves or other exposed surfaces, and possibly also through the uptake and translocation of PCDDs and PCDFs through the root system.[12] Cows that consume contaminated grain, vegetation, or soil may in turn exhibit elevated levels of PCDDs and PCDFs, especially in their fat and milk. As a result, human ingestion of milk and beef can represent a significant pathway of exposure to PCDDs and PCDFs emitted from MSW incinerators.

Studies of the risks posed by MSW incinerator emissions often conclude that exposures to PCDDs and PCDFs via the food chain are likely to be more significant than exposures via direct inhalation.[13,14] For this reason, predictions of PCDD and PCDF concentrations in plants, milk, beef, and other sources of food are often necessary to conduct a full evaluation of the public health risks associated with an MSW incinerator. This chapter presents a general approach for estimating the possible extent of bioaccumulation of PCDDs and PCDFs in cow milk and beef. Although this approach was developed specifically for predicting PCDD/PCDF levels in milk and beef, it can be extended to other persistent organic chemicals, such as hexachlorobenzene, and to other components of the human diet, including pork, poultry, and lamb.

II. THE SINGLE COMPARTMENT ELIMINATION MODEL

Due to their lipophilic nature, PCDDs and PCDFs tend to bioaccumulate in the fatty tissues of exposed animals.[15-18] When the magnitude of exposure is essentially constant, concentrations in fat are expected to increase until they reach a relatively stable plateau known as the steady-state concentration. In dairy cows, PCDDs and PCDFs accumulate in both adipose tissue (which constitutes at least 10% of most types of beef) and milk fat. The exposure period necessary to reach steady-state concentrations in adipose tissue is believed to be much longer than the time required to reach such a level in milk.[19]

Ideally, steady-state measurements from controlled, long-term experiments should be used to determine the relationship between the level of exposure to PCDDs and PCDFs and the concentration of PCDDs and PCDFs in beef and milk. The exposure periods examined in experiments conducted to date, however, are too short to permit the direct measurement of steady-state PCDD and PCDF concentrations in either milk or beef. For example, the only measurement of the bioaccumulation of 2,3,7,8-TCDD in the body fat of nonlactating cattle involved an exposure of 28 d.[20] Concentrations approaching steady-state levels in adipose tissue are not likely to be reached for at least 6 months.[19,20]

Given the lack of adequate long-term studies, researchers have applied a single compartment elimination model to estimate steady-state PCDD and PCDF concentrations in animal fat and tissue.[20,21] The use of this particular model is based in part on studies conducted by Piper et al.,[22] Fries and Morrow,[17] and Rose et al.,[18] which indicate that the model can adequately describe the kinetics of 2,3,7,8-TCDD elimination in rats.

The single compartment model is based on a relatively simple concept: the rate of accumulation within a compartment is equal to the rate of uptake minus the rate of elimination. If the compartment is defined as an animal, or a part of an animal, this concept can be expressed mathematically as follows:

$$dC_a/dt = k_1C_f - k_2C_a \tag{1}$$

where:

C_a = concentration of contaminant in the animal at time t
C_f = concentration of contaminant in food
k_1 = rate constant for uptake
k_2 = rate constant for elimination
t = time

Equation 1 can be solved in terms of the contaminant concentration in the animal (C_a) at time t:

$$C_a = (k_1/k_2)\, C_f\, [1 - \exp(-k_2t)] + C_ao\, [\exp(-k_2t)] \tag{2}$$

where C_ao = initial concentration of contaminant in the animal. At steady-state, the contaminant concentration in the animal is constant over time:

$$dC_a/dt = 0 = k_1C_f - k_2C_ass \tag{3}$$

where C_ass = steady-state concentration of contaminant in the animal. Therefore:

$$C_f = k_2C_ass/k_1 \tag{4}$$

By substituting Equation 4 into Equation 2, the concentration at any time (C_a) can be related to the steady-state value (C_ass) as follows:

$$C_a = C_ass\, [1 - \exp(-k_2t)] + C_ao\, [\exp(-k_2t)] \tag{5}$$

Rearranging for C_ass, Equation 5 becomes:

$$C_a ss = \frac{C_a - C_a O\,[\exp(-k_2 t)]}{[1 - \exp(-k_2 t)]} \qquad (6)$$

Equation 6 allows for the calculation of the steady-state concentration of a contaminant ($C_a ss$) given the period of exposure (t), the level present at time t (C_a), the initial concentration ($C_a o$), and the elimination rate constant (k_2).

The elimination rate constant can be estimated using dissipation studies that identify the chemical half-life in the compartment. Assuming that the dissipation is a first-order process, the elimination rate constant can be calculated based on the following equation:

$$C/C_o = \exp(-k_2\, t_{1/2}) \qquad (7)$$

where:

C/C_o = chemical concentration in milk at time t divided by the initial concentration
k_2 = rate constant for elimination
$t_{1/2}$ = elimination half-life

Since $t_{1/2}$ is defined as the time at which the initial concentration has been reduced by half, $C/C_o = 0.5$. Rearranging and solving for k_2:

$$k_2 = -\ln(0.5)/t_{1/2} = 0.693/t_{1/2} \qquad (8)$$

Using a k_2 value estimated from Equation 8, Equation 6 can serve as the basis for estimating the steady-state concentration PCDDs and PCDFs in meat and milk.

The single compartment model is likely to provide only a relatively rough estimate of the extent to which a chemical may accumulate in an animal. In reality, the processes that determine chemical accumulation in animal tissues are highly complex and influenced by a variety of factors that are poorly understood or poorly characterized. Nonetheless, the single compartment model does offer a reasonable approach for estimating steady-state concentrations in animal tissues when only limited experimental data are available.

III. BIOCONCENTRATION AND BIOTRANSFER FACTORS

Bioconcentration factors (BCFs) represent a measure of the extent to which a chemical accumulates within an exposed animal. The BCF for a chemical in animal tissue is defined as follows:

$$BCF = C_a ss/C_f \qquad (9)$$

where:

BCF = bioconcentration factor [(mg/kg)/(mg/kg)]
$C_a ss$ = steady-state concentration in animal tissue (mg/kg)
C_f = concentration in feed (mg/kg)

For example, if a cow were administered feed with a PCDD/PCDF concentration of 5 ppb, and the resulting steady-state concentration of PCDDs and PCDFs in adipose tissue was 50 ppb, the BCF would be equal to 10.

A similar parameter, the biotransfer factor (BTF), can also be used to evaluate the extent of chemical accumulation within an animal.[23] The BTF relates the concentration of a chemical in the animal to total chemical dose:

$$\text{BTF} = C_a\text{ss}/D \tag{10}$$

where:

BTF	=	biotransfer factor [(mg/kg)/(mg/d)]
$C_a$ss	=	steady-state concentration in animal tissue (mg/kg)
D	=	total dose of chemical (mg/d)

The BTF value can be more useful than the BCF value for calculating contaminant concentrations in livestock when exposures occur through a variety of pathways (e.g., consumption of feed, ingestion of water, and inhalation). For example, consider a case in which a cow is exposed to 10 mg/d PCDDs and PCDFs via ingestion of feed and 5 mg/d through consumption of water. If the BTF value is known to be 3 (mg/kg)/(mg/d), the resulting steady-state concentration of PCDDs and PCDFs in the animal tissue is estimated to be 45 mg/kg.

The chemical and physical properties of a substance influence the magnitude of its BCF and BTF values. Since there is considerable variability among some properties of the different forms of PCDDs and PCDFs, their BCF and BTF values might be expected to vary as well. For example, the BCF value for PCDDs with four chlorine atoms (tetrachlorodibenzo-*p*-dioxins) may be very different from that for PCDDs with eight chlorine atoms (octachlorodibenzo-*p*-dioxins). Furthermore, BCF and BTF values may depend on the position as well as the number of chlorine atoms on the PCDD or PCDF molecule. A detailed assessment of the accumulation of PCDDs and PCDFs in milk and beef should consider the possibility of such differences in BCF and BTF values.

IV. THE BIOACCUMULATION OF PCDDs AND PCDFs IN MILK

The bioaccumulation of PCDDs and PCDFs in milk has been examined by Jensen and Hummel[24] and Firestone et al.[21] Jensen and Hummel[24] examined only 2,3,7,8-TCDD; Firestone investigated the behavior of hexa-, hepta-, and octachlorinated dibenzo-*p*-dioxins (HxCDD, HpCDD, and OCDD, respectively).

A. THE JENSEN AND HUMMEL (1982) STUDY

Jensen and Hummel[24] administered 5 parts per trillion (ppt) 2,3,7,8-TCDD in feed to three lactating cows for 14 consecutive days. The 2,3,7,8-TCDD concentration in the feed was then increased to 15 ppt for the next 14 d. Similarly, increases to 50, 150, and 500 ppt were made at the end of consecutive 14-d periods. The cows remained on feed containing 500 ppt 2,3,7,8-TCDD for an additional 21 d.

Milk was taken from the cows periodically during the feeding experiment and analyzed for 2,3,7,8-TCDD. The highest 2,3,7,8-TCDD concentration measured in milk from any cow was 89 ppt following 16 d of exposure to feed containing 500 ppt 2,3,7,8-TCDD. A review of the data, however, indicates that steady-state concentrations of 2,3,7,8-TCDD had not been reached in milk in at least two of the three cows when exposure was terminated.

The decline in 2,3,7,8-TCDD concentration in milk from two of the cows was measured following the completion of the experiment. Using these data, Jensen and Hummel[24] calculated an elimination half-life of 41 d for 2,3,7,8-TCDD, corresponding to an elimination rate constant of 0.017/d. This elimination rate constant can be used to estimate a steady-state concentration of 2,3,7,8-TCDD in milk produced by each cow by applying the single compartment model. The calculations are summarized in Table 1.

TABLE 1
Calculation of Steady-State 2,3,7,8-TCDD Concentrations in Milk[24]

Cow	C_m [a]	t [b]	C_mo [c]	k_2 [d]	C_mss [e]
A	79	21	16	0.017	226
B	89	16	19	0.017	313
C	68	21	14	0.017	194

[a] Observed 2,3,7,8-TCDD concentration in milk at time t (ppt).
[b] Time (days).
[c] 2,3,7,8-TCDD concentration in milk at the start of the 21-d feeding experiment (ppt).
[d] Rate constant for elimination (1/d).
[e] Calculated steady-state 2,3,7,8-TCDD concentration in milk (ppt).

TABLE 2
BCF and BTF Values for 2,3,7,8-TCDD in Milk

Cow	C_mss [a]	C_f [b]	BCF [c] milk	BTF [d] milk
A	226	500	0.45	0.028
B	313	500	0.63	0.039
C	194	500	0.39	0.024

[a] Calculated steady-state 2,3,7,8-TCDD concentration in milk (ppt).
[b] Concentration in feed (ppt).
[c] Bioconcentration factor [(mg/kg)/(mg/kg)].
[d] Biotransfer factor [(mg/kg)/(mg/d)].

The BCF for 2,3,7,8-TCDD in milk from each cow can be calculated by dividing the steady-state concentration by the concentration in the feed. The BTF value can be calculated by dividing the BCF by the feed rate, 16 kg/d, since PCDDs and PCDFs were administered only in the feed. The BCFs and BTFs for 2,3,7,8-TCDD in cow milk are presented in Table 2.

The BTF for 2,3,7,8-TCDD in milk can be estimated as the average of the BTF values calculated for the individual cows, or about 0.03 [(mg/kg)/(mg/d)]. The average BCF value is approximately 0.5.

It should be noted that there is considerable uncertainty associated with the use of the Jensen and Hummel[24] data to estimate 2,3,7,8-TCDD concentration in cow milk. The data collected in that study indicate that the pattern of 2,3,7,8-TCDD uptake varied among the cows tested, and prediction of steady state using simple mathematical models may not be accurate.

B. THE FIRESTONE ET AL. (1979) STUDY

The BTF values for the more highly chlorinated PCDD congeners can be derived from a study conducted by Firestone et al.[21] In this experiment, three dairy cows were administered feed containing technical grade pentachlorophenol (tPCP) contaminated with detectable levels of several chlorinated dioxins. The concentration of PCDDs in the feed can be calculated as follows:

$$C_f = (D_p)\,(C_p)\,(W)/(10^6)\,(F) \qquad (11)$$

where:

C_f = PCDD concentration in feed (ppm)
D_p = dose of tPCP to the cow [10 mg/(kg-d)]

TABLE 3
Concentration of PCDDs in Feed in the Firestone et al. (1979) Experiment[21]

Congener[a]	Conc in tPCP (ppm)	Conc in feed (ppb)
1,2,4,6,8,9-HxCDD	2.0	0.75
1,2,3,6,8,9-HxCDD	7.8	2.9
1,2,3,6,7,8-HxCDD	10	3.7
1,2,3,4,6,7,9-HpCDD	75	28
1,2,3,4,6,7,8-HpCDD	205	77
OCDD	690	260

[a] The numerical prefix indicates the location of the chlorine atoms on the PCDD molecule. Those congeners with chlorines in the 2, 3, 7, and 8 positions are known as "laterally substituted".

TABLE 4
Calculation of Steady-State PCDD Concentrations in Milk Fat[21]

Congener	C_{fat}[a]	t[b]	k_2[c]	C_{fat}^{SS}[d]
1,2,3,6,7,8-HxCDD	19	70	0.014	31
1,2,3,4,6,7,8-HpCDD	39	70	0.015	61
OCDD	24	70	0.017	35

[a] Concentration in milk fat (ppb).
[b] Time (d).
[c] Rate constant for elimination (1/d).
[d] Calculated steady-state concentration in milk fat (ppb). Since data on PCDD concentrations in milk fat at the beginning of the experiment were not available, initial concentrations ($C_{fat}o$) were assumed to be zero.

C_p = PCDD concentration in tPCP (ppm)
W = average weight of the cows tested (599 kg)
F = assumed feed rate for lactating cows (16 kg/d) (actual feed rate not reported)

The concentration of PCDDs present in the tPCP at levels exceeding 1 ppm are summarized in Table 3. PCDFs were not considered in the experiment.

Three PCDD congeners (1,2,3,6,7,8-HxCDD, 1,2,3,4,6,7,8-HpCDD, and OCDD) were detected in the milk fat. Although the cows were administered tPCP for 70 d, it is not clear that steady-state levels of these congeners in milk were reached during the experiment. The single compartment model can be applied, however, to estimate steady-state concentrations in milk fat using elimination rate constants calculated by Firestone et al. The estimation of steady-state concentrations is summarized in Table 4.

Interestingly, the elimination rate constants derived by Firestone et al.[21] for the HxCDD, HpCCD, and OCDD congeners (0.014, 0.015, and 0.017, respectively) are very similar in magnitude to the value estimated for 2,3,7,8-TCDD by Jensen and Hummel[24] (0.017/d). It should also be noted that only those congeners with chlorine atoms in the 2, 3, 7, and 8 positions (the "lateral" positions) of the PCDD molecule were detected in the milk samples. According to the authors, "This suggests that dioxin congeners with low or zero biological activity (lateral ring positions incompletely filled with halogen atoms) are rapidly metabolized or excreted, whereas the biologically active congeners with halogens in all lateral ring positions are relatively stable after absorption by the animal."[21] These findings are in agreement with other reports that non-2,3,7,8-substituted PCDDs accumulate in fat to a much lesser extent than PCDDs that are substituted in the 2, 3, 7, and 8 positions.[25]

TABLE 5
Steady-State PCDD Concentrations in Cow Milk

Congener	$C_{fat}ss^a$	$C_m ss^b$
1,2,3,6,7,8-HxCDD	31	1.2
1,2,3,4,6,7,8-HpCDD	61	2.4
OCDD	35	1.4

[a] Calculated steady-state concentration in cow milk fat (ppb).
[b] Calculated steady-state concentration in cow milk (ppb).

TABLE 6
BCF and BTF Values for PCDDs in Cow Milk

Congener	$C_m ss^a$	$C_f^{\,b}$	BCF[c] milk	BTF[d] milk
1,2,3,6,7,8-HxCDD	1.2	3.7	0.32	0.020
1,2,3,4,6,7,8-HpCDD	2.4	77	0.031	0.002
OCDD	1.4	260	0.0054	0.0003

[a] Calculated steady-state concentration in cow milk (ppb).
[b] Concentration in feed (ppb).
[c] Bioconcentration factor [(mg/kg)/(mg/kg)].
[d] Biotransfer factor [(mg/kg)/(mg/d)].

Assuming that cow milk contains 4% fat,[26] the corresponding steady-state PCDD concentrations in milk have been estimated and are shown in Table 5. BCF values were calculated by dividing the steady-state milk values for each congener by the concentrations administered with the feed (C_f). Corresponding BTF values were determined by dividing the BCF values by an assumed feed rate of 16 kg/d for lactating cows. The BCFs and BTFs are listed in Table 6.

C. SUMMARY

The Jensen and Hummel[24] and Firestone et al.[21] experiments provide a basis for estimating steady-state BCFs and BTFs for tetra-, hexa-, hepta-, and octachlorinated dibenzo-*p*-dioxins in milk. No data, however, are currently available on the accumulation of pentachlorodibenzo-*p*-dioxins (PnCDDs) or any of the PCDFs in milk. To extend the analysis to these compounds, the following assumptions can be made:

- A BTF or BCF value midway between the values calculated for laterally substituted TCDD and HxCDD can be assigned to 2,3,7,8-PnCDDs. This assumption is based on the general decreasing trend in BCFs and BTFs estimated for the laterally substituted TCDD, HxCDD, HpCDD, and OCDD.
- PCDFs have the same BTF and BCF values as the corresponding PCDDs. This assumption is based on similarities in structure and biological activity between the two groups of compounds.[27,28]

Using these assumptions, BCF and BTF values for all dioxin and furan congeners were calculated and are summarized in Table 7. Based on the findings of Firestone et al.,[21] the BTFs and BCFs for non-2,3,7,8-substituted congeners are assumed to be negligible.

V. THE BIOACCUMULATION OF PCDDs AND PCDFs IN BEEF

Three studies[20,21,29] have examined the bioaccumulation of PCDDs and PCDFs in the body

TABLE 7
BCF and BTF Values for PCDDs and PCDFs in Milk

Congener	BCF[a] milk	BTF[b] milk
2,3,7,8-TCDD/TCDF	0.5	0.03
2,3,7,8-PnCDDs/PnCDFs	0.41	0.025
2,3,7,8-HxCDDs/HxCDFs	0.32	0.02
2,3,7,8-HpCDDs/HpCDFs	0.031	0.002
OCDD/OCDF	0.0054	0.0003

[a] Bioconcentration factor [(mg/kg)/(mg/kg)].
[b] Biotransfer factor [(mg/kg)/(mg/d)].

TABLE 8
Concentrations of PCDDs in Milk Fat and Body Fat[21]

	Conc (ppb) day 69			Conc (ppb) day 169		
	HxCDD	HpCDD	OCDD	HxCDD	HpCDD	OCDD
Milk fat	19	39	24	4.3	6.9	3.0
Body fat	13	24	32	2.5	6.6	5.6

fat of cattle. Jensen et al.[20] studied the bioaccumulation of 2,3,7,8-TCDD in nonlactating heifers. Firestone et al.[21] and Parker et al.[29] examined various PCDDs and PCDFs in lactating dairy cows and nonlactating heifers, respectively.

A. THE FIRESTONE ET AL. (1979) STUDY

Firestone et al.[21] administered PCDD-contaminated technical grade pentachlorophenol (tPCP) to three dairy cows for a period of 70 d. Average PCDD levels in both milk fat and body fat were measured on the 69th d of exposure and again 100 d after the administration of PCDD-contaminated tPCP had ceased. The results of these analyses are presented in Table 8.

As shown in Table 8, the PCDD concentrations in the milk fat were found to be similar to those in the body fat at both day 69 and 169. This suggests that, for lactating cows, the BTF for PCDDs to body fat may be approximately the same as that for PCDDs to milk fat. In general, levels of PCDDs in the body fat of beef cattle would be expected to be higher than those in dairy cows, since PCDD would be released into the milk taken from the cows.[19]

B. THE JENSEN ET AL. (1981) STUDY

Jensen et al.[20] evaluated both the tissue distribution of 2,3,7,8-TCDD administered to cattle and the rate of 2,3,7,8-TCDD dissipation from body fat. For 28 d, 7 heifers were fed rations contaminated with 24 ppt of 2,3,7,8-TCDD. The tissue distribution of 2,3,7,8-TCDD was studied by sacrificing three of the treated heifers and analyzing samples of muscle, fat, liver, and kidney. Dissipation of 2,3,7,8-TCDD in body fat was measured by analyzing samples from the remaining four calves at intervals after discontinuance of 2,3,7,8-TCDD in the diet.

At the end of the 28-d feeding period, concentrations of 2,3,7,8-TCDD in body fat ranged from 66 to 95 ppt, with an average of approximately 92 ppt. These concentrations cannot be considered steady state, however, due to the relatively short period of exposure. Jensen et al.,[20] therefore, used data from the dissipation study to calculate an elimination half-life of approximately 116 d, corresponding to a 2,3,7,8-TCDD elimination rate constant (k_2) of 0.006/d. This value can be used in the single compartment model to estimate a steady-state 2,3,7,8-TCDD concentration of 550 ppt in body fat, a value 6 to 7 times higher than the concentration detected

TABLE 9
BCF and BTF Values for 2,3,7,8-TCDD in Beef

$C_{fat}ss$[a]	C_f[b]	BCF fat[c]	BCF beef[d]	BTF fat[e]	BTF beef[f]
550	24	22.9	4.6	3.8	0.8

[a] Calculated steady-state concentration of 2,3,7,8-TCDD in beef fat (ppt).
[b] Concentration of 2,3,7,8-TCDD in feed (ppt).
[c] Bioconcentration factor for 2,3,7,8-TCDD in beef fat [(mg/kg)/(mg/kg)].
[d] Bioconcentration factor for 2,3,7,8-TCDD in beef [(mg/kg)/(mg/kg)].
[e] Biotransfer factor for 2,3,7,8-TCDD in beef fat [(mg/kg)/(mg/d)].
[f] Biotransfer factor for 2,3,7,8-TCDD in beef [(mg/kg)/(mg/d)].

TABLE 10
Concentrations of PCDDs and PCDFs in Contaminated Pentachlorophenol (PCP) Used in the Parker et al. (1980) Experiment[29]

Congener (identification tentative)	Conc (ppm)	
	aPCP	tPCP
1,2,3,7,8,9-HxCDD or	<1	5
1,2,3,6,7,8-HxCDD	<1	10
1,2,3,4,6,7,9-HpCDD	<10	132
1,2,3,4,6,7,8-HpCDD	<10	278
OCDD	<20	1500
1,2,3,4,6,8-HxCDF or	<5	16
1,2,4,6,7,8-HxCDF or	<5	28
1,2,4,6,8,9-HxCDF	<5	13
1,2,3,4,6,7,8-HpCDF	<5	28
1,2,3,4,6,8,9-HpCDF	<5	99
OCDF	<10	90

after 28 d. Using a different weighting of the data, Jensen et al.[20] estimated the steady-state 2,3,7,8-TCDD residue to be 594 ppt. Jensen et al.[20] noted that the steady-state concentration estimated in the body fat of the heifers "... is close to the 540 ppt found in the fat of rats fed 22 ppt of TCDD for two years." (The rat study cited by Jensen et al.[20] was conducted by Kociba et al.[30]) Hence, the extent of 2,3,7,8-TCDD bioaccumulation in the fat of cattle appears to be similar to that in the adipose tissue of rats.[20]

The BCF for beef fat was estimated by dividing the calculated steady-state fat concentrations by the concentration in the feed. The BTF for beef fat was then determined by dividing the BCF by the feed rate of 6 kg/d for the nonlactating heifers. The BTF and BCF values for beef were calculated by assuming a fat content of 20% in meat consumed by humans[31] and are presented in Table 9.

C. THE PARKER ET AL. (1980) STUDY

Parker et al.[29] fed PCDD/PCDF-contaminated pentachlorophenol (PCP) to three groups of three heifers. The heifers were fed mixtures of two grades of PCP: tPCP and analytical (aPCP). As shown in Table 10, the tPCP contained much higher PCDD/PCDF concentrations than the aPCP.

The three groups of heifers were exposed to different mixtures of aPCP and tPCP. The PCP administered to group A consisted of 10% tPCP/90% aPCP; group B, 35% tPCP/65% aPCP; group C, 100% tPCP. The period of exposure for all three groups was 160 d. The dose of PCP to each group remained constant at 15 mg/(kg/d) for the final 118 d of the feeding period. The

TABLE 11
Concentrations of PCDDs and PCDFs in Feed Used in the Parker et al. (1980) Experiment[29]

Congener	Feed conc (ppb)[a]		
	A (10% tPCP)	B (35% tPCP)	C (100% tPCP)
1,2,3,7,8,9-HxCDD	0.38	1.3	3.8
1,2,3,6,7,8-HxCDD	0.75	2.6	7.5
1,2,3,4,6,7,9-HpCDD	9.9	35	99
1,2,3,4,6,7,8-HpCDD	21	73	210
OCDD	113	395	1130
1,2,3,4,6,8-HxCDF	1.2	4.2	21
1,2,4,6,7,8-HxCDF	2.1	7.4	21
1,2,4,6,8,9-HxCDF	0.98	3.4	9.8
1,2,3,4,6,7,8-HpCDF	2.1	7.4	21
1,2,3,4,6,8,9-HpCDF	7.4	6	74
OCDF	6.8	24	68

[a] PCDD and PCDF concentrations in aPCP were assumed to be negligible. If actual levels in aPCP are close to the detection limits, BCF and BTF values may be overestimated.

TABLE 12
Concentrations of PCDDs and PCDFs in Body Fat of Cattle from the Parker et al. (1980) Experiment[29]

Congener	Fat conc (ppb)[a]		
	A (10% tPCP)	B (35% tPCP)	C (100% tPCP)
1,2,3,7,8,9-HxCDD	ND	ND	0.7
1,2,3,6,7,8-HxCDD	1	10	16
1,2,3,4,6,7,8-HpCDD	7	12	52
OCDD	13	36	61
1,2,3,4,6,7,8-HpCDF	0.5	4.2	6.9
OCDF	4	5	7

[a] Nd: nondetect.

weight and feed rate for the heifers were not specified in the experiment; hence, a body weight of 300 kg and a feed rate of 6 kg/d were assumed for this analysis. Using these assumptions, the concentration of each PCDD and PCDF congener in the feed for the three groups of animals was estimated and is summarized in Table 11.

At the end of the 160-d period, the animals were sacrificed, and levels of PCDDs and PCDFs in blood, liver, and body fat were measured. The dissipation of PCDDs and PCDFs from fat was not examined. The levels of PCDDs and PCDFs detected in the adipose tissue of the heifers at the end of the 160-d period are summarized in Table 12.

As observed by Firestone et al.,[21] only those PCDDs and PCDFs chlorinated in the lateral (2,3,7,8) positions were found to accumulate in animal fat. This phenomenon is attributed by Parker et al.[29] "... to the fact that an open lateral position provides an opportunity for the metabolism of dioxins and furans to more polar compounds which can be excreted".

Although the cows were exposed for a relatively long period (160 d), it is not clear that steady-state PCDD/PCDF concentrations were reached. Since an examination of the dissipation of PCDD/PCDF in the fat was not conducted, an elimination half-life for the various congeners cannot be estimated from data generated in the experiment. It is therefore necessary to assume an elimination half-life in order to apply the single compartment model.

TABLE 13
Steady-State Concentrations of PCDDs and PCDFs in Body Fat of Cattle

	Steady-state fat conc (ppb)[a]		
Congener	**A (10% tPCP)**	**B (35% tPCP)**	**C (100% tPCP)**
1,2,3,7,8,9-HxCDD	ND	ND	1.1
1,2,3,6,7,8-HxCDD	1.6	16	26
1,2,3,4,6,7,8-HpCDD	11	19	84
OCDD	21	58	99
1,2,3,4,6,7,8-HpCDF	0.8	6.8	11
OCDF	6.5	8.1	11

[a] Nd: nondetect.

TABLE 14
BCF Values for PCDDs and PCDFs in Body Fat of Cattle

	BCF[a]			
Congener	**A (10% tPCP)**	**B (35% tPCP)**	**C (100% tPCP)**	**Mean**
1,2,3,7,8,9-HxCDD	—	—	0.3	0.3
1,2,3,6,7,8-HxCDD	2.1	6.1	3.5	3.9
1,2,3,4,6,7,8-HpCDD	0.52	0.26	0.41	0.40
OCDD	0.19	0.15	0.088	0.14
1,2,3,4,6,7,8-HpCDF	0.38	0.92	0.52	0.61
OCDF	0.95	0.34	0.16	0.48

[a] Bioconcentration factor [(mg/kg)/(mg/kg)].

Studies using dairy cows found half-lives for laterally substituted TCDD, HxCDD, HpCDD, and OCDD in milk fat to be very similar, ranging from 41 d for TCDD and OCDD to 51 d for HxCDD.[21,24] Assuming that the dissipation of laterally substituted PCDDs and PCDFs in the adipose tissue of nonlactating cattle is also similar, then the half-life value of 116 d[20] for 2,3,7,8-TCDD can be applied to other 2,3,7,8-substituted PCDD congeners as well. This allows the use of the single compartment model to calculate steady-state concentrations for the PCDD congeners studies by Parker et al.[29]

An elimination rate constant of 0.006/d corresponds to a half-life of 116 d. Using this value, the steady-state PCDD and PCDF concentrations in adipose tissue were calculated and are listed in Table 13. Dividing the steady-state fat concentrations by the feed concentrations yields the BCF values for each congener examined. These BCF values are presented in Table 14.

The low BCF value shown in Table 14 for 1,2,3,7,8,9-HxCDD is questionable for several reasons. First, this congener was not detected in liver samples, which exhibited much higher levels than the fat for all other PCDD and PCDF congeners. Furthermore, all other data from this and other experiments indicate that biotransfer of PCDDs and PCDFs to cow milk and beef tends to decrease with increasing chlorination. Finally, the BCF value for 1,2,3,7,8,9-HxCDD is based on a single data point. Accordingly, only the BCF value for 1,2,3,6,7,8-HxCDD is used to represent the HxCDD congeners.

Assuming a fat content of 20% in beef,[31] the BCFs for beef were calculated and are shown in Table 15. BTFs for beef for the various PCDD and PCDF congeners were calculated by dividing the appropriate BCF values by an assumed feed rate of 6 kg/d for heifers, and are also presented in Table 15.

TABLE 15
BCF and BTF Values for PCDDs and PCDFs in Body Fat and Beef

	BCF[a]		BTF[a]	
Congener	**Fat**	**Beef**	**Fat**	**Beef**
2,3,7,8-HxCDDs	3.9	0.78	0.65	0.13
2,3,7,8-HpCDDs	0.40	0.08	0.067	0.013
OCDD	0.14	0.028	0.023	0.005
2,3,7,8-HpCDFs	0.61	0.12	0.10	0.020
OCDF	0.48	0.096	0.080	0.016

[a] Bioconcentration factor [(mg/kg)/(mg/kg)].
[b] Biotransfer factor [(mg/kg)/(mg/d)].

It is interesting to note that the PCDFs examined in the Parker et al.[29] study exhibit somewhat higher BCF and BTF values than do the PCDDs. This difference is more pronounced for the octachlorinated species (where the BCF for the dibenzofuran is about 3.5 times greater than the BCF for the dibenzo-*p*-dioxin) than for the heptachlorinated species (where the HpCDF BCF is greater than the HpCDD BCF by a factor of approximately 1.5). It is not clear why the structure of the PCDFs would cause them to have a higher bioaccumulation potential than the PCDDs. Since experiments involving HxCDF, PnCDF, and TCDF have not been conducted, it is also unknown whether the difference between the octa- and heptachlorinated species would also be observed in the less highly chlorinated PCDD and PCDF.

D. SUMMARY

BTF and BCF values for tetra-, hexa-, hepta-, and ocatachlorinated dibenzo-*p*-dioxins and for hepta- and octachlorinated dibenzofurans in beef were estimated from data obtained by Jensen et al.[20] and Parker et al.[29] However, data on the accumulation of pentachlorodibenzo-*p*-dioxins and tetra-, penta-, and hexachlorinated dibenzofurans in beef are currently unavailable. To allow for a consideration of these compounds in the risk assessment, the following assumptions can be made:

- The BTF and BCF values for 2,3,7,8-PnCDDs lie midway between the values for 2,3,7,8-TCDD and 2,3,7,8-HxCDDs. This assumption is based on the general decreasing trend in BCFs and BTFs estimated for the laterally substituted TCDD, HxCDD, HpCDD, and OCDD.
- For those PCDFs for which no data are available, PCDF, BCF, and BTF values are assumed to be equal to those of the corresponding PCDD. This assumption is based on the similarities in structure and biological activity between the two groups of compounds.[27,28]

Using these assumptions, BCFs and BTFs for all PCDDs and PCDFs were estimated and are summarized in Table 16.

VI. CONCLUSIONS

Three general observations can be made following a review of literature on the biotransfer of PCDDs and PCDFs to milk and beef:

TABLE 16
BCF and BTF Values for PCDDs and PCDFs in Beef

Isomer	BCF[a] beef	BTF[b] beef
2,3,7,8-TCDDs	4.6	0.8
2,3,7,8-PnCDDs	2.7	0.47
2,3,7,8-HxCDDs	0.78	0.13
2,3,7,8-HpCDDs	0.08	0.013
OCDDs	0.028	0.005
2,3,7,8-TCDFs	4.6	0.8
2,3,7,8-PnCDFs	2.7	0.47
2,3,7,8-HxCDFs	0.78	0.13
2,3,7,8-HpCDFs	0.12	0.020
OCDFs	0.096	0.016

[a] Bioconcentration factor [(mg/kg)/(mg/kg)].
[b] Biotransfer factor [(mg/kg)/(mg/d)].

- The extent of PCDD and PCDF accumulation in fat appears to be inversely related to the extent of chlorination. This observation is limited to PCDDs with four or more chlorine atoms; data are not available for less chlorinated congeners.
- PCDDs and PCDFs chlorinated in the lateral (2, 3, 7, 8) positions appear to accumulate to a much greater extent than nonlaterally substituted compounds. This conclusion is supported by data from animal[25] and human studies.[32,33]
- The estimated steady-state BCFs and BTFs for PCDDs appear to be greater for beef fat than for milk fat. This may be due to the greater excretion of PCDDs from lactating cows.

These observations cannot be confirmed using data from field studies[34] since such data do not necessarily represent steady-state conditions, and actual levels of exposure in the field are rarely known with any certainty.

It should be emphasized that there is considerable uncertainty associated with the estimation of steady-state concentrations of PCDDs and PCDFs in milk and beef as a result of limitations in available data. Experiments conducted to date have been too short to provide direct measurements of the steady-state concentrations of PCDDs and PCDFs, and the accuracy of the single compartment elimination model used to predict these concentrations has not been established. It is believed, however, that the approach described in this chapter provides reasonable estimates of the steady-state concentrations of PCDDs and PCDFs in milk and beef.

REFERENCES

1. **Pazderova-Vejlupkova, J., Nemcova, M., Pickova, J., and Jirasek, L.,** The development and prognosis of chronic intoxication by TCDD in men, *Arch. Environ. Health,* 36, 5, 1981.
2. **Hatch, M. C.,** Reproductive effects of the dioxins, in *Public Health Risks of the Dioxins,* Lawrence, W. W., Ed., Rockefeller University Press, Los Altos, CA, 1984, 255.
3. International Agency for Research on Cancer, *IARC Monographs on the Evaluation of the Carcinogenic Risk of Chemicals to Humans, Chemicals, Industrial Processes and Industries Associated with Cancer in Humans,* Vol. 1 to 29, Suppl. 4, 1982, 238.
4. **Bellin, S. S. and Barnes, D. G.,** Health hazard assessment for chlorinated dioxins and dibenzofurans other than 2,3,7,8-TCDD, *Toxicol. Ind. Health,* 1, 235, 1985.
5. **Stehl, R., Pagenfuss, R., Bredeweg, R., and Roberts, R.,** The stability of pentachlorophenol and chlorinated dioxins to sunlight, heat, and combustion, in *Chlorodioxin-Origin and Fate,* Blair, E. H., Ed., American Chemical Society, Washington, D.C., 1973, 119.

6. **Berlincioni, M. and di Domenico, A.,** Polychlorodibenzo-*p*-dioxins and polychlorodibenzofurans in the soil near the municipal incinerator of Florence, Italy, *Environ. Sci. Technol.,* 21, 1063, 1987.
7. **Czuczwa, J. and Hites, R.,** Environmental fate of combustion-generated polychlorinated dioxins and furans, *Environ. Sci. Technol.,* 18, 444, 1984.
8. **Firgerio, A.,** The Seveso case: TCDD in animals from contaminated areas, *Methodol. Surv. Biochem.,* 7, 161, 1978.
9. **Stalling, D. L., Smith, L., Petty, J., Hogan, J., Johnson, J., Rappe, C., and Buser, M.,** Residues of polychlorinated dibenzo-*p*-dioxins and dibenzofurans in Laurentian Great Lakes fish, in *Human and Environmental Risks of Chlorinated Dioxins and Related Compounds,* Tucker, R. E., Young, A. L., and Gray, A. P., Eds., Plenum Press, New York, 1983, 221.
10. **Petty, J.,** Composition of polychlorinated dibenzofurans and dibenzo-*p*-dioxin residues in sediments of the Hudson and Housatonic Rivers, in *Chlorinated Dioxins and Dibenzofurans in the Total Environment,* Choudhry, G., Keith, L., and Rappe, C., Eds., Butterworths, Boston, 1983, 203.
11. **Wipf, H. K., Homberger, E., Neimer, N., Ranalder, V., Vetter, W., and Veiheumeir, J.,** TCDD levels in soil and plant samples from the Seveso area, in *Chlorinated Dioxins and Related Compounds — Impact on the Environment,* Hutzinger, O., Frei, R., Merian, E., and Pocchiari, F., Eds., Pergamon Press, Oxford, 1982, 115.
12. **Cocucci, S., Di Gerolamo, F., Verderio, A., Cavallaro, A., Colli, G., Gorni, A., Invernizzi, G., and Luciani, L.,** Absorption and translocation of tetrachlorodibenzo-*p*-dioxin by plants from polluted soil, *Experientia,* 35, 482, 1979.
13. **Olie, K., Van Den Berg, M., and Hutzinger, O.,** Formation and fate of PCDD and PCDF from combustion processes, *Chemosphere,* 12, 627, 1983.
14. **Travis, C. C. and Hattemer-Frey, H. A.,** Human exposure to 2,3,7,8-TCDD, *Chemosphere,* 16, 2331, 1987.
15. **Van Miller, J. P., Marlar, R. J., and Allen, J. R.,** Tissue distribution and excretion of tritiated tetrachlorodibenzo-*p*-dioxin in nonhuman primates and rats, *Food Cosmet. Toxicol.,* 14, 31, 1976.
16. **Thalken, C. E. and Young, A. L.,** Long-term field studies of a rodent population continuously exposed to TCDD, in *Human and Environmental Risks of Chlorinated Dioxins and Related Compounds,* Tucker, R. E., Young, A. L., and Gray, A. P., Eds., Plenum Press, New York, 1983, 357.
17. **Fries, G. F. and Morrow, G. S.,** Retention and excretion of 2,3,7,8-tetrachlorodibenzo-*p*-dioxin by rats, *J. Agric. Food Chem.,* 23, 265, 1975.
18. **Rose, J. Q., Ramsey, J. C., Wentzler, T. H., Hummel, R. A., and Gehring, P. J.,** The fate of 2,3,7,8-tetrachlorodibenzo-*p*-dioxin following single and repeated oral doses to the rat, *Toxicol. Appl. Pharmacol.,* 36, 209, 1976.
19. **Fries, G. F.,** Record of Proceedings, Maine Department of Environmental Protection Dioxin Workshop, Augusta, February 4, 1986.
20. **Jensen, D. J., Hummel, R. A., Mahle, N. H., Kocher, C. W., and Higgins, H. S.,** A residue study on beef cattle consuming 2,3,7,8-TCDD, *J. Agric. Food Chem.,* 29, 265, 1981.
21. **Firestone, D., Clower, M., Bonsetti, A. P., Teske, R. H., and Long, P. E.,** Polychlorodibenzo-*p*-dioxin and pentachlorophenol residues in milk and blood of cows fed technical pentachlorophenol, *J. Agric. Food Chem.,* 27, 1171, 1979.
22. **Piper, W. N., Rose, J. Q., and Gehring, P. J.,** Excretion and tissue distribution of 2,3,7,8-tetrachlorodibenzo-*p*-dioxin in the rat. *Environ. Health Perspect.,* 5, 241, 1973.
23. **Travis, C. C. and Arms, A. D.,** Bioconcentration of organics in beef, milk, and vegetation, *Environ. Sci. Technol.,* 22, 271, 1988.
24. **Jensen, D. J. and Hummel, R. A.,** Secretion of TCDD in milk and cream following the feeding of TCDD to lactating dairy cows, *Bull. Environ. Contam. Toxicol.,* 29, 440, 1982.
25. **Rappe, C.,** Chemistry and analysis of polychlorinated dioxins and dibenzofurans in biological samples, in *Biological Mechanisms of Dioxin,* Kimbrough, A. and Poland, R. D., Eds., Cold Spring Harbor Laboratory, Cold Spring Harbor, NY, 1984.
26. **Hodgman, C. D., Weast, R. C., Wallace, C. W., and Selby, S. M.,** *CRC Handbook of Chemistry and Physics,* Chemical Rubber Company, Cleveland, 1954.
27. Ontario Ministry of the Environment, Scientific Criteria Document for Standard Development, No. 4-84, Polychlorinated Dibenzo-*p*-Dioxins (PCDDs) and Polychlorinated Dibenzofurans (PCDFs), Toronto, 1985.
28. **Safe, S.,** Determination of 2,3,7,8-TCDD toxic equivalent factors (TEFs): support for the use in the *in-vitro* AHH induction assay, *Chemosphere,* 16, 791, 1987.
29. **Parker, C. E., Jones, W. A., Matthews, H. B., McConnell, E. E., and Haas, J. R.,** The chronic toxicity of technical and analytical pentachlorophenol in cattle, *Toxicol. Appl. Pharmacol.,* 55, 359, 1980.
30. **Kociba, R. J., Keyes, D. G., Beyer, J. E., Carreon, R. M., Wade, C. E., Dittenber, P. A., Kalnins, R. P., Frauson, L. E., Park, C. N., Bernard, S. D., Hummel, R. A., and Humiston, C. G.,** Results of a two-year chronic toxicity and oncogenicity study of 2,3,7,8-tetrachlorodibenzo-*p*-dioxin in rats, *Toxicol. Appl. Pharmacol.,* 46, 229, 1978.
31. **West, B. K. and Merrill, A. L.,** Composition of Foods, Agriculture Hand. No. 8, U.S. Department of Agriculture, Washington, D.C., 1975.

32. **Ryan, J. J.,** Variation of dioxins and furans in human tissues, *Chemosphere,* 15, 1585, 1986.
33. **Rappe, C., Choudhary, G., and Keith, L. H.,** *Chlorinated Dioxins and Dibenzofurans in Perspective,* Lewis Publishers, Chelsea, MI, 1986.
34. **Beck, H., Eckart, K., Mathar, W., and Wittkowski, R.,** PCDD and PCDF body burden from food intake in the Federal Republic of Germany, *Chemosphere,* 18, 417, 1989.

Chapter 9

HEALTH RISK ASSESSMENT FOR WASTE-TO-ENERGY PROJECTS IN CALIFORNIA

John A. Lowe, Wyatt Dietrich, and Michael T. Alberts

TABLE OF CONTENTS

I. INTRODUCTION

The generation of electricity from incinerating municipal solid waste (MSW) in waste-to-energy (WTE) facilities is being considered as an alternative waste management practice by counties and cities in California whose landfills will reach capacity in the near future. Approximately 11 WTE projects are under consideration in California. Most of these projects are in various stages of the planning or permitting processes. One facility is currently in operation, and two projects are under construction with operation scheduled in 1989. Public concern about environmental contamination associated with volatile organic compounds, such as benzene, vinyl chloride, and trichloroethylene, migrating from sanitary landfills to groundwater supplies, or being emitted into the atmosphere has diminished public acceptance of this traditional waste disposal process. However, WTE facilities have their own set of issues that generate apprehension, notably the emission of toxic air contaminants (TACs) when refuse is incinerated. The presence of polychlorinated dibenzo-*p*-dioxins (PCDDs) and polychlorinated dibenzofurans (PCDFs) in WTE emissions have heightened public concern about the potential health impacts associated with these facilities.

Concerns about the potential adverse effects of WTE emissions have significantly affected decisions to site and permit WTE facilities in California. Significant regulatory and scientific uncertainties exist in evaluating the impacts on public health from exposure to TACs emitted from WTE facilities. Regulatory uncertainties result because there are no state or federal standards governing emissions or ambient air quality of TACs associated with the incineration of MSW. Scientific uncertainties result from using health risk assessment as the primary analytical and decision-making technique for evaluating potential impacts on public health from exposure to TACs emitted from WTE facilities.

Health risk assessments in California have become important tools for developing project-specific TAC emission limitations and making siting decisions for proposed WTE facilities. The purpose of such assessments is to quantitatively estimate the health risk to the surrounding community as a result of exposure to facility-emitted pollutants. The increased use of health risk assessments for this purpose results largely from the lack of regulatory standards for WTE facilities, as the health risk assessment process, as applied to WTE facilities, has evolved largely in response to concerns raised during the regulatory review process. This chapter outlines the evolution of health risk assessments for WTE facilities in California and describes the methodology used in and technical issues inherent to a health risk assessments for WTE with a case study from a California WTE facility.

II. CALIFORNIA REGULATORY PROCESS

The primary objective of a WTE project is to reduce the total amount of residential, commercial, and industrial waste currently disposed of in landfills and to use MSW as an energy resource. Costs associated with construction and operation of a WTE facility are partially recovered by the sale of electrical power produced by the WTE project.

To be implemented in California, a WTE project needs the following elements:

- A guaranteed supply of MSW for the life of the project
- A sound, long-term market for the energy produced, usually in terms of an electricity sales contract with a local or regional utility company

- A proven WTE technology with a demonstrated record of performance
- Availability of a landfill for ash residue disposal
- Required environmental permits and regulatory approval of a health risk assessment.

Health risk assessment has become one of the most important aspects in regulatory approval for proposed WTE projects in California. In the absence of ambient standards for TACs, the health risk assessment process is used to quantify potential health impacts and set TAC emission limits for WTE projects on a case-by-case basis.

Public concern in California regarding TACs resulted in the adoption of Assembly Bill 1807, known as the Tanner Process, by the State Legislature in 1984. This law requires the California Air Resources Board (ARB) and California Department of Health Services (DHS) to prioritize candidate TACs for evaluation and regulation. In 1986, ARB and DHS recommended that dioxin (2,3,7,8-dibenzo-*p*-dioxins and dibenzofurans) be listed as TACs based on the agencies' findings that dioxin is a carcinogen.[1]

Prior to 1986, health risk assessments for proposed WTE projects were prepared at the discretion of local air pollution control districts (APCDs) as part of the air quality permitting process. The APCDs usually requested that health risk assessments be technically reviewed by ARB and DHS. Because the regulatory review process did not have defined time limitations, final permit approval by the local APCD could take 2 to 3 years. There were also no guidelines concerning the appropriate level of detail required in health risk assessments. Early assessments considered the inhalation pathway only; health risk assessments that considered other exposure pathways, such as ingestion and dermal exposure, called multipathway health risk assessments, were rarely required before 1985.

In January 1986, a new state law (Assembly Bill 3989) was enacted that required preparation of a health risk assessment and an environmental impact report (EIR) for all WTE projects in California. Environmental groups supported passage of this law, since all WTE projects would receive careful scrutiny and adequate technical review. The WTE industry also supported this law, because it standardized the permitting process and established formal requirements for timely agency review of health risk assessment. Technical review of the health risk assessment by a state agency is mandatory within 45 d for ARB review and 90 d for DHS review. The *Air Toxics Source Assessment Manual for California Air Pollution Control Districts*,[2] which was finalized in 1987, provides methods for performing simple, screening-level assessments and limited guidance on performing detailed multipathway assessments.

A consequence of Assembly Bill 3989 has been a more direct linkage between health risk assessments and EIRs for proposed WTE projects in California, since the results of the health risk assessment are included in the draft EIR. The public is encouraged to comment on health risk assessments and EIRs during the public review period established under the California Environmental Quality Act (CEQA). Public comments received concerning the health risk assessment and other environmental issues are addressed in the final EIR. The lead agency preparing the EIR will determine if the health risk associated with a proposed project constitutes a "significant" environmental impact. This determination takes into account the technical findings of the ARB, the DHS, and the risk management policy of the local air pollution control district (APCD).

The Monterey Bay Unified Air Pollution Control District (MBUAPCD) is one regulatory agency in California that has adopted regulations (Rule 1000) defining an acceptable cancer risk level for new or modified sources.[3] Under Rule 1000, an increased lifetime risk of $<1 \times 10^{-5}$ (or a 1 in 100,000 chance of developing cancer from lifetime exposure to facility-emitted contaminants) is considered acceptable. Risk levels exceeding 1×10^{-5} would result in permit denial.[3]

The South Coast Air Quality Management District (SCAQMD) currently has under consideration similar rules to regulate both cancer risk and cancer burden (proposed Rules 1401 and

223).[4] Cancer burden is the estimated increase in the number of cancer cases due to facility emissions of TACs. The proposed rules stipulate that new or modified existing sources seeking an authority-to-construct permit cannot pose a cancer risk greater than 1×10^{-5} for the maximum exposed individual.[4] This risk level is considered acceptable only when the best available control technology for toxics (T-BACT) is used. Without installation of T-BACT, the maximum exposed individual risk must be less than 1×10^{-6} (1 in 1 million).[4] The rule further stipulates that cancer burden, or excess cancer cases associated with exposure to a carcinogen, cannot exceed 0.5 for a population exposed to a 1×10^{-6} risk.[4]

The MBUAPCD Rule 1000 and the SCAQMD Proposed Rules 1401 and 223 represent local regulatory agency efforts to define acceptable risk. It is evident from these and other rules that a trend is emerging which sets the upper-bound of acceptable cancer risk for the maximally exposed individual at 1×10^{-5}.

III. WTE FACILITY PROCESS

The health risk assessment case study presented in this chapter was drawn from the permitting process for the Milliken WTE project. The Milliken WTE facility was proposed for siting in Ontario, California. This facility, however, failed to obtain the approval of the San Bernadino County Board of Supervisors in late 1987. This facility was similar to the Stanislaus WTE facility, which began operation in 1989 in San Joaquin Valley. These two facilities are representative of air emission levels that could result from the operation of modern WTE facilities. These facilities were of similar design, although the Milliken facility was designed to process twice as much waste as the Stanislaus facility. The Milliken facility was selected to present a case study of how health risk assessments for WTE facilities are performed.

A. WTE PROCESS AND POLLUTION CONTROL DEVICES

Both of these facilities are massburn. In massburn operations, overhead cranes transfer unprocessed refuse from an enclosed storage pit to the furnace feed hopper. Refuse is fed to an inclined, movable grate furnace where it is dried, volatilized, and ignited. Large, noncombustible items are removed from the storage pit for recycling or landfilling before burning. The burned-out residue and noncombustible portion of the refuse travels down the grate and are collected in an ash discharge area. Ferrous metals are recovered before landfill disposal of the ash and noncombustibles. The ash residues are generally equivalent to 20 to 30% of the mass of the refuse incinerated.

The flue gas generated during the combustion process contains a variety of criteria pollutants, including nitrous oxides (NO_x), carbon monoxide (CO), particulates, and sulfur dioxinde (SO_2) and noncriteria pollutants, such as toxic metals and organic compounds. A variety of pollution control options are available for both the criteria and TACs.

Criteria pollutant emissions must be controlled for WTE facilities to meet state and federal air quality standards. A state-of-the-art method of controlling NO_x emissions from WTE facilities is the Thermal DeNOx technology, which consists of injecting ammonia into the flue gas for reaction with NO_x. SO_2 emissions are controlled through a dry scrubber that uses a lime slurry injection. Particulate matter entrained in the flue gas is primarily inert solids but may contain metals and organics adsorbed onto particle surfaces or incorporated into the particle matrix. A fabric filter baghouse is used to control particulate emissions. The baghouse is also effective at controlling acid gas emissions when used downstream of a scrubber. The lime slurry is carried in the flue gas to the baghouse where it deposits onto the bag surface with the other particulate matter. The lime coating is then available to react with any acid gases that may have not been neutralized in the scrubber.

B. WTE FACILITY PERMITTING PROCESSES

The permitting process in California for WTE facilities is described briefly in Section II. The major difference in the permitting processes for these two facilities is that the Milliken WTE facility was subject to mandatory review by state agencies, the ARB and the California DHS required by legislation passed in 1986. The Milliken WTE health risk assessment was also subject to the requirements of SCAQMD proposed Rule 1401, which specified acceptable individual cancer risk levels and population burden (additional cancer cases in an exposed population) for any new sources emitting TACs.[5]

Many of the permit requirements for the Stanislaus WTE facility were completed prior to the passage of state law mandating review of the health risk assessment by ARB and DHS; however, these agencies had considerable advisory roles in the development of the health risk assessment for this facility.

C. LAND USE CONSIDERATIONS

The area surrounding a WTE facility plays a large role in defining the scope of the health risk assessment. The Stanislaus WTE facility site, located near Crows Landing, CA, is in the San Joaquin Valley, which is one of the largest commercial agricultural areas in the country. The surrounding area close to the facility is largely rural, consisting of small towns, cropland, and orchards. An important consideration in the health risk assessment for this facility was evaluation of potential exposures through the food chain. Because of the rural nature of the area, it was possible that a high proportion of an individual's diet was derived from locally produced food items or from backyard gardens.

Evaluation of food chain exposures was also addressed in the health risk assessment for the Milliken WTE facility. Land use surrounding the facility site included a large concentration of commercial dairies. Previously, food chain models used in health risk assessments accounted for the exposure of animals to facility emissions deposited on pasture. However, a model was developed for the Milliken health risk assessment that attempted to evaluate exposures to livestock confined to dairies and feedlots. The food chain analysis also accounted for the possibility that residents might obtain a large percentage of their dairy products contaminated by emissions from the facility.[5]

IV. RISK ASSESSMENT PROCESS

Risk assessment is defined by the National Research Council (NRC) as, "The characterization of the potential adverse health effects of human exposures to environmental hazards. The risk assessment presents summary judgments of the existence and magnitude of a potential public health problem, and a characterization of the uncertainties inherent in the process of inferring risk."[6] Risk assessment (referred to here as "health risk assessment") is a systematic study of the possible relationships between exposure to toxic substances in the environment and the occurrence of adverse health effects. The NRC[6] describes four steps in a health risk assessment:

- Hazard identification
- Exposure assessment
- Dose-response (or toxicity) assessment
- Risk characterization

Hazard identification is the process of determining whether exposure to a substance can cause an increase in the incidence of an adverse health effect. In this step, chemicals of concern to be considered in the health risk assessment are identified. This selection process takes into account the amount of a chemical emitted from a facility and its observed toxicological hazards. Indicator chemicals are typically designated to represent a family of chemicals that may be emitted. For example, benzo(a)pyrene usually serves as the indicator chemical for emitted polycyclic aromatic hydrocarbons (PAHs), and 2,3,7,8-tetrachlorodibenzo-*p*-dioxin (TCDD) serve as the indicator for PCDDs and PCDFs.

Exposure assessment is the estimation of the magnitude, frequency, duration, and route of exposure to a substance. The magnitude of exposure is determined by measuring or estimating the amount of chemical to which individuals are exposed through inhalation, ingestion, or skin (dermal) contact during a specified time period. The frequency and duration of exposure are largely determined by how long the facility will operate and the activities of individuals living near it. The resulting exposure estimates are then evaluated in light of the uncertainties present in the assumptions, data, and methods used in the exposure assessment.

Dose-response assessment characterizes the relationship between the chemical dose and the incidence of an adverse effect in exposed populations. This relationship is often depicted by extrapolating adverse effects associated with the relatively high doses administered to laboratory animals in toxicity studies to the low levels of exposure to which humans are most likely to be exposed. Dose-response assessments for WTE facilities typically use agency-derived numerical estimates of chemical toxicity.

Risk characterization involves combining the results of the dose-response and exposure assessments to provide numerical estimates of health risk. Additionally, risk characterization summarizes the nature and weight of the evidence supporting those numerical estimates.

V. RISK ASSESSMENT METHODOLOGY FOR WTE FACILITIES

Exposure assessment for a WTE facility is based on computerized mathematical models. These models require data on facility emissions, meteorology, and local terrain features, including the presence and size of bodies of water used for drinking water or fishing, farm land, home gardens, feedlots, and pasture land. Various models are then used to calculate the distribution of chemicals in air, water, soil, crops, and livestock (meat and milk). The populations potentially exposed to these chemicals are estimated using census data as well as information on where individuals are living (i.e., in urban vs. rural areas) and the amount of food and water consumed by the average exposed individual. This process produces human intake (or exposure) estimates in units of milligrams of chemical per kilograms of human body weight per day.

Health impacts were determined using modeled air emission, evaluating the environmental transport and fate of emitted pollutants, and evaluating all direct and indirect pathways of human exposure. Individual and population health risks resulting from facility emissions were quantified through a series of sequential processes beginning with estimating facility emissions and subsequent ambient air concentrations. The following sections discuss the modeling methods, data, and assumptions used in health risk assessments for the WTE facility and present estimates of the magnitude of human health risk from exposure to facility-emitted pollutants. An overview of the risk assessment process used in the Milliken WTE projects is presented in Figure 1. The steps involved in this process are

- Screening and sensitivity analyses
- Emissions estimation
- Atmospheric dispersion and particle deposition modeling
- Environmental fate and transport evaluation
- Exposure assessment
- Dose-response assessment
- Risk characterization

A. SCREENING AND SENSITIVITY ANALYSIS

An initial screening and sensitivity analysis using health conservative (or upper-limit)

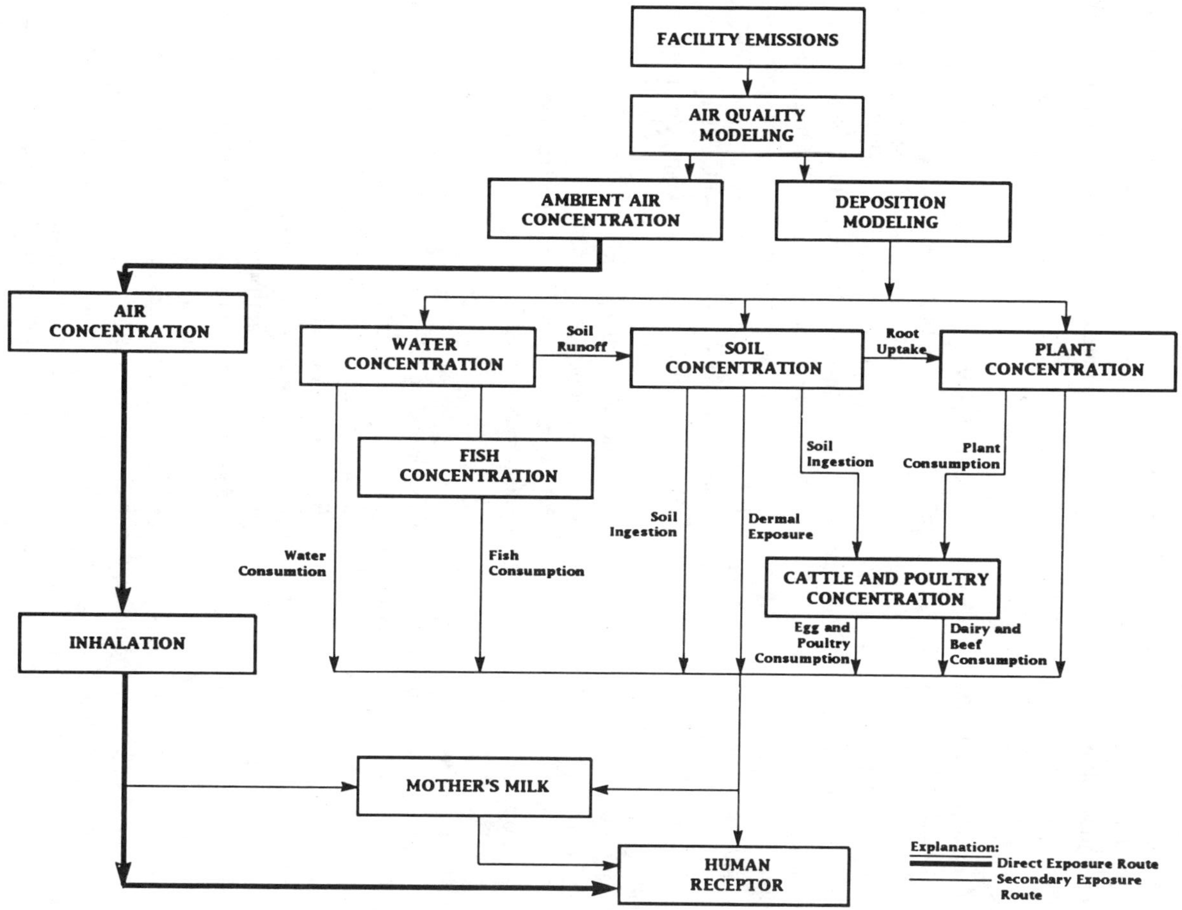

FIGURE 1. Environmental fate and exposure pathways.

assumptions, methods, and data was done for all model inputs and variables associated with evaluating both the deposition of and human exposure to emitted pollutants. Screening analyses are designed to reduce the scope of the assessment to a more manageable size. Health conservative values and assumptions intentionally overestimate the magnitude of health risks to the public. Performing a screening analysis allows researchers to focus on the most important issues and variables, since all variables and inputs shown to have a minimal impact on human exposure are excluded from further analysis. Conversely, inputs and variables shown to substantially affect risk estimates are evaluated in detail. A sensitivity analysis identified those assumptions and parameters that contribute most to be estimated risks, allowing further study to be focused on those elements and assumptions.

B. EMISSIONS ESTIMATION

Facility emissions of TACs, including metals and organics, are estimated using source test data from operating facilities similar in design and technology to the proposed facility (Table 1). The selection of source test data for emissions estimation is critical in the risk assessment process and has typically been an area which involves extensive review by regulatory agencies.

Emissions estimates for selected pollutants of concern for the Milliken facility are based on source test data obtained from newer, modern facilities which all have a dry scrubber and fabric filter baghouse for particulate control. This type of pollution control scheme, now regarded as the best available control technology (BACT), is much more effective at retaining toxic pollutants than the older, less efficient control units. Case 1 represents a more plausible, upper-bound estimate of emissions. Case 2 estimates represent a worst-case scenario. These emissions estimates are presented in Table 2.

The major problems associated with estimating emissions from proposed WTE facilities are (1) selecting appropriate test data for PCDD/PCDF emissions; (2) making appropriate assumptions for estimating chromium emissions; and (3) establishing procedures for estimating emissions from off-design operation. These issues are discussed below.

1. Dioxin/Furan Emissions

Special attention was given to emission of PCDDs and PCDFs due largely to concerns about the potential toxicity of these compounds. Health risk assessment for these compounds can be complicated, since numerous PCDDs and PCDFs are emitted from a WTE facility. PCDDs and PCDFs are organized into congener groups distinguished by the degree of chlorination. Within each congener group, the compounds are further speciated into isomers according to the position each chlorine atom occupies on the ring structure. The relative quantity of each congener formed during combustion and the placement of chlorine atoms on the ring structure may vary slightly from facility to facility according to design, operating conditions, and waste feed. Speciation of these compounds in WTE emissions is crucial in assessing risk because of the wide range in potency among the different congener and isomer groups.

The toxicity of the different PCDDs and PCDFs emitted from a WTE facility are expressed in terms of toxic equivalent factors (TEFs). This methodology estimates the toxicity of PCDD/PCDF congeners relative to the known toxicity of the most toxic isomer, 2,3,7,8-TCDD, using one of several weighing schemes.[7] Health risk assessments for WTE facilities proposed in California use the California DHS scenario #4 scheme.[8] This scheme assumes that PCDDs and PCDFs not chlorinated at the 2, 3, 7, and 8 positions or that do not have at least one ring position open are not carcinogenic. The DHS weighing scheme is presented in Table 3.

The PCDD/PCDF emissions from the proposed Milliken WTE facility were derived from source test data collected at two facilities (Marion County, Oregon, and Wurzburg, West Germany) that use a similar stoker combustion and advanced air pollution control systems as those proposed for the Milliken WTE facility. To address the uncertainty in estimating dioxin emissions, two emissions scenarios were used in the health risk assessments. The first scenario,

TABLE 1
Emissions Data from Modern WTE Facilities Using a Dry Scrubber/Baghouse[a]

Pollutant	Marion County (ng/nm^3)	Wurzburg (ng/nm^3)	Hogdalen (ng/nm^3)	Quebec (ng/nm^{3b})	Average (ng/nm^3)	Case 1 (ng/nm^{3c})	Case 2 (ng/nm^{3d})
Arsenic	NR[e]	7.3	NR	36	21.5	7.3	22
Beryllium	2.5	NR	22	NR	12.3	2.5	13
Cadmium	3,400	7,300	108	292	2,775	3,400	3,001
Chromium[f]	200	652	NR	569	474	200	483
Cr(VI)[g]	20	93	NR	81	65	20	66
Nickel	3,200	295	NR	1,136	1,544	3,200	1,627
PCBs	NR	NR	NR	8.1	8.1	8.1	12
PAHs[h]	NR	NR	43	15	29	15	42
C1-Benzene	NR	NR	NR	292	292	82	423
C1-Phenols	NR	NR	NR	426	426	246	618
Vinyl-C1	NR	700	NR	NR	700	700	1,015
2,3,7,8-TCDD	0.3	2.1	NR	NR	1.2	0.3	1.7

[a] All data are corrected to 12% CO_2. When a range of data are given, the average value is used.
[b] Quebec City data includes all operating modes except 200°C.
[c] C1-benzenes and C1-phenols area based on the 140°C wet-dry system.
[d] California Air Resources Board, 95% on-design, 5% off-design scenario. For metals, on-design is the average of the data, and off-design is the highest value shown. For organics, on-design is the average, and off-design is a factor of 10 times the average. Methodology prescribed by ARB. Case 1 emissions of cadmium and nickel are greater than in Case 2 and are used for both cases.
[e] Not reported.
[f] Total chromium.
[g] PAH emissions are adjusted to reflect that only 12% of the emissions are carcinogenic based on testing at Chicago Northwest, Charlottetown, and Nurnberg.
[h] Assumed 1/7 of total chromium is hexavalent unless specifically measured as prescribed by ARB and SCAQMD.

TABLE 2
Emissions Data for the Proposed Milliken WTE Facility[5]

Pollutant	Case 1 emission rate (g/s)	Case 2 emission rate (g/s)
Antimony	5.7×10^{-5}	5.7×10^{-5}
Arsenic	5.4×10^{-7}	1.6×10^{-6}
Beryllium	1.8×10^{-7}	9.6×10^{-7}
Cadmium	2.5×10^{-4}	2.5×10^{-4}
Total chromium	1.5×10^{-5}	3.6×10^{-5}
Hexavalent chromium	1.5×10^{-6}	4.9×10^{-6}
Copper	4.5×10^{-4}	4.5×10^{-4}
Lead	1.9×10^{-3}	1.9×10^{-3}
Manganese	1.4×10^{-4}	1.4×10^{-4}
Mercury	2.1×10^{-2}	2.1×10^{-2}
Hydrofluoric acid[a]	6.0×10^{-2}	6.0×10^{-2}
Nickel	2.4×10^{-4}	2.4×10^{-4}
Selenium	3.6×10^{-5}	3.6×10^{-5}
Tin	2.7×10^{-3}	2.7×10^{-3}
Vanadium	2.0×10^{-4}	2.0×10^{-4}
Zinc	1.4×10^{-1}	1.4×10^{-1}
Hydrochloric acid[a]	6.6	6.6
Polynuclear aromatic hydrocarbons (PAHs)[b]	1.1×10^{-6}	3.1×10^{-6}
Polychlorinated biphenyls	6.0×10^{-7}	8.8×10^{-7}
Vinyl chloride	5.2×10^{-5}	7.5×10^{-5}
Chlorinated benzenes	6.0×10^{-6}	3.1×10^{-5}
Chlorinated phenols	1.8×10^{-5}	4.6×10^{-5}
2,3,7,8-TCDD1[c,d]	2.2×10^{-8}	
2,3,7,8-TCDD2[c,e]		1.3×10^{-7}

[a] 90% control efficiency.
[b] Total carcinogenic PAH compounds as benzo(a)pyrene.
[c] Toxic equivalent 2,3,7,8-TCDD emissions using DHS scenario 4 weighting scheme.
[d] Based on Marion County.
[e] Average of Marion County and Wurzburg; CARB 95% on-design, 5% off-design scenario.

TABLE 3
California Department of Health Services Toxic Equivalent Weighing Scheme for PCDDs and PCDFs[8]

Homolog	Isomer	Dioxin potency	Furan potency
Tetra	2,3,7,8	1	1
	Non-2,3,7,8	0	0
Penta	2,3,7,8	1	1
	Non-2,3,7,8	0	0
Hexa	2,3,7,8	0.03	0.03
	Non-2,3,7,8	0	0
Hepta	2,3,7,8	0.03	0.03
	Non-2,3,7,8	0	0
Octa	All	0	0

TCDD1, is the expected value based on the average of three source tests. The second scenario, TCDD2, is the worst-case emissions estimate. The TCDD2 estimate was derived from the Wurzburg test data, which had the highest emissions of the three facilities tested.

2. Chromium Emissions

The emission rate of chromium is another technical issue that significantly influences health risk estimates. Chromium exists in two valence states, trivalent (Cr^{3+}) and hexavalent (Cr^{6+}), both of which are emitted from a WTE facility. However, only Cr^{6+} is considered carcinogenic. Determination of the carcinogenic potency and assessment of cancer risk from exposure to chromium is based on the Cr^{6+} oxidation state.[9]

In the emissions testing conducted at the Marion County facility, both total and hexavalent chromium were measured at the inlet and outlet of the air pollution control equipment. Because the quantity of hexavalent chromium measured at the outlet of the baghouse was at or below the detection limit, the detection limit concentration was used to estimate hexavalent chromium emissions for the most plausible scenario estimate. Based on the outlet detection limit, Cr^{6+} was 10% of total chromium. However, based on the inlet sampling, Cr^{6+} was less than 0.1% of total chromium. Since oxidation did not appear to occur in the scrubber or baghouse, estimating the Cr^{6+} emission rate using the detection limit (10% of total chromium) was considered to be a conservative assumption.[5]

Upper-bound, worst-case emission rates assume that 1/7 of the total chromium present was in the hexavalent form.[5] This assumption has been recommended by regulatory agencies in California and is consistent with the assumption concerning chromium speciation used by EPA to develop the cancer potency slope for chromium.

3. Off-Design Emissions

Estimation of off-design emission rates remains highly uncertain. Off-design operation can be defined as failure to maintain combustion gas temperatures at or above 1800°F for 2 s. The data required to evaluate emissions associated with off-design operation are unavailable, because source testing does not usually measure off-design emission rates.

Modern WTE facility design incorporates auxiliary burners designed to maintain adequate combustion temperatures within the stoker. The set point for the burners is typically 1800°F, which would presumably prevent the occurrence of off-design operation. However, the use of auxiliary burners as a mitigation measure has not been considered acceptable for permitting purposes, since adequate test data demonstrating the effect of auxiliary fuel on PCDD/PCDF emissions are not available. Modeling studies suggest that auxiliary fuel use would decrease dioxin formation by promoting the destruction of precursor compounds.[10]

To estimate off-design emission, two assumptions concerning the annual frequency of occurrence and the magnitude of resulting emissions were made. The procedure for estimating off-design emission rates, which was developed in conjunction with regulatory agencies, assumes that a specific fraction of operating time was off-design. The magnitude of off-design emissions was based on emissions from the incinerator that had the highest emission rate estimates. For the Milliken WTE risk assessment, it was recommended that the Toronto, Canada, incinerator test data be used for estimating the magnitude of the emissions and that the annual frequency of occurrence of off-design operation by 5% of annual operating time.[5] This 95% on-design, 5% off-design assumption was considered conservative based on the limited data available from existing WTE facilities.

C. PARTICLE DEPOSITION MODELING

Deposition of particles onto the ground is used to evaluate exposures through the food chain. Pollutant deposition is assumed to be proportional to concentration of pollutants in air. Hence, deposition is estimated using a proportionality constant referred to as a deposition velocity. The deposition velocity is expressed in units of centimeters per second. The methods for evaluating deposition are critical and greatly influence the results of a risk assessment. The assumptions used and the values selected for deposition velocities tend to over predict the magnitude of particle deposition from WTE facilities.

Deposition of particles emitted from WTE facilities is a function of particle size, density, meteorological conditions, and terrain. Deposition is not constant over time changes with variations in wind speed, stability, and vegetative canopy. All of these factors are considered in estimating the rate of deposition of facility-emitted pollutants.

Existing techniques for estimating the rate of deposition are not adequate to provide a realistic simulation. The most commonly used dispersion models were not specifically designed to simulate dispersion of fine particulate matter. EPA-preferred dispersion models typically have a deposition algorithm in which the user specifies a settling velocity computed from Stokes' law and a reflection (or resuspension) coefficient dependent on settling velocity. This type of deposition model, however, is not recommended for particles with a diameter of less than 20 μm for two reasons: (1) the dominant mechanism of deposition for particles with diameters less than 20 μm is not gravity but diffusion and (2) the reflection coefficient for particles with a settling velocity of 3 to 5 cm/s (i.e., particles that are 10 to 20 μm in diameter) is essentially 100%. Hence, the dispersion model used in many regulatory applications assumes 100% reflection, which means that particles do not deposit onto environmental surfaces. It essentially treats emitted particles as gases. The main problem with this approach is that it is not mass conservative, because the model does not subtract mass from the plume. Thus, particulate emitted from a WTE facility are assumed to both deposit from the plume (which is estimated by using a deposition velocity) and are allowed to disperse throughout the air as well. This approach also neglects the influence of meteorology and variations in particle size and terrain in estimating deposition.

However, modern WTE facilities equipped with a dry scrubber and baghouse filter emit a very small fraction of particles in the 20 to 50 μm size range; most emitted particles are less than 20 μm in diameter. In the absence of approved models and techniques, several alternative models have been developed for simulating particle deposition. The most common method for simulating dispersion and deposition of particulate matter involves multiplying the modeled concentration in air by a chemical-specific deposition velocity, typically 1 to 2 cm/s. This approach is used in screening analyses but is considered conservative, since it tends to overestimate deposition and ground-level air concentrations.

A more realistic approach utilizes the work of Sehmel and Hodgson[11] to estimate a deposition velocity that is dependent on particle size, meteorological conditions, and surface roughness (terrain effects). They also developed[11] a series of nomographs that relate particle diameter and density to wind speed, stability, and surface roughness. These nomographs have been incorporated into a computer model that was developed by ARB. This model uses hourly meteorological data and a size distribution of the emitted particles to obtain an hourly, size-dependent estimate of deposition velocity. These values are then used to compute deposition fluxes on an hourly basis, which are averaged over 1 year to determine annual average deposition rates. This method is also not mass conservative, but it does provide deposition velocities that are more realistic and reflect changing weather conditions. For a typical WTE facility equipped with a baghouse, the deposition velocity for emitted pollutants can range from 0.05 to approximately 1 cm/s, depending on the particle size distribution used in the model.

Another approach to deposition modeling is currently under development.[12] In this approach, Sehmel and Hodgson's nomographs are used to calculate a deposition rate, but a site-specific reflection coefficient is specified by the user. The difficulty of this approach is selection of a justifiable reflection coefficient, as this task involves an extensive research effort that is not feasible for health risk assessments prepared to support regulatory permitting. To date, this method has not been used or proposed for use in California.

Facility emissions in the Milliken WTE health risk assessment were modeled using the Industrial Source Complex-Short Term (ISCST) dispersion model. Deposition of pollutants onto environmental surfaces was modeled using a modified version of ISCST developed by the Radian Corporation[5] (based on Sehmel and Hodgson[11]), in which deposition is calculated as a function of particle size and meteorological conditions. Particle size distributions were estimated from tests conducted at the WTE facility in Wurzburg, West Germany. Two size distributions were used in the modeling to account for the effects of fine particulate enrichment

TABLE 4
Particle Size Distribution Used in Estimating Deposition Velocities[54]

Particle size (μm)	Mass fraction (%)[a]	Area (%)
0.1	13.25	47.1
0.18	13.25	26.2
0.33	13.25	14.3
0.60	13.25	7.9
1.0	4.0	1.4
1.6	3.0	0.7
3.2	10.0	1.1
5.0	8.0	0.6
7.3	8.0	0.4
10.8	6.0	0.2
40.0	8.0	0.1

[a] Original data listed 53.7% by mass less than 0.6 μm. Mass was distributed over the range of 0.1 to 0.6 using a log normal distribution.

of organic compounds. Metals were assumed to be an integral part of the flyash and to be evenly distributed on a mass basis. Semi-volatile organics were assumed to adsorb to the surface of particles during cooling of the flue gas and, thus, would distribute according to surface area. The surface area-weighted distribution (Table 4) indicates that over 99% of the total available surface area is in the 0 to 2 μm range. As a result of the different weighing schemes, metals had an estimated deposition rate of 0.6 cm/s, while the semi-volatile organics had an estimated rate of 0.06 cm/s.

Estimated health risks were developed for two exposure scenarios. These scenarios describe the best estimate of upper-bound risk to the maximally exposed individual. These results represent the bounds of estimated risks, given the ranges of values for the data inputs. A deposition velocity of 1 cm/s, recommended by the California Air Resources Board, provides an upper-bound estimate of particle deposition. The lower bound estimate (0.6 cm/s for trace metals and 0.06 cm/s for semi-volatile organic compounds) accounts for the distribution of organic emissions onto particle surface areas.[5]

D. ATMOSPHERIC DISPERSION MODELING

Dispersion models typically used to estimate ground-level air concentrations are Gaussian plume models included in the Users' Network of Applied Models for Air Pollution (UNAMAP) series developed for the EPA and used for air quality impact analyses required under the Clean Air Act.[13] ISCST is part of this collection. These models are used to predict ambient air concentrations based on meteorological conditions (wind speed, direction, and stability class) and stack characteristics of the facility (stack height, stack temperature, and exit velocities of emitted pollutants).

For a given emission rate, the dispersion model will estimate the ground-level air concentration of all emitted pollutants of concern expressed in units of micrograms of pollutant per cubic meter of air. In most instances, a unit emission rate (e.g., 1 g/s) is assumed. The actual concentration of each pollutant emitted is determined by multiplying the concentration estimated when a unit emission rate was assumed ($\mu g/m^3/g/s$) by the actual emission rate. Pollutant concentrations in air are used to estimate inhalation exposure.

Ambient air concentrations of pollutants emitted from the proposed Milliken WTE facility were estimated using the ISCST model, an EPA-recommended atmospheric dispersion model.[13] Annual averages of hourly surface wind speed, direction, cloud cover, and temperature obtained from the Ontario Airport were used to estimate the annual ground-level concentrations in the vicinity of the proposed facility. A high degree of conservatism is built into the model to ensure

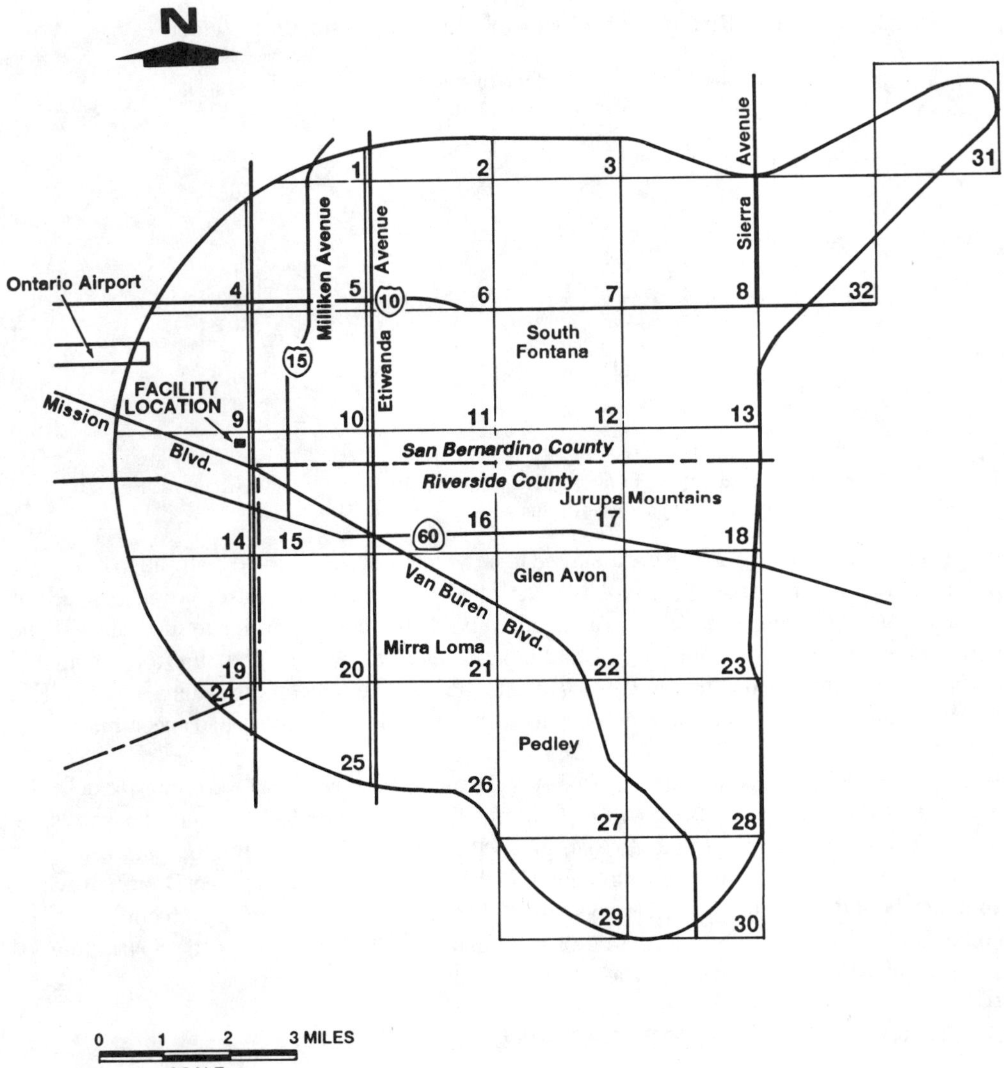

FIGURE 2. Study area and population sectors for the Milliken WTE facility.

that the estimated annual ground-level concentrations are overestimated. This conservatism is due, in part, to the Gaussian-based models that tend to overpredict ground level concentrations by a factor of two or greater.[5]

A technique for evaluating population exposures is to divide the geographical region surrounding the proposed WTE facility into sectors with a specific population in each sector. The area around the proposed Milliken WTE facility was divided into the 32 sectors shown in Figure 2. Ground-level air concentrations and deposition rates for all sectors are listed in Table 5. Estimated air concentrations for the center of each sector are assumed to represent the concentration to which all individuals living or working in that sector are exposed. The point of maximum ground-level air concentration was estimated to be 0.17 $\mu g/m^3$ and to occur at the point 500 m ENE of the facility.[5] Population at other locations are potentially exposed to emissions from the facility, but at lower levels that at the point of maximum ground-level concentration. The boundary of the population at risk was set at a point where ground-level air concentrations decreased to 1/10 of the maximum ground-level air concentration.[5]

TABLE 5
Modeled Ground-Level Air Concentrations and Deposition Velocities for the Milliken WTE Study Area

Sector	Land use[b]	Population (2060)	GLC[a] (μg/m³/g/s)[c]	Deposition flux (μg/m²/s/g/s)[d]
1	RES	192	0.018	0.0000285
2	RES	3,000	0.015	0.0000250
3	RES	19,309	0.014	0.0000224
4	C/I	6,383	0.011	0.0000182
5	C/I	6,383	0.023	0.0000379
6	C/I	6,383	0.023	0.0000377
7	RES	38,298	0.014	0.0000221
8	RES	44,681	0.039	0.0000622
9	C/I	6,383	0.013	0.0000211
10	C/I	6,383	0.059	0.0000964
11	C/I	6,383	0.056	0.0000894
12	C/I	6,383	0.029	0.0000466
13	RES	25,532	0.014	0.0000220
14	RES	19,149	0.011	0.0000177
15	RUR	6,383	0.082	0.0001300
16	RES	15,958	0.022	0.0000347
17	RES	15,958	0.011	0.0000183
18	RUR	3,192	0.008	0.0000126
19	RUR	6,383	0.012	0.0000187
20	RUR	1,277	0.087	0.0001400
21	RUR	19,149	0.066	0.0001050
22	RES	15,958	0.016	0.0000260
23	RES	31,915	0.015	0.0000239
24	RUR	6,383	0.017	0.0000274
25	RUR	6,383	0.035	0.0000553
26	RES	31,915	0.030	0.0000483
27	RES	25,532	0.045	0.0000711
28	RES	9,600	0.015	0.0000242
29	RES	22,340	0.022	0.0000345
30	RES	35,745	0.028	0.0000445
31	RES	27,300	0.021	0.0000342
32	RES	68,830	0.021	0.0000335

Note: Average rural ground-level concentration equals 0.038 μg/m³/g/s. Average rural deposition flux equals 0.0000598 μg/m²/s/g/s.

[a] Ground-level concentration.
[b] RES = residential; C/I = commercial/industrial; RUR = rural. See Figure 2 for the location of each sector.
[c] Microgram per cubic meter per gram per second.
[d] Microgram per meter squared per second per gram per second.

E. ENVIRONMENTAL FATE AND TRANSPORT ANALYSIS

To assess the extent of human exposure to facility-emitted contaminants, the environmental fate of those pollutants and the potential pathways of human exposure must be evaluated. Potential pathways of exposure are depicted in Figure 1. To monitor the environmental fate of pollutants emitted from WTE facilities, it is necessary to evaluate the rate at which chemicals enter a given medium and the pathway processes that affect chemical concentration in that medium. Pathway variables and chemical-specific parameters used in the environmental fate evaluation are listed in Table 6.[5]

A screening-level compartment model was used in the Milliken health risk assessment to simulate the transport of emitted chemicals through the environment. This model was developed largely from the procedures presented in the NRCs Regulatory Guide 1.109,[14] which has been

TABLE 6
Key Parameters Used in the Environmental Fate Analysis

Description	Input value
Soil concentrations	
Soil density	1500 kg/m^3
Mixing depth	15 cm
TCDD half-life in soil	12 years
PAH half-life in soil	20 years
Degradation of other pollutants in soil	No degradation
Deposition velocity	0.006, 0.0006 m/s
Edible plant concentrations	
Elimination rate constant	0.045/d
Edible crop density	2 kg/m^2
Crop intercept fraction	0.051
Livestock concentrations	
Feed rate	16 kg/d
Milk production rate	27 kg/d
Feed exposure period	2 d
Inhalation rate	158 m^3/d
Feed surface area	1 m^2
Water ingestion rate	150 kg/d
Pasture grass intercept fraction	0.55/d
Pasture grass yield	0.28 kg/m^2
Mother's milk concentrations	
Female fat content	18 kg
TCDD elimination rate constant	9.49×10^{-4}/d

used to evaluate the potential health impacts from routine releases of radionuclides to the environment. This model provides conservative estimates of the various processes affecting chemical fate and transport. The model is termed conservative because predicted environmental concentrations (and, hence, human health risks) are higher than would actually be encountered in the real world.

Assuming human exposure to facility-emitted pollutants from all other pathways other than direct inhalation requires estimating pollutant concentrations in soil, vegetation, and water due to deposition.

1. Assessing Pollutant Concentrations in Soil

Soil concentrations are used to calculate the concentration of chemicals in animals and humans due to direct ingestion of soil as well as pollutant concentrations in plants due to root uptake. Soil concentrations associated with emissions from the Milliken WTE facility were estimated assuming that 70 years of facility emissions were deposited onto soil surfaces at the start of facility operation and mixed uniformly to a depth of 15 cm.[14] To provide a conservative estimate, pollutant concentrations for most chemicals were not reduced by loss due to removal processes, such as volatilization, resuspension, leaching, and biodegradation, since these processes operate concurrently with deposition to yield equilibrium soil concentrations. The concentration in the soil in the absence of any removal processes was determined by the following equation:

$$C_s = D \times T \times 0.001/d \times M \quad (1)$$

where:

C_s = Concentration in soil (mg/kg)
T = Facility lifetime (70 years)
D = Deposition flux (μg/m^2/year)
d = Density of soil (1500 kg/m^3)
M = Mixing depth in soil (0.15 m)
0.001 represents the conversion from μg to mg

Exceptions to the above assumption were benzo(a)pyrene and TCDD. TCDD and benzo(a)pyrene levels in soil were estimated a half-life of 12[15] and 20 years,[16] respectively (Table 7). The calculation used to determine steady state concentrations in the soil, assuming both deposition and removal processes, was derived from NRC Regulatory Guide 1.109[14] and is as follows:

$$C_s = D \times 0.001 \times [1 - \exp(-k \times T)]/d \times M \times k \quad (2)$$

where :

C_s = Concentration in soil (mg/kg)
T = Facility lifetime (70 years)
D = Deposition flux (μg/m^2/year)
k = First-order elimination rate constant (per year^{-1})
d = Density of soil (1500 kg/m^3)
M = Mixing depth in soil (0.15 m)
0.001 represents the conversion from μg to mg

Table 8 presents estimated soil concentrations for pollutants emitted from the Milliken facility (70-year accumulation).

2. Assessing Pollutant Concentrations in Plants Due to Deposition

Backyard gardens, which are subject to deposition of facility emissions, present a potential source of human exposure. Produce grown in backyard gardens may be contaminated via two pathways: root uptake and direct deposition of pollutants onto outer plant surfaces. Pollutant concentrations in crops were assessed by estimating deposition onto above-ground foliage and accumulation through roots from the soil according to the following assumptions:

- Deposition was assumed to occur during the growing season only
- Pollutant concentration in crops was corrected for the fraction of deposited particles that are intercepted by foliage using an interception fraction (r)
- Pollutant concentration in crops was corrected for particle loss from plant surfaces caused by weathering using an elimination rate constant (k_e)
- Pollutant concentration in crops due to root uptake was assumed to be a linear proportion of the pollutant concentration in soil
- The growing season for all crops was assumed to be 60 d[17]
- Vegetative yield for all crops was assumed to be 2 kg/m^2 [5]

a. Interception Fractions

Interception fractions account for the fact that not all of the airborne material depositing within a unit area will accumulate on edible plant surfaces. Intercept fractions are plant-specific and take into account the change in plant surface area due to growth. Baes et al.[18] estimated interception fractions for pasture grasses (including grains), leafy vegetables, and exposed

TABLE 7
Chemical-Specific Variables Used in the Environmental Fate Evaluation and Exposure Assessment

Chemical	Soil elimination rate constant (years^{-1})	Root uptake from soil (crop/soil ratio)	Dermal uptake factor	Elimination rate constant—humans (days^{-1})	Elimination rate constant—livestock (days^{-1})
PCDDs/PCDFs[26]					
Best estimate	0.0693	0.01	0.03	5.13×10^{-4}	6.0×10^{-3}
Upper-bound estimate	0.0578	0.10	0.03	5.13×10^{-4}	6.0×10^{-3}
PAHs[17,20,21]					
Best estimate	0.0346	0.005	0.03	0.1	0.1
Upper bound estimate	0.0115	0.10	0.03	0.1	0.1
Trace metals[22]					
Arsenic	—	0.015	—	—	—
Cadmium	—	0.25	—	—	—
Chromium	—	0.005	—	—	—
Mercury	—	0.38	—	—	—
Lead	—	0.018	—	—	—
Nickel	—	0.06	—	—	—
Beryllium	—	0.004	—	—	—

TABLE 8
Soil Concentrations of Metals Due to Facility Emissions and Existing Background Levels

Compound	70-year soil conc due to facility emissions[a] (mg/kg)	Background soil conc[b] (mg/kg)
Antimony	2.344×10^{-5}	<2.0
Arsenic	3.168×10^{-7}	4.7
Beryllium	1.091×10^{-7}	1.8
Cadmium	1.467×10^{-4}	1.4
Chromium	5.867×10^{-7}	25.0
Copper	2.640×10^{-4}	15.0
Lead	1.085×10^{-3}	68.0
Manganese	8.214×10^{-5}	200—3,000
Mercury	1.232×10^{-2}	<0.05
Nickel	2.347×10^{-5}	22.0
Selenium	2.112×10^{-5}	<0.2
Tin	1.584×10^{-3}	2—200[c]
Vanadium	1.173×10^{-4}	43.0
Zinc	8.214×10^{-2}	50.0

[a] Based on rural average deposition rate.
[b] California Energy Commission.[55]
[c] Piscator.[56]

produce (fruits and vegetables). The interception fraction (r) for hay and pasture grasses was derived from the following equation:[18]

$$r = 1 - \exp(-2.88\ y) \qquad (3)$$

where y equals productivity (standing crop biomass) in kilograms per square meter. Assuming that annual standing crop biomass of pasture grasses is 0.28 kg/m^2,[17] the interception fraction would be 0.55. For grains, the interception fraction was assumed to be 0, since the grain portion is protected by husks and would not be contaminated by emitted pollutants.

The interception fraction for leafy vegetables and exposed fruits is based on a geometrical model, which assumes that the fractional surface area represented by these crops is a function of the distance between plants, the number of plants per unit area, and plant size.[18] Interception fractions calculated for the individual plants were weighted by their relative importance in U.S. agriculture (acreage of that crop planted). The average interception fraction for mature leafy vegetables and exposed fruits was assumed to be 0.15 and 0.052, respectively.[18]

To simplify the analysis, the intercept fractions for vegetation consumed humans were reduced to a single value which was considered to be representative of all types of vegetation in the human diet. The assumptions used to estimate this single value are

- The interception fraction for different crops was weighted by the proportion of that crop that is grown in home gardens
- The proportion of a crop from home gardens was assumed to be proportional to the surface area of crop available to intercept facility-emitted particles
- The intercept fraction for root vegetables was assumed to be 0

The percentage of crops originating from homegrown sources differs with different levels of urbanization (with families classified as urban, rural-nonfarm, and rural-farm). Table 9 presents the weighted interception fractions for different levels of homegrown crop consumption.

TABLE 9
Weighted Intercept Fractions (r)

Crop	Urban family	Rural/nonfarm family	Rural/farm family	Intercept fraction (r)
	Percentage of homegrown crops (C_h)			
Grains	0.007	0.009	0.008	0
Leafy vegetables	0.26	0.24	0.22	0.15
Other vegetables	0.40	0.35	0.34	0.052
Root vegetables	0.33	0.40	0.43	0
r[a]	0.06	0.054	0.051	

[a] $R = \Sigma (C_h \times r)$.

b. Weathering Processes

Following deposition, environmental processes being to remove particulate from plant surfaces. Weathering processes include wind, water (rainfall and irrigation), and dilution through growth. Chemical- and plant-specific weathering half-life values should be used when available in the literature. In the absence of measured values, Baes et al.[18] recommends the use of a 14-d weathering half-life in estimating food chain exposures.

Since removal by weathering tends to occur in an exponential fashion, a weathering rate constant (kc) can be derived from the following equation:

$$k_c = 1n2/T_{1/2} \tag{4}$$

where 1n2 is the natural log of 2, and $T_{1/2}$ is the weathering half-life in days. For a half-life of 14 d, the weathering rate constant is 0.05/d.

The concentration of pollutants on vegetation due to deposition can be estimated using the following equation:

$$C_d = D \times 0.001 \times r \times [1 - \exp(-k_c \times S)]/Y \times k_c \tag{5}$$

where:

C_d = Concentration in plants from deposition (mg/kg)
D = Deposition flux ($\mu g/m^2/d$)
r = Interception fraction (unitless)
k_c = Weathering rate constant (per day)
S = Length of the growing season (days)
Y = Plant yield or standing crop biomass (kg/m^2)

3. Assessing Pollutant Concentration in Plants Due to Root Uptake

Accumulation of pollutants by crops from root uptake is calculated using root uptake factors, which relate the concentration of pollutants in plants to the concentration of pollutants in soil. The calculation of root uptake from soil is

$$C_r = C_s \times U_r \tag{6}$$

where:

C_r = Concentration in plants from root uptake (mg/kg)

C_s = Concentration in soil (mg/kg)
U_r = Root uptake factor (unitless)

Briggs et al.[19] demonstrated that the root uptake and translocation of a chemical decreases as its octanol-water partition coefficient (K_{ow}) increases. Wang and Mersez[20] noted that low molecular weight PAHs appeared to be absorbed more readily than high molecular weight PAHs, and most of the PAH residues were concentrated in the skins of the crops. Chemical-specific root uptake factors should be used in an exposure assessment. In cases where field study data are not available, the translocation of chemicals from soil to crops can be approximated using the regression equation developed by Travis and Arms:[21]

$$\log U_r = 1.588 - 0.578 \log K_{ow} \tag{7}$$

Root uptake values for several trace metals[18] and PAHs[20,22-24] are summarized in Table 7. Wipf et al.[25] studied root uptake of TCDD following the explosion of the chlorophenol production plant at Seveso, Italy. These root uptake factors for TCDD, however, represent a worst-case scenario, because elevated soil concentrations were present compared to concentrations resulting from deposition of the relatively small amounts of particulate emitted from WTE facilities. Since TCDD is very lipophilic ($K_{ow} = 6.85$), root uptake may not be the major pathway of vegetative contamination.[26,27] Helling et al.[27] reported a root uptake factor of 0.017 for TCDD. Using the above regression equation, a root uptake value of 0.004 is obtained for TCDD. Generally, the accumulation of pollutants from soil represent a negligible contribution to concentrations in crops compared with direct deposition.

4. Assessing Pollutant Concentrations in Livestock

Cattle are exposed to facility-emitted pollutants through the consumption of contaminated feed, water, and soil and via inhalation. Ingestion of contaminated soil and vegetation represents a major pathway of exposure for terrestrial organisms, including beef and dairy cows.[26] Consumption of lipophilic chemicals, such as PCBs, PAHs, and TCDD by animals can result in their bioaccumulation in animal tissues and milk and their transferral to humans via ingestion of contaminated meat and milk products.

Quantifying chemical concentrations in animal tissues involves the use of a bioconcentration factor (BCF) specific for each chemical. BCFs are defined as the equilibrium concentration of chemical in a tissue (mg/kg) divided by the equilibrium concentration of chemical in food (mg/kg).[21] A large BCF indicates that the chemical is likely to accumulate in living organisms and that the food chain will be an important source of human exposure.

Concentrations of pollutants in beef and dairy products resulting from the ingestion of vegetation, soil, and water by livestock are calculated as:

$$C_T = (BCF_f)(F)(C_D) \tag{8}$$

where:

C_T = Concentration of chemical in tissue (or milk) (mg/kg)
BCF_f = Bioconcentration factor (fat basis) [(mg/kg)/(mg/kg)]
F = Fat content of tissue (kg fat/kg tissue)
C_D = Concentration of pollutant in the diet [vegetation, soil, and water (mg/kg)]

The concentration in beef or milk fat due to ingestion of contaminated feed, soil, and water is estimated by weighing the concentration in each dietary source by the fraction of total dietary intake that originates from that source, summing the resulting concentrations, and multiplying by the appropriate chemical-specific BCF. Pollutant concentrations in soil and pasture grass consumed by cows are estimated in the same manner as described in previous sections. Concentrations of pollutants in animal tissues are calculated assuming that the total fat content of beef is 45%.[28]

Feedlot and dairy cattle are fed hay or fodder that is assumed to be grown outside the area and stored under cover shielded from deposition. When cattle are fed, however, the feed is placed in open troughs where pollutants emitted from a WTE facility could contaminate it. The following assumptions were used to estimate the concentration of pollutants in this food source: feed is exposed to deposition for 48 h; the exposed surface area is 1 m^2 per bale of hay, and bale weight is 36 kg.

The concentration of pollutants in feed was estimated from the following equation:

$$C_f = D \times E \times 1\ m^2 \times 0.001/W_b \tag{9}$$

where:

C_f = Concentration in livestock feed (mg/kg)
D = Deposition flux (μg/m^2/s)
E = Exposure period (sec)
W_b = Bale weight (kg)
A factor of 0.001 corrects μg to mg

Drinking water in a dairy or feedlot is assumed to be supplied in circular tanks. Surveys of dairies in San Bernardino County showed that these tanks are approximately 10 to 12 ft in diameter, 1.5 to 2 ft deep, and serve 100 to 200 cattle.[5] The rate at which a dairy cow consumes water is a function of the ambient temperature. Using these data and assuming an average ambient temperature during the retention time of the water in the tank, the resulting pollutant concentration in drinking water was determined using the following equation:

$$C_w = D \times S_w \times R_w \times 0.001/W_w \tag{10}$$

where:

C_w = Concentration in water (mg/kg)
D = Deposition flux (μg/m^2/s)
S_w = Surface area of water tank (m^2)
R_w = Retention time of water in the tank (sec)
W_w = Weight of water in the tank (kg)
A factor of 0.001 corrects μg to mg

Cows can also be exposed through inhalation of contaminated particulates that are retained in the lung. The average inhalation rate for a cow is 158 m^3 of air per day.[29] Since the absorption efficiency for inhaled pollutants is assumed to be 100%, exposure from inhalation was obtained by multiplying the ground-level air concentration (mg/m^3) times the respiration rate (158 m^3/d).[5] Exposure due to ingestion was determined by multiplying the concentration of facility-emitted pollutants in feed, soil, pasture grass, and water by the average daily consumption rate of each dietary source (kilograms per day).

5. Assessing Pollutant Concentrations in Fish

Fish in nearby rivers can accumulate organic and trace metal pollutants in their tissues at levels thousands of times greater than the pollutant concentration in water. Pollutants can enter a water body from direct deposition and surface runoff from contaminated soil. The amount of pollutants entering the water via deposition is expressed as:

$$I_d = D \times A \times 86{,}400 \times 0.001 \quad (11)$$

where:

I_d = Input to the water from deposition (mg/d)
D = Deposition flux ($\mu g/m^2/s$)
A = Surface area of the river (m^2)
A factor of 0.001 is used to convert from µg to mg
A factor of 86,400 s/d converts deposition flux to a daily basis

A simplified approach[5] for estimating the quantity of soil runoff into a body of water assumes that the quantity of runoff from contaminated soil equals the average daily sediment loading to the river as estimated from U.S. Geological Survey (USGS) data.[30] The total area of the drainage basin was determined by inspecting topographical maps of the site surrounding the facility. The ratio of the sediment load to the drainage basin was defined as the runoff factor. The volume of contaminated soil runoff into the river was computed by multiplying the area of soil receiving deposition by the runoff factor. The quantity of pollutants entering via soil runoff (milligrams per day) are estimated by multiplying the pollutant concentration in soil (milligrams per kilogram) times the surface runoff factor (kilograms per day). The concentration of pollutants in water is then determined by dividing the sum of the pollutant mass input into the water from soil runoff and water deposition (milligrams per kilogram) by the river flow rate (kg/day).

Pollutant concentrations in fish are computed by multiplying the concentration of pollutant in water by a pollutant-specific BCF for fish. The BCF for a chemical in fish is defined as equilibrium concentration of pollutant in fish tissue divided by the equilibrium concentration of pollutant in water. Pollutant concentrations in water are typically calculated without taking into account removal processes, such as sediment impaction and burial, volatilization, photodegradation, and adsorption to particulate. BCF values should be obtained from the literature when possible or estimated from a chemical's solubility or octanol/water partition coefficient (K_{ow}) using the methods described in Kenaga and Goring.[31]

F. EXPOSURE ASSESSMENT

The objectives of an exposure assessment are to identify the relevant pathways of human exposure and to determine the total dose of pollutant taken in that is attributable to facility emissions. The following exposure pathways were determined to be potentially significant sources of human exposure:

- Inhalation
- Ingestion of contaminated soil
- Ingestion of locally or home-grown fruits and vegetables
- Ingestion of locally produced meat and dairy products
- Ingestion of mother's milk by infants
- Dermal absorption

Exposure scenarios are created that define the location of individuals relative to the site, frequency and duration of exposure, and the pathways of exposure. A common exposure scenario that serves as a basis for risk management is the maximally exposed individual (MEI). An MEI is a hypothetical individual located at the point of maximum TAC exposure to emissions from a WTE facility for 70 years. This conceptual individual is assumed to be born, breast-fed, educated, employed, and reside at this location. The rationale for evaluating health risks of an MEI is that estimated risks for the MEI represent the highest risk to which any individual may be exposed due to facility emissions. Initially, the MEI is assumed to be exposed to maximum pollutant concentrations (as predicted from dispersion modeling) continuously for 70 years. This assumption is consistent with numerical indicators of chronic toxicity, which also assume a continuous lifetime exposure. However, in selecting the MEI for a more realistic assessment, land use surrounding the proposed facility is considered. For example, if receptors are located in commercial business areas, exposure duration is assumed to be 8 h/d, 5 d/week, for 40 years.

Several of the model input and chemical-specific parameters used in assessing human exposure can have different values depending on the method of estimation or variation in experimental results. To account for the variability, two exposure scenarios were developed. The first scenario represents assumptions associated with deriving a plausible, upper-limit risk estimate. The second scenario represents worst-case assumptions. Key parameters used in the human exposure assessment for both scenarios are listed in Table 10.

1. Inhalation Exposures

The risk associated with inhaling ambient concentrations of particle-absorbed pollutants was determined using modeled ground-level air concentrations. Assuming that pollutants adsorbed to inhaled particles are completely absorbed into the body through the lung, inhalation exposure was calculated as a function of the concentration of pollutants in air, respiration rate, and body weight as follows:

$$E_i = C_a \times R \times 0.001/BW \quad (12)$$

where:

E_i = Exposure from inhalation (mg/kg/d)
C_a = Pollutant concentration in air (μg/m^3)
R = Adult respiration rate (m^3/d)
BW = Adult body weight (kg)
0.001 converts from μg to mg

An adult respiration rate of 20 m^3/d and body weight of 70 kg were assumed.[33]

This approach provides a conservative estimate of human exposure via inhalation, since it assumes that the MEI is continuously exposed to the maximum ambient ground-level air concentration for 70 years. This fact, combined with the 100% absorption of inhaled particulate, tends to overestimate inhalation exposure.

2. Ingestion Exposures

Ingestion of soil, backyard food crops, beef and dairy products from locally raised livestock, and fish from nearby streams or rivers, were considered in the health risk assessment. To account for differences in the bioavailability from these different matrices, chemical intake (mg/kg/d) was multiplied by a chemical-specific gastrointestinal (GI) absorption factor to determine the amount of chemical ingested that is actually retained in the human body. Pollutant-specific GI factors for both exposure scenarios are listed in Table 11. These adjusted dose values were used in estimating health risks from ingestion pathways.

a. Soil Ingestion

Pollutants in the soil can be accumulated into food crops or ingested directly. The quantity of soil ingested by humans varies with age, with maximum ingestion occurring in children 0 to 6 years old.[32] Soil ingestion was age-weighted to account for different levels of consumption in

TABLE 10
Key Parameters Used in the Exposure Assessment

	Case 1	Case 2
General		
Body weight	60 kg	60 kg
Exposure period		
Rural population	30 years	70 years
Residential population	30 years	70 years
Inhalation		
Respiration rate		
Residential population	18 m^3/d	18 m^3/d
Rural population	20 m^3/d	20 m^3/d
Fruit and vegetable consumption		
Percentage that is locally grown produce		
Rural population	17%	40%
Residential population	3%	15%
Cleansing efficiency	50%	25%
Consumption of mother's milk		
Infant milk consumption		
Rural population	850 ml/d	850 ml/d
Residential population	850 ml/d	850 ml/d
Infant body weight	5 kg	5 kg
Duration of feeding period	1 year	1 year
Ingestion of contaminated soil		
Soil consumption rate	0.07 g/d	0.07 g/d
Skin contact with soil		
Skin surface area	2940 cm^2	2940 cm^2
Dust on skin	1.5 $mg/cm^2/d$	1.5 $mg/cm^2/d$
Gastrointestinal absorption		
Pollutant-specific	100% absorption with the exception of TCDD	

TABLE 11
Gastrointestinal Absorption Efficiencies for Ingested Contaminants

Chemical	Absorption (%) (literature value)	Absorption best-estimate exposure scenario (%)	Absorption upper-bound exposure scenario (%)	Ref.
Arsenic	100	100	100	57
Beryllium	1	1	10	58
Cadmium	1—6	1	6	59
Chromium	3—11	3	11	60
Lead	10	10	10	61
Mercury	NA[a]	NA[b]	NA	
Nickel	10	10	100	62
PCDDs/PCDFs				
Adsorbed to fly ash or soil	5—10	5	15	63—66
Accumulated in crops	50	50	50	50
Accumulated in meat or milk	70—86	70	86	67—68
PAHs				
Adsorbed to fly ash or soil	NA	50	100	
Accumulated in crops	NA	50[c]	100	69,70

TABLE 11 (continued)
Gastrointestinal Absorption Efficiencies for Ingested Contaminants

Chemical	Absorption (%) (literature value)	Absorption best-estimate exposure scenario (%)	Absorption upper-bound exposure scenario (%)	Ref.
Accumulated in meat, milk, or fish	NA	100	100	

[a] Data not available.
[b] GI absorption was not estimated. Mercury is assumed to be emitted as a vapor.
[c] Absorption of 50% reported in rats gavaged with benzo(a)pyrene in starch solution or animal feed.[67]

infants, toddlers, children, and adults. Hawley[33] computed ingestion rates for different age groups, which were weighted to provide a single, average soil ingestion rate of 0.07 g/d. Exposure to absorbed pollutants resulting from the ingestion of contaminated soil was calculated using the following equation:

$$E_s = C_s \times S_c \times GI_s/BW \tag{13}$$

where:

E_s = Exposure from soil ingestion (mg/kg/d)
C_s = Concentration in soil (mg/kg)
S_c = Age-weighted soil consumption rate (70 μg/d)
GI_s = Gastrointestinal absorption factor for pollutants adsorbed to soil (unitless)
BW = Adult body weight (kg)

b. Crop Ingestion

Exposure through ingestion of contaminated plants may result if the local population consumes crops which are grown in the vicinity of the proposed WTE facility. The fraction of food that is grown or produced locally must be determined, as fruits and vegetables obtained from retail markets are assumed not to be affected by facility emissions and are not considered in the assessment. Home-grown produce provides a larger contribution to the diets of rural-farming and rural-nonfarming families than to the diets of urban families. Of all U.S. families 32% are designated as urban, 9% are rural-nonfarming, and 3% are rural-farming, and 56% of U.S. families are classified as nongardening.[34] Estimation of the contribution of home-grown produce for different levels of urbanization has not been performed specifically for households in California, so these nationwide values were applied to California residents in the Milliken health risk assessment.

The EPA reported that the median daily consumption of six crop food groups (cereals/grains, potatoes, leafy vegetables, legumes, root vegetables, and fruits) totals 738 g per person per day.[34] These consumption values were adjusted by percentage of annual consumption that was obtained from home-grown sources to derive consumption rates for home-grown produce in grams per person per day. The percentage of annual consumption that is home-grown was obtained from U.S. Department of Agriculture (USDA) food consumption surveys[35] and are reported by the EPA.[34] The percentage of total annual consumption from home-grown sources was calculated for urban and rural-nonfarming families. Since rural-farming households constitute an extremely small fraction of the population, this group was not included in the estimation of home-grown crop consumption rates. These data show that home-grown crop consumption for an urban household equals 29.4 g per person per day, partitioned as 19.4 g/d

TABLE 12
Average Percent of Annual Consumption that is Home-Grown for Various Foods: Urban and Rural Nonfarm Households[34,35]

Food group	Percentage	
	Urban	Rural/nonfarm
Milk, cream, cheese	0	3.99
Fats, oils	≤0.21	1.63
Flour, cereal	≤0.13	0.82
Meat	0.78	6.14
Poultry, fish	2.85	11.90
Eggs	0.62	9.14
Sugar, sweets	1.61	4.95
Potatoes, sweet potatoes	1.21	14.62
Fresh vegetables	7.55	35.65
Fresh fruits	3.24	13.20
Juice (vegetable, fruit)	0.68	4.13
Dried vegetables, fruits	2.63	6.9

TABLE 13
Adjusted Home-Grown Produce Consumption[34]

Food group	Percent consumption from home-grown sources				
	Median consumption (g/d)	Urban households (%)	Urban consumption (g/d)	Rural households (%)	Rural consumption (g/d)
Cereals	157.2	0.13	0.20	0.82	1.29
Potatoes	141.7	1.21	1.70	14.62	20.72
Leafy vegetables	99.7	7.55	7.53	35.65	35.54
Legumes	98.0	7.55	7.40	35.65	34.94
Root vegetables	109.7	7.55	8.28	35.65	39.11
Fruits	131.8	3.24	4.27	14.20	17.40
Total	738.1		29.4		149.0

of above-ground produce (cereals, leafy vegetables, legumes, and garden fruits) and 10.0 g/d of root vegetables (potatoes, other root vegetables). Home-grown crop consumption for a rural-nonfarming household totals 149.0 g per person per day, partitioned as 89.2 g/d above-ground produce and 59.8 g/d root vegetables. Table 12 presents the percentage of annual consumption from home-grown sources developed by USDA.[35] Table 13 presents the EPA median consumption values as adjusted by these percentages.[34]

Food preparation activities, such as washing, blanching, cooking, peeling, and canning, can remove surface particulate from contaminated foliage. Removal of pesticide residues (DDT, carbaryl, and parathion) from green beans, spinach, and broccoli by blanching ranged from 48 to 84%, while washing and blanching together resulted in the removal of 60 to 97%.[36,37] Peeling tomatoes and potatoes removed virtually all pesticide residues.[38,39] A study of PAH residues in lettuce grown near known emission sources (a highway and an aluminum smelter) reported that washing considerably reduced the foliar levels of high molecular weight PAHs.[38] The amount of benzo(a)pyrene removed by washing ranged from 67 to 95%, with a mean removal of 87%. The mean removal of six PAHs was 50%. Washing had little effect on removal of the low molecular weight PAHs (phenanthrene and fluoroanthrene). The authors suggested that low molecular weight PAHs, when in the vapor phase, penetrate the waxy layer of lettuce. Higher molecular weight PAHs remain exposed on foliar surfaces and could be readily removed.[38] A cleansing removal factor of 25% was used in the Milliken assessment.[39]

The fractional absorption of pollutants by the human body is accounted for through the use of GI absorption factors (Table 11). Exposure from ingestion of home-grown crops was calculated as follows:

$$E_c = [(C_d \times E_r \times GI_p) + (C_r \times GI_t)] \times V_a + (C_r \times V_r \times GI_t)/BW \tag{14}$$

where:

E_c = Exposure from crop consumption (mg/kg/d)
C_d = Concentration in crops from deposition (mg/kg)
E_r = Efficiency factor for particle removal from crop surfaces by food preparation (unitless) (25%)
GI_p = Gastrointestinal absorption factor for pollutants adsorbed to particulate on crop outer surfaces (unitless)
C_r = Concentration in crops from root uptake (mg/kg)
GI_t = Gastrointestinal absorption factor for pollutant absorbed into crop tissue (unitless)
V_a = Average consumption of above-ground crops (kg/d)
V_r = Average consumption of root crops (kg/d)
BW = Adult body weight (kg)

c. Meat and Milk Ingestion

The average consumption of beef and dairy products, based on the average of rural and urban family data, is 5.82 lb of meat and 14.84 qt of milk per family per week.[35] The average per-capita beef and dairy product consumptions for purposes of estimating exposure is 105.4 and 557 g per person per day. These consumption rates were exposed on a fat basis, since the BCFs for cow tissue and cow milk are based on accumulation of pollutants in the fatty portions of these tissues. The fat content of beef and dairy products consumed by humans was assumed to be 15 and 7.8%, respectively.[40] These fat-adjusted consumption rates were then summed into a single rate for exposure assessment and multiplied by the fraction of meat and dairy products that is produced in the local area (Table 13). Human exposure via ingestion of contaminated beef and dairy products was calculated as:

$$E_{bd} = C_{bd} \times I_{bd} \times GI/BW \tag{15}$$

where:

E_{bd} = Exposure from consumption of locally produced beef and dairy products (mg/kg/d)
C_{bd} = Concentration of pollutants in beef and dairy products (mg/kg)
I_{bd} = Ingestion rates for beef and dairy products (mg/kg)
GI = Chemical-specific GI absorption factors for pollutants in beef and dairy products (unitless)
BW = Adult body weight (kg)

d. Fish Ingestion

Human exposure from consumption of contaminated fish can be calculated in the same manner as for consumption of beef and dairy products using Equation 21 and an average fish consumption rate of 6.5 g/d.[19]

3. Mother's Milk Exposures

Another potential pathway of human exposure is ingestion of contaminated breast milk by infants. The level of pollutants present in human breast milk depends on the level of exposure and the rate of elimination from the mother's body.[42] The elimination rate constant, k_e, was determined from the following phramacokinetic equation assuming a first order rate relationship:[42]

$$Ke = 1n2/T_{1/2} \tag{16}$$

where $T_{1/2}$ equals the half-life of a chemical in the human body, and 1n2 is the natural log of two. Selection of a $T_{1/2}$ value that is representative of exposure to a mixture of PCDDs and PCDFs involves consideration of the estimated half-lives of individual compounds and their relative contribution to individual exposure. To estimate a representative half-life in adipose tissue, it was assumed that the half-life for all PCDDs was 5.8[43] years and the half-life for all PCDDs was 2.5 years.[44,45] Assuming that dose rate for humans is proportional to the emission rate of PCDDs and PCDFs, then PCDFs represent 63.4% of total exposure, while PCDDs represent 36.6%. Using this ratio, a $T_{1/2}$ of 3.7 years was used to represent the mixture of PCDDs and PCDFs present in humans. Metals were not assumed to accumulate in breast milk at toxicologically significant levels.

Adipose tissue levels have been found to equal breast milk levels when both are expressed on a fat basis.[46] Hence, the concentration of pollutants in breast milk can be estimated from adipose tissue levels assuming that the average female contains 30% body fat[47] and that the average fat content of human breast milk is approximately 4%.[48] Infant exposure depends on breast milk consumption rate, which was assumed to be 850 ml/d[49] for 1 year. The absorption of TCDD by the infant is assumed to be 100%. Since cancer potency estimates are based on 70-year exposure period, an individual's lifetime exposure was normalized by multiplying by 1/70 to account for the decreased exposure period as an infant. Exposure from consumption of contaminated breast milk was estimated according to the following equation:

$$E_{bm} = C_m \times I_m \times GI/BW \tag{17}$$

where:

E_{bm} = Exposure from consumption of breast milk (mg/kg/d)
C_m = Pollutant concentration in breast milk (mg/ml)
I_m = Ingestion rate of breast milk (ml/d)
GI = Gastrointestinal absorption factor (100%)
BW = Infant body weight (5 kg)

4. Dermal Exposures

Dermal exposure can result from direct deposition of pollutants onto skin or contact with surface dust. Pollutant concentrations on the skin are based on the rate of dust accumulation per skin surface area and the amount of exposed skin.[40] Absorption of trace metals through the skin was assumed to be unlikely. The absorption of TCDD following contact with soil was studied by Poiger and Schlatter.[50] Depending on the soil concentration, TCDD absorption through the skin ranged from 0.05 to 2.0%, yielding uptake factors ranging from 0.0005 to 0.02.[50]

Comparison of dermal exposure from airborne pollutants vs. contact with surface dust indicates that airborne emissions are insignificant relative to exposure via contact with surface dust. Therefore, deposition of airborne emissions were ignored in the following calculation of exposure

$$E_d = D_s \times S \times C_s \times U/BW \tag{18}$$

where:

E_d = Exposure from dermal contact (mg/kg/d)
D_s = Dust loading on the skin (mg/cm^2)
S = Exposed surface area of skin (cm^2)
C_s = Concentration of pollutant in soil (mg/mg)
U = Uptake factor (unitless)
BW = Adult body weight

G. DOSE-RESPONSE ASSESSMENT

Estimating cancer risks in health risk assessments are very sensitive to the procedures used for determining the dose-response relationship of chemical carcinogens. Typically, individual cancer risk is the factor considered most carefully in the health risk assessment for WTE emissions, as noncancer health effects generally are insignificant in comparison with cancer risks. The following sections briefly describe the procedures for evaluating potential adverse health effects from exposure to WTE emissions.

1. Cancer Risks

Evidence of chemical carcinogenicity comes from two sources: lifetime studies with laboratory animals and human (epidemiological) studies in which excess cancer risk has been associated with exposure to chemical carcinogen. If a carcinogenic response occurs at the exposure levels studies, it is assumed that a carcinogenic response will occur at all doses. Exposure to any level of a carcinogen is considered to have a finite risk of inducing cancer (i.e., carcinogens do not have thresholds).

Since risks at low levels of exposure cannot be measured directly in either animal or epidemiological studies, mathematical models are used to extrapolate from high to low doses. There is no universally acceptable scientific basis for any model that relates cancer risk with the extremely low concentrations present in the environment to which humans are most likely to be exposed. The linearized, multistage model procedure for low dose extrapolation is recommended by both the U.S. EPA and the State of California.[51,52] Use of the multistage model leads to a plausible, upper-limit risk estimate.

The model incorporates a procedure for estimating the largest possible slope at low extrapolated doses that is consistent with experimental dose-response data (the larger the slope, the higher the estimate of cancer risk). The animal test data used for the extrapolation is the most sensitive species, based on the assumption that man is equally sensitive as the most sensitive animal species. The risk estimates made with this model should be regarded as conservative, representing the most plausible, upper limit of risk. That is, the true risk is not likely to be higher than the estimate and is most likely lower.

Numerical estimates of cancer potency are presented as carcinogenic potency slopes (CPS). Under an assumption of dose-response linearity at low doses, the potency slope defines the cancer risk due to continuous lifetime exposure of one unit of carcinogen concentration. The EPAs Carcinogen Assessment group (CAG) has ranked the relative carcinogenic potencies of the 55 chemicals suspected to be human carcinogens, while the California Department of Health Services (DHS) has evaluated the carcinogenicity of 7 airborne contaminants. Table 14 lists the cancer potency factors used in the Milliken health risk assessment.

Cancer risk assessment involves calculating upper-bound estimates of individual cancer risk as well as the population risk (excess number of cases within a defined population). The individual cancer risk estimate of the MEI is most frequently used as a basis for risk management. Individual cancer risk is calculated as the product of exposure to a pollutant (mg/kg/day) and the CPS for that pollutant, (mg/kg/day). In some cases, different cancer potency slopes are developed separately for inhalation and ingestion exposure pathways. Cancer risks

TABLE 14[2,42]

Relative Carcinogenic Potencies of Chemicals Evaluated by the EPA Carcinogenic Assessment Group or California Department of Health Services[a]

	Potency slopes (mg/kg/d^{-1})			
	Best-estimate scenario		Upper-bound scenario	
Chemical	Inhalation	Ingestion	Inhalation	Ingestion
Arsenic	50	15	50	15
Benzo(a)pyrene	6.11	23[b]	6.11	23
Beryllium	4.86	4.86	4.86	4.86
Cadmium	6.1	6.1	40	40
Chromium VI	41	41	500	500
Nickel	1.7	1.7	1.7	1.7
Dioxins/furans	1.3×10^5	2.6×10^5[b]	1.56×10^5	3.1×10^5

[a] Animal slopes are 95% upper-bound slopes on the linearized multistage model. They are calculated based on animal oral studies, except for those indicated by I (animal inhalation), W (human occupational exposure), and H (human drinking water exposure). Human slopes are point estimates based on the linear, nonthreshold model. Not all of the carcinogenic potencies presented in this table represent the same degree of certainty. All values are subject to change as new evidence becomes available. Values are upper-bound in the sense that the true value, which is unknown, is not likely to exceed the upper-bound and may be much lower, with a lower-bound approaching zero. Thus, the use of the slope estimate in risk evaluation requires an appreciation for the implication of the upper-bound concept as well as the "weight of evidence" for the likelihood that the substance is a human carcinogen.

[b] These ingestion slopes have been corrected for partial gastrointestinal absorption of pollutants.

from exposure to multiple carcinogens and multiple pathways of exposure are assumed to be additive based on EPA carcinogen risk assessment guidelines.[51] Population burden is estimated as the product of individual cancer risk estimated at various receptors surrounding a facility times the populations assumed to be located at each receptor.

2. Noncancer Risks

Potential effects of chronic exposure to the noncarcinogenic compounds present in WTE emissions are assessed by comparing exposure levels with chronic acceptable daily intake (ADI) levels. Substances that cause systemic toxicity, (i.e., toxic effects other than cancer) appear to do so through mechanisms that include physiological thresholds. Thus, a certain dose of compound must be present before toxic effects will be observed. ADIs represent the level of chemical intake that is not expected to produce adverse effects over a lifetime of exposure. ADIs are calculated by dividing a NOEL, NOAEL, or LOAEL dose (mg/kg/day) by an uncertainty or safety factor that typically ranges from 10 to 10,000. The acronyms "NOEL", "NOAEL", "LOAEL" are defined as follows:[53]

- NOEL: No observed effect level. The dose at which there are no statistically or biologically significant increases in the frequency or severity of effects between the exposed and control populations.
- NOAEL: No observed adverse effect level. The dose at which there are no statistically or biologically significant increases in the frequency or severity of adverse effects between the exposed and control populations (Effects which are not considered adverse may occur at this dose level.)
- LOAEL: Lowest observed adverse effect level. The lowest dose of chemical in a study or group of studies that produces statistically or biologically significant increases in the frequency or severity of adverse effects between the exposed and control populations.

In many cases, these ADIs are also called reference doses (RfDs). RfDs are defined as estimates of lifetime daily exposure of a noncarcinogenic substance for the general human population which appears to be without an appreciable risk of deleterious effects. Historically, this concept was denoted by the term acceptable daily intake (ADI); however, the term acceptable implies that exceeding the set level is unacceptable. This term is gradually being replaced by RfD, implying greater flexibility in the characterization of health risks from exposure to noncarcinogens. For most purposes, ADI and RfD are equivalent.

H. RISK CHARACTERIZATION

Risk characterization involves estimating the magnitude of potential adverse effects and making summary judgments about the nature of the health risk threat to the public. Characterization of risks involves combining the results of the dose-response and exposure assessments to provide numerical estimates of health risk. Additionally, risk characterization summarizes the nature and weight of the evidence supporting the numerical estimates. The uncertainties in the evidence and how these uncertainties influence the results are also discussed in this portion of the risk assessment. A description of how risk characterization is presented in the health risk assessment case study for the Milliken WTE facility in Section VI.

The numerical results of this case study are estimated cancer risks from exposure of the maximally exposed individual from the carcinogens in the emissions. The cancer risks range from a plausible, upper-bound estimate of 3.1×10^{-7} to a worst-case estimate of 3.4×10^{-5}. Tables 15 and 16 show the contributions to total cancer risk from each pathway and each pollutant. Some observations about risk estimates obtained from this methodology are that emissions estimation and deposition modeling provide the greatest sources of uncertainty in the results. A particularly critical assumption is the speciation of chromium emissions. An assumption that all emitted chromium is the hexavalent species results in chromium driving the cancer risk estimate. The food chain pathway is a significant contributor to risks in the upper-bound scenario due to the deposition modeling methods.

VI. UNCERTAINTY IN HEALTH RISK ASSESSMENTS

The health risk assessment provides information needed for risk management decisions by evaluating the possibility for adverse health effects resulting from operation of a proposed WTE facility. As the results of a health risk assessment are reviewed by state and local regulatory agencies, technical issues arise, such as data selection, formulation of assumptions, and selection of exposure scenarios to be used in risk management. The selection of data inputs and models is influenced by several factors, including agency standards for data quality, data availability under the stringent permitting deadlines, and differences in judgment concerning the balance of conservatism and realism in the assessment methodology. In a regulatory application, the health risk assessment methodology is typically selected primarily to assure that risk management is based on a health conservative assessment.

Numerous assumptions and values are incorporated in the risk assessment. In several cases, the values for these inputs can be expressed as ranges. These differences in risk estimates that resulted from variability in assumptions and inputs are assessed using different exposure scenarios that reflects these ranges in values for assumptions and data inputs. The California Air Pollution Control Officers Association recommends using two different scenarios for risk management in their guidance on health risk assessments. One scenario typically represents a best estimate of risks, while the second represents a worst-case assessment. Supposedly, if the risk estimates from both scenarios are vastly different, it becomes important to determine the reasons for the differences and to understand how changes in the assumptions affect the predicted level of risk. It should be recognized that both scenarios provide conservative estimates of health risks due to the screening-level nature of the exposure assessment methods and the assumptions inherent in cancer risk assessment methods.

TABLE 15
Best Estimate Cancer Risks for the Milliken WTE Facility Case Study (case 1, rural)

	Pathway								
Pollutant	**Inhalation**	**Soil**	**Plant**	**Fish**	**Dairy**	**Dermal**	**Water**	**Mother's milk**	**Total**
Arsenic	2.349E-10	1.298E-11	1.664-10	0.000E+00	0.000E+00	2.453E-12	0.000E+00	0.000E+00	4.167E-10
Beryllium	2.621E-11	1.499E-12	1.857E-11	0.000E+00	0.000E+00	2.738E-13	0.000E+00	0.000E+00	4.650E-11
Cadmium	4.423E-08	1.466E-10	1.879E-09	0.000E+00	0.000E+00	4.619E-10	0.000E+00	0.000E+00	4.671E-08
Chromium	1.189E-09	7.227E-12	9.262E-11	0.000E+00	0.000E+00	1.242E-11	0.000E+00	0.000E+00	1.301E-09
Nickel	1.334E-09	7.371E-12	9.447E-11	0.000E+00	0.000E+00	1.393E-11	0.000E+00	0.000E+00	1.450E-09
PAHs	1.281E-10	2.657E-12	2.102E-10	0.000E+00	4.065E-11	5.021E-13	0.000E+00	1.087E-10	1.604E-10
PCB	5.286E-12	2.921E-13	3.744E-12	0.000E+00	1.678E-12	5.521E-14	0.000E+00	0.000E+00	1.105E-11
TCDD	1.275E-07	3.975E-12	1.238E-10	0.000E+00	2.565E-09	3.552E-10	0.000E+00	1.567E-07	2.874E-07
Total	1.448E-07	1.963E-11	3.721E-10	0.00E-00	2.573E-09	3.564E-10	0.000E+00	1.567E-07	3.125E-07

TABLE 16
Upper-Bound Cancer Risk Estimates for the Milliken WTE Case Study (case 2, rural)

	Pathway								
Pollutant	**Inhalation**	**Soil**	**Plant**	**Fish**	**Dairy**	**Dermal**	**Water**	**Mother's milk**	**Total**
Arsenic	2.349E-10	1.298E-11	1.954E-10	0.000E+00	0.000E+00	2.453E+12	0.000E+00	0.000E+00	4.458E-10
Beryllium	2.621E-11	1.449E-12	2.181E-11	0.000E+00	0.000E+00	2.738E+13	0.000E+00	0.000E+00	4.975E-11
Cadmium	2.900E-07	1.602E-08	2.413E-07	0.000E+00	0.000E+00	3.029E+09	0.000E+00	0.000E+00	5.503E-07
Chromium	1.450E-08	8,012E-10	1.206E-08	0.000E+00	0.000E+00	1.514E+10	0.000E+00	0.000E+00	2.752E-08
Nickel	1.334E-09	7.3791109	1.110E-09	0.000E+00	0.000E+00	1.393E+11	0.000E+00	0.000E+00	2.531E-09
PAHs	1.281E-10	2.657E-12	3.864E-10	0.000E+00	4.065E-11	5.021E+13	0.000E+00	1.587E-10	7.171E-10
PCB	5.286E-12	2.921E-13	4.398E-12	0.000E+00	1.678E-12	5.521E+14	0.000E+00	0.000E+00	1.171E-11
TCDD	7.015E-07	1.744E-08	1.414E-06	1.570E-08	2.782E-07	0.000E+00	0.000E+00	2.243E-06	4.670E-06
Total	1.762E-05	3.179E-07	0.000E-00	5.940E-06	2.985E-07	0.000E+00	0.000E+00	2.365E-06	3.407E-05

The utility of an assessment based on a worst case is that if the predicted risk approaches predetermined levels of regulatory concern, further analysis using more sophisticated, data-intensive methodologies may be needed to arrive at a more realistic prediction of health risk. There are problems, however, in presenting a worst-case assessment for use in risk management, particularly in an environment where a regulatory level of acceptable risk does not exist. In an effort to assure health-conservative decision making, risk managers may rely heavily on the worst-case scenario. This practice may bias risk management decision making if worst-case estimates are not presented with some prediction of the probability of their occurrence. Another viewpoint is that the worst-case scenario requires that an individual be born, breast-fed, reside, and work continuously at the point of maximum impact, and consume food and soil from the immediate vicinity that are contaminated at maximum levels. Viewed in this fashion, the worst-case becomes a remote event that is not particularly valuable in risk management.

The analysis of food chain exposure pathways in WTE health risk assessments are typically based on a screening level compartment model. This type of model relies on single values for input parameters that represent either plausible or upper-bound conditions. This approach provides a limited evaluation of uncertainties in exposures through the food chain. Duration and frequency of exposure are represented by the maximum exposed individual. Population mobility considerations are typically not addressed in quantitative estimates of exposure.

Several input parameters for calculating pollutant concentrations in soil and crops are not directly available in the technical literature. Plausible values for these parameters (including half-life in soil and crop/soil transfer ratios) must be estimated from data available in the literature, and available data for all relevant pathways of exposure are insufficient. For example, data on the bioaccumulation of facility-emitted pollutants in the food chain are inadequate. Standard risk assessment methodologies and input values do not exist. Since standard methods for routine evaluations include specified protocols, data sets for rate functions and model input parameters, and assumption used in risk assessments are lacking, results of risk assessments done for various proposed projects are often not readily comparable.

VII. DISCUSSION

The regulatory guidance for health risk assessment developed by California agencies has implications for risk management. Currently, only broad guidelines are provided for many aspects of health risk assessment, requiring that appropriate technical approaches be developed in conjunction with the permit applicant and agency staffs. The guidelines favor a deterministic approach, in some cases specifying parameter values. The single parameter approach also does not consider uncertainty in presenting health risk estimates. The cumulative effect of numerous upper-bound or worst-case parameters may lead to large overestimates of actual health risks.

As the use of waste-to-energy facilities for managing municipal solid waste increases, the importance of detailed, formal analysis to support controversial risk management decision will also increase. Advanced modeling features that may be used in future WTE health risk assessments include plume depletion in air dispersion modeling, quantification of the atmospheric fate of pollutants, population mobility, microenvironmental analysis, and stochastic analysis.

The static representation of potentially exposed populations, with estimates of risks being based on a maximally exposed individual, is a common feature in health risk assessments for WTE facilities. However, individuals move from their primary residence to commute to work, go to school, or relocate to a residence far from the impacts from the facility under study. Additionally, individuals typically spend a portion of the day in particular microenvironments, such as an automobile or the work place. Acceptable modeling approaches for population mobility and microenvironmental analyses have not been adopted in health risk assessments for WTE facilities. The current approach of assuming a 70-year continuous lifetime exposure to a single maximum concentration is also a significant source of conservatism in health risk assessments for WTE facilities that needs to be evaluated.

Health risk assessment for WTE facilities commonly have used deterministic models that produce single-value estimates of exposure and cancer risk. Risk management in California is typically based on single-value estimates. Current regulatory guidelines make no provision for the use of uncertainty analysis in risk management. Sensitivity analyses can be used to identify data gaps and problems for future study by ranking parameters by their sensitivity (i.e., determining which parameters produce large variations in estimated risk levels when varied over the plausible range of values). However, the preferred approach to evaluation uncertainty is stochastic simulation.

Carefully designed stochastic analyses can be used to estimate accuracy, overall bias, and precision in model predictions when model validation studies are not available. The major challenges to using stochastic analysis in risk management include preparing realistic parameter distributions and assisting risk managers in the process of usefully interpreting the additional information derived from this type of uncertainty analysis.

It is likely that health risk assessment will advance technically and local agency risk management guidelines will be more clearly defined in California during the next several years. Public comment and input on health risk assessments will likely be merged with the EIR process for WTE projects to provide full disclosure of the impacts associated with air toxic emissions. Health risk assessment will be an important analytical tool in permitting individual proposed WTE projects and evaluating cumulative health impacts from proposed multiple WTE projects within a regional airshed.

REFERENCES

1. Report to the Scientific Review Panel on Chlorinated Dibenzodioxins and Dibenzofurans, California Air Resources Board, State of California, Sacramento, 1986.
2. Air Toxics Source Assessment Manual for California Air Pollution Control Districts and Applicants for Air Pollution Control District Permits, Final Report, California Air Pollution Control Officers Association, Sacramento, 1987.
3. Rule 1000, Monterey Bay Unified Air Pollution Control District Regulations, Monterey, CA, 1987.
4. Proposed Rules 1401 and 223, South Coast Air Quality Management District, Los Angeles, 1987.
5. Milliken Waste-to-Energy Project Health Risk Assessment, Vol. 1, Submitted to the South Coast Air Management District, El Monte, CA, by Radian Corporation, Sacramento, 1987.
6. National Research Council, *Risk Assessment and Risk Management in the Federal Government: Managing the Process,* National Research Council, National Agency Press, Washington, D.C., 1983.
7. **Barnes, D. G., Bellin, J., and Cleverly, D.,** Interim procedures for estimating risks associated with exposures to mixtures of chlorinated dibenzodioxins and dibenzofurans (CDDs and CDFs), *Chemosphere,* 15, 1895, 1986.
8. Health Effects of 2,3,7,8-Tetrachlorodibenzo-*p*-Dioxin and Related Compounds, California Department of Health Services, Epidemiological Studies Section, Berkeley, 1986.
9. Health Assessment Document for Chromium, EPA 600/8-83-014f, U.S. Environmental Protection Agency, Washington, D. C., 1984.
10. **Shaub, W. M. and Tsang, W.,** Dioxin formation in incinerators, *Environ. Sci. Technol.,* 17, 721, 1983.
11. **Sehmel, G. A. and Hodgson, W. H.,** A Model for Predicting Dry Deposition of Particulates and Gases to Environmental Surfaces, DOE Report PLN-SA-6721, Pacific Northwest Laboratories, Richland, WA, 1978.
12. **Tesche, T. W., Honrath, R. A., Kaphai, R., and Dietrich, R. W.,** Improved Dry Deposition Estimates for Air Toxics Risk Assessments, presented at 80th Ann. Meeting of the Air Pollution Control Assoc., New York, 1987.
13. Guidelines on Air Quality Modeling (Revised), EPA 450/2-78-027R, Office of Air Quality Planning and Standards, U.S. Environmental Protection Agency, Research Triangle Park, NC, 1986.
14. Calculation of Annual Doses to Man from Routine Releases of Reactor Effluents for Purposes of Evaluating Compliance with 10 CFR 50, Appendix I, Revision 1, Regulatory Guide 1.109, Nuclear Regulatory Commission, Washington, D.C., 1977.

15. Health Assessment Document for Polychlorinated Dibenzo-*p*-dioxins, EPA 600/8-84-014f, Office of Health and Environmental Assessment, U.S. Environmental Protection Agency, Washington, D.C., 1985.
16. **Herbes, S. E. and Schwall, L. R.,** Microbial transformation of polycyclic aromatic hydrocarbons in pristine and petroleum-contaminated sediments in the vicinity of coal-coking wastewater discharge, *Appl. Environ. Microbiol.,* 41, 20, 1978.
17. AIRDOS-EPA: A Computerized Methodology for Estimating Environmental Concentrations and Dose to Man from Airborne Releases of Radionuclides, Office of Radiation Programs, EPA 520/1-79-009, U.S. Environmental Protection Agency, Washington, D.C., 1979.
18. **Baes, C. F., Sharp, R. D., Sjoreen, A. L., and Shor, R. W.,** A Review and Analysis of Parameters for Assessing Transport of Environmentally Released Radionuclides Through Agriculture, ORNL-5786, Oak Ridge National Laboratory, Oak Ridge, TN, 1984.
19. **Briggs, G. G., Bromilow, R. H., and Evans, A. A.,** Relationships between lipophilicity and root uptake and translocation of non-ionized chemicals by barley, *Pest. Sci.,* 13, 495, 1982.
20. **Wang, D. T. and Meresz, O.,** Occurrence and potential uptake of polynuclear aromatic hydrocarbons of highway traffic origin by proximally grown food crops, in *Polynuclear Aromatic Hydrocarbons, Sixth International Symposium,* Cooke, M., Dennis, A. J., and Fisher, G. L., Eds., Battelle Press, Columbus, OH, 1982.
21. **Travis, C. C. and Arms, A. D.,** Bioconcentration of organics in beef, milk, and vegetation, *Environ. Sci. Technol.,* 22, 3, 271, 1988.
22. **Sims, R. C. and Overcash, R.,** Fate of polynuclear aromatic compounds (PNA) in soil-plant systems, *Residue Rev.,* 88, 2, 1982.
23. **Edwards, N. T.,** Polycyclic aromatic hydrocarbons (PAHs) in the terrestrial environment — a review, *J. Environ. Qual.,* 12, 427, 1983.
24. **Ellwardt, P.,** Variations in content of polycyclic aromatic hydrocarbons in soils and plants by using municipal wastes components in agriculture, in Soil Organic Matter Studies, Vol. II, IAEA-SM-211/31, International Atomic Energy Agency, Vienna, 1977.
25. **Wipf, H. K., Homberger, E., Neuner, N., Ranalder, V. B., Vettev, H., and Vuillemier, J. P.,** TCDD levels in soil and plant samples from the Seveso area, in *Chlorinated Dioxins and Related Compounds: Impact on the Environment,* Hutzinger, O., Frei, R. W., Meriam, E., and Pocchiari, F., Eds., Pergamon Press, New York, 1982.
26. **Travis, C. C. and Hattemer-Frey, H. A.,** Human exposure to 2,3,7,8-TCDD, *Chemosphere,* 10, 2331, 1987.
27. **Helling, C. S., Isensee, A. R., Woolson, E. A., Ensor, P. D. J., Jones, G. E., Plimmer, J. R., and Kearney, P. C.,** Chlorodioxins in pesticides, soils, and plants, *J. Environ. Qual.,* 2, 171, 1973.
28. Composition of Foods, Agricultural Hand. No. 8, U.S. Department of Agriculture, Washington, D.C., 1963.
29. **Swenson, M. J.,** *Duke's Physiology of Domestic Animals,* 8th Ed., Comstock Publishing, London, 1980.
30. Water Resources Data — California, U.S. Geological Survey, National Technical Information Center, Richmond, VA, 1983.
31. **Kenaga, E. E. and Goring, C. A. I.,** Relationship between water solubility, sorption, octanol-water partitioning, and the concentration of chemicals in biota, in *Aquatic Toxicology,* Eaton, J. E., Parrish, P. R., and Hendricks, A. C., Eds., American Society for Testing Materials, Philadelphia, 1980.
32. **Kimbrough, R. D., Falk, F., Stehr, P., and Fries, G.,** Health implications of 2,3,7,8-tetrachlorodibenzodioxin (TCDD) contamination of residential soil, *J. Toxicol. Environ. Health,* 14, 47, 1983.
33. **Hawley, J. K.,** Assessment of health risk from exposure to contaminated soil, *Risk Anal.,* 5, 289, 1985.
34. Dietary Consumption Distributions of Selected Food Groups for the U.S. Population, EPA 560-11-80-012, Office of Pesticides and Toxic Substances, U.S. Environmental Protection Agency, Washington, D.C., 1980.
35. Food Consumption of Households in the United States, Seasons and Year, 1965-1966, Agricultural Research Service, U.S. Department of Agriculture, Washington, D.C., 1972.
36. **Farrow, R. P., Elkins, E. R., Rose, W. W., Lamb, F. C., Ralls, J. W., and Mercer, W. A.,** Canning operations that reduce insecticide levels in prepared foods and in solid food wastes, *Residue Rev.,* 29, 73, 1969.
37. **Geisman, J. R.,** Reduction of pesticide residues in food crops by processing, *Residue Rev.,* 54, 43, 1975.
38. **Larsson, B. and Sahlberg, G.,** Polycyclic aromatic hydrocarbons in lettuce, in *Polynuclear Aromatic Hydrocarbon, Sixth International Symposium,* Cooke, M., Dennis, A. J., and Fisher, G. L., Eds., Battelle Press, Columbus, OH, 417, 1982.
39. **Gunther, F. A., Barkley, J. H., and Westlake, W. E.,** Sampling and processing techniques for determining dislodgeable pesticide residues on leaf surfaces, *Bull. Environ. Contam. Toxicol.,* 12, 641, 1974.
40. **Schaum, J.,** Risk Analysis of TCDD Contaminated Soil, EPA 600/8-84-02, U.S. Environmental Protection Agency, Washington, D.C., 1984.
41. Superfund Public Health Evaluation Manual, EPA 540/1-86/060, Office of Solid Waste and Emergency Response, U.S. Environmental Protection Agency, Washington, D.C., 1986.
42. **Tuey, D. B.,** Toxicokinetics, in *Introduction to Biochemical Toxicology,* Hodgson, E. and Guthrie, F. E., Eds., Elsevier, Amsterdam, 1980.
43. **Poiger, H. and Schlatter, C.,** Pharmacokinetics of 2,3,7,8-TCDD in man, *Chemosphere,* 15, 1489, 1986.

44. **Kuroki, H. and Masuda, Y.,** Determination of polychlorinated dibenzofuran isomers retained in patients with Yusho, *Chemosphere,* 10, 771, 1978.
45. **Kashimoto, T., Miyata, H., Kunita, S., Tung, T., Hsu, S., Chang, K., Tang, S., Chi, G., Nakagawa, J., and Yamamoto, S.,** Role of polychlorinated dibenzofuran in Yusho (PCB poisoning), *Arch. Environ. Health,* 36, 321, 1981.
46. **Rappe, C., Bergquist, P. A., Hansson, M., Kjeller, L. O., Lindstrom, G., Marklund, S., and Nygren, M.,** Chemistry and analysis of polychlorinated dioxins and dibenzofurans in biological samples, in *Banbury Report 18, Biological Mechanisms of Dioxin Action,* Poland, A. and Kimbrough, R., Eds., Cold Spring Harbor Laboratory, Cold Spring Harbor, NY, 1984, 17.
47. *Report of the Task Force on Reference Man,* ICRP Publ. No. 23, International Commission on Radiological Protection and Measurement, 1975.
48. **Butte, N. F., Garza, C., Stuff, J. E., Smith, E. O., and Nichols, B. L.,** Effect of maternal diet and body composition on lactational performance, *Am. J. Clin. Nutr.,* 39, 296, 1984.
49. **Whitehead, R. G. and Paul, A. A.,** Infant growth and human milk requirements, *Lancet,* 2, 161, 1981.
50. **Poiger, H. and Schlatter, C. H.,** Influence of solvents and adsorbents on dermal and intestinal absorption of TCDD, *Food Cosmet. Toxicol.,* 18, 477, 1980.
51. U.S. Environmental Protection Agency, Guidelines for carcinogen risk assessment, *Fed. Reg.,* 51, 33992, 1986.
52. Guidelines for Chemical Carcinogen Risk Assessments and Their Scientific Rationale, California Department of Health Services, State of California, Sacramento, 1985.
53. **Dourson, M. L. and Stara, J. F.,** Regulatory history and experimental support of uncertainty (safety) factors, *Regul. Toxicol. Pharmacol.,* 3, 224, 1983.
54. **Eschenroeder, A., Petito, C., and Wolff, S.,** An Analysis of Health Risks for the Irwindale Resource Recovery Facility, California Energy Commission, Sacramento, CA, 1986.
55. Los Angeles County Soil Data, Biomass Study No. 500-81-037, California Energy Commission, State of California, Sacramento, 1985.
56. **Piscator, M.,** Manganese, in *Handbook on the Toxicology of Metals,* Friberg, L., Nordberg, G. F., and Vouk, V. B., Eds., Elsevier, Amsterdam, 1979.
57. Health Assessment Document for Inorganic Arsenic, EPA 600/8-83-014a, U.S. Environmental Protection Agency, Washington, D.C., 1983.
58. **Furchner, J. E., Richmond, C. R., and London, J. E.,** Comparative metabolism of radionuclides in mammals. VIII. Retention of beryllium in the mouse, rat, monkey, and dog, *Health Phys.,* 24, 293, 1973.
59. **Friberg, L., Elinder, C. G., Kjellstrom, T., and Nordberg, G. F.,** *Cadmium and Health: A Toxicological and Epidemiological Appraisal,* CRC Press, Boca Raton, FL, 1985.
60. **Donaldson, R. M. and Barreras, R. F.,** Intestinal absorption of trace quantities of chromium, *J. Labor. Clin. Med.,* 68, 484, 1966.
61. **Heard, M. M. and Chamberlain, A. C.,** Effect of minerals and food on uptake of lead from the gastrointestinal tract in humans, *Hum. Toxicol.,* 1, 411, 1982.
62. **Horak, E. and Sunderman, F. W.,** Fecal excretion of nickel by healthy adults, *Clin. Chem.,* 12, 429, 1973.
63. **Van den Berg, M., Olie, K., Hutzinger, O.,** Uptake and selective retention in rats of orally administered chlorinated dioxins and dibenzofurans from fly ash and fly ash extract, *Chemosphere,* 12, 537, 1983.
64. **Van den Berg, M., Devroom, E., Van Grevenbrook, M., Olie, K., and Hutzinger, O.,** Bioavailability of PCDDs and PCDFs adsorbed on fly ash in rat, guinea pig and syrian hamster, *Chemosphere,* 14, 865, 1983.
65. **Van den Berg, M., Van Grevenbrook, M., Olie, K., and Hutzinger, O.,** Bioavailability of PCDDs and PCDFs on fly ash after semi-chronic oral ingestion by the rat, *Chemosphere,* 15, 509, 1985.
66. **Van den Berg, M., Devroom, E., Olie, K., and Hutzinger, O.,** Bioavailability of PCDDs and PCDFs on fly ash after semi-chronic oral ingestion by the guinea pig and syrian golden hamster, *Chemosphere,* 15, 519, 1986.
67. **Rose, J. Q., Ramsey, J. C., Wentzler, T. H., Hummel, R. A., and Gehring, P. J.,** The fate of 2,3,7,8-tetrachlordibenzo-*p*-dioxin following single and repeated oral doses to the rat, *Toxicol. Appl. Pharmacol.,* 36, 209, 1976.
68. **Piper, W. N., Rose, R. Q., and Gehring, P. J.,** Excretion and tissue distribution of 2,3,7,8-tetrachlordibenzo-*p*-dioxin in the rat, *Environ. Health Perspect.,* 5, 241, 1973.
69. **Grimmer, G.,** *Environmental Carcinogens: Polycyclic Aromatic Hydrocarbons,* CRC Press, Boca Raton, FL, 1983.
70. **Chang, L. J.,** The fecal excretion of polycyclic aromatic hydrocarbons following their administration to the rat, *J. Biol. Chem.,* 151, 93, 1943.

Chapter 10

ANALYSIS OF UNCERTAINTIES IN HEALTH RISK ESTIMATES FOR MUNICIPAL WASTE COMBUSTORS

Mike Marchlik

TABLE OF CONTENTS

I. INTRODUCTION

In 1985, Americans disposed of approximately 250 million tons of municipal solid waste (MSW), 95% of which was disposed of in landfills.[1] As existing landfills reach capacity and the number of viable sites for new landfills decreases, there is increasing pressure on municipal officials to evaluate other disposal options, including incineration. The potential health risks posed by the emissions from municipal waste combustors (MWCs) has emerged as a major factor affecting the decision to build these facilities. The decision to build an MWC is made even more difficult in light of the large uncertainties associated with any estimate of the chronic health effects on the population surrounding a given facility.

Increasing attention is being centered on reducing uncertainties in the risk assessment process and deriving more realistic estimates of human exposure and risk.[2] The parameters and assumptions associated with any environmental transport and exposure model are inherently uncertain. According to the Office of Science and Technology Policy,[3] health risk assessments should not only characterize the potential risk to affected populations but should also express risk estimates in such a way that the assumptions made to derive those estimates and the uncertainties associated with such estimates are evident to the decision maker. Uncertainty is defined as a quantification of the state of knowledge about the input variables and results of a health risk assessment. The purpose of this chapter is to discuss the sources of uncertainty in assessing health risks from MWCs, the impact of this uncertainty on the decision-making process, and methods available to explicitly evaluate uncertainties in the risk assessment process.

II. IMPACT OF UNCERTAINTY ON THE DECISION-MAKING PROCESS

Before examining sources of uncertainty and methods available to quantify uncertainty, it is important to evaluate the impact that uncertainty has on the decision-making process. Figure 1 presents a simplified diagram of the decision faced by a local government in siting and operating an MWC. This decision tree is a graphical representation of the choices available to a decision maker and the potential outcomes of those choices.

In this simplified tree, two choices are presented: (1) build an MWC or (2) expand an existing landfill. While there are several other alternatives that could also be examined, only two choices are shown here for the sake of clarity. Associated with each choice are two alternative health risk estimates provided for consideration by decision makers. If study "A" is correct, operating the proposed incinerator would result in 100 additional cancer cases among the population living near the proposed MWC, while expanding the landfill would result in one only additional cancer case. Study "B," however, predicts that one additional cancer case would result from operating the proposed incinerator, while expansion of the landfill would cause 100 additional cancer cases.

To choose the best course of action, decision makers must assess the likelihood that either risk estimate is an accurate prediction of future health effects from each proposed alternative. The situation depicted in Figure 1 is not uncommon, except that decision makers are often presented with two different estimates of the health risk for an incinerator but no estimate of the health risk from alternative waste disposal options. If uncertainties are not explicitly addressed and accounted for in the risk assessment process, it is difficult to determine the true range of uncertainty in risk estimates.

The magnitude of the uncertainty associated with each risk estimate can be evaluated by examining the results of various health risk assessments prepared for proposed MWCs. Analyses of the health risks from the same facility as estimated in a risk assessment sponsored by project proponents will often differ by a factor of 10, 100, or sometimes even more from

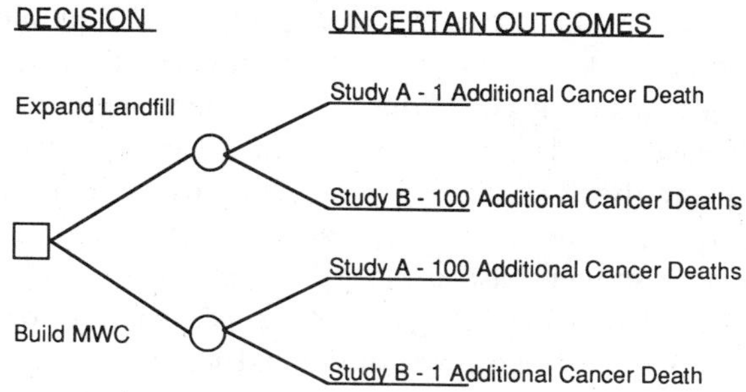

FIGURE 1. Sample decision tree.

estimates derived in a study commissioned by project opponents. These differences are often the result of alternative interpretations of uncertain scientific processes and data and the application of different air quality models and risk assessment methodologies.

A situation that is of particular concern in evaluating risk analyses is the use of single value, "worst-case" estimates. Worst-case evaluations are often prepared in the interest of being health conservative. This approach has been criticized for inappropriately making risk management decisions in the risk analysis stage of the problem.[4] More significantly, worst-case analyses can greatly overestimate the true risk and make the alternatives appear to be the better choice from a health perspective, especially if the alternatives were not subject to a similar worst-case evaluation. Models tend to incorporate conservative (potentially unrealistic) assumptions that are likely to overestimate the true risk to the public.[2] For example, one conservative assumption is typically made in health risk assessments for MWCs: all exposed individuals live at the point of maximum impact. The result estimated intake of pollutant is much higher than if the modeler were to assume that only a fraction of exposed individuals actually live at the point of maximum deposition impact.

In worst-case analyses, a conservative (i.e., health protective) value for each model input is selected. In addition, when simplifying assumptions, such as assuming that all exposed individuals reside at the point of maximum impact, are used, they also tend to be highly conservative. The effect of calculating the risk estimate based exclusively on conservative input values is that the probability associated with the risk estimate is more unlikely than the uncertainty associated with the individual input variables. This outcome can be illustrated using the following example.

The rules of combining probabilities can be applied to calculate the probability of rolling four dice at once and rolling all sixes. The probability of rolling a six with one die is 1/6 or 16.7%. Similarly, the chance of rolling a six with all four dice is (1/6)(1/6)(1/6)(1/6) or 0.0008%. Similarly, if a model consists of four inputs, each of which is assigned a value with a corresponding probability of 5%, the corresponding output value does not have a probability of 5% as one might guess but of (0.05)(0.05)(0.05)(0.05) or 0.00000625%. Thus, it is not difficult to realize that performing a worst-case analysis with a model that has a large number of inputs will produce results that are extremely unlikely to occur. A similar problem exists if all input parameters are assigned low probability or "optimistic" values. In this case, the results could greatly underestimate the true risk.

While worst-case risk analyses were the norm in the past, they are gradually giving way to more complete analyses of the full range of possible inputs. This transformation is promoted by advances in the available tools and increased experience in the use of uncertainty analyses. In addition, the courts are less receptive to the use of conservative assumptions and have advised the U.S. Environmental Protection Agency (EPA) in a number of cases to consider a more detailed analysis of uncertainty when performing regulatory analyses.[5]

The value of an uncertainty analysis increases with the costs of an incorrect decision. In the language of decision analysis, this is termed opportunity loss.[6] Returning to the example in Figure 1, if the decision is made to expand the landfill based on the results of study B and the results predicted in study A actually come true, then the opportunity loss would be to cause 99 excess cancer cases in the exposed population. In real life, a community is faced with a decision that will likely have many effects on its citizens from the cost of garbage disposal to potential health consequences. If an incorrect decision is made, the result could be a significant increase in morbidity or an unnecessary increase in waste disposal costs.

There is no universal solution for making the optimum decision. Two primary requirements for making the "correct" decision, however, are to understand and account for uncertainties when analyzing the problem and to make a fully informed decision that acknowledges potential losses associated with various alternatives.[7] The remainder of this chapter discusses the uncertainty affecting the assessment of human health risks from the incineration of MSW and methods that can be used to understand and account for this uncertainty in the risk assessment process.

III. SOURCES OF UNCERTAINTY

In performing probabilistic risk assessments for nuclear power plants, Vesely and Rasmussen[8] have differentiated the sources of uncertainty into model uncertainty, parameter uncertainty, and completeness uncertainty. Model uncertainty pertains to uncertainty about the form of the mathematical equations used to represent the process of interest. Parameter uncertainty refers to the variability inherent in the input variables or in the measurement of these variables. Completeness uncertainty considers whether all sources of risk have been accounted for and adequately described. All three types of uncertainty affect the results of risk assessments performed for MWCs.

In performing a risk assessment, the types of uncertainty that can affect the results should be identified and quantified or at least qualitatively discussed. The three categories of uncertainty listed above and their relationship to MWC risk assessments are discussed in more detail below.

A. MODEL UNCERTAINTY

Estimating health risks from MWC emissions requires the use of several models. For instance, atmospheric transport models estimate the ambient air concentrations and deposition rates for each contaminant of concern. A number of models are available to estimate these variables, each of which may produce different results from the same basic input data. Many factors, such as the type of terrain surrounding the proposed site (e.g., flat vs. complex), influence the choice of atmospheric model. For the most part, however, air quality model selection depends upon the discretion of the analyst or the specifications of the permitting agency.

A variety of models exist to quantify human exposure to pollutants emitted from point sources, including MWCs.[9-12] When considering various models, the user should identify differences in model design, the underlying assumptions inherent to each model, and the data requirements associated with establishing parameter distributions.[2] To validate a selected model, investigators can compare model predictions to results obtained from other, similar models or to measured data.

B. PARAMETER UNCERTAINTY

The second type of uncertainty, parameter uncertainty, addresses the variance inherent in the data used to perform a health risk assessment. The necessary inputs rely on a mixture of experimental data, the analysts' judgment about the quality of those data, and simplifying assumptions. Uncertainty about the values used as inputs may result from measurement error, referred to as statistical error, and from a lack of understanding about the underlying physical and chemical processes affecting the variable.

When performing a detailed, multimedia health risk assessment, the number of variables can be very large. For example, a model developed for use by the California Energy Commission has over 40 input variables, each of which represents a source of uncertainty.[13] If variability exists in the observed data for an input, it is necessary to examine why that variability exists and to determine if the data reflects the full range of variability associated with the model input.

If sufficient measured data are available, the correct statistical distribution can be developed for each parameter.[14] In reality, however, limited measured data are typically available. As a result, a reasonable value for many parameters must be estimated indirectly using various predictive equations. A good example of parameter uncertainty is encountered when performing a risk assessment that includes exposures from home gardens. An important variable crucial to quantifying exposure from this pathway is root uptake of pollutants from contaminated soil. Root uptake factors vary according to whether the edible portion of the plant is a root, such as carrots; a leafy portion, such as lettuce; an exposed fruit, such as tomatoes; or a protected fruit, such as corn.[15] Variability may also be attributed to species-specific factors (i.e., tomatoes may accumulate a contaminant at a faster rate than green peppers even though both are exposed fruits).

For many contaminants, there are little data on plant uptake and what data do exist may be applicable to only one or two species. There are numerous cases where no measured data are available for a specific contaminant. The accepted practice in the absence of measured data is to utilize published regression equations that relate the quantity of interest, such as root uptake factors, to a common physicochemical property, such as the octanol-water partition coefficient (K_{ow}). Correlations between the K_{ow} value of a chemical and its potential to bioaccumulate in living organisms have been developed for several animal groups. For example, Chiou et al.,[16] Kenaga,[17] Könemann and van Leeuwen,[18] Lu and Metacalf,[19] Neely et al.,[21] Travis and Arms,[22] and Travis et al.[23] showed that a linear relationship exists between the bioconcentration of organics in fish, cows, swine, and humans and their K_{ow} values. When using these approaches, it is necessary to account for the uncertainty in using regression equations and the uncertainty in the measured physical parameter involved (in this case, the octanol-water partition coefficient). It is also important to examine the implication of a worst-case analysis. For example, the worst-case value may be for a species that does not represent the typical vegetables grown by local residents. The result would be that the value chosen is not representative of the actual exposures that will be experienced.

C. COMPLETENESS UNCERTAINTY

Completeness uncertainty is closely coupled to model uncertainty. The completeness of a risk assessment relates to the detail of the models used and the ability of the models to account for all contributors to risk. In a risk assessment, completeness is affected by the number and types of exposure pathways examined and the level of detail on the model equations. The location of the facility and the demographics of the surrounding population should guide the selection of pathways evaluated. The detail of the models used should be guided by the availability of data that can serve as inputs for the model. For instance, in a metropolitan area such as New York City, the only pathways of interest may be inhalation, dust ingestion, and dermal absorption. Conversely, evaluating only these few pathways in a locality where a commercial dairy or slaughterhouse is located near the proposed facility would be insufficient. This this cases, human exposure via ingestion of contaminated food items should also be evaluated.

A second aspect of completeness applies to sources of contamination. For example, there are several routes by which vegetation can be contaminated, including direct deposition of contaminants onto plant surfaces, root uptake of contaminants that have accumulated in the soil over time, and uptake of pollutant vapors by aerial plant parts, also known as air-to-leaf transfer.[24] While not every route may be significant, for completeness all routes should be evaluated when developing the initial exposure model.

Uncertainty about the completeness of a risk assessment does not lend itself to direct quantification and measurement. As a result it is difficult to ascertain its effects on the final results. One way to reduce completeness uncertainty is to evaluate the model carefully during the development stage. At times, models may be simplified for ease of use or due to lack of suitable data. As the model is developed, an attempt may be made to quantify the effect of these limiting assumptions and to qualify the results obtained from the model as a result of these investigations.

IV. CHARACTERIZING UNCERTAINTY

Of the three types of uncertainty discussed, parameter uncertainty is the most readily quantified. Parameter uncertainty analyses use an estimated frequency distribution for each model parameter to produce a frequency distribution of model results. These analyses are also valuable for determining the contribution of each parameter to over all uncertainty.[2] Uncertainty about a specific variable can be characterized as a range of values or as a probability distribution where probabilities are assigned to the range of values. Distributions may be discrete or continuous depending on the nature of the variable. Familiar distributions, such as normal or log normal distributions, are often used to characterize the uncertainty arising from measurement error. Log normal distribution are frequently used to represent the uncertainty in biological processes, such as plant uptake factors.

V. REVIEW OF UNCERTAINTY ANALYSIS TECHNIQUES

As discussed previously, the use of single value, conservative inputs in risk assessment models produces results that tend to be extremely unlikely. This is especially true in situations where there is a great deal of uncertainty about the "correct" value to be assigned to the input variables. There are a variety of techniques available to perform uncertainty analyses, such as the Monte Carlo and Latin Hypercube approaches. The objective of utilizing these techniques is to propagate the uncertainty of input data through the model to obtain an estimate of the uncertainty in model output.

A Monte Carlo analysis is often used for conducting uncertainty analyses on risk assessment models. This technique calculates a single output value, or model situation, from a set of randomly selected input variable values.[2] The advantages of using a Monte Carlo analysis include ease of use, applicability to broad range of model equations, a general lack of restraints on the form of the input distributions used or the functional form of the model equations, and the ability to characterize the output distribution. The disadvantages of other, available methods include restraints on the functional form of the model, restrictions on the distributions used to characterize the uncertainty of inputs, and the inability to fully characterize the output distribution.

The following sections examine the application of Latin Hypercube Sampling (LHS), a variation of the Monte Carlo approach, to the health risk assessments for MWCs. The primary disadvantage of using a Monte Carlo approach is the requirement that the uncertainty model be iterated several hundred or several thousand times, depending on the variance of the input distributions. LHS greatly reduces the number of computer runs necessary by constraining the sampling of the input distributions.[25-27] This is particularly advantageous when dealing with complex exposure models that have a large number of input variables, as the computer simulations would become impractical and expensive using a Monte Carlo approach.

VI. LATIN HYPERCUBE SAMPLING

The first step in any uncertainty analysis is to perform a sensitivity analysis. A sensitivity analysis identified those variables that have the greatest impact on model results. In a sensitivity analysis, each parameter is varied over its range of plausible values while all other model variables are held constant to observe the effect each parameter has on model outcome.[2] Since all parameters are assumed to have the same distribution, the most influential parameters on model output are the most sensitive. Hence, a sensitivity analysis allows the uncertainty analysis to focus on the most sensitive parameters only, thereby minimizing the costs of conducting an uncertainty analysis. Based on the results of the sensitivity analysis, data gathering efforts can be restricted to parameters that have a major impact on human exposure.

Once a smaller list of input variables has been identified, the next step is to determine the proper distribution of parameter values for each parameter included in the uncertainty analysis. A reasonable range of values for each parameter can be estimated indirectly using predictive techniques or derived from measured data reported in the literature. In some cases, sufficient data exist to develop frequency distributions that adequately characterize the uncertainty about a given input variable. In other instances, insufficient data are available necessitating a subjective interpretation of those limited data.[28] Traditional choices for parameter distributions include normal, log normal, and uniform for continuous parameters and binomial or Poisson for discrete variables.[29]

Once the distributions are developed, the LHS technique is used to generate an input sample be generated using a computerized code, such as that developed by Iman et al.[30] This matrix of inputs is then read one row at a time and used as input for a single model simulation, and the results recorded. After all the samples have been executed, it is possible to use the output data generated for the outputs to produce frequency distributions of the results.

The LHS approach produces a k by n sample matrix where k is the number of input variables, and n is the number of samples to be taken from each input distribution. As a general rule, the larger the number of samples, the more effective the results will be in reproducing the true distribution of the output variables. The number of samples should be a least equal to the number of inputs. A schematic of LHS sampling is shown in Figure 2. Each input that is assigned a probability distribution function is sampled n times. These samples are ranked in order of value and a second random sample is drawn from these values and combined with random values from each of the other input distributions. The results is the n by k sample matrix. The model is then run n times, using a different set of values from the input matrix each time. After all n output values are calculated, they can be rank ordered in value and arranged in a probability density function.

Iman et al.[30] developed a FORTRAN computer code that produces sample matrices from a set of input probability distributions. This LHS code incorporates an enhancement to the process described here by incorporating an algorithm designed to reduce the occurrence of random correlations among the input variables, which can also be used to induce a specified correlation between variables.

VII. INCORPORATING UNCERTAINTY ANALYSIS INTO THE RISK ASSESSMENT PROCESS

Uncertainty analysis, to be most effective, should not be considered as a discrete step in the risk assessment process but as an element in each step from initial model selection to estimation of final results. For example, when selecting an atmospheric transport or an exposure model, an attempt should be made to qualitatively assess the impact of choosing these models over alternative formulations. If at all possible, use alternative models to quantify these differences. This information should be carried through the risk assessment process and incorporated into

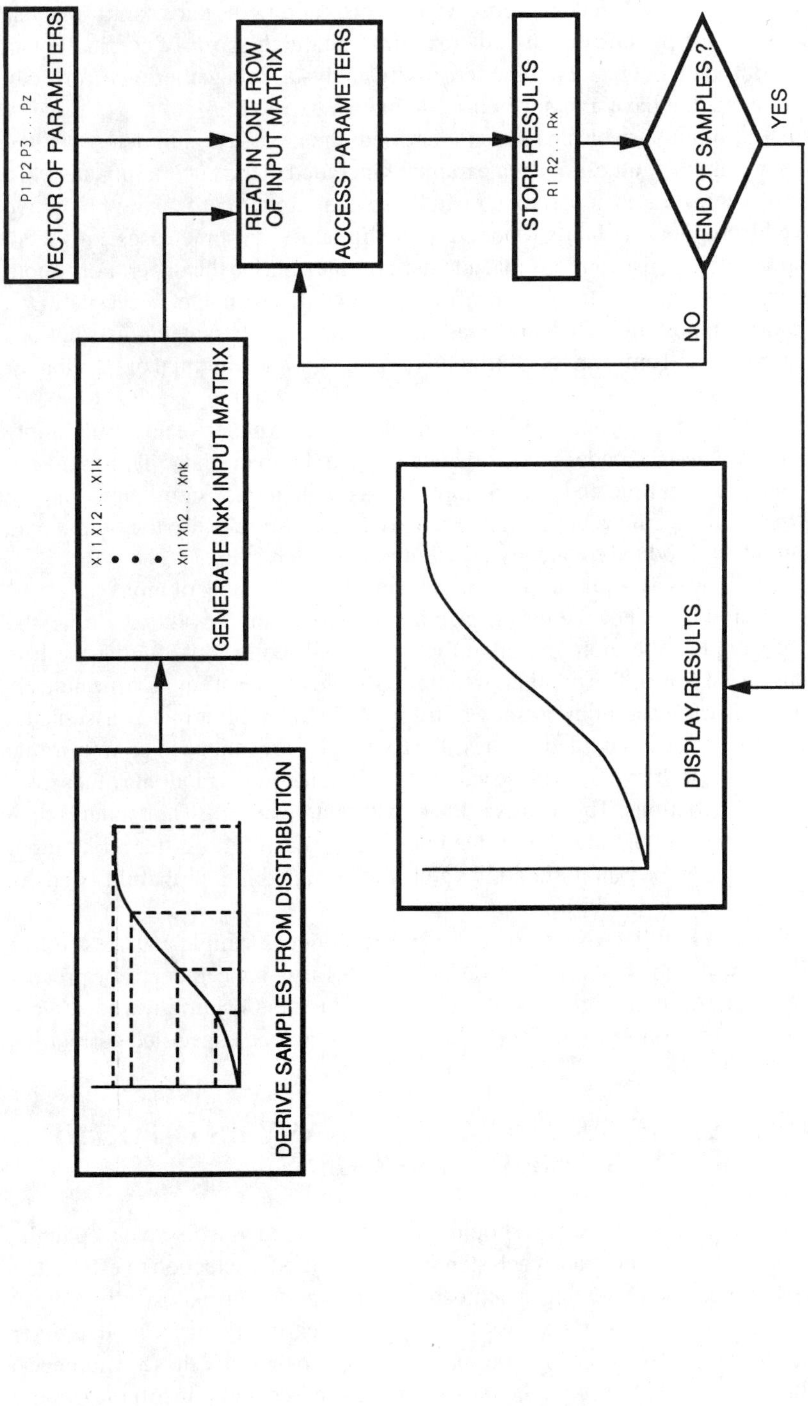

FIGURE 2. Illustration of the LHS model process.

the results. When developing new models, steps should be taken to evaluate and assess the impact of using simplifying assumptions and selecting from alternative model formulations.

It is very important to state what assumptions were made when presenting the results of an uncertainty analysis. If not properly stated, the results will often be mistaken as a full statement of the range of uncertainty about the analysis, while in fact, the results present the range of uncertainty about a single estimate of the risk. Once the results are obtained, it is possible to perform a variety of analyses in support of the decision-making process. These analyses are discussed below.

The frequency distributions produced as a result of an uncertainty analysis can be used to calculate more reasonable (i.e., probable) health risk estimates as opposed to deriving worst-case risk estimates. For example, the median value can be presented as a best estimate of risk, while the 95th percentile value can be chosen as a conservative estimate. A second use of the results is the determination of sensitive model inputs. The sensitivity analysis is conducted by performing a regression analysis of the input sample matrix against the output distributions.[26]

When a regression is performed on all the uncertain inputs against the outputs, the inputs that have the greatest influence on the results will have the largest absolute value for the measured slope since the regression coefficient may be negative or positive. Varying the values of these parameters will produce the greatest impact on the results. For a model with several outputs, the importance of a variable may change depending on the output variable of interest. Performing a sensitivity analysis also allows limited resources to be more cost-effectively funneled to investigate those variables that have the greatest impact on health risk estimates and to insure that the uncertainty about these variables is adequately described or, if possible, reduced through the collection of better information.

VIII. CONCLUSIONS

Results of an uncertainty analysis demonstrate that estimating the potential health risks associated with operating an MWC is difficult to predict with any certainty. Use of single-valued estimates, especially worst-case analyses, does not provide adequate information to those who must decide whether to build an MWC. It is important, therefore, to provide the decision makers with an estimate of the range of possible health impacts so that an optimal decision can be made.

Techniques to calculate the value of information are available using the outputs generated by an uncertainty analysis. This analysis can be used to guide a decision maker in choosing whether to make a choice given the current level of uncertainty present in the analysis or to spend additional funds in an effort to reduce the uncertainty. This could consist of collecting better site specific data, conducting a more in-depth review of the literature, or sponsoring research projects aimed at collecting better information.

REFERENCES

1. Kidder, Peabody, and Company, Engineering — Construction Resource-Recovery Outlook, Industrial Analysis, March 5, 1986.
2. **Hoffman, H. O. and Gardner, R. H.,** Evaluation of uncertainties in radiological risk assessment models, in *Radiological Assessment,* NUREG/CR-3332, Till, J. E. and Meyer, H. R., Eds., U.S. Nuclear Regulatory Commission, Washington, D.C., 1983.
3. Office of Science and Technology Policy, Chemical carcinogens: a review of the science and its associated principles, *Fed. Reg.,* 50, 10372, 1985.
4. **Eschenroeder, A., Guldberg, P., Kellermeyer, D., Smith, A., and Wolff, S.,** *An Analysis of Health Risks from the Irwindale Resource Recovery Facility,* Vol. 3, prepared for the California Energy Commission, Sacramento, 1986.

5. **Graham, J. D.,** Cancer in the Courtroom: Risk Assessment in the Post-Benzene Era, Environmental Risk Management — Is Analysis Useful?, Proc. APCA Specialty Int. Conf., Chicago, 1986.
6. **Raiffa, H.,** *Decision Analysis—Lectures on Choices Under Uncertainty,* Addison-Wesley, Reading MA, 1968.
7. **Marchlik, M. and Anderson, S.,** Municipal Solid Waste Management: Evaluating the Alternatives, presented at Society for Risk Analysis Annual Meeting, Washington, D.C., November 1988.
8. **Vesely, W. E. and Rasmussen, D. M.,** Uncertainties in nuclear probabilistic analyses, *Risk Anal.,* 4, 313, 1984.
9. **Dacre, J. C., Rosenblatt, D. H., and Cogley, D. R.,** Preliminary pollutant limit values for human health effects, *Environ. Sci. Technol.,* 14, 778, 1980.
10. **MacKay, D. and Paterson, S.,** Fugacity revisited, *Environ. Sci. Technol.,* 16, 654A, 1982.
11. **Cohen, Y.,** Organic pollutant transport, *Environ. Sci. Technol.,* 20, 538, 1980.
12. **Travis, C. C. and Hattemer-Frey, H. A.,** An overview of food chain impacts from municipal waste combustion, Proc. Conf. on Risk Assessment for Municipal Waste Combustion: Deposition, Food Chain Impacts, and Research Needs, Cincinnati, June 8-9, 1989.
13. Envirosphere Co., environmental Pathways Uncertainty and Screening Model User's Guide, Final Report to California Energy Commission, June 1988.
14. **Johnson, N. L. and Kotz, S.,** *Distributions in Statistics: Continuous Univariate Distributions,* Vol. I, Houghton-Mifflin, Boston, 1970.
15. **Schaffer, S.,** Environmental Transfer Parameters for Five Selected Organic Priority Pollutants: Benzo(a)Pyrene, DDT, PCBs, Trichloroethylene, and Dioxins, presented at Natl. Conf. on Hazardous Wastes and Environmental Emergencies, Cincinnati, May 14-16, 1985.
16. **Chiou, C. T., Freed, V. H., Schmedding, D. W., and Kohnert, R. L.,** Partition coefficient and bioaccumulation of selected organic chemicals, *Environ. Sci. Technol.,* 11, 475, 1977.
17. **Kenaga, E. E.,** Correlations of bioconcentration factors in aquatic and terrestrial organisms with their physical and chemical properties, *Environ. Sci. Technol.,* 14, 553, 1980.
18. **Könemann, H. and van Leeuwen, K.,** Toxicokinetics in fish: accumulation and elimination of six chlorobenzenes by guppies, *Chemosphere,* 9, 3, 1980.
19. **Lu, P.-Y. and Metcalf, R. L.,** Environmental fate and biodegradability of benzene derivatives as studies in a model aquatic ecosystem, *Environ. Health Perspect.,* 10, 269, 1975.
20. **MacKay, D.,** Correlation of bioconcentration factors, *Environ. Sci. Technol.,* 16, 274, 1982.
21. **Neely, W. B., Branson, D. R., and Blau, G. E.,** Partition coefficient to measure bioconcentration potential or organic chemicals in fish, *Environ. Sci. Technol.,* 8, 1113, 1974.
22. **Travis, C. C. and Arms, A. D.,** Bioconcentration of organics in beef, milk, and vegetation, *Environ. Sci. Technol.,* 22, 271, 1988.
23. **Travis, C. C., Hattemer-Frey, H. A., and Arms, A. D.,** Relationship between dietary intake of organic chemicals and their concentrations in human adipose tissue and breast milk, *Arch. Environ. Contam. Toxicol.,* 17, 473, 1988.
24. **Travis, C. C. and Hattemer-Frey, H. A.,** Uptake of organics by aerial plant parts: a call for research, *Chemosphere,* 17, 277, 1988.
25. **Cranwell, R. M. and Helton, J. C.,** Uncertainty analysis for geologic disposal of radioactive waste, Proc. Symp. on Uncertainties Associated with the Regulation of the Geologic Disposal of High-Level Radioactive Waste, CONF-810372, Kocher, D. C., Ed., Gatlinburg, TN, March 9-13, 1981.
26. **Iman, R. L., Helton, J. C., and Campbell, J. E.,** An approach to sensitivity analysis of computer models. II. Ranking of input variables, response surface validation, distribution effects, and technique synopsis, *J. Qual. Technol.,* 13, 233, 1980.
27. **McKay, M. D., Conover, W. J., and Beckman, R. J.,** A comparison of three methods for selecting values of input variables in the analysis of output from a computer code, *Technometrics,* 21, 239, 1979.
28. **Morgan, M. G., Henrion, M., Morris, S. C., and Amareal, D. A.,** Uncertainty in risk assessment, *Environ. Sci. Technol.,* 19, 662, 1985.
29. **McKay, M. D.,** Sensitivity and uncertainty analysis using a statistical sample of input variables, in *Uncertainty Analysis,* Ronen, Y., Ed., CRC Press, Boca Raton, FL, 1988.
30. **Iman, R. L., Davenport, J. M., and Ziegler, D. K.,** Latin Hypercube Sampling Program User's Guide, SAND 79-1473, Sandia National Laboratory, Albequerque, 1980.

Chapter 11

AN UNCERTAINTY ANALYSIS OF FOOD CHAIN EXPOSURES TO POLLUTANTS EMITTED FROM MUNICIPAL WASTE COMBUSTORS*

Greg D. Belcher and Curtis C. Travis

TABLE OF CONTENTS

* Research was sponsored by the Environmental Protection Agency under an Interagency Agreement applicable under Martin Marietta Energy Systems, Inc., Contract No. DE-AC05-840R21400 with the U.S. Department of Energy.

I. INTRODUCTION

The incineration of municipal solid waste (MSW) produces residual amounts of pollutants that are emitted from municipal waste combustors (MWCs). Constant operation of the waste facility increases the concentration of these pollutants in air and soil near the facility. Since the use of MWCs as a waste management alternative has increased, public concern over possible environmental and human health effects has also intensified.

When models are used as the basis for making estimates of exposure in risk assessment, the question arises as to the sensitivity of model predictions to uncertainties in model input parameters.[1] A first step in the direction of answering this question is the identification of model parameters which made the greatest contribution to overall model error. These represent the parameters which must be quantified more accurately than other variables. The second step is to determine probability distributions for each model parameter and to propagate these distributions through the model to obtain a characterization of uncertainty in model output. Both of these steps will be performed for two pollutants released by MWCs: cadmium and 2,3,7,8-tetrachlorodibenzo-*p*-dioxin (TCDD), commonly referred to as dioxin. These pollutants were chosen as representative of two chemical classes: metals and organics.

Probabilistic simulation techniques allow more realistic exposure and risk estimates to be made and reduce reliance on worst-case assumptions. The purpose of this chapter is to perform a sensitivity and uncertainty analysis of the multimedia impacts of MWC using the terrestrial food chain (TFC) model and a probablistic simulation technique. The specific exposure pathways for which probablistic distributions were developed are those pathways included in the TFC model. The distributions generated from this analysis reflect the variability (uncertainty) in the input parameters used in the TFC model.

II. THE TERRESTRIAL FOOD CHAIN MODEL

The TFC model is a multimedia transport model that uses average atmospheric air concentration values ($\mu g/m^3$) and annual deposition rates (g/m^2/year) to calculate the daily pollutant intake by humans through the TFC. The model estimates soil concentrations and uptake into vegetation, animals, and humans. Pollutants are assumed to enter the food chain via the following routes: (1) root uptake, (2) direct deposition, (3) air-to-plant transfer, and (4) direct

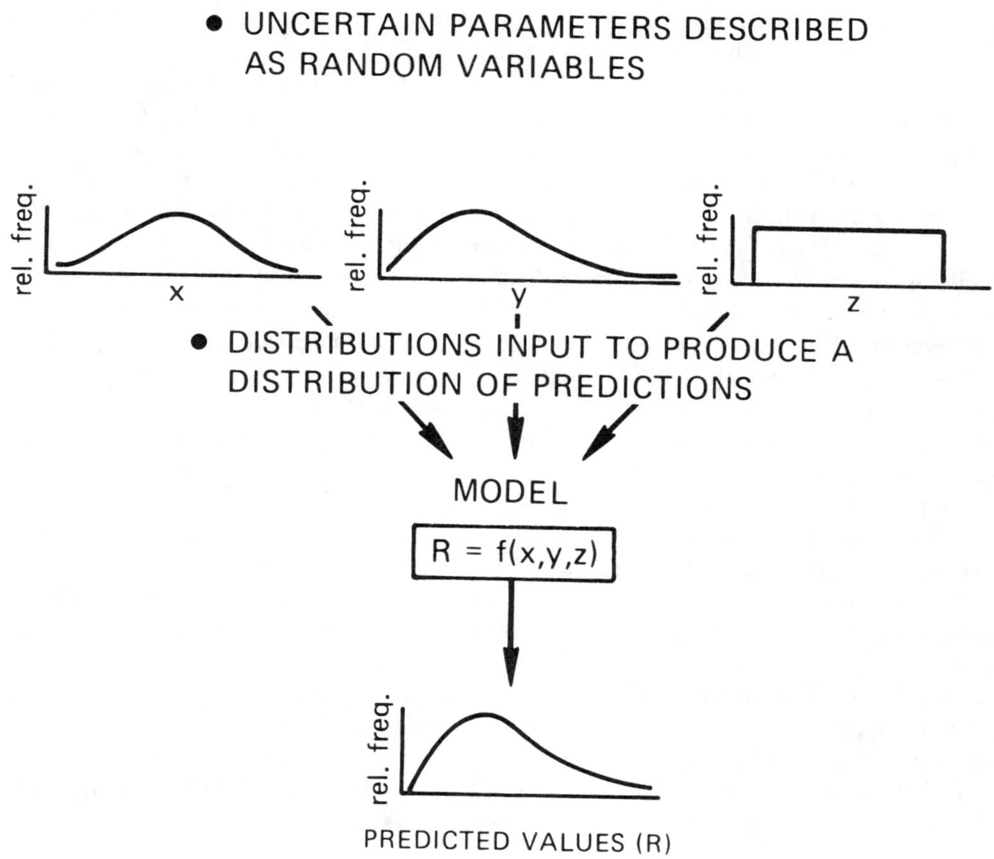

FIGURE 1. General approach to uncertainty analysis.

ingestion of contaminated soil by humans or animals. For a detailed description of the model, see Chapter 10.

III. UNCERTAINTY ANALYSIS

When models such as the TFC model are used to estimate the extent of human exposure, the question arises as to the accuracy of model predictions. The use of conservative assumptions (i.e., exposure estimates based on worst-case scenarios) have predominated in the past.[1] However, as the emphasis has moved to eliminate conservative assumptions and increase the "realism" of model predictions, uncertainty analysis has become an essential requirement in determining the reliability of model predictions.[2] An uncertainty analysis of the multimedia impacts of MSW combustion was performed, so that reliance on worst-case assumptions can be minimized and a comparison of risks associated with alternative waste disposal options can be made.

Figure 1 describes the methodology of an uncertainty analysis. As the figure shows, a model exists as a function of a series of input parameters. Each of these parameters can be described as a random variable with an appropriate probability distribution. Random samples are taken from these distributions and are entered into the model to produce a single estimate of model output. This process is repeated many times to produce a distribution of model predictions. Hence, as opposed to a point estimate that has no description of model variability, an uncertainty analysis characterizes a range of plausible values for each input parameter from which a best estimate (i.e., mean, median, etc.) can be ascertained.

A computer code called PRISM was used to evaluate the uncertainty associated with model predictions resulting from model input parameter variability. PRISM uses a systematic sampling method to generate random model parameter values and iteratively simulates model predictions for each parameter set. PRISM analyzes the results to determine means, variances, and important statistical relationships between model predictions and input parameters.[3]

The systematic sampling methodology employed by PRISM is a Latin Hypercube Sampling (LHS) approach. This technique generates a sample from a given distribution by dividing the total distribution into a number of equal segments and selecting a value at random from within each segment. This constrained random sampling procedure forces sampling from the tails of the distribution, resulting in a smaller sample being more representative of the distribution than the larger, unconstrained samples produced by Monte Carlo or simple random sampling techniques.[4] In fact, when efficiency is defined as the minimum number of samples needed to adequately represent the probability distribution of each parameter, a LHS approach appears to be the most efficient systematic sampling methodology. A LHS approach requires approximately ten times fewer samples than simple random sampling techniques.[3] For each chemical considered, 150 repetitions of the LHS process were performed.

To perform an uncertainty analysis of the multimedia impacts of combusting MSW as assessed by the TFC model, appropriate probability distributions were determined for each model input parameter. These distributions were propagated throughout the model to obtain a characterization of model output. Although the user can only enter normal distributions, the PRISM code allows conversion of those distributions to the type of distribution (i.e., log normal, uniform, etc.) specified by the user.

The following steps were used to determine the appropriate distribution for each model parameter.

1. Previously published model parameter uncertainty distributions were used when available. (For some parameters used in the TFC model, input distributions for use in other stochastic models have already been created.)
2. Data from primary source literature were used to obtain a range of plausible values and determine the appropriate probability distribution.
3. The range of parameter values was plotted on probit paper to determine if the values adhered to a log normal distribution.
4. If data limitations made the proper distribution type undiscernible, a log normal distribution was assumed.

It is important to note that parameter distribution characterization is a rather subjective process, especially for parameters for which limited literature data are available. While distribution selection is an important component of the uncertainty analysis process, in most cases the mean and variance are the most critical moments in stochastic modeling.[5]

A. SENSITIVITY ANALYSIS

The first step in any uncertainty analysis is to determine which model parameters have the greatest contribution to overall model output variability. The most effective process to determine parameter influence is to use a sensitivity analysis, which quantifies the magnitude of effect a given input parameter has on model outcome.[6] A sensitivity analysis was performed using the PRISM model, which ranks each input parameter according to the influence it has on model output. Each parameter was assumed to be normally distributed with a standard deviation equal to 1% of the mean.[7] Since all parameters are assumed to have the same probability and standard deviation, those parameters that have the greatest impact on model outcome will be

TABLE 1
Percent Contribution of the Most Influential Parameter to Variability in Total Daily Intake

Chemical	Most influential parameter	Influence (%)
Cadmium	Dy	57.0%
Dioxin	Dy	55.8%

TABLE 2
Percent of Parameter Influence on Total Daily Intake (no variation in Dy and Cy)

Cadmium[a]		Dioxin[a]	
Precip	17.11%	Ba (dairy)	22.06%
Evap	6.64	Ca (dairy)	9.88
Irigat	4.41	Ba (beef)	8.66
Cp (fruits)	4.13	Qs (beef)	6.49
Kd	3.70	Bd	5.33
Tc	3.67	Qs (dairy)	5.00
Theat	3.23	Fa (dairy)	3.33
Rp (legdr)	2.90	Bg	2.39
Qf (dairy)	2.60		
Fa (pork)	2.33		
Ca (pork)	2.33		
Br (potatoes)	2.20		
Ba (poultry)	2.12		
Total	57.37%	Total	63.14%

[a] Parameters are formally defined in Sections IV and V.

ranked as the most sensitive parameters when varied over their range of plausible values. Results show that variability in the annual dry deposition rate, Dy, is the primary contributor to uncertainty in model estimates to total daily intake for cadmium and dioxin, accounting for 57 and 56 % of the variability in human exposure, respectively (Table 1).

A sensitivity analysis was also performed assuming no uncertainty in the annual dry deposition rate, Dy, and the atmospheric air concentration, Cy. This scenario is conceivable for a site-specific analysis in which there might be little known variability in these parameters for a given, fixed distance from an incinerator. The remaining input parameters were assumed to have a normal distribution with the standard deviation set at 1% of the mean. Parameters that contribute more than 2% to total model output variability were designated as the most influential. Thirteen variables met this criterion for cadmium, and they account for almost 60% of the variation in total daily intake of cadmium. Eight parameters satisfied this criterion for dioxin, and those eight contribute over 60% of the variation in the total daily intake of dioxin. The individual contributions of each parameter to this variation are summarized in Table 2.

The major contributors to the variance of cadmium intake levels are precipitation, evapotranspiration, and irrigation. These parameters are important in determining equilibrium soil concentrations of cadmium. Since ingestion of contaminated vegetation is the main route of exposure to cadmium, and root uptake is responsible for almost 95% of the exposure of this pathway (see Section VI), these variables have the greatest influence on variation in cadmium intake.

The primary sources of variation in dioxin intake are the biotransfer factor and the daily human consumption rate for dairy products. A critical characteristic of dioxin is that due to its high lipophilicity, it readily accumulates in living organisms. Travis and Hattemer-Frey[8] estimated that consumption of beef and dairy products account for 77% of human exposure to background levels of dioxin. Therefore, the biotransfer factor for dairy products and the daily human consumption rate for dairy products are two parameters which have the most influence on variability in intake of dioxin.

IV. TECHNICAL BASIS FOR CHEMICAL-INDEPENDENT PARAMETERS

TFC model parameters are either chemical-independent or chemical-specific. Chemical-independent parameter values and distributions are summarized in Table 3. The technical derivation and/or justification of each parameter distribution is described below.

A. Bd: SOIL BULK DENSITY

Hoffman and Baes[9] summarized the range of values for Bd, which has been extensively measured and documented in the literature. They concluded that soil bulk density appears to be log normally distributed with a mean of 1.40 g/cm^3.

B. Ca: ADULT DAILY CONSUMPTION OF MEATS

Wet weight (WW) consumption rates reported by Yang and Nelson[10] were converted to a dry weight (DW) basis using conversion factors reported by Baes et al.[11] and Watt and Merrill.[12] Distributions were assigned to daily intake rates in grams DW per day (g DW/d) for beef, beef liver, pork, poultry, lamb, milk, and eggs. The respective mean consumption rates are as follows: 33.9, 6.4, 10.7, 11.9, 1.7, 40.9, 7.0, respectively. Log normal distributions were assumed for all meat groups.

C. Cp: ADULT DAILY CONSUMPTION OF PLANT FOOD GROUPS

Again, WW adult consumption rates reported by Yang and Nelson[10] were converted to a DW basis using conversion factors given by Baes et al.[11] Distributions were designated log normal and were characterized for leafy vegetables, potatoes, root vegetables, garden fruits, dried legumes, fresh legumes, and grains. The respective mean consumption rates (g DW/d) for these plant groups are 2.6, 30.0, 3.3, 11.0, 3.3, 0.8, and 177.9.

D. Cs: SOIL INGESTION RATE FOR ADULTS

Soil ingestion data for adults is sparse. To obtain a distribution for this parameter, estimates reported by Eschenroeder et al.,[13] Lagoy,[14] Kimbrough et al.,[15] and the Cleverly group[16] were used to calculate a mean of 0.037 g/d for incidental soil ingestion by adults. Due to data limitations, probability plots were unable to ostensibly discern a distribution type, and a log normal distribution was assumed.

E. E: EXPOSURE DURATION ADJUSTMENT FACTOR

This parameter was incorporated into the TFC model to effectively distribute pollutant exposure resulting from the intentional or incidental ingestion of soil over the entire life of an individual for carcinogenic pollutants. However, E is set equal to 1.0 when evaluating toxic effects other than carcinogenesis, exposure to metals, and when determining the adult human total daily pollutant intake.

F. Evap: AVERAGE ANNUAL EVAPOTRANSPIRATION

A range of values was extrapolated from Baes et al.,[11] who reported annual average evapotranspiration rates for 69 selected cities across the U.S. A mean of 64.81 cm/year was calculated. Due to the limited data, a log normal distribution was assumed.

TABLE 3
Chemical-Independent Parameters

Parameter	n	Distribution[a]	Mean	Standard deviation	Range	Ref.
Bd	299	L	1.40	0.15	0.93—1.84	9
Ca[b]						
Beef	4	L	33.9	2.2	31.7—35.8	10, 11
Beef liver	4	L	6.4	0.7	5.4—7.0	10, 11
Lamb	4	L	1.7	0.2	1.4—1.8	10, 11
Pork	4	L	10.7	1.4	8.5—11.6	10, 11
Poultry	4	L	11.9	1.7	10.2—14.1	10, 11
Dairy products	4	L	40.9	5.5	33.0—45.3	10, 11
Eggs	4	L	7.0	1.0	6.2—8.1	10—12
Cp[b]						
Potatoes	4	L	30.0	1.9	27.4—32.0	10, 11
Leafy vegetables	4	L	2.6	0.3	2.4—3.0	10, 11
Fresh Legumes	4	L	0.8	0.2	0.6—1.1	10, 11
Dried Legumes	4	L	3.3	0.8	2.4—4.4	10, 11
Root vegetables	4	L	3.3	0.3	3.0—3.6	10, 11
Garden fruits	4	L	11.0	1.0	9.7—12.0	10, 11
Grain	4	L	177.9	4.5	171.2—180.7	10, 11
Cs	4	L	0.037	0.043	0.0044—0.10	19—22
Evap	69	L	64.81	17.98	32.50—100.00	10
Fa						
Beef	4	L	0.0526	0.0056	0.0462—0.0599	17
Beef Liver	4	L	0.0526	0.0056	0.0462—0.0599	17
Lamb	4	L	0.0526	0.0056	0.0462—0.0599	17
Pork	4	L	0.0526	0.0056	0.0462—0.0599	17
Poultry	4	L	0.0719	0.0197	0.0439—0.0893	17
Dairy products	4	L	0.0403	0.0057	0.0347—0.0482	17
Eggs	4	L	0.0644	0.0185	0.0427—0.0870	17
Fa						
Potatoes	4	L	0.0931	0.0405	0.0503—0.1288	17
Leafy vegetables	4	L	0.0931	0.0405	0.0503—0.1288	17
Fresh legumes	4	L	0.0931	0.0405	0.0503—0.1288	17
Dried legumes	4	L	0.0931	0.0405	0.0503—0.1288	17
Root vegetables	4	L	0.0931	0.0405	0.0503—0.1288	17
Garden fruits	4	L	0.0765	0.0286	0.0500—0.1171	17
Grain	4	L	0.0044	0.0041	0.0011—0.0102	17
Irigat	69	L	32.99	27.08	12.50—100.00	11
Precip	69	L	88.18	34.74	18.06—164.19	18
Qf						
Beef	7	L	3.60	1.36	2.12—5.74	19—22
Beef liver	7	L	3.60	1.36	2.12—5.74	19—22
Lamb	3	L	1.58	0.07	1.50—1.64	19—22
Dairy products	7	c	10.37	2.42	7.28—15.18	19—21
Qg						
Beef	7	L	4.05	1.53	2.39—6.46	19—22
Beef liver	7	L	4.05	1.53	2.39—6.46	19—22
Lamb		c	0.0			19
Pork	2	L	2.07	0.050	2.03—2.10	21, 22
Poultry	5	L	0.053	0.13	0.038—0.066	21, 22
Dairy products	7	L	4.03	0.94	2.83—5.90	19—21
Eggs	5	L	0.053	0.13	0.038—0.066	21, 22
Qs						
Beef	4	L	0.39	0.27	0.1—0.72	21
Beef liver	4	L	0.39	0.27	0.1—0.72	21
Lamb	3	L	0.04	0.03	0.016—0.07	23
Pork	3	L	0.034	0.034	0.0—0.0688	23, 24

TABLE 3 (continued)
Chemical-Independent Parameters

Parameter	n	Distribution[a]	Mean	Standard deviation	Range	Ref.
Poultry	3	L	0.001	0.001	0.0—0.002	23, 24
Dairy products	5	L	0.41	0.24	0.1—0.72	21
Eggs	3	L	0.001	0.001	0.0—0.002	23, 24
Rf	12	n	0.47	0.30	0.02—0.82	9
Rg		c	0.0			25
Rp						
Potatoes		c	0.0			
Leafy vegetables	9	L	0.16	0.10	0.08—0.38	11, 19
Fresh legumes	2	L	0.008	0.004	0.005—0.01	11, 19
Dried legumes	2	L	0.008	0.004	0.005—0.01	11, 19
Root vegetables		c	0.0			
Garden fruits	6	L	0.05	0.05	0.004—0.008	11, 19
Grain		c	0.0			
Tc	2	L	65.00	49.50	30.00—100.00	22
Tf	2	L	0.123	0.058	0.082—0.164	11
Tg	2	L	0.164	0.083	0.082—0.247	11
Theta	299	n	0.22	0.07	0.03—0.40	9
Tp	2	L	0.164	0.083	0.082—0.247	11
Yf	10	L	0.31	0.25	0.02—0.75	9
Yg	48	L	0.30	0.09	0.14—0.45	19
Yp						
Potatoes	2	L	0.48	0.106	0.405—0.555	11, 19
Leafy vegetables	9	L	0.177	0.086	0.091—0.353	11, 19
Fresh legumes	2	L	0.104	0.38	0.077—0.130	11, 19
Dried legumes	2	L	0.104	0.38	0.077—0.130	11, 19
Root vegetables	5	L	0.334	0.142	0.090—0.434	11, 19
Garden fruits	6	L	0.107	0.093	0.012—0.253	11, 19
Grain	48	L	0.30	0.09	0.14—0.45	11, 19
Z2	L	20.0	5.00	15.0—25.0	22	

[a] Distribution types: L = log normal, n = normal, c = constant.
[b] Distribution types for Ca and Cp based on McKone and Ryan.[21]

G. Fa: FRACTION OF MEATS CONSUMED BY HUMANS THAT ORIGINATES FROM THE CONTAMINATED AREA

The U.S. Environmental Protection Agency (EPA)[17] estimated the annual average percent of adult dietary consumption that is homegrown for various food groups on a seasonal basis for rural/nonfarm, rural/farm, and urban households. Distributions, which were determined from the reported percentages for spring, summer, fall, and winter for all U.S. households, proved to be log normal. Distributions were assigned for the following meat groups: beef, beef liver, lamb, pork, poultry, dairy, and eggs. The respective mean fraction that is locally produced for these meat groups were as follows: 0.05, 0.05, 0.05, 0.05, 0.07, 0.04, and 0.06. The available data combined beef, beef liver, lamb, and pork, making their resulting distributions the same.

H. Fp: FRACTION OF PLANTS CONSUMED BY HUMANS THAT ORIGINATES FROM THE CONTAMINATED AREA

Data were taken from the U.S. EPA[17] based on the average percent of seasonal consumption which is homegrown for various plant food groups. Distributions were determined from the reported percentages for spring, summer, fall, and winter for all U.S. households and were assumed to be log normal. Distributions were assigned for the following plant groups: potatoes, leafy vegetables, fresh legumes, dried legumes, root vegetables, garden fruits, and grains. The mean fraction of these plant groups that was assumed to be locally grown were 0.09, 0.09, 0.09, 0.09, 0.08, 0.004, respectively. Again, the available data combined potatoes, leafy vegetables, legumes, and root vegetables together, making the distributions for those plant groups the same.

I. Irigat: AVERAGE ANNUAL IRRIGATION

A range of values was extrapolated from Baes et al.,[11] who reported average annual irrigation values for 69 selected cities across the U.S. Due to data limitations, probability plots were unable to discern a definite distribution type for this parameter, so it was assumed to have a log normal distribution with a mean of 32.99 cm/year.

J. Precip: AVERAGE ANNUAL PRECIPITATION

Average annual precipitation values for 69 selected cities across the U.S. were taken from the U.S. Bureau of The Census.[18] A probit plot of these data showed that this parameter appears to be log normally distributed with a mean of 88.18 cm/year.

K. Qf: QUANTITY OF LOCALLY GROWN FORAGE CROPS CONSUMED BY ANIMALS

The average quantity of forage consumed by beef cattle, dairy cattle, and sheep is 3.60, 10.37, and 1.58 kg DW/d, respectively.[19-21] All distributions were log normally distributed. It was assumed that all forage consumed by agricultural animals was locally grown and that pigs and chickens do not to ingest forage as part of their typical diet.[22,23]

L. Qg: QUANTITY OF LOCALLY GROWN GRAINS CONSUMED BY ANIMALS

The average amount of locally grown grain consumed by beef cattle, dairy cattle, pigs, and chickens is 4.05, 4.03, 2.07, and 0.053 kg DW/d, respectively.[19-21] All distributions were log normally distributed, and sheep were assumed not to ingest grain.[19]

M. Qs: QUANTITY OF SOIL CONSUMED BY ANIMALS

The average amount of soil ingested by beef cattle, dairy cattle, pigs, sheep, and chickens is estimated to be 0.39, 0.41, 0.034, 0.04, and 0.001 kg DW/d, respectively.[21,23,24] Log normal distributions were assumed.

N. Rf: INTERCEPTION FRACTION FOR FORAGE CROPS

Hoffman and Baes[9] summarized measured intercept fraction values for forage taken from 12 reports and showed that the probability distribution of reported values more nearly approximates a normal instead of log normal distribution. The mean (unitless) of reported values is 0.47.

O. Rg: INTERCEPTION FRACTION FOR GRAINS

Since grains have a protective outer husk, they were considered similar to protected produce and assumed not to be contaminated via deposition of facility-emitted pollutants.[25,26] Thus, this variable was assumed to have a constant value of 0.0.

P. Rp: INTERCEPTION FRACTION FOR PLANTS CONSUMED BY HUMANS

Interception fractions for food crops consumed by humans were calculated from the relationships for leafy vegetables and exposed produce listed in Baes et al.[11] Potatoes, root vegetables, and grains were assumed not to be contaminated via direct deposition and, therefore, have an intercept fraction of 0. Intercept fractions for garden fruits and legumes were extrapolated from the value given for exposed produce.[11] Fresh weight standing crop biomass values, which were used to calculate intercept fractions, were taken from Shor et al.[19] A probability plot of intercept fraction values for leafy vegetables showed the range of reported values adheres to a log normal distribution. Hence, a log normal distribution was assumed for all plant groups modeled. The mean intercept fraction (unitless) for each plant group contaminated via direct deposition is as follows: leafy vegetables (0.16), fresh and dried legumes (0.008), garden fruits (0.05).

Q. Tc: TOTAL TIME PERIOD OF DEPOSITION

The total time period of deposition is equivalent to the operating lifetime of the combustor. The U.S. EPA[16] estimated that the average life span of an MWC ranges from 30 to 100 years. A log normal distribution with a mean of 65 years was assumed for this analysis.

R. Tf: LENGTH OF THE GROWING SEASON FOR FORAGE CROPS

Baes et al.[11] quantified the average time between successive hay harvests (60 d) and successive grazings by cattle (30 d). These values were divided by the number of days in 1 year to obtain a range of values for Tf. A log normal distribution with a mean of 0.123 years was assumed.

S. Tg AND Tp: LENGTH OF THE GROWING SEASON FOR GRAINS AND PLANTS CONSUMED BY HUMANS

Reported values for the average time between successive hay harvests, 60 and 90 d,[11] were divided by the number of days in 1 year to obtain a range of values for Tg and Tp. Both parameters were assumed to be log normally distributed with a mean of 0.164 years.

T. Theta: SOIL VOLUMETRIC WATER CONTENT

Values for theta have been extensively reviewed in the theoretical literature and are summarized by Hoffman and Baes,[9] who determined that theta is normally vs. log normally distributed based on probability plots of 299 data points. The mean of the distribution is 0.22 ml/cm^3.

U. Yf: YIELD OR STANDING CROP BIOMASS FOR FORAGE CROPS

Hoffman and Baes[9] summarized the available data on forage yields from ten studies and determined that Yp was log normally distributed with a mean of 0.31 kg DW/m^2.

V. Yg: YIELD OR STANDING CROP BIOMASS FOR GRAINS

Shor et al.[19] reported grain crop yields for the 48 contiguous states. Average standing crop biomass values for each state were plotted on a probit plot. Results show that Yg is log normally distributed with a mean of 0.30 kg DW/m^2.

W. Yp: YIELD OR STANDING CROP BIOMASS FOR PLANTS CONSUMED BY HUMANS

Fresh weight values of Yp reported by Shor et al.[19] were converted to a DW basis using the conversion factors from Baes et al.[11] A log normal distribution was determined based on a probability plot for leafy vegetables. Mean yields (kg DW/m^2) for each food crop consumed by humans are as follows: potatoes (0.48), leafy vegetables (0.177), root vegetables (0.338), garden fruits (0.107), legumes (0.104).

X. Z: SOIL DEPTH

Values for soil depth normally range from 15 to 25 cm.[16] A log normal distribution with a mean of 20 cm was assumed for this analysis.

TABLE 4
Chemical-Specific Parameters for Cadmium

Parameter	n	Distribution[a]	Mean	Standard deviation	Range	Ref.
Ba						
Beef	8	L	0.003	0.005	0.0—0.012	24
Beef liver	6	L	0.135	0.129	0.045—0.3793	24
Lamb	6	L	0.0270.026	0.003—0.06	24	
Pork	6	L	0.028	0.048	0.0—0.12	24
Poultry	12	L	12.45	11.94	1.36—36.57	24
Dairy products	3	L	0.0055	0.002	0.0036—0.0076	24
Eggs	3	L	0.6496	0.1836	0.4894—0.8500	24
Bf	3	L	0.39	0.17	0.22—0.55	11, 34, 35
Bg	8	L	0.05	0.07	0.003—0.22	35
Br						
Potatoes	4	L	0.09	0.03	0.06—0.14	24, 35
Leafy vegetables	6	L	1.18	0.72	0.04—2.38	24, 35
Fresh legumes	4	L	0.24	0.32	0.03—0.70	24, 35
Dried legumes	4	L	0.24	0.32	0.03—0.70	24, 35
Root vegetables	11	L	1.98	4.757	0.0072—18.76	24, 35
Garden fruits	5	L	1.16	1.36	0.0019—3.14	24, 35
Grain	8	L	0.05	0.07	0.003—0.22	35
Bv						
Potatoes		c	0.0			
Leafy vegetables		c	0.0			
Dried legumes		c	0.0			
Fresh legumes		c	0.0			
Root vegetables		c	0.0			
Garden fruits		c	0.0			
Grain		c	0.0			
Bvf		c	0.0			
Bvg		c	0.0			
Cy	na[b]	L	4.0E-6	5.31	1.3E-7—1.3E-4	28, 33, 36-44
Dy	na	L	7.3E-6	7.61	1.1E-7—4.7E-4	28, 33, 36-44
Kd	3	L	148.13	238.08	6.55—423.0	11, 35
Kp	4	L	31.88	14.83	18.07—52.51	11, 35
Ksgv		c	0.0			

[a] Distribution types: L = log normal, c = constant.
[b] na = not applicable.

V. TECHNICAL BASIS FOR CHEMICAL-SPECIFIC PARAMETERS

Tables 4 and 5 summarize the distributions and give reference for each chemical-specific parameter.

A. Ba: BIOTRANSFER FACTOR FOR ANIMAL TISSUES

A traditional measure of a chemical's potential to accumulate in animal tissue is the bioconcentration factor (BCF), which is defined as the equilibrium concentration of pollutant in an organism or tissue (μg/g) divided by the equilibrium concentration of pollutant (μg/g) in food for terrestrial organisms. This concept, however, is not readily applied to humans and other terrestrial organisms because the amount of chemical in various food items (and, hence, total

TABLE 5
Chemical-Specific Parameters for Dioxin

Parameter	n	Distribution[a]	Mean	Standard deviation	Range	Ref.
Ba						
Beef	3	L	0.3089	0.4285	0.0550—0.80	20, 45, 46
Beef liver	3	L	0.8097	1.1256	0.0286—2.10	20, 45-47
Lamb[b]	1	L	0.0213			26
Pork[b]	1	L	0.5610			26
Poultry[b]	1	L	118.96			26
Dairy products	3	L	0.030	0.008	0.024—0.039	46
Eggs[b]	1	L	90.03			26
Bf	6	L	0.033	0.034	0.00003—0.083	20, 21, 48
Bg	6	L	0.033	0.034	0.00003—0.083	20, 21, 48
Br						
Potatoes	6	L	0.033	0.034	0.00003—0.083	20, 21, 48
Leafy vegetables	6	L	0.033	0.034	0.00003—0.083	20, 21, 48
Fresh legumes	6	L	0.033	0.034	0.00003—0.083	20, 21, 48
Dried legumes	6	L	0.033	0.034	0.00003—0.083	20, 21, 48
Root vegetables	6	L	0.033	0.034	0.00003—0.083	20, 21, 48
Garden fruits	6	L	0.033	0.034	0.00003—0.083	20, 21, 48
Grain	6	L	0.033	0.034	0.00003—0.083	20, 21, 48
Bv						
Potatoes		c	0.0			
Leafy vegetables	3	L	1.02E4	7.65E2	9.61E3—1.10E4	49-52
Dried legumes	3	L	1.02E4	7.65E2	9.61E3—1.10E4	49-52
Fresh legumes	3	L	1.02E4	7.65E2	9.61E3—1.10E4	49-52
Root vegetables		c	0.0			
Garden fruits	3	L	1.02E4	7.65E2	9.61E3—1.10E4	49-52
Grain		c	0.0			
Bvf	3	L	1.02E4	7.65E2	9.61E3—1.10E4	49-52
Bvg		c	0.0			
Cy	na[c]	L	9.1E-11	13.46	4.4E-13—1.9E-8	53
Dy	na	L	3.5E-11	15.49	1.3E-13—9.8E-9	53
Kd	3	L	3.1E4	8.48E3	2.3E4—4.0E4	54, 55
Kp	2	L	27.06	12.84	17.98—36.14	8, 24
Ksgv	2	L	0.0624	0.0061	0.0578—0.0693	49, 54

[a] Distribution types: L = log normal, c = constant.
[b] Estimated based on background concentration of TCDD in these tissues.
[c] na = not applicable.

daily intake of pollutants) can vary markedly. Thus, for risk assessment purposes, it is more efficacious to examine the biotransfer factor (BTF), which is defined as the equilibrium concentration of pollutant in an organism or tissue (mg/kg) divided by the average daily intake of pollutant (mg/d).[20]

If bioconcentration values were reported as a BCF, they were converted to a BTF by dividing the BCF by the dry feed ingestion rate (kg/d) for the animal group.[20] If only one BCF or BTF was available, a distribution of BTFs was created by dividing by the minimum, mean, and maximum dry feed ingestion rates. The dry feed ingestion rates used to convert to BTFs were determined by adding the quantity of grain (Qg) and forage (Qf) by each animal group. Log normal distributions were assumed.[21]

B. Bf, Bg, Bp: SOIL-TO-PLANT BIOCONCENTRATION FACTORS FOR ALL PLANT GROUPS

Soil-to-plant BCFs represent a chemical's potential to take up pollutants from contaminated soil through the root system and are defined as the equilibrium concentration of pollutant in upper plant parts (μg pollutant/g plant tissue DW) divided by the equilibrium concentration of pollutant in soil (μg pollutant/g soil). A log normal distribution was assumed for Bf, the soil-to-plant BCF for forage crops; Bg, the soil-to-plant BCF for grains; and Bp, the soil-to-plant BCF food crops consumed by humans.[21]

C. Bvf, Bvg, Bvp: AIR-TO-PLANT BIOCONCENTRATION FACTORS FOR ALL PLANT GROUPS

Air-to-plant BCFs are defined as the equilibrium concentration of organic in upper plant parts (μg pollutant/g plant tissue DW) divided by the equilibrium concentration of organic in air as a vapor (μg pollutant/g air). By definition, since metals do not exist in the vapor phase, they were assumed not to contaminate vegetation via the air-to-leaf transfer pathway. Thus, this parameter was set constant to zero for cadmium. Since measured data for air-to-leaf transfer factors are sparse, air-to-plant BCFs for dioxin were estimated from the following geometric mean regression equation developed by Hattemer-Frey and Travis[26]:

$$Bv = 5.0 \times 10^{-6} K_{ow} \div H \qquad (1)$$

where K_{ow} = octanol water partitioning coefficient and H = Henry's Law Constant.

Air-to-plant BCFs were used to model the uptake of facility-emitted pollutants by the following plant groups: leafy vegetables, legumes, and garden fruits. It was assumed that protected produce, including potatoes and other root vegetables, whose edible portions either grow underground or are protected by pods, shells, or nonedible peels were not contaminated via this pathway. A log normal distribution was assumed based on a probability plot of nine air-to-plant transfer values calculated for dioxin using the above equation and a range of K_{ow} values.

D. Cy: ATMOSPHERIC AIR CONCENTRATION

This parameter reflects the average annual atmospheric air concentration of the pollutant due to direct emissions from a MWC and is recorded in terms of μg pollutant/m^3 air. The methodology used to determine Cy is described in Hattemer-Frey and Travis[26] and Lowe et al.[27] Average annual atmospheric concentrations resulting from a unit release of 1 g of pollutant per second within a 50-km radius area were calculated for 24 operating MWCs located across the U.S. using the ISCLT air dispersion model.[28] Facility characteristics and locations can be found in Table 6.

Next, a range of emission rates was determined for each chemical (Table 7). These distributions of emissions were merged with the exposure concentration distributions based on a unit emission rate to obtain a distribution of atmospheric concentrations for each chemical. Since atmospheric concentrations were determined for an arbitrary MWC based on data from MWCs throughout the U.S., estimates for Cy derived using the above-mentioned procedure can be expected to vary more than site-specific estimates of average annual atmospheric concentrations. Log normal distributions were assumed.[26]

E. Dy: ANNUAL DRY DEPOSITION RATE

The methodology used to determine distributions for the annual dry deposition rate (g/m2/year) for each pollutant is the same as the procedure used to determine distributions for Cy. The ISCLT air dispersion model was used to estimate average annual dry deposition rates for cadmium and dioxin within a 50-km radius of the 24 MWCs listed in Table 6. These distributions (based on a unit emission rate of 1 g/s) were merged with distributions of emissions for each chemical to obtain a distribution of actual deposition rates (Table 7). Again, since this procedure estimates annual dry deposition rates for an arbitrary MWC based on actual facility data, site-specific estimates can be expected to vary less than these rates. Log normal distributions were assumed.[26]

TABLE 6
Operating Characteristics and Location of the 24 MWCs Used to Determine Unit Emission Results for Cy and Dy[26,56]

Location	Stack height (m)	Stack diameter (m)	Exit gas velocity (m/s)	Exit gas temp (°K)
Livingston, MT	5	1.8	3.5	420
Ogden, UT	46	1.5	13.4	477
Collegeville, MN	37	3.6	0.7	350
Waukesha, WI	32	2.1	9.2	420
Stuttgart, AZ	11	0.8	9.3	420
Palestine, TX	15	0.9	5.5	350
Pinellas Co., FL	49	3.7	9.2	420
Lewisburg, TN	15	1.2	6.6	350
Baltimore, MD	96	1.8	13.4	506
Euclid, OH	38	0.8	7.3	505
Harrisburg, PA	49	0.6	8.3	420
Huntington, NY	38	2.1	12.1	494
Durham, NH	35	1.8	11.5	455
Bellingham, WA	11	1.1	6.2	420
Meredith, NH	24	0.9	11.1	830
Pittsfield, NH	8	0.6	21.2	420
Oswego Co., NY	43	0.6	22.1	420
Johnsonville, SC	9	2.1	1.8	420
Ames, IA	61	2.4	1.3	505
Haverhill, MA	99	2.1	19.0	435
Waxahachie, TX	27	0.9	11.8	350
Lakeland, FL	61	3.7	1.3	400
Sitka, AK	24	1.0	4.0	420
Osceola, AR	9	1.2	5.6	420
Geometric mean	26.5	1.4	6.6	435

TABLE 7
Predicted Emission Rates

Chemical	Geometric mean (g/s)	Geometric standard deviation
Cadmium[a]	1.06E-3	0.76
Dioxin[b]	2.39E-8	1.24

[a] Emission rates for cadmium based on estimated values for 11 proposed MWCs in the U.S.[28,33,36-44,57]

[b] Emission rates for dioxin based on actual emissions from existing MWCs in the U.S.[26,56]

F. Kd: SOIL-TO-WATER PARTITIONING COEFFICIENT

Kd represents the ratio of the concentration of pollutant in soil to concentration of pollutant in water at equilibrium (ml/g). Baes et al.[11] assigned a log normal distribution for this variable based on distributions determined for cesium and strontium. Hence, a log normal distribution was used in this analysis.

TABLE 8
Breakdown of Exposure Pathways for Cadmium (percent contributions to average daily intake)

Pathway	Contribution (%)
Direct ingestion of contaminated soil	12.9
Consumption of contaminated vegetation	
Atmospheric deposition	3.3
Root uptake	62.8
Air-to-plant transfer	0.0
Consumption of contaminated meat and dairy products	
Soil-animal-human route	13.3
Plant-animal-human route	7.7

TABLE 9
Contribution to Daily Intake of Cadmium by Vegetation Food Groups

Source	Total intake (%)
Garden fruits	32.3
Root vegetables	19.6
Potatoes	6.9
Leafy vegetables	4.7
Dried legumes	1.7
Grain	0.6
Fresh legumes	0.3

G. Kp: PLANT SURFACE LOSS CONTENT

The equation for calculating pollutant loss from plant surfaces due to weathering is

$$\ln 2/T_{1/2}\ (\text{years}) \qquad (2)$$

where ln 2 is the natural log of two, and $T_{1/2}$ is the half-life of the pollutant on outer plant parts. This parameter was assigned a log normal distribution based on results reported by Miller and Hoffman.[29]

H. Ksgv: SOIL LOSS CONTENT

This parameter represents pollutant loss from soil due to degradation (biotic and abiotic) and volatilization (1/years). Since metals do not decay in soil, this parameter was set constant at zero for cadmium. Equation 2 can be used to calculate Ksgv values for organics by defining $T_{1/2}$ as the half-life of pollutant in soil (years) vs. the half-life of pollutant on outer plant surfaces. A log normal distribution for Ksgv was assumed.

VI. RESULTS

A. CADMIUM

Table 8 shows that cadmium intake due to emissions from MWCs is due mainly to the consumption of contaminated vegetation (66%), of which 63% is due to root uptake and 3% to

TABLE 10
Contribution to Daily Intake of Cadmium by Meat Food Groups

Source	Total intake (%)
Beef liver	12.3
Dairy products	3.7
Poultry	3.1
Beef	1.4
Pork	0.3
Eggs	0.2
Lamb	<0.1

TABLE 11
Distribution of Daily Cadmium Intake Values (μg/d)

Geometric mean	0.00211
Geometric standard deviation	0.771
Minimum	0.00009
Maximum	0.0719

atmospheric deposition. Table 9 shows that contaminated garden fruits and root vegetables are the primary plant groups contributing to total daily intake of cadmium, accounting for 32 and 20% of total daily intake, respectively.

The consumption of contaminated meat and dairy products is also an important pathway of cadmium exposure, accounting for 21% of total daily intake by humans. Table 10, which lists the contribution to daily intake of each meat group, shows that ingestion of beef liver contributes most to human exposure, accounting for 12% of total daily intake.

The goal of the uncertainty analysis was to determine a probability distribution for the daily human intake of cadmium due to emissions from MWCs. Figure 2 shows that these values fit a log normal distribution, and their statistical summary is given in Table 11. Figure 3 depicts the 95% confidence intervals for the daily intake of cadmium from the various exposure pathways, with inhalation included for comparison purposes. Inhalation accounts for only 4% of total daily intake of cadmium by individuals living near an arbitrary MWC. The 95% confidence interval is less than a factor of 23 times the best estimate of cadmium daily intake from the food chain. The range of uncertainty in total daily intake from the food chain across the interval is about 2.7 orders of magnitude.

Previous studies have shown that much of the overall uncertainty in estimates of human exposure to pollutants emitted from MWCs is attributable to uncertainty in the deposition component and the concentration value when modeling an "arbitrary" incinerator using cross-site data.[21,30] Hence, for comparison purposes model runs were made treating Dy and Cy as constants. Figure 4 depicts the 95% confidence interval for the total daily intake of cadmium due to emissions from MWCs based on these point estimates. The 95% confidence interval is less than a factor of 2.1 times the best estimate of cadmium daily intake from the food chain. The range of uncertainty in daily intake from the food chain across the interval is approximately four times the lower bound.

B. DIOXIN

Table 12 shows that the primary route of human exposure to dioxin from the food chain due to emissions from MWCs is the consumption of contaminated meat and dairy products, which

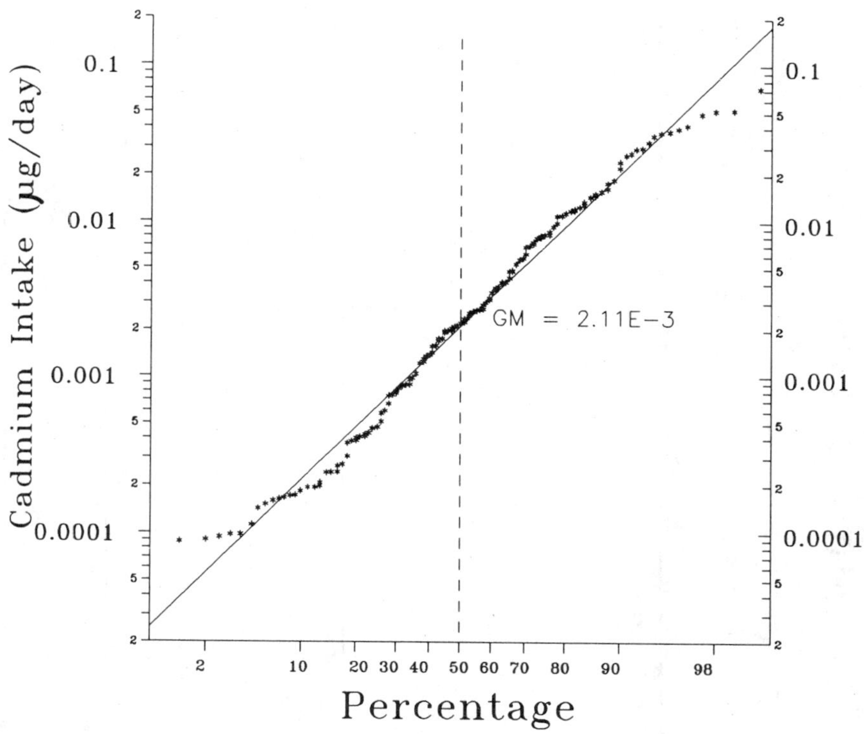

FIGURE 2. Predicted cadmium daily intake values.

accounts for 89% of total daily intake. This pathway can be further subdivided into the soil-animal route (53%) and the plant-animal route (35%). Table 13, which gives the contributions of each individual meat group, shows that beef and beef liver contribute the most to total daily intake, accounting for 34 and 24% of total daily intake of dioxin, respectively.

While soil ingestion by humans accounts for about 3% of dioxin daily intake, the consumption of contaminated vegetation contributes about 8%. The contaminated vegetation pathway can be subdivided into the air-to-plant transfer component (6%), the atmospheric deposition component (2%), and the root uptake component (0.1%). Table 14 shows that consumption of garden fruits contributes more to total daily intake of dioxin (4%) than the other plant groups consumed by humans.

Again, a probability distribution for total daily intake of dioxin due to emissions of the contaminant from an arbitrary MWC was determined. Figure 5 shows that these values fit a log normal distribution, and their statistical summary is listed in Table 15. Figure 6 depicts the 95% confidence intervals for the daily intake of dioxin from the various exposure pathways. Human exposure via inhalation is included for comparison purposes. Inhalation accounts for only 9% of the total daily intake of dioxin. The 95% confidence interval for total daily intake of dioxin from the food chain is approximately a factor of 40 times the best estimate. The estimates of dioxin intake across this interval range almost 3.2 orders of magnitude.

For comparison purposes, model runs were made in which Dy and Cy were held constant. Figure 7 depicts the 95% confidence interval for the total daily intake of dioxin due to emissions from MWCs based on these point estimates, which reflect no variability in the deposition rate and the air concentration value. The 95% confidence interval is less than a factor of 1.6 times the best estimate of dioxin daily intake from the food chain. The range of uncertainty across the interval is approximately 2.2 times the lower bound.

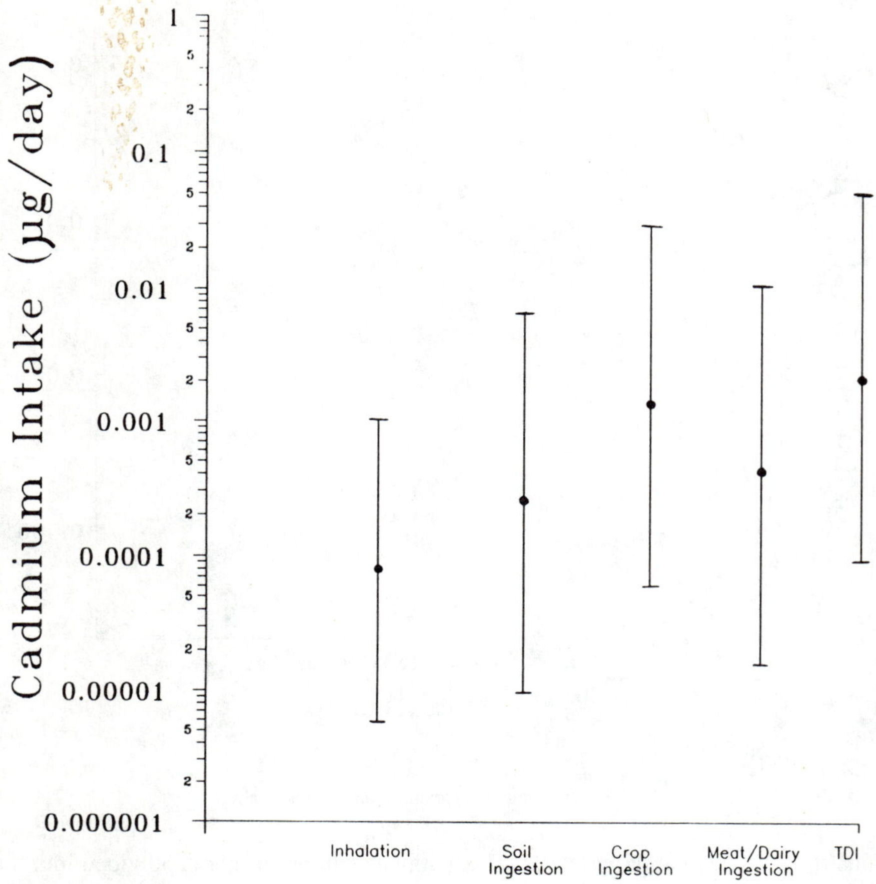

FIGURE 3. Intake of cadmium from various exposure pathways. Error bars represent 95% confidence intervals. TDI refers to total daily intake from the food chain pathways.

VII. CONCLUSIONS

The food chain has been shown to be the primary pathway of human exposure to a large class of organics.[8,25,31,32] A sensitivity and uncertainty analysis was performed on the TFC model to identify the input parameters that have the most influence on model outcome (i.e., human daily intake estimates), so that reliance on worst-case assumptions can be minimized. Three main pathways of human exposure were analyzed: (1) ingestion of contaminated soil, (2) consumption of contaminated vegetation, and (3) consumption of contaminated meat and dairy products. Three routes of vegetative contamination were considered: (1) atmospheric deposition, (2) uptake from the soil, and (3) air-to-plant transfer.

Results showed that cadmium intake by humans is due mainly to the consumption of contaminated plants, especially garden fruits and root vegetables, and that the primary pathway of vegetation contamination is root uptake. Human exposure to dioxin occurs primarily via the ingestion of contaminated meat and dairy products, especially beef.

Using dry deposition rates and air concentration values for an arbitrary MWC based on cross-site data, distributions of cadmium, and dioxin daily intake values were determined. Cadmium values ranged from 3.9×10^{-5} to 7.2×10^{-2} µg/d with a geometric mean of 2.1×10^{-3}. The 95%

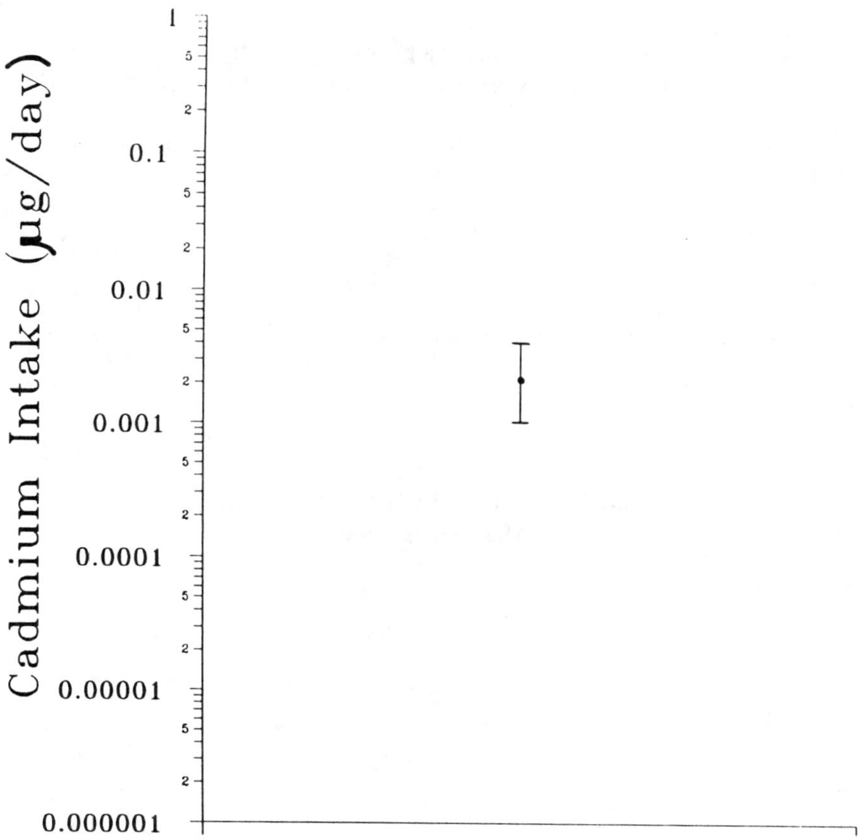

FIGURE 4. 95% confidence interval for total daily intake of cadmium from the food chain using point estimates for the dry deposition rate (Dy) and annual average air concentration (Cy).

confidence interval was less than a factor of 23 times the best estimate for cadmium exposure. Dioxin intake values varied from 2.9×10^{-10} to 9.25×10^{-7} μg/d with a geometric mean of 2.8×10^{-8}. The 95% confidence interval was approximately a factor of 40 times the best estimate for total daily intake. Inhalation was shown to be a minor source of cadmium and dioxin exposure when compared to contributions from the food chain.

When distributions of total daily intake were determined based on point estimates for the dry deposition rate and average annual air concentration value, uncertainty in intake estimates dramatically declined. Results from this analysis showed that the 95% confidence intervals for total daily intake of cadmium and dioxin were less than a factor of 2.1 times the best estimate. Also, total uncertainty in the daily intake estimates for both chemicals from all pathways was less than 0.5 orders or magnitude. These results support previous findings, indicating that only a small portion of the variability in human intake estimates is attributable to uncertainty in the FC model.[30] Thus, the results presented here strongly suggest that much of the overall uncertainty in estimates of human exposure to pollutants emitted from MWCs is attributable to uncertainty in the deposition rate and the air concentration value. When determining human daily intake estimates based on a specific site in which deposition rates and concentration values experience limited variation, uncertainty in estimates of human exposure will likely be reduced,[33] with only a small degree of variability attributable to uncertainty in the TFC model.

TABLE 12
Breakdown of Exposure Pathways for Dioxin (percent contributions to average daily intake)

Pathway	Contribution (%)
Direct ingestion of contaminated soil	3.3
Consumption of contaminated vegetation	
Atmospheric deposition	1.6
Root uptake	0.1
Air-to-plant transfer	6.4
Consumption of contaminated meat and dairy products	
Soil-animal-human route	53.3
Plant-animal-human route	35.3

TABLE 13
Contribution to Daily Intake of Dioxin by Meat Food Groups

Source	Total intake (%)
Beef	34.1
Beef liver	23.8
Poultry	14.4
Dairy products	8.4
Eggs	6.3
Pork	1.6
Lamb	<0.1

TABLE 14
Contribution to Daily Intake of Dioxin by Vegetation Food Groups

Source	Total intake (%)
Garden fruits	4.4
Leafy vegetables	2.0
Dried legumes	1.3
Fresh legumes	0.3
Potatoes	0.1
Grain	<0.1
Root vegetables	<0.1

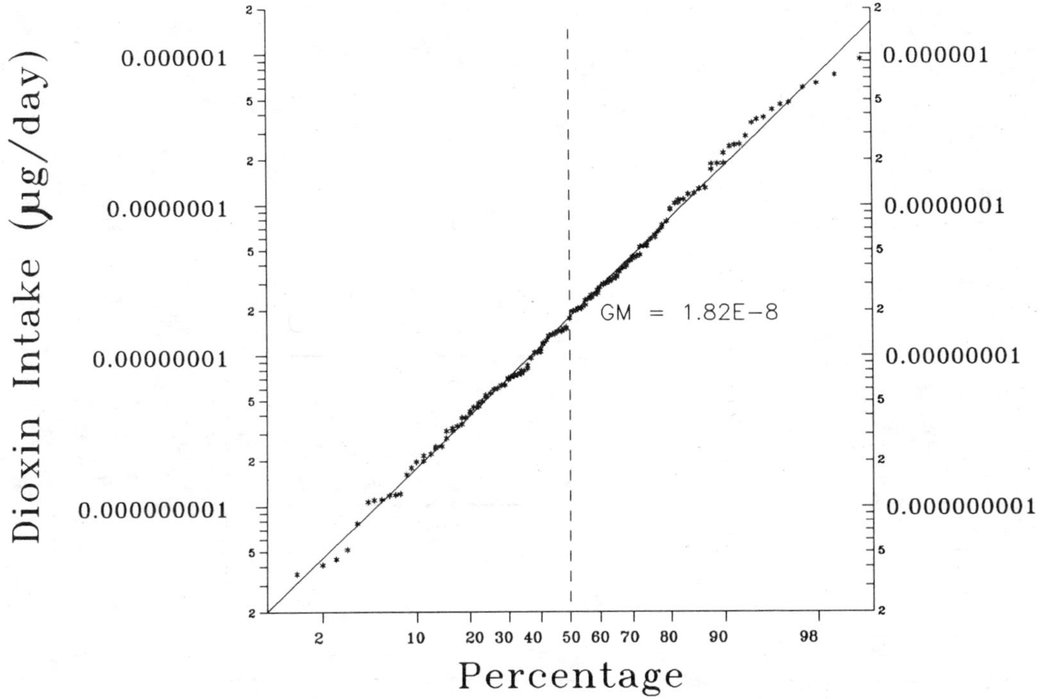

FIGURE 5. Predicted dioxin daily intake values.

TABLE 15
Distribution of Daily Dioxin Intake Values (μg/d)

Geometric mean	1.82×10^{-8}
Geometric standard deviation	0.782
Minimum	2.86×10^{-10}
Maximum	9.25×10^{-7}

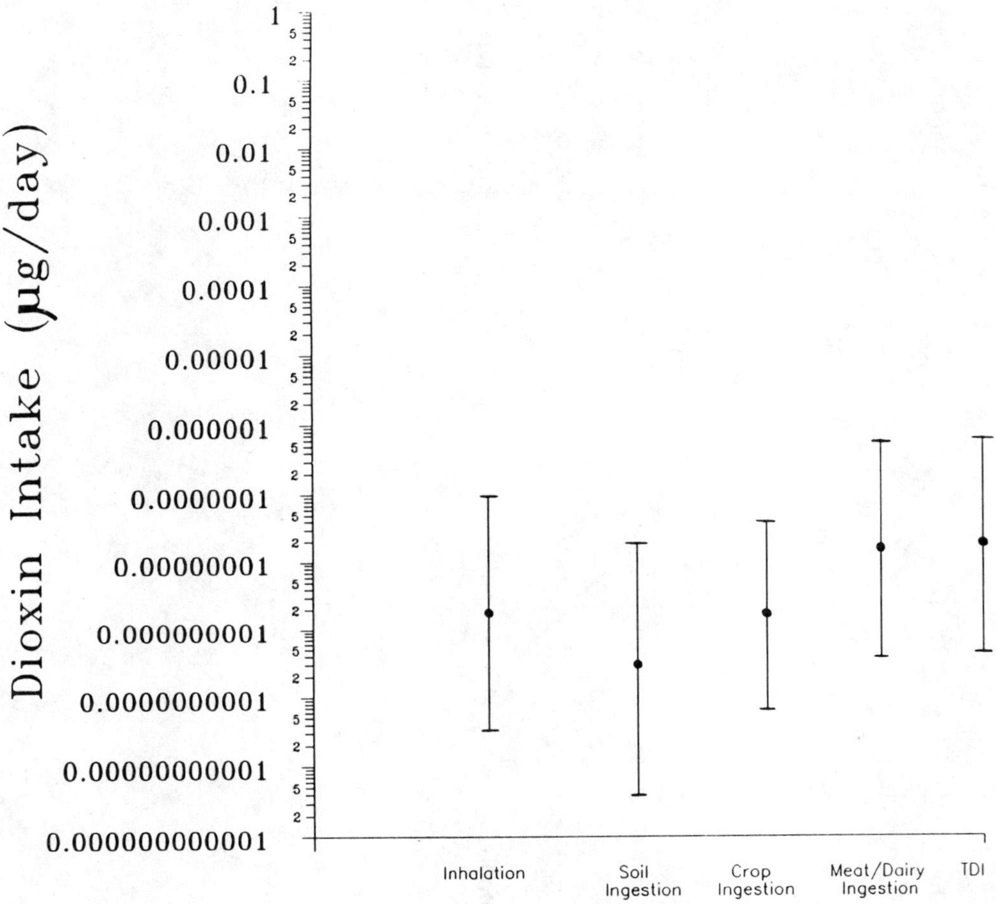

FIGURE 6. Intake of dioxin from various exposure pathways. Error bars represent 95% confidence intervals. TDI refers to total daily intake from the food chain pathways.

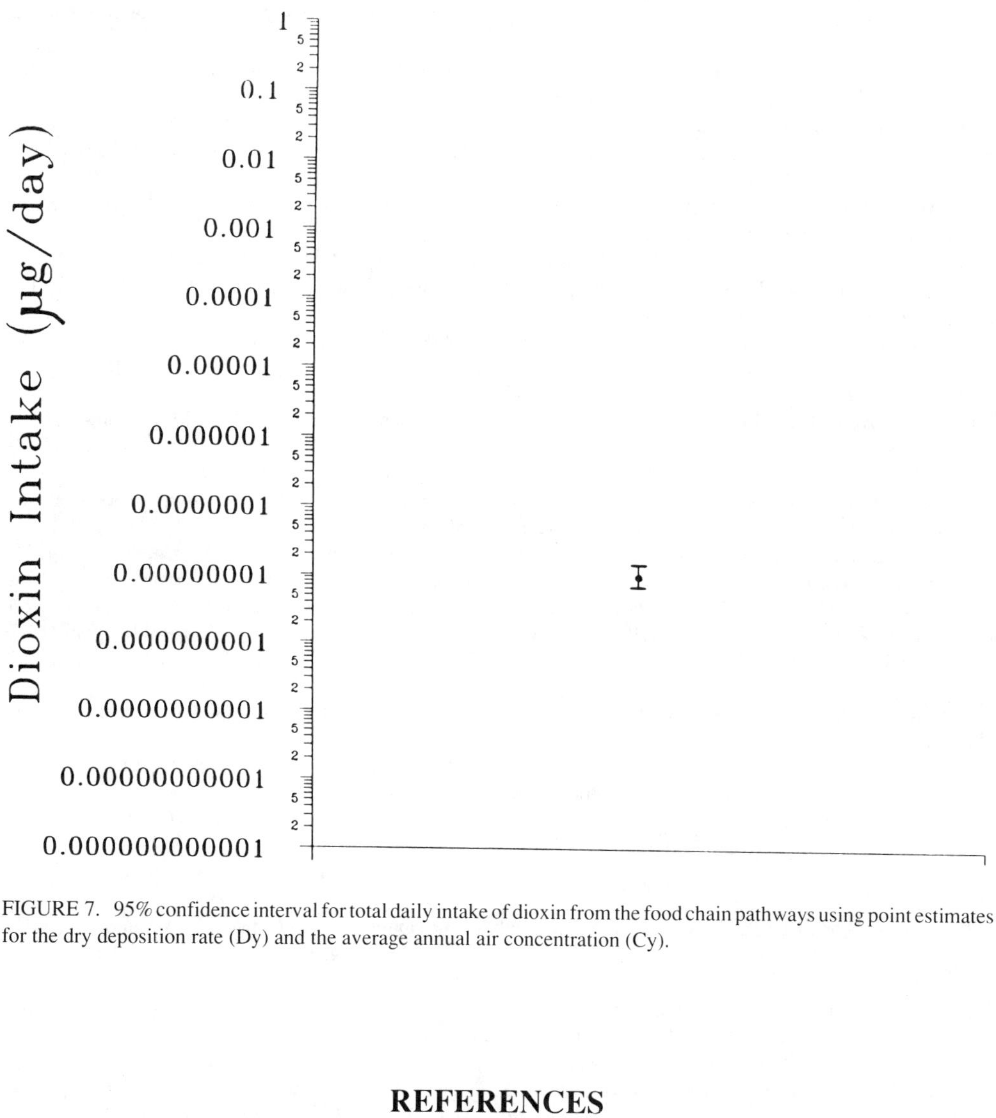

FIGURE 7. 95% confidence interval for total daily intake of dioxin from the food chain pathways using point estimates for the dry deposition rate (Dy) and the average annual air concentration (Cy).

REFERENCES

1. **Hoffman, F. O. and Gardner, R. H.,** Evaluation of uncertainties in radiological assessment models, in Radiological Assessment: A Textbook on Environmental Dose Analysis, NUREG/CR-3332, Till, J. E. and Meyer, H. R., Eds., 1983, Oak Ridge National Laboratory, Oak Ridge, TN, chap. 11.
2. **Ronen, Y.,** The role of uncertainties, in *Uncertainty Analysis,* Ronen, Y., Ed., CRC Press, Boca Raton, FL, 1988, chap. 1.
3. **Gardner, R. H., Rojder, B., and Bergstrom, U.,** PRISM, A Systematic Method for Determining the Effect of Parameter Uncertainties on Model Predictions, Studvisk Energiteknik AB Report STUDSVIK/NW-83/555, Nykoping, Sweden, 1983.
4. **Ronen, Y.,** Sensitivity and uncertainty analysis using a statistical sample of input values, in *Uncertainty Analysis,* Ronen, Y., Ed., CRC Press, Boca Raton, FL, 1988, 4.
5. **Gardner, R. H.,** person communication, 1989.
6. **Ronen, Y.,** Uncertainty analysis based on sensitivity analysis, in *Uncertainty Analysis,* Ronen, Y., Ed., CRC Press, Boca Raton, FL, 1988, 2.
7. **Gardner, R. H., O'Neill, V., Mankin, J. B., and Carney, J. H.,** A comparison of sensitivity analysis and error analysis based on a stream ecosystem model, *Ecol. Model.,* 12, 177, 1981.
8. **Travis, C. C. and Hattemer-Frey, H. A.,** Human exposure to 2,3,7,8-TCDD, *Chemosphere,* 17, 277, 1987.

9. **Hoffman, F. O. and Baes, C. F.,** A Statistical Analysis of Selected Parameters for Predicting Food Chain Transport and Internal Dose of Radionuclides, Report ORNL/NUREG/TM-882, Oak Ridge National Laboratories, Oak Ridge, TN, 1979.
10. **Yang, Y. and Nelson, C. B.,** An estimation of daily food usage factors for assessing radionuclide intakes in the U.S. population, *Health Phys.*, 50, 245, 1986.
11. **Baes, C. F., Sharp, R. D., and Sjoreen, R. W.,** A Review and Analysis of Parameters for Assessing Transport of Environmentally Released Radionuclides Through Agriculture, Report ORNL-5786, Oak Ridge National Laboratory, Oak Ridge, TN, 1984.
12. **Watt, B. K. and Merrill, A. L.,** Composition of Foods, Agricultural Handb. No. 8, U.S. Department of Agriculture, Washington, D.C., 1963.
13. **Eschenroeder, A., Guldberg, P., Hahn, J., Kellermeyer, D., Smith, A., Wolff, S., and Ziemer, S.,** An Analysis of Health Risks from the Irwindale Resource Recovery Facility, prepared for the California Energy Commission, Sacramento, 1986.
14. **Lagoy, P. K.,** Estimated soil ingesting rates for use in risk assessment, *Risk Anal.*, 7, 355, 1987.
15. **Kimbrough, R. D., et al.,** Health implications of 2,3,7,8-tetrachlorodibenzodioxin (TCDD) contamination of residential soil, *J. Toxicol. Environ. Health,* 14, 47, 1984.
16. **Cleverly, D. Fradkin, L., Bruins, R., McGinnis, P., Dawson, G., and Bond, R.,** Methodology for the Assessment of Health Risks Associated with Municipal Waste Combustor Emissions, Report EPA-530/SW-87-021g, U.S. Environmental Protection Agency, Cincinnati, 1986.
17. *Dietary Consumption Distributions of Selected Food Groups for the U.S. Population,* Office of Toxic Substances, U.S. Environmental Protection Agency, Washington, D.C., 1980.
18. *Statistical Abstract of the United States: 1987,* 107th ed., U.S. Bureau of the Census, Washington, D.C., 1986.
19. **Shor, R. W., Baes, C. F., III, and Sharp, R.,** Agricultural Production in the United States by County: A Compilation of Information from the 1974 Census of Agriculture for Use in Terrestrial Food-Chain Transport and Assessment Models, Report ORNL-5768, Oak Ridge National Laboratory, Oak Ridge, TN, 1982.
20. **Travis, C. C. and Arms, A. D.,** Bioconcentration of organics in beef, milk, and vegetation, *Environ. Sci. Technol.*, 22, 271, 1988.
21. **McKone, T. E. and Ryan, P. B.,** Human exposure to chemicals through food chains, *Environ. Sci. Technol.*, 23, 1154, 1989.
22. **Ng, Y., et al.,** Transfer Coefficients for Assessing the Dose from Radionuclides in Meat and Eggs, Rep. NUREG/CR02967, U.S. Government Printing Office, Washington, D.C., 1980.
23. **Fries, G. F.,** Assessment of potential residues in food derived from animals exposed to TCDD-contaminated soil, *Chemosphere,* 16, 2123, 1987.
24. *Risk Assessment Study of the Dickerson Site,* Maryland Power Plant and Review Division, Montgomery Co., MD, 1988.
25. **Travis, C. C., Holton, G. A., Etnier, E. L., Cook, C., O'Donnell, F. R., Hetrick, D. M., and Dixon, E.,** Assessment of inhalation and ingestion population exposures from incinerate hazardous wastes, *Environ. Int.*, 12, 533, 1986.
26. **Hatemer-Frey, H. A. and Travis, C. C.,** An overview of food chain impacts from municipal waste combustion, in *Proceedings of the U.S. EPA/ORNL Workshop on Risk Assessment for Municipal Waste Combustion: Deposition, Food Chain Impacts, Uncertainty, and Research Needs,* Plenum Press, New York, 1989.
27. **Lowe, J. A., Dietrick, R. W., and Alberts, M. T.,** Health risk assessments for waste-to-energy projects in California, in *Health Effects of Municipal Incineration,* Travis, C. C. and Hattemer-Frey, H. A., Eds., CRC Press, Boca Raton, FL, 1991, chap. 9.
28. Industrial Source Complex (ISC) Dispersion Model User's Guide, Vol. 1, Rep. EPA-450/4-79-030, U.S. Environmental Protection Agency, Washington, D.C., 1979.
29. **Miller, C. and Hoffman, F.,** An examination of the environmental half-time for radionuclides deposited on vegetation, *Health Phys.*, 45, 731, 1983.
30. **Belcher, G. D. and Travis, C. C.,** The food chain as a source of human exposure from municipal waste combustion, paper presented at the International Conference on Municipal Waste Combustion, Hollywood, FL, 1989.
31. **Travis, C. C. and Arms, A. D.,** The food chain as a source of toxic chemical exposure, in *Toxic Chemicals, Health and the Environment,* Lave, L. B. and Upton, A. C., Eds., Plenum Press, New York, 1987, 75.
32. **Hattemer-Frey, H. A. and Travis, C. C.,** Pentachlorophenol: environmental partitioning and human exposure, *Arch. Environ. Contam. Toxicol.*, 18, 482, 1989.
33. Exposure Estimation, San Diego Energy Recovery (SANDER) Project, Appendix C, Part 2, Public Health Rep. 06-24-88/4254e, California Energy Commission, Sacramento, 1988.
34. **Heffron, C. L., et al.,** Cadmium and zinc in growing sheep fed silage corn grown on municipal sludge amended soil, *J. Agric. Food Chem.*, 18, 58, 1980.
35. Environmental Profiles and Hazard Indices for Constituents of Municipal Sludge, Cadmium, Office of Water Regulations and Standards, U.S. Environmental Protection Agency, Washington, D.C., 1985.

36. Health Risk Assessment for a Resource Recovery Facility in Montgomery County, Maryland, submitted to the Montgomery County government, prepared by Weston, Inc., 1988.
37. San Diego Energy Recovery (SANDER) Project, Revised Health Risk Assessment, submitted to the California Energy Commission, prepared by Signal Environmental Systems, Inc., 1987.
38. Risk Assessment for the Proposed Trash-to-Steam Municipal Solid Waste Incinerator at the U.S. Naval Base in Philadelphia, Pennsylvania, submitted to the Philadelphia Public Health Advisory Commission, prepared by Clement Associates, Inc., 1986.
39. Stanislaus Waste-to-Energy Facility Health Risk assessment, submitted to the Stanislaus County Air Pollution Control District, prepared by the Stanislaus Waste Energy Co., 1985.
40. Final Environmental Impact Report of the Los Angeles City Energy Recovery (LANCER) Project, submitted to the Department of Public Works, Bureau of Sanitation, prepared by Cooper Engineers, Inc., 1985.
41. **Smith, A. H.,** *Health Risk Assessment for the Brooklyn Navy Yard Resource Recovery Facility,* University of California, Berkeley, 1988.
42. Bloomington Incinerator Project Risk Assessment, prepared by the Westinghouse Electric Corporation, Waste Technology Services Division, Pittsburgh, 1986.
43. Assessment of Health Risks Associated with Trace Elements and Organic Emissions from the Proposed SPADRA Refuse-to-Energy Project, Pomona, California, HDR Techserv, Inc., Santa Barbara, 1987.
44. Draft Site and Technology Specific Impact Addendum to the Generic Environmental Impact Statement for the North Hempstead Solid Waste Management Facility Project, Health Risk Assessment, prepared by Malcolm Pirnie, White Plains, NY, 1987.
45. **Garten, C. T. and Tradbalka, J. R.,** Evaluation of models for predicting terrestrial food chain behavior of xenobiotics, *Environ. Sci. Technol.,* 17, 590, 1983.
46. **Washburn, S. T.,** The accumulation of chlorinated dibenzo-*p*-dioxins and dibenzofurans in beef and milk, in *Health Effects of Municipal Waste Inceration,* Travis, C. C. and Hattemer-Frey, H. A., Eds., CRC Press, Boca Raton, FL, 1991, chap. 8.
47. **Travis, C. C., Yambert, M. W., and Arms, A. D.,** Food Chain Exposure from Municipal Waste Incineration, Oak Ridge National Laboratory, Oak Ridge, TN, 1989.
48. **Isensee, A. R. and Jones, G. E.,** Absorption and translocation of root and foliage applied 2,4-dichlorophenol, 2,7-dichlorodibenzo-*p*-dioxin, and 2,3,7,8-tetrachlorodibenzo-*p*-dioxin, *J. Agric. Food Chem.,* 19, 1210, 1971.
49. Superfund Public Health Evaluation Manual, EPA/540/1-86/060, Office of Emergency Response and Remedial Response, U.S. Environmental Protection Agency, Washington, D.C., 1986.
50. **McKone, T. E.,** The Use of Environmental Health-Risk Analysis for Managing Toxic Substances, presented at the 78th Annu. Meet. of the Air Pollution Control Assoc., Detroit, 1985.
51. **Crosby, D.,** The degradation and disposal of chlorinated dioxins, in *Dioxins in the Environment,* Kamrin, M. A., and Rodgers, P. W., Eds., Hemisphere Press, Washington, D.C. 1985.
52. **Briggs, G.,** Theoretical and experimental relationships between soil adsorption, octanol-water partition coefficients, water solubilities, bioconcentration factors, and the parachor, *J. Agric. Food Chem.,* 29, 1050, 1981.
53. Municipal Waste Combustion Study, Emission Data Base for Municipal Waste Combustors, EPA/530-SW-87-021b, U.S. Environmental Protection Agency, Offices of Solid Waste and Emergency Response, Air and Radiation, and Research and Development, Washington, D.C., 1987.
54. Preliminary Risk Calculations for Deposited Contaminants from Municipal Waste Combustor Emissions: Polychlorinated Dibenzo-*p*-Dioxins (PCDD) and Polychlorinated Dibenzofurans (PCDF), U.S. Environmental Protection Agency, Office of Health and Environmental Assessment, Environmental Criteria and Assessment Office, Cincinnati, 1987.
55. Environmental Profiles and Hazard Indices for Constituents of Municipal Sludge: Chlorinated Dioxins, Office of Water Regulations and Standards, U.S. Environmental Protection Agency, Washington, D.C., 1985.
56. Municipal Waste Combustion Study, Characterization of the Municipal Waste Combustion Industry, EPA/530-SW-87-021h, Offices of Solid Waste and Emergency Response, Air and Radiation, and Research and Development, U.S. Environmental Protection Agency, Washington, D.C., 1987.
57. Milliken Waste-to-Energy Project Health Risk Assessment, submitted to the South Coast Air Quality Management District, prepared by Radian Corporation, Sacramento, 1987.

Chapter 12

METHODS OF ESTIMATING TOXIC EQUIVALENTS FOR POLYCHLORINATED DIBENZODIOXINS AND DIBENZOFURANS

John A. Lowe

TABLE OF CONTENTS

I. INTRODUCTION

An important concern in quantitative health risk assessments for municipal solid waste (MSW) incinerators or waste-to-energy (WTE) facilities is the need to evaluate potential adverse human health effects from exposure to mixtures of polychlorinated dibenzo-*p*-dioxins (PCDDs) and polychlorinated dibenzofurans (PCDFs). This task is difficult because the toxicities of the various dioxin and furan congeners vary widely. Regulation of only one selected compound, such as 2,3,7,8-tetrachlorinated-*p*-dioxin (2,3,7,8-TCDD), may not be protective of human health, since the concentration of this compound in stack emissions may be minimal relative to the concentration of other PCDD/PCDF congeners, or this compound may be only slightly more toxic than the other, nonregulated compounds. [A *congener* is a member of a particular chemical family (e.g., there are 75 different chlorinated dioxin congeners). The term *homolog* refers to a group of structurally similar compounds that have the same number of chlorine atoms. There are eight chlorinated dioxin homologs, monochlorinated (i.e., dioxins that have a single chlorine atom) through octachlorinated (i.e., dioxins that have eight chlorine atoms). *Isomers* are compounds that belong to the same homolog class. For example, 2,3,7,8-tetrachlorinated-*p*-dioxin (TCDD) is one of 22 isomers belonging to the tetrachlorinated dioxin group.][1]

Public concern about possible adverse effects to human health and the environment from exposure PCDDs and PCDFs emitted from MSW incinerators present the need to assess and compare the toxicity of these compounds collectively, rather than focusing on a few, well-studied isomers. Although there are 75 PCDDs and 135 PCDFs, only a few of these compounds have undergone appreciable toxicity testing, with TCDD receiving the most extensive evaluation. Similar toxic responses have been reported for a variety of PCDDs and PCDFs, suggesting that there is a common mechanism of action related to the observed structure-activity relationships between the different PCDD/PCDF congeners.[2,3] Current knowledge of the structure-activity relationships between different PCDDs and PCDFs allow these compounds to be ranked relative to the known toxicity of the most potent isomer, TCDD.

Various ranking schemes have been developed in the U.S. and other countries for expressing the potency of PCDD/PCDF mixtures relative to TCDD.[1,4-6] The challenge inherent to assessing the risks associated with exposure to PCDDs and PCDFs emitted from MSW incinerators is summarized here:

- Variability in the biological potency of different PCDDs and PCDFs is not accurately know.
- Toxicological data available for characterizing the biological potency of different PCDDs and PCDFs are limited.
- The analytical data characterizing the relative abundance of PCDD and isomers.

Despite these limitations it is imperative that accounts for various dioxin and furan congeners and isomers and their probable distribution in MSW emissions be developed for and applied to quantitative risk assessments for MSW incinerators.

II. TOXICOLOGICAL CHARACTERIZATION OF PCDDs AND PCDFs

The evaluation of PCDD and PCDF toxicity in a quantitative risk assessment is based largely on the animal toxicological and *in vitro* data. Development of a toxicological database for PCDDs and PCDFs has occurred in response to specific public health problems and not with the intent of systematically studying the effects of exposure to PCDD/PCDF mixtures present in the environment. Hence, there are a few PCDD and PCDF species whose potency is well known, while the toxicity of most PCDDs and PCDFs is poorly characterized. The comparative toxicity of PCDDs and PCDFs relative to TCDD (acknowledged as the most toxic of all PCDDs and PCDFs) is presented in Table 1. Table 1 shows that comparisons of the toxicity of various PCDD/PCDF congeners must be made indirectly, since long-term bioassays in laboratory animals have been performed for 2,3,7,8-TCDD and two hexachlorodibenzodioxins only. As a result, potency estimates for most PCDDs and PCDFs are based largely on the results of acute toxicity tests and *in vitro* assays of enzyme induction, since detailed evaluation of the toxicity of most PCDDs and PCDFs is restricted by the absence of a comprehensive toxicological database encompassing all PCDDs and PCDFs.

Various PCDDs and PCDFs have produced similar toxic effects in experimental animals, including a wasting syndrome, lymphoid involution, hepatotoxicity, edema, urinary tract hyperplasia, hyperlipidemia, teratogenicity, adverse reproductive effects, and enzyme induction, suggesting a common mechanism of toxicity.[2,3] Several studies have demonstrated that a correlation exists between the structure-activity relationships of PCDDs and PCDFs and their ability to bind to a cystolic protein receptor and induce the microsomal monooxygenase aryl hydrocarbon hydroxylase (AHH).[2,3,8-10] These studies showed that dioxins that bind to the protein receptor and induce AHH have two common properties: (1) Halogen atoms occupying at least three, and for maximal potency, four of the lateral ring positions (i.e., the 2, 3, 7, and 8 positions). (2) At least one ring position is unsubstituted. For example, the octochloroderivative is inactive, but heptachloro-derivatives have some activity.[2]

It is generally agreed that PCDDs and PCDFs that are chlorinated at the 2, 3, 7, or 8 positions are more likely to exhibit toxic effects at lower doses than other isomers,[1] which encompasses the 15 tetra-, penta-, hexa-, and heptachlorinated dioxins and furans. This factor demonstrates the necessity of evaluating human exposure to PCDDs and PCDFs collectively rather than focusing on a single congener. Various approaches to estimate the toxicity of a PCDD/PCDF mixture relative to the known toxicity of 2,3,7,8-TCDD have been proposed. Using one of these ranking schemes, the toxicity of a mixture of PCDDs and PCDFs is then expressed in terms of toxicity equivalent factors (TEFs). The basis of the various methods for developing TEFs are discussed below.

III. COMPARISON OF APPROACHES FOR ESTIMATING THE RELATIVE TOXICITIES OF PCDDs AND PCDFs

Recognizing the importance of the structure-activity relationships demonstrated in various *in vivo* and *in vitro* experiments, several approaches have been proposed for expressing the potency of PCDD/PCDF mixtures in terms of TEFs (Table 2). The rationale for the approaches listed in Table 2 is summarized below. Note that the four approaches outlined in Table 2 do not represent the comprehensive list of weighing schemes that have been proposed or are applied in current risk assessments.

A. THE U.S. ENVIRONMENTAL PROTECTION AGENCY INTERIM APPROACH

The interim approach of the U.S. Environmental Protection Agency (EPA) to developing TEFs[1,4] assigns weighing factors based on the following criteria:

1. Definitive human evidence.
2. Lacking definitive data on human carcinogenicity, carcinogenic potencies based on long-term animal studies take precedence over all other types of data. (Since long-term bioassays are costly and time-consuming, they have been performed for 2,3,7,8-TCDD and a mixture of 1,2,3,4,7,8- and 1,2,3,7,8,9-hexachlorodibenzo-*p*-dioxins [HxCDDs] only.)

TABLE 1
Toxicity of Dioxins and Furans Relative to the Toxicity of 2,3,7,8-TCDD

Compound	Guinea pig (LD_{50})	Carcinogenicity	Reproductive/ teratogenic effects	Receptor binding	AHH enzyme induction in: Animal cells	AHH enzyme induction in: Human cells	Cell keratinization	Flat (XB) cell assay	*In vitro* immunotoxicity
PCDDs									
Mono-Tri	<0.0001[a]	—	—	0.001—0.01[a]	<0.001[b]	—	0.01[a]	—	0.005[c]
2,3,7,8-TCDD	1[d]	1[e]	1[f,g]	1[a]	1[a]	1[h]	1[a]	1[i]	1[c,j]
TCDDs	<0.001[d]	—	<0.001[k]	<0.01—0.16[a]	<0.001—0.02[l]	—	<0.001—0.01[a]	—	—
2,3,7,8-PeCDD	0.67[d]	—	—	1[a]	0.02—0.2[l]	—	0.5[a]	—	—
PeCDDs	0.002[d]	—	—	—	<0.001[l]	—	—	—	—
2,3,7,8-HxCDDs	0.03[d]	0.04[e]	0.01[f]	0.05[a]	0.001—0.1[l]	—	0.005[a]	—	—
HxCDDs	—	—	—	—	<0.001[l]	—	—	—	—
2,3,7,8-HpCDDs	0.004[d]	—	—	—	0.002—0.004[b,l]	—	—	—	—
HpCDDs	0.002[d]	—	—	—	<0.001[b]	—	—	—	—
OCDD	—	—	<0.00001[k]	—	<0.001[b]	—	—	—	—
PCDFs									
Mono-Tri	—	—	—	≤0.001—0.02[m,n]	<0.001[m]	<0.001[m]	0.001[a]	—	—
2,3,7,8-TCDF	0.28; 0.5[d]	—	0.03—0.13[g,k]	0.2[n]; 0.3[a]; 0.4[o]	0.01—0.4[b,h,n]	0.4[h]	0.05[a]	0.1[i]	0.1[j]; 1.0[e]
TCDFs	—	—	—	0.001—0.05[a,m]	≤0.001[m]; 0.04[h]	0.4[h]	—	—	—
2,3,7,8-PeCDF	—	—	—	0.13[m]; 0.6[n]; 0.7[a]	<0.3[m]; 0.4[h]	0.8[h]	—	—	—
1,2,4,6,7-PeCDF	—	—	—	0.15[n]	0.002[n]	—	—	—	—
PeCDFs	—	—	—	0.001—0.1[a,m]	≤0.001—2.0[h,m,p]	0.6[h]	—	—	—
2,3,7,8-HxCDFs	0.017[d]	—	—	0.04—0.5[a,n]	0.05—0.2[h,n]	0.9[h]	—	—	—
HxCDFs	—	—	—	0.001[a,n]	0.001[h]; 0.002[n]	—	—	—	—
2,3,7,8-HpCDFs	—	—	—	—	0.004[l]	—	—	—	—
HpCDFs	—	—	—	<0.001[n]	<0.001[b]	—	—	—	—

[a] Knutson and Poland.[18]
[b] Bradlaw et al.[19]
[c] Greenlee et al.[20]
[d] McKinney and McConnell;[21] Moore et al.[22]
[e] U.S. EPA.[23]
[f] Murray et al.;[24] Schwetz et al.;[25] Weber et al.[26]
[g] Hassoun et al.[27]
[h] Nagayama et al.[28,29]
[i] Bradlaw et al.[30]
[j] Gierthy and Crane.[31]
[k] Dencker et al.[32]
[l] Weber et al.[26]
[m] Bandiera et al.[33]
[n] Bandiera et al.[8]
[o] Poland et al.[34]
[p] Poland et al.[35]

TABLE 2
Selected Approaches for Estimating the Relative Toxicity of PCDDs and PCDFs

Congener	NY[a]	CDHS[b]	EPA[c]	I-TEF[d]
Mono through DiCDDs	0	0	0	0
TriCDDs	0	0	0	0
2,3,7,8-TCDD	1	1	1	1
Other TCDDs	0	0	0.01	0
2,3,7,8-PeCDDs	1	1	0.5	0.5
Other PeCDDs	0	0	0.005	0
2,3,7,8-HxCDDs	0.03	0.04	0.04	0.1
Other HxCDDs	0	0.004	0.004	0
2,3,7,8-HpCDDs	0	0.04	0.001	0.01
Other HpCDDs	0	0	0.00001	0
OCDD	0	0	0	0.001
2,3,7,8-TCDF	0.33	1	0.1	0.1
Other TCDFs	0	0	0.001	0
2,3,7,8-PeCDFs	0.33	1	0.1	0.5 or 0.05
Other PeCDFs	0	0	0.001	0
2,3,7,8-HxCDFs	0.01	0.04	0.01	0.1
Other HxCDFs	0	0	0.0001	0
2,3,7,8-HpCDFs	0	0.04	0.001	0.01
Other HpCDFs	0	0	0.00001	0
OCDF	0	0	0	0.001

[a] Eadon et al.[6]
[b] California Department of Health Services.[5]
[c] U.S. EPA.[1]
[d] NATO/CCMS.[7]

3. For compounds where carcinogenic effects have not been demonstrated, information on systematic effects is considered.
4. If neither carcinogenic nor systematic effects have been observed, data from *in vitro* test systems are used, with greatest reliance placed on data from receptor binding studies and oxidative enzyme induction studies, since these *in vitro* effects correlate most strongly with observed *in vivo* toxicity (e.g., thymic atrophy and body weight loss).
5. Data on other systematic effects, such as body or organ weight losses (i.e., the wasting syndrome) and hepatotoxicity, are used for qualitative support only.

Evaluation of systemic effects data in developing TEFs proceeds in the following manner:

- Data on reproductive effects take precedence due to the significance of this endpoint in humans and because the doses producing reproductive effects and carcinogenicity in laboratory animals are similar in the case of 2,3,7,8-TCDD.
- Data on acute lethality (i.e., LD_{50}s) are accorded low priority, since a correlation between LD_{50}s and chronic effects has not been established for PCDDs and PCDFs.

Although definitive human evidence is preferred by the Agency when assessing the potential human carcinogenicity of PCDDs and PCDFs, such data are scarce. As a result, the EPA relies on demonstrated effects in animals obtained from short-term animal studies and *in vitro* assays to determine the relative carcinogenic potency for most PCDDs and PCDFs. TEFs for 2,3,7,8-TCDD (1.0), the 2,3,7,8-HxCDDs (0.04), and 2,3,7,8-tetrachlorodibenzofuran (2,3,7,8-TCDF) (0.1) are based on the results of chronic animal bioassays. The TEF for 2,3,7,8-TCDF is 0.1, because it was shown to be one to two orders of magnitude less toxic than 2,3,7,8-TCDD in reproductive toxicity tests and one order of magnitude less toxic in *in vitro* assays.

The TEF of 0.5 for the 2,3,7,8-substituted pentachlorodibenzodioxins (2,3,7,8-PeCDDs) is the arithmetic mean of the carcinogenic potencies for 2,3,7,8-TCDD (1.0) and 2,3,7,8-HxCDDs (0.04). Receptor binding, cell keratinization, and enzyme induction data generally support this value. The TEF of 0.1 for 2,3,7,8-substituted pentachlorodibenzofurans (2,3,7,8-PeCDFs) is based on responses observed in *in vitro* oxidative enzyme induction tests. *In vitro* tests also showed that the 2,3,7,8-hexachlorodibenzofurans (2,3,7,9-HxCDFs) are typically one tenth as potent as PeCDFs; hence, they were assigned a TEF of 0.01. Enzyme induction assays demonstrated that 2,3,7,8-heptachlorodibenzo-*p*-dioxins (HpCDDs) and heptachlorodibenzofurans (HpCDFs) are 1000 times less potent than 2,3,7,8-TCDD. Limited *in vitro* data indicate that the non-2,3,7,8-substituted HpCDDs and HpCDFs are 10 to 100 times less potent than the 2,3,7,8-substituted isomers. Hence, these compounds were given a TEF weighing factor of 0.00001.

Octachlorodibenzo-*p*-dioxin (OCDD) and octachlorodibenzofuran (OCDF) were not considered toxic and were assigned a TEF of 0. For the sake of uniformity and because this approach is relatively insensitive to small changes in the TEFs, all non-2,3,7,8-substituted isomers are considered to be 100 times less potent than their 2,3,7,8-substituted counterparts.

B. THE STATE OF NEW YORK APPROACH

The State of New York developed a weighing scheme for evaluating the toxicity of mixtures of PCDDs and PCDFs in soot resulting from a polychlorinated biphenyl (PCB) transformer fire.[6,11] This weighing scheme is based largely on the results of acute toxicity (LD_{50}) tests done on guinea pigs using soot as the test material. From these studies, investigators concluded that the soot was 1/60th as toxic as 2,3,7,8-TCDD. The LD_{50} data for the soot, published LD_{50} values for 2,3,7,8-TCDD and TCDF, and analytical data from soot samples were used to estimate TEFs for the PCDDs and PCDFs present in the soot. Based on the limited availability of acute toxicity data for PCDDs and PCDFs, the following assumptions were made in estimating the TEFs:

- The ratio of the LD_{50} value of a particular PCDF congener and 2,3,7,8-TCDF is the same as the ratio of the LD_{50} value of the correspondingly numbered PCDD congener and 2,3,7,8-TCDD. This assumption was made because in the early 1980s investigators did not have acute toxicity data for PCDFs other than 2,3,7,8-TCDF. This assumption yielded the development of TEFs based on the available acute toxicity data for PCDDs.
- LD_{50} values for PCDDs and PCDFs lacking chlorines at all four laterally substituted positions (i.e., the 2, 3, 7, and 8 positions) were assumed to be sufficiently high that their influence could be neglected (i.e., these isomers were given a TEF of 0). This assumption is supported by LD_{50} data for 2,7-dichlorodibenzo-*p*-dioxin, 2,3,7-trichlorodibenzo-*p*-dioxin, and 1,2,4,7,8-PeCDD, all of which have LD_{50} values 450 times higher than the LD_{50} value for 2,3,7,8-TCDD.[12]
- The LD_{50} values for 2,3,7,8-PeCDD congeners and 2,3,7,8-TCDD were the same in acute toxicity tests done on guinea pigs.[12] Hence, both of the congeners were assigned a TEF of 1.0.
- TEFs for the 2,3,7,8-HxCDD and HxCDD congeners was determined to be 0.3, because these congeners had LD_{50} values that were 30-fold greater than the LD_{50} value obtained for 2,3,7,8-TCDD.[12]
- The LD_{50} values observed for all hepta- and octachlorinated dioxins and furans were sufficiently greater than the LD_{50} value observed for 2,3,7,8-TCDD, allowing these congeners to be assigned a TEF value of 0.

C. THE CALIFORNIA DEPARTMENT OF HEALTH SERVICES APPROACH

The California Department of Health Services (CDHS) evaluated the toxicity of 2,3,7,8-TCDD under a statutory program for regulating toxic air contaminants in California. Since carcinogenicity was the adverse effect of greatest concern to CDHS, TEFs are based on the potential carcinogenicity of PCDDs and PCDFs. To estimate cancer risk associated with exposure to mixtures of PCDDs and PCDFs requires knowledge of the proportion of each PCDD and PCDF isomer within the mixture and the carcinogenic potency of those isomers. In response to uncertainties associated with inferring cancer risks from a poorly characterized mixture of chemicals, most of which have not been tested for carcinogenicity, CDHS has developed four different scenarios for evaluating the carcinogenic potency of mixtures of PCDDs and PCDFs:[5]

Scenario 1 assumes that all measured PCDDs and PCDFs with four to eight chlorines are carcinogenic and all are equipotent to 2,3,7,8-TCDD.

Scenario 2 assumes that only the tetra-, penta-, hexa-, and hepta-congeners with chlorines in the 2, 3, 7, and 8 positions (i.e., laterally substituted compounds) are carcinogenic and equally as potent as 2,3,7,8-TCDD, except for HxCDD, whose potency is derived from its long-term animal bioassay.

Scenario 3 assumes that all laterally substituted tetra-, penta-, hexa-, hepta-congeners are carcinogenic and have the same potency as for Scenario 2. Non-2,3,7,8-substituted isomers are assumed to be 100-fold less carcinogenic than their corresponding 2,3,7,8-substituted isomers based on *in vitro* and short-term studies.

Scenario 4 assumes that only 2,3,7,8-substituted PCDD/PCDF isomers of the tetra-, penta-, hexa-, and hepta-congeners are carcinogenic. Laterally substituted tetra- and penta-congeners are considered equipotent to 2,3,7,8-TCDD, while the 2,3,7,8-substituted hexa- and hepta-congeners are considered equipotent to HxCDD.

CDHS typically recommends Scenario 4 for estimating TEFs for MSW incinerator emissions in California.

D. THE INTERNATIONAL TOXIC EQUIVALENT FACTOR (I-TEF) APPROACH

The North Atlantic Treaty Organization's Committee on Challenges of Modern Society (NATO/CCMS) Pilot Study on International Information Exchange on Dioxins and Related Compounds was initiated in 1986 to address several public health issues related to PCDDs and PCDFs. One group in this study addressed risk assessment methodologies related to exposure to complex mixtures of PCDDs and PCDFs.[7] This activity was intended to resolve discrepancies among existing TEF schemes and to develop an updated TEF scheme (I-TEF) that would provide a consistent basis for risk assessment and regulation.

The I-TEF approach does not represent a significant advance in the evaluation of PCDD and PCDF mixtures compared with the three approaches discussed previously. The distinguishing feature of the I-TEF scheme, however, is that NATO/CCMS obtained an international consensus on methods of estimating TEFs and validating the TEF approach for characterizing the toxicity of PCDD/PCDF mixtures in the absence of more definitive data.

The development of I-TEF weighing scheme is guided by three principles:

1. The scheme should be simple and practicable, since a complex scheme suggests greater precision and sophistication in estimating toxicity of mixtures than can be scientifically supported.

2. The scheme focuses on the PCDD and PCDF congeners that are preferentially accumulated in animal and human tissues, principally the laterally substituted congeners.
3. The TEFs developed through this scheme should reflect the differences in toxicity of different congeners for a variety of toxicological endpoints. (Some of the previously described schemes have been based on a single toxicological endpoint.)

The I-TEFs listed in Table 2 were estimated using the following criteria:

- Evidence for carcinogenicity derived from long-term animal bioassays.
- In the absence of long-term evidence, data from studies evaluating reproductive effects were considered, since toxicological data indicate that doses resulting in reproductive effects correspond to the range of doses that induced carcinogenic effects.
- Lacking long-term data, other subchronic effects, such as thymus atrophy, body weight loss, or general toxicity, were considered.
- Data from acute lethality (e.g., LD_{50}) studies.
- Data characterizing *in vivo* and *in vitro* effects, such as receptor binding and enzyme induction.

The main differences between the I-TEF scheme and other schemes are

- 2,3,4,7,8-PeCDF is assigned a TEF of 0.5 while 1,2,3,7,8-PeCDF is assigned a TEF of 0.01, reflecting the greater toxicity of 2,3,4,7,8-PeCDF as demonstrated in different test systems and the lesser toxicity of 1,2,3,7,8-PeCDF exhibited in teratogenicity studies.
- OCDD and OCDF are assigned a TEF of 0.001, reflecting recent findings that subchronic exposure of laboratory animals to OCDD leads to accumulation of OCDD in body tissues and the appearance of incipient 2,3,7,8-TCDD-like toxicity.
- Non-2,3,7,8-substituted congeners are assigned a TEF of 0 reflecting the focus on congeners that accumulate in body tissues.

In most environmental samples, non-2,3,7,8-substituted congeners are present in higher concentrations than 2,3,7,8-substituted compounds. When tissue samples from living organisms are analyzed, however, only the 2,3,7,8-substituted congeners are generally detected.[13,14] Similarly, background level of PCDDs and PCDFs routinely detected in human biological tissues also almost entirely contain 2,3,7,8-substituted congeners.[15] Since 2,3,7,8-substituted compounds appear to pose the greater long-term threat to human health, full attention is centered on these compounds.

Status of the I-TEF Approach: Many groups have documented shortcomings with the proposed I-TEF concept, including the fact that the correlation between enzyme induction response, which serves as the basis for some of the TEF factors, and several toxic endpoints induced by PCDDs and PCDFs remains unclear. Despite this limitation, most countries including Austria, Canada, Denmark, Sweden, West Germany, and Switzerland, support the proposed I-TEF approach as a reasonable method of assessing the risks associated with exposure to complex mixtures of PCDDs and PCDFs. Analytical tests that can more accurately and less expensively determine the dose-response relationship for various PCDD/PCDF isomers should continue. Additional research needs in this area include:

- Evaluation of receptor-mediated mechanisms of toxicity (e.g., elucidation of intracellular events contributing to enzyme induction and better explanation of the distinct differences in response observed in various species).
- Further exploration of the relationship between short-term toxicity (e.g., enzyme induction and subchronic effects) and carcinogenicity and other chronic effects.

IV. APPLICATION TO RISK ASSESSMENT

The approximate toxicity of a mixture of PCDDs and PCDFs emitted from MSW incinerators is estimated by multiplying the concentration of each isomer present in stack emissions by its corresponding TEF value. A release of 1 PCDDs and PCDFs expressed in toxic equivalents is assumed to have the same toxic effect as a release of 1 g 2,3,7,8-TCDD.

The use of any TEF weighing scheme requires extensive characterization of the PCDDs and PCDFs present. Data on the 2,3,7,8-substituted and non-2,3,7,8-substituted isomers are ideally required. Such extensive laboratory analyses of stack emissions from MSW facilities, however, are not typically available. A review of source test data from 24 MSW incinerators showed that most tests contained only 2,3,7,8-TCDD isomer-specific emissions data.[16] In cases where only congener-specific analytical data are reported (i.e., the data do not distinguish between 2,3,7,8-substituted and non-2,3,7,8-substituted isomers) or only total PCDD and PCDF concentrations are reported, the following three approaches have been suggested to estimate the relative concentration of 2,3,7,8-substituted congeners:

1. Assume that all emitted PCDDs and PCDFs have the same toxicity as 2,3,7,8-TCDD.
2. Assume that all isomers within a congener class have the same toxicity as the 2,3,7,8-substituted isomer. For example, all TCDDs are equally as potent as 2,3,7,8-TCDD.
3. Assume the PCDDs and PCDFs are emitted as a perfect isomeric distribution (i.e., the occurrence of each congener in a mixture has equal probability). For example, the concentration of 2,3,7,8-TCDD is 1/22 or 4.5% of total TCDD emitted.[17]

As expected, the first approach is the most conservative and provides the highest estimated 2,3,7,8-TCDD toxic equivalent emission rates. The U.S. EPA, however, does not recommend this approach, since available long- and short-term toxicity data suggest that this approach is overly conservative. The second approach is a slight variation of the first, which can be applied if congener-specific analytical data are available. This approach is also likely to substantially overestimate actual emission rates.

The third approach has been reviewed to evaluate its ability to predict the concentrations 2,3,7,8-substituted isomers. Jones et al.[16] statistically analyzed PCDD and PCDF emissions from MSW incinerators to determine the ratio of 2,3,7,8-substituted isomers to its parent congener. The results of that study showed that:

- 2,3,7,8-TCDD emissions tend to be lower than the perfect isomeric distribution
- Emissions of the remaining 2,3,7,8-substituted PCDD isomers and 2,3,7,8-TCDF were generally greater than the perfect isomeric distribution
- Emissions of the remaining 2,3,7,8-substituted PCDF isomers tend to be less than the perfect isomeric distribution

V. CASE STUDY

A comparison of PCDD/PCDF emission rates for the Wurzburg, West Germany, resource recovery facility expressed in toxic equivalents using the four weighing schemes presented in Table 2 are listed in Table 3. These emissions data were selected because they included measurable levels of all PCDDs and PCDFs isomers so that the full range of TEFs could be applied. Using the same actual emissions data, the four TEF schemes provide 2,3,7,8-TCDD toxicity equivalent emission rates ranging from 0.47 to 3.4 ng/m^3. The TEF scheme developed by the State of California[5] yielded the highest emission rate, due to the high values assigned to the 2,3,7,8-substituted PCDFs. The New York[6] and I-TEF schemes yielded comparable toxic equivalent emission rates, both of which were higher than the U.S. EPA TEF emission rate. The

TABLE 3
Comparison of 2,3,7,8-TCDD Toxic Equivalent Emission Rates Using Different Weighing Schemes

Congener/isomer	Actual emission rate (ng/m³)[a]	2,3,7,8-TCDD toxic equivalent emission rate			
		CDHS[b]	Eadon[c]	EPA[d]	I-TEF[e]
2378-TCDD	0.02200	0.02200	0.02200	0.02200	0.02200
Other TCDDs	2.52210	0.00000	0.00000	0.02522	0.00000
12378-PeCDD	0.26130	0.26130	0.26130	0.05226	0.13065
Other PeCDDs	3.01060	0.00000	0.00000	0.00602	0.00000
123478-HxCDD	0.10225	0.00307	0.00307	0.00409	0.01023
123678-HxCDD	0.20450	0.00613	0.00613	0.00818	0.02045
123789-HxCDD	0.14769	0.00443	0.00443	0.00591	0.01477
Other HxCDDs	3.68100	0.00000	0.00000	0.00147	0.00000
1234678-HpCDD	3.32881	0.09986	0.00000	0.00333	0.03329
Other HpCDDs	2.94253	0.00000	0.00000	0.00003	0.00000
OCDD	10.17000	0.00000	0.00000	0.00000	0.01017
2378-TCDF	0.44300	0.44300	0.14619	0.04430	0.04430
Other TCDFs	18.92700	0.00000	0.00000	0.01893	0.00000
12378-PeCDF	1.36330	1.36330	0.44989	0.13633	0.06817
23478-PeCDF	1.02250	1.02250	0.33743	0.10225	0.51125
Other-PeCDFs	12.16700	0.00000	0.00000	0.01217	0.00000
123478-HxCDF	0.57941	0.01738	0.00579	0.00579	0.05794
123678-HxCDF	0.69302	0.02079	0.00693	0.00693	0.06930
234678-HxCDF	0.95433	0.02863	0.00954	0.00954	0.09543
123789-HxCDF	0.12497	0.00375	0.00125	0.00125	0.01250
Other HxCDFs	7.45280	0.00000	0.00000	0.00075	0.00000
1234678-HpCDF	2.30631	0.06919	0.00000	0.00231	0.02306
1234789-HpCDF	0.09089	0.00273	0.00000	0.00009	0.00000
Other HpCDFs	0.37490	0.00000	0.00000	0.00000	0.00000
OCDF	0.88000	0.00000	0.00000	0.00000	0.00088
Total toxic equivalent emission rate		3.36807	1.25395	0.46915	1.12350

[a] Nanograms per cubic meter.
[b] California Department of Health Services.[5]
[c] Eadon et al.[6]
[d] U.S. EPA.[1]
[e] NATO/CCMS.[7]

higher TEF values assigned to 2,3,7,8-substituted PCDDs and PCDFs by the I-TEF scheme account for the higher emission rates relative to the EPA scheme. Although the U.S. EPA approach assigned TEF values to both 2,3,7,8-substituted, and non-2,3,7,8-substituted isomers, these values are generally lower than the corresponding values adopted by the other schemes. It should be observed, however, that OCDD, non-2,3,7,8-substituted PCDFs, and OCDF are the most common congeners present in stack emissions. All of these compounds are assumed to be nontoxic (i.e., are assigned low or zero TEF values).

VI. DISCUSSION

TEF weighing schemes have been developed using different rationales that account for the varying toxic effects associated with PCDDs and PCDFs. The weighing scheme developed by CDHS is intended to account for the potential carcinogenicity of PCDDs and PCDFs based on data from laboratory animal studies. The New York State weighing scheme attempts to account for differences in toxic effects by considering differences in acute toxicity. The U.S. EPA and I-TEF weighing schemes also account for carcinogenicity; however, they address data gaps in the cancer testing database with the use of *in vitro* data. Selecting a scheme that is the "best" in terms of toxicological justification depends on how the risk assessor regards the uncertainties in the PCDD and PCDF toxicological database. Given these uncertainties, use of any particular TEF weighing scheme should be accompanied by justification of its toxicological rationale.

The accuracy of an assessment of the risks associated with exposure to PCDDs and PCDFs depends upon several factors, of which the uncertainty in the TEF approach is only one and possibly not the most significant factor. Uncertainties in emission rate estimates, assessing potential exposures, bioavailability, interspecific extrapolation, and dose-response modeling are likely to provide as much or more uncertainty than the application of TEF weighing scheme.[7] The case study results showed that the uncertainties associated with applying various TEF weighing schemes are appreciable (a sevenfold range); however, this uncertainty should be considered small when compared to other, probably more significant sources of uncertainty.

There is an international consensus that the TEF approach is an appropriate scientific policy tool for use in the absence of more definitive toxicity data. This approach rests on a rational but not on a definitive base of scientific data. Given all its limitations, the TEF approach provides a practical, interim method for the scientific and regulatory community to communicate the significance of PCDDs and PCDFs in the environment.[7]

REFERENCES

1. Interim Procedures for Estimating Risks Associated with Exposures to Mixtures of Chlorinated Dibenzo-*p* Dioxins and Dibenzofurans (CDDs and CDFs), Risk Assessment Forum, EPA/625/3-87-012, U.S. Enviro. mental Protection agency, Washington, D.C., 1987.
2. **Poland, A. and Knutson, J. C.,** 2,3,7,8-tetrachlorodibenzo-*p*-dioxin and related halogenated aromatic hydrocarbons: examination of the mechanism of toxicity, *Ann. Rev. Pharmacol.*, 22, 517, 1982.
3. **Neal, R. A.,** Mechanisms of the biological effects of PCBs, polychlorinated dibenzo-*p*-dioxins and polychlorinated dibenzofurans in experimental animals, *Environ. Health Perspect.*, 60, 41, 1985.
4. **Barnes, D. G., Bellin, J., and Cleverly, D.,** Interim procedures for estimating risks with exposures to mixtures of chlorinated dibenzodioxins and dibenzofurans (CDDs and CDFs), *Chemosphere*, 15, 1895, 1986.
5. Health Effects of 2,3,7,8-Tetrachlorodibenzo-*p*-Dioxin and Related Compounds, Epidemiological Studies and Surveillance Section, California Department of Health Services, Berkeley, 1986.
6. **Eadon, G., Kaminsky, L., Silkworth, J., Aldous, K., Hilker, D., O'Keefe, P., Smith, R., Gierthy, J., Hawley, J., Kim, N., and DeCaprio, A.,** Calculation of 2,3,7,8-TCDD equivalent concentrations of complex environmental contaminant mixtures, *Environ. Health Perspect.*, 70, 221, 1986.
7. Pilot Study on International Information Exchange on Dioxins and Related Compounds: International Toxicity Equivalency Factor (I-TEF) Method of Risk Assessment for Complex Mixtures of Dioxins and Related Compounds, NATO/CCMS Rep. No. 176, North Atlantic Treaty Organization, Committee on the Challenges to Modern Society, U.S. Environmental Protection Agency Contract No. 68-02-4254, Washington, D.C., 1988.
8. **Bandiera, S., Sawyer, T., Romkes, M., Zmudzka, B., Safe, L., Mason, G., Keys, B., and Safe, S.,** Polychlorinated dibenzofurans (PCDFs): effects of structure on binding to the 2,3,7,8-TCDD cystolic receptor protein, AHH induction and toxicity, *Toxicologist*, 32, 131, 1984.
9. **Okey, A. B., Villa, L. M., and Iverson, F.,** Ah receptor in primate liver: binding of 2,3,7,8-tetrachlorodibenzo-*p*-dioxin and carcinogenic aromatic hydrocarbons, *Can. J. Physiol. Pharmacol.*, 62, 1292, 1984.
10. **Safe, S., Sawyer, T., Mason, G., Bandiera, S., Keys, B., Romkes, M., Piorski, J., Zmudzka, B., and Safe, L.,** Polychlorinated dibenzofurans: quantitative structure-activity relationships, *Chemosphere*, 14, 675, 1985.
11. **O'Keefe, P. W., Silkworth, J. B., Gierthy, J. F., Smith, R. M., DeCaprio, A. P., Turner, J. N., Eadon, G., Hilker, D. R., Aldous, K. M., Kaminsky, L. S., and Collins, D. N.,** Chemical and biological investigations of a transformer accident at Binghamton, New York, *Environ. Health Perspect.*, 60, 201, 1985.
12. **McConnell, E. E., Moore, J. A., Haseman, J. K., and Harris, M. W.,** The comparative toxicity of chlorinated dibenzo-*p*-dioxins in mice and guinea pigs, *Toxicol. Appl. Pharmacol.*, 44, 335, 1978.

13. **Keuhl, D. W., Cook, P. M., and Batterman, A. R.,** Uptake and depuration studies of PCDDs and PCDFs in fresh water fish, *Chemosphere,* 15, 2023, 1986.
14. **Van den Berg, M., de Broom, E., van Grrvenbroek, M., Olie, K., and Hutzinger, O.,** Bioavailability of PCDDs and PCDFs absorbed on fly ash in rat, guinea pig, and Syrian golden hamster, *Chemosphere,* 14, 865, 1985.
15. **Rappe, C., Anderson, R., Bergqvist, P. A., Brohede, C., Hansson, M., Keller, L. O., Lindstrom, G., Marklund, S., Nygren, M., Swanson, S. E., Tysklind, M., and Wiberr, K.,** Overview on environmental fate of chlorinated dioxins and dibenzofurans: sources, levels, and isomeric patterns in various matrices, *Chemosphere,* 16, 1603, 1987.
16. **Jones, K. H., Walsh, J., and Alston, D.,** The statistical properties of available worldwide MSW combustion dioxin/furan emissions data as they apply to the conduct of risk assessments, *Chemosphere,* 16, 2183, 1987.
17. **Olie, K., Lustenhouwer, J. W. A., and Hutzinger, O.,** Polychlorinated dibenzo-*p*-dioxins and related compounds in incinerator effluents, in *Chlorinated Dioxins and Related Compounds,* Hutzinger, O., Frei, R. W., Merian, E., and Pocchiari, F., Eds., Pergamon Press, New York, 1983, 227.
18. **Knutson, J. and Poland, A.,** Keratinization of mouse teratoma cell line XB produced by 2,3,7,8-tetrachlorodibenzo-*p*-dioxin: an *in vitro* model of toxicity, *Cell,* 22, 27, 1980.
19. **Bradlaw, J. and Casterline, J. L.,** Induction of enzyme activity in cell culture: a rapid screen for detection of planar polychlorinated organic compounds, *J. Assoc. Off. Anal. Chem.,* 62, 904, 1979.
20. **Greenlee, W. F., Osborne, R., Dold, K. M., Hudson, L. G., and Toscana, W. A., Jr.,** Toxicology of chlorinated aromatic hydrocarbons in animals and humans: *in vitro* approach to toxic mechanisms, *Environ. Health Perspect.,* 60, 69, 1985.
21. **McKinney, J. and McConnell, E.,** Structural specificity and the dioxin receptor, *Pergamon Ser. Environ. Sci.,* 5, 367, 1982.
22. **Moore, J. A., McConnell, E. E., Dalgard, D. W., and Harris, M. W.,** Comparative toxicity of three halogenated dibenzofurans in guinea pigs, mice, and rhesus monkeys, *Ann. N.Y. Acad. Sci.,* 320, 151, 1979.
23. Ambient Water Quality Criteria for 2,3,7,8-Tetrachlorodibenzo-*p*-Dioxin, EPA-440/5-84/007, U.S. Environmental Protection agency, Washington, D.C., 1984.
24. **Murray, F. J., Smith, F. A., Ntischke, K. D., Humiston, C. G., Kociba, R. J., and Schwetz, B. A.,** Three-generation reproduction study of rats given 2,3,7,8-tetrachlorodibenzo-*p*-dioxin in the diet, *Toxicol. Appl. Pharmacol.,* 50, 241, 1979.
25. **Schwetz, B. A., Norris, J. M., Sparschu, G. L., Rowe, V. K., Gehring, P. J., Emerson, J. L., and Gerbig, C. G.,** Toxicity of chlorinated dibenzo-*p*-dioxins, *Environ. Health Perspect.,* 5, 87, 1973.
26. **Weber, H., Lamb, J. C., Harris, M. W., and Moore, J. A.,** Teratogenicity of 2,3,7,8-tetrachlorodibenzofuran (TCDF) in mice, *Toxicol. Lett.,* 20, 183, 1984.
27. **Hassoun, E., D'Argy, R., and Dencker, L.,** Teratogenicity of 2,3,7,8-tetrachlorodibenzofuran (TCDF) in mice, *J. Toxicol. Environ. Health,* 14, 337, 1984.
28. **Nagayama, J., Kiyohara, C., Masuda, Y., and Kuratsune, M.,** Inducing potency of aryl hydrocarbon hydroxylase activity in human lymphoblastoid cells and mice by polychlorinated dibenzofurans, *Environ. Health Perspect.,* 59, 107, 1985.
29. **Nagayama, J., Kiyohara, C., Masuda, Y., and Kuratsune, M.,** Genetically mediated induction of aryl hydrocarbon hydroxylase activity in human lymphoblastoid cells by polychlorinated dibenzofuran isomers and 2,3,7,8-tetrachlorodibenzo-*p*-dioxin, *Arch. Toxicol.,* 56, 230, 1985.
30. **Bradlaw, J., Garthoff, L. H., Hurley, N. E., and Firestone, D.,** Comparative induction of aryl hydrocarbon hydroxylase activity *in vitro* by analogues of dibenzo-*p*-dioxin, *Cosmet. Toxicol.,* 18, 627, 1980.
31. **Gierthy, J. F. and Crane, D.,** *In vitro* bioassay for dioxin-like activity based on alterations in epithelial cell proliferation and morphology, *Fundam. Appl. Toxicol.,* 5, 754, 1985.
32. **Dencker, L., Hassoun, E., and Dargy, R.,** Fetal thymus organ culture as an *in vitro* model for the toxicity of 2,3,7,8-tetrachlorodibenzo-*p*-dioxin and its congeners, *Mol. Pharmacol.,* 27, 133, 1985.
33. **Bandiera, S., Sawyer, T. W., Campdell, M. A., Fujita, T., and Safe, S.,** Competitive binding of the cystolic tetrachlorodibenzo-*p*-dioxin receptor, *Biochem. Pharmacol.,* 32, 3803, 1983.
34. **Poland, A., Greenlee, W. E., and Kende, A. S.,** Studies on the mechanism of action of ten chlorinated dibenzo-*p*-dioxins and related compounds, *Ann. N.Y. Acad. Sci.,* 320, 214, 1979.
35. **Poland, A., Glover, E., and Kende, A. S.,** 3,4,3′,4′- tetrachloroazoxybenzene and azobenzene: potent inducers of aryl hydrocarbon hydroxylase, *Science,* 194, 627, 1976.

Chapter 13

MANAGEMENT OF ASH FROM MUNICIPAL SOLID WASTE COMBUSTION

Paul C. Chrostowski and Shawn L. Sager

TABLE OF CONTENTS

I. INTRODUCTION

The combustion of municipal solid waste (MSW) typically results in a volume reduction of the waste material of approximately 10 to 20%. The ash which is transported through the bottom of the stoker grate is known as bottom ash, while the fraction of ash which escapes through the stack is known as fly ash. The function of particulate control devices such as baghouses, electrostatic precipitators, and cyclones is to trap fly ash which may be transported via the exhaust gas.

Fly ash consists of 70 to 95% inorganic matter. The primary constituents are silicon, iron, aluminum, calcium, magnesium, and sodium.[1] The remaining 5 to 30% contains potentially toxic organic compounds such as polychlorinated dibenzodioxins (PCDDs), polychlorinated dibenzofurans (PCDFs), polynuclear aromatic hydrocarbons (PAHs), and toxic heavy metals and metalloids, such as arsenic, cadmium, chromium, nickel, and lead. Typically, the bottom and fly ash from a facility are combined and transported to a landfill for disposal. If the ash were to be disposed of in an environmentally unsound manner, water could filter through the ash, generating a leachate, which may be contaminated with materials present in the ash. If uncontrolled, some of the toxic materials present in the leachate may be transferred to other environmental compartments.

The purpose of this chapter is

- To define the characteristics of ash generated from MSW combustion, including the amount of ash generated, the physical properties of ash, the metal and organic chemical composition of ash, and the sources of metals and organics in ash
- To discuss the potential for pollutants to leach from the ash and contaminate the environment along with implications for ash management
- To explain the different types of leaching tests currently used both in a regulatory and a chemical context
- To discuss some important current issues associated with ash management such as data gaps, landfill design, and exposure routes

II. CHARACTERISTICS OF ASH GENERATED FROM MSW COMBUSTION

The quantity and characteristics of ash produced during combustion are a function of the composition of the MSW, combustion conditions, and the type of combustion and pollution control equipment used. Most processes that produce ash, such as incineration, tend to yield a lower volume of residues than other, nonash processes, such as composting. In these cases, volume reduction is greater than mass reduction due to the tendency for heavier elements, such as iron, to remain in the ash while lighter elements such as carbon are released in a gaseous form. Table 1 presents an overview of ash production and disposal at MSW facilities in the U.S.[2]

In addition to these generic estimates of ash production, it is instructive to examine the actual amount of ash generated by various operating facilities. Table 2 presents information on the quantities of fly and bottom ash generated at various facilities in the U.S. As of 1987, 7547 tons (wet weight) of ash were being disposed of in the U.S. on a daily basis.[3] Of this total, 361 tons were fly ash, 2930 tons were bottom ash, and 4255 tons were combined fly and bottom ash.

Although fly ash is composed primarily of oxides and silicates of iron, calcium, magnesium, and sodium, the trace amounts of heavy metals and other organics, such as PCDDs and PCDFs, have received the greatest attention due to their potential toxicity. While few characterizations of the inorganic compounds present in MSW ash are reported in the literature, numerous data on trace metals are available. Table 3 summarizes the metal content of ash generated from MSW incinerators (in columns labeled 1 through 6),[3-8] coal-fired power plants (column labeled 7,8),[9,10]

TABLE 1
Summary of Municipal Solid Waste Incinerator Statistics[2]

	All facilities	Modular	Conventional	RDF[a]
No. of operating facilities	102	51	48	3
No. with heat recovery	55	31	24	NA
Total MSW managed (tons/d)	33,541	5,296	26,018	2,227
Total amount of ash generated (tons/d)				
Ash disposed	7,547	1,943	5,374	231
Dry-weight basis	5,191	965	4,027	199
Bottom ash as disposed	2,930	1,630	1,175	125
Fly ash as disposed	361	143	114	105
Combined ash as disposed	4,255	169	4,085	0
Number of facilities with				
Disposal method				
On-site landfill	23	12	9	2
Off-site landfill	73	36	36	1
Other/unknown	6	3	3	0
Type of landfill				
Monofill	36	16	18	2
Codisposal	41	19	21	1
Other /unknown/not LF[b]	25	16	9	0
Percent of Ash disposed of				
Disposal method				
On-site landfill	22.78	3.37	28.61	50.39
Off-site landfill	77.22	96.63	71.39	49.61
Other/unknown	0	0	0	0
Type of landfill				
Monofill	35.50	25.97	38.30	50.39
Codisposal	14.75	46.14	5.70	49.61
Other/unknown/ not LF	46.75	27.89	55.32	0.00

[a] RDF = Refuse-derived fuel.
[b] LF = Landfill.

and sewage sludge incinerators (column labeled 9).[11] The wide range of variability among the ash samples from MSW incinerators indicates that it is incorrect to speak of a "typical" MSW incineration ash. For example, lead concentrations range from less than 250 to 40,000 ppm. This variability points out the importance of obtaining site-specific information when performing health risk analyses or environmental impact assessments on ash. Furthermore, with the exception of lead, zinc, and arsenic levels, which are somewhat higher in MSW ash than in coal ash, there do not appear to be substantial overall differences between ash from MSW incineration, sewage sludge incineration, or coal-fired power plants. Some of the differences between coal and MSW ash may be due to combustion conditions. When refuse-derived fuel (RDF) and coal were burned under similar conditions, similar levels of trace metals were found in the ash.[4,12]

Few studies have been conducted on nonmetals or anions in MSW ash. Chloride content of fly ash was reported to be 0.30 to 1.12% for one facility and 0.116 to 0.153% for another facility.[7] NUS[3] reported that the chloride content of fly ash ranged from 1169 to 11,200 ppm.

Even fewer studies have been conducted on the organic content of ash. The limited studies done have focused on PCDDs and PCDFs due to their potential for adverse human health effects. Other organic chemicals often found in MSW ash which are suspected to cause adverse health effects include PAHs, chlorinated phenols, chlorinated benzenes, and polychlorinated biphenyls (PCBs). The organic content of ash is, like the metal content, highly variable, depending on

TABLE 2
Amount of Ash Generated at Various Municipal Solid Waste Combustors in the U.S.

Plant	Capacity (TPD)[b]	Amount of ash generated[a]			Total amount of ash (TPD)	Type of facility	Pollution control device used
		Bottom	Fly	Combined			
Prudhoe Bay, AK	NA[c]	62.1	NC[d]	62.1	68.31	NA	NA
Sitka, AK	25	9.8	1.9	11.7	12.87	Modular	ESP[e]
Batesville, AR	50	9.1	NC	9.1	10.01	Modular	NA
Reduction Center, DC	NA	129.9	NC	129.9	142.89	NRR[f]	ESP
Lakeland, FL	300	10.9	43.6	54.5	59.95	RDF	Cyclone
Pinellas County, FL	2,000	227.3	NC	227.3	250.03	MB[g]/waterwall	ESP
Oahu, HI	NA	229.9	NC	229.9	252.89	NA	NA
Ames, IA	200	36.9	24.6	61.5	67.65	RDF	ESP/cyclone
Heyburn, ID	NA	30.3	NC	30.3	33.33	Modular	NA
Chicago NW, IL	1,600	79.7	11.2	90.9	99.99	MB/waterwall	ESP
Louisvelle, KY	NA	36.4	NC	36.4	40.04	NRR	WS[h]
Framingham, MA	NA	57.5	1.6	59.1	65.01	NRR	Baghouse
Pittsfield, MA	240	64.8	1.8	66.6	73.26	Modular	Gravel bed
Saugus, MA	1,500	272.7	NC	272.7	299.97	MB/waterwall	ESP
Baltimore, MD	2,250	409.1	NC	409.1	450.01	MB/waterwall	ESP
Auburn, ME	200	85	1.6	86.6	92.26	Modular	Baghouse
Fort Leonard Wood,MI	75	1.5	NC	1.5	1.65	Modular	NA
New Hanover County, NC	200	16.7	NC	16.7	18.37	MB/waterwall	ESP
Auburn, NH	NA	0.4	NC	0.4	0.44	Modular/NRR	NA
Canterbury, NH	NA	0.1	NC	0.1	0.11	Modular/NRR	NA
Meredith, NH	NA	1.9	NC	1.9	2.09	Modular/NRR	NA
Oneida County, NY	200	88.4	NC	88.4	97.24	Modular	ESP
North Dayton, OH	NA	154.5	NC	154.5	169.95	NRR	ESP
South Dayton, OH	NA	154.4	NC	154.4	169.95	NRR	ESP
Harrisburg, PA	720	63.6	27.3	90.9	99.99	MB/waterwall	ESP
Lewisville City, TN	60	12.1	NC	12.1	13.31	Modular	WS
Nashville, TN	720	129.3	NC	129.3	142.23	MS/waterwall	ESP
Waxahachie, TX	50	27.6	NC	27.6	30.36	Modular	WS
Harrisonburg, VA	100	20.5	NC	20.5	22.55	MB/refractory	ESP

Salem, VA	100	24.2	NC	24.2	26.62	Modular	NA
Bellingham, WA	NA	36.4	NC	36.4	40.04	Modular	NA
Sheboygan, WI	NA	32.7	1.1	33.8	37.18	MB/refractory	WS

[a] Amount of ash generated in metric tons.
[b] Tons per day.
[c] Data not available.
[d] Data not collected.
[e] Electrostatic precipitator.
[f] No resource (heat) recovery.
[g] Massburn facility.
[h] Wet scrubber.

TABLE 3
Metal Content of Municipal Solid Waste Ash (micrograms per gram)[a]

	1	1	1	2	2	3	4	5	6	7,8	9
Antimony	17	—	—	—	—	3,000	—	139—760	—	2.2—13	4—163
Arsenic	1,250	<1,200	<620	<1,200	<560	100	2.9—50	9.4—74	1.3—12	0.5—312	0.9—22
Cadmium	3	<17	<9	<17	<8	2,000	0.2—100	<5—125	1.9—35	0.1—18	1.3—127
Chromium	70	230	320	320	1,500	200	12—1,500	1,070—1,900	13—129	3.4—437	346—9,696
Cobalt	100	<24	<19	<91	<11	—	1.7—9.1	26—54	—	4.9—79	8.2—69
Copper	100	550	2,000	650	1,700	3,000	40—5,900	300—2,000	—	3.7—616	485—3,225
Lead	500	<470	<250	<470	<230	30,00	31—36,600	18,000—54,000	330—17,500	1.3—252	82—1,333
Manganese	190	1,200	1,600	1,000	1,100	—	14—3,130	2,700—8,500	—	79—769	342—3,095
Nickel	1,000	<200	<110	<200	<96	100	13—12,910	580—960	19—175	2—258	40—4,200
Selenium	62	<780	<420	<780	<380	—	0.1—50	1.4—11	—	0.4—19	0.04—0.9
Thorium	19	—	—	—	—	—	—	11—20	—	30—68	—
Uranium	9	—	—	—	—	—	—	—	—	3.9—9.7	—
Vanadium	800	36	33	51	18	100	13—150	110—166	—	12—570	23—387
Zinc	250	—	—	—	—	80,000	92—46,000	78,000—13,600	710—2,030	16—2,300	400—8000

(1) Taylor et al.[4] (2) Francis and White.[5] (3) Vogg et al.[6] (4) NUS.[3] (5) Greenberg et al.[7] (Fly ash only). (6) State of Washington.[8] (7) EPRI[9] (coal ash). (8) Furr et al.[10] (coal ash). (9) Furr et al.[11] (sewage sludge ash).

[a] All combined ash unless otherwise noted.

TABLE 4
Concentration of PCDDs and PCDFs in Ash from Municipal Solid Waste Incinerators (nanograms/gram)[3]

Isomer	Fly ash	Combined ash	Bottom ash
MCDD	2.0	NR[a]	ND[b]
DCDD	0.4—200	NR	ND
T_3CDD	1.1—82	NR	ND
T_4CDD	ND—250	0.14—14	<0.04—0.65
PCDD	ND—722	1.9—50	ND—3.0
H_6CDD	ND—5,565	1.4—78	ND—2.3
H_7CDD	ND—3,030	0.84—89	ND—29
OCDD	ND—3,152	0.84—89	ND—29
2378-TCDD	0.1—42	0.02—0.78	<0.04—0.07
Total PCDDs	5.23—10,883	6.2—350	ND—110
MCDF	41	NR	1.1
DCDF	ND—90	NR	0.63
T_3CDF	0.7—550	NR	ND
T_4CDF	ND—410	2.3—91	0.15—1.4
PCDF	ND—1,800	1.6—37	0.07—6.2
H_6CDF	T[c]—2,353	1.2—35	ND—2.5
H_7CDF	T—666	0.62—36	ND—6.9
OCDF	ND—362	0.18—8.4	ND—3.7
2378-TCDF	0.1—5.4	0.41—12	ND—10
Total PCDFs	3.7—3,187	6.1—154	ND—65

[a] Not reported.
[b] Below detection limit.
[c] Trace = 0.01 < T < 0.1.

the ash and on the method used to analyze for organic content. Hrudey and Perry[13] analyzed six samples of ash using six methods and found that the organic carbon content in all samples ranged from 0.005 to 5.86%. Eiceman et al.[1] identified almost 100 organic compounds in fly ash, most of which were branched or straight chain hydrocarbons or related compounds with generally low toxicity. The total extractable organic content ranged from 1 to 30 ppm (micrograms per gram) depending on the sample, with PAHs constituting 1 to 500 ppb (= nanograms per gram) and tetrachlorinated dibenzo-*p*-dioxins (TCDDs) 2 to 10 ppb.[1] Clement et al.[14] tested various extraction solvents for their ability to extract organics from ash and found benzene to be the most efficient extractant. Mean values extracted by benzene were 3566 ppb for alkanes, 509 ppb for PAHs, and 165 ppb for PCDDs. These data indicate that results obtained by extraction with solvents other than benzene, such as methylene chloride, cyclohexane, or methanol, may underestimate the organic content of ash. NUS[3] measured the PCDD/PCDF content of ash (Table 4). These data indicate that PCDD/PCDF levels are generally higher in fly ash than in bottom or combined ash.

The physical characteristics of ash vary considerably due to the heterogeneous nature of the waste burned. Coal fly ash is primarily composed of regular, spherical particles with diameters between 10 and 70 μm, whereas MSW fly ash is composed of irregular particles with diameters between 50 and 500 μm.[15] Most of the particle size distributions developed for fly ash have been based on air pollution control requirements rather than residual disposal requirements; therefore, they represent the condition of the ash as it leaves the stack. Most of these measurements present a size distribution considerably smaller than that shown by Surgi.[15] For example, Ferraro and Parenteau[16] report that 100% of the ash leaving the stack at the Maine Energy Recovery Company is less than 20 μm in diameter and the majority (52.8%) is less than 6 μm in diameter.

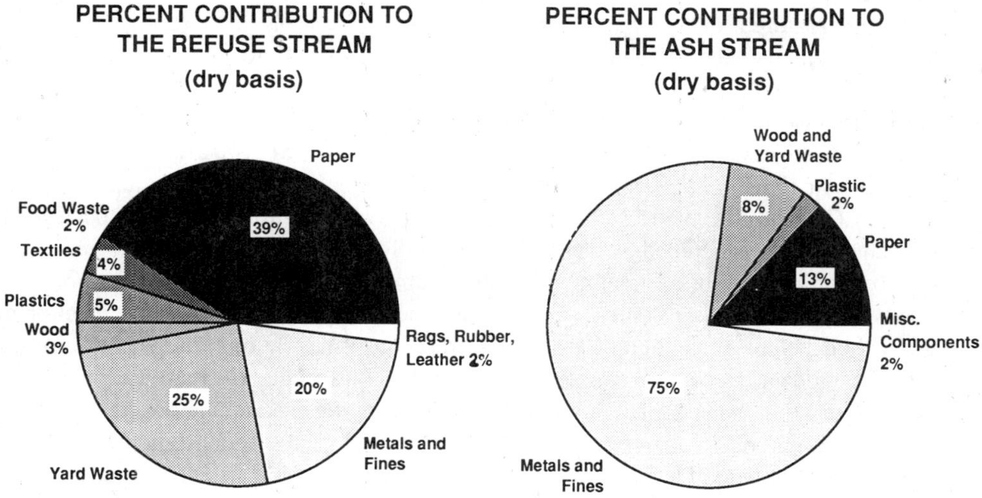

FIGURE 1. Comparison of the percent contribution of materials to the refuse stream and the ash stream.

The ash is primarily derived from the inorganic component of the solid waste incinerated. Kaiser[17] reported that MSW typically contains 31 to 35% inorganic matter on a dry weight (DW) basis. The major contributors to the inorganic fraction are glass and metallics, although paper can contribute up to 29% and rubber and leather up to 25%. Figure 1 compares the contribution of materials percent to the refuse and ash streams, respectively. With the exception of zinc, metals of potential human health or environmental concern are present in trace quantities (up to 999 ppm) in MSW feedstock.[18] Most of the heavy metals in their metallic form are either separated from the waste before it is burned or end up in the bottom ash. Some representative sources of metallic material are lead from can solder, cadmium and chromium from metal plating processes, and lead, cadmium, and mercury in batteries. Metals in combined form often end up in the fly ash. The primary source of these materials stems from their use as colorants and stabilizers in inks and plastics.

Most organic compounds present in ash are products of incomplete combustion. Although the scientific controversy surrounding their formation has not been entirely resolved, most scientists agree that PCDDs and PCDFs are formed during the combustion process and adsorb onto the ash. Sources of chlorine, both inorganic (such as sodium chlorine) and organic (such as polyvinyl chloride plastics) have been implicated in the formation of PCDDs, PCDFs, and their precursors, chlorobenzenes and chlorophenols. PAHs are ubiquitous products of incomplete combustion and will form whenever organic compounds are burned at less than optimal conditions.

III. LEACHATE GENERATION AND ASH MANAGEMENT

Since there currently is no market demand for MSW ash, the large quantities of ash generated daily by MSW incineration must be disposed of in land treatment facilities, or landfills. During the active period of a landfill, precipitation can fall onto the material present in the landfill and dissolve part of the waste as it percolates through the fill. The contaminated water which passes through the landfill is known as leachate. Proper management of both the ash and the leachate from landfills is necessary to avoid environmental contamination and human health impacts. In the past, sanitary landfills were operated under the theory of natural renovation. This theory is based on the premise that leachate from a sanitary landfill would be biologically degraded by soil microbes and physically modified by soil filtration, thus rendering it harmless. The numerous cases of groundwater contamination as a result of leachate releases from sanitary landfills which have been reported over recent years demonstrate that better management techniques are required.

Modern ash management techniques are essentially divided into three types: chemical treatment, leachate confinement and control, and separation of ash from other types of waste. Before discussing each of these processes, it is pertinent to define some of the terms used in ash management. *Codisposal* refers to the management practice where MSW and ash are comingled and disposed of together. A *monofill* is a landfill that contains only solid waste combustion ashes and residues. Currently, 35.5% of MSW ash is disposed of in monofills, 17.45% is codisposed, and the remainder is not handled by landfilling or its ultimate disposal treatment is unknown.[3]

The most common chemical treatment of ash is stabilization by a pozzolanic reaction with lime used in acid gas emission controls. Pozzolanic processes generally involve the reaction of fine-grained pozzolans (e.g., aluminum silicates present in the ash), lime, and water to form a concrete-like product. Nonreacting materials in the ash are trapped in the matrix formed from the chemical reaction between the aluminum silicates from the ash and the calcium from the lime. The U.S. Environmental Agency[19] (EPA) has discussed the use and testing of pozzolanic processes in the treatment of hazardous waste. This information could also be applied to MSW ash. Although many resource recovery facilities use lime in their emission control processes, not all do so under the optimum conditions necessary for the formulation of a pozzolan in the residues. The pozzolanic reaction typically requires a reaction time of 7 to 28 d, about 40% water content, thorough mixing of components, and application of pressure (analogous to the shear forces exerted on the mass in a cement mixer). Failure to attain these conditions will result in an imperfect pozzolan which could lead to leachate generation. At this time, few other chemical treatments are used. One exception is a proprietary process using phosphoric acid developed by Wheelabrator Environmental Systems.[15]

The primary method used for ash management in the U.S. is leachate confinement and control. Before discussing these technologies, it is instructive to examine the chemical composition of leachate generated from ash disposal. Since monodisposal of MSW ash is a relatively new technology, there is a limited database from which to derive information concerning leachate composition. NUS[3] reported that seven organic compounds, primarily oxygenated and sulfonated hydrocarbons, were found in nine leachate samples from monofills at concentrations as high as 0.4 mg/l. Only minute quantities (ppt) of PCDDs and PCDFs were found in field leachate samples. A summary of inorganic concentrations in leachate from monofills is given in Table 5 along with similar information for leachate from codisposal landfills and landfills containing materials which have not been thermally treated.[3] These data show that on the average, inorganic concentrations are higher in MSW landfill leachate than in monofill or codisposal landfill leachates.

The most comprehensive study of leachate composition from an operating facility is that conducted by Malcolm Pirnie, Inc. at the Peekskill plant in Westchester County, New York.[20] The design and characteristics of both the resource recovery plant and landfill are given by Cundari and Lauria.[21] Leachate from the monofill was evaluated 29 times over a 12-month period. Results of these analyses, given in Table 6, show that lead and nickel were present in the highest concentrations, while much smaller quantities of arsenic, hexavalent chromium, and mercury were measured.

IV. LANDFILL DESIGN

Landfilling is currently the method of choice for disposal of MSW ash residues and is likely to remain so in the future. Ash generated from the combustion of MSW can be disposed of in landfills with design characteristics ranging from a monofill with a clay liner to a landfill with a double liner and leachate collection system. The purpose of any liner system is to retard the movement of leachate from the fill to the subsurface environment.

TABLE 5
Inorganic Concentrations in Leachate from Monofills, Co-Disposal Landfills, and Municipal Solid Waste Landfills (micrograms/liter)[3]

Element	Monofills	Co-disposal	MSW[a]
Arsenic	0.005—0.218	0.008—0.046	ND[b]—70.2
Barium	1.0	0.29—0.89	ND—12.5
Cadmium	ND—0.044	ND—0.011	ND—0.19
Chromium	0.006—1.53	ND—0.013	ND—33.4
Copper	0.022—24	ND	ND—10
Lead	0.012—2.92	0.01—0.27	ND—14.2
Manganese	0.103—4.57	1.29—11.3	ND—1,400
Mercury	0.001—0.008	ND	ND—0.064
Nickel	ND—0.412	ND—0.24	ND—7.5
Selenium	0.0025—0.037	NA[c]	NA
Silver	0.07	NA	NA
Zinc	ND—3.3	0.09—1.21	ND—1,000

[a] No thermal treatment.
[b] Below detection limit.
[c] Not analyzed.

TABLE 6
Westchester County Monofill Leachate Concentrations[a] (milligrams/liter)[20]

Element	Range	Average
Arsenic	—	<0.005
Barium	<0.01—0.30	0.06
Cadmium	<0.001—0.13	0.073
Chromium^{+6}	—	<0.02
Total chromium	<0.01—0.29	0.05
Copper	0.03—0.11	0.06
Lead	0.005—0.71	0.434
Mercury	—	<0.0005
Nickel	0.04—0.47	0.281
Selenium	<0.005—0.008	0.005
Silver	<0.01—0.08	0.05
Zinc	0.029—0.384	0.089

[a] Summary of 29 measurements taken from October 1986 to October 1987.

There is limited information on the field performance of synthetic liners and their anticipated failure rates due in part to the relatively short period during which synthetic liners have been used. Several laboratory and field studies revealed that synthetic liners are performing up to their specifications after 9 years of continuous exposure to leachate.[22-24] The liners studied are made of materials that include chlorosulfonated phlyetylene (CSPE), high density polyethylene (HDPE), and polyvinyl chloride (PVC) materials. In developing a series of models for investigating the impact of various policy decisions involving the regulation of landfills, the U.S. EPA assumed that the average time it takes a synthetic liner to fail because of aging is 25 to 45 years.[25]

If a landfill has a synthetic liner underlain by clay and a leachate collection system, as long as the synthetic liner is intact, leachate will be diverted to the collection system and removed from the landfill. If a break occurs in the liner, however, only a portion of the leachate will be collected. The remaining leachate will seep through the clay liner and contaminate the surrounding environment. The quantity of leachate likely to seep through the clay depends on the size and location of the break and the depth of the clay liner.

There are three major types of synthetic liner failure: physical, biological, and chemical. Numerous forms of physical failure exist. Puncturing or tearing the liner during installation, for example, is one potential source. Other forms of physical failure include creep and differential settling. Creep is a term used to describe a liner undergoing increasing deformation due to a sustained local stress load. Differential settling is a localized structural stress due to improper subgrade design or construction. Careful geological analysis prior to installation and greater liner thickness and elasticity can substantially decrease the potential for differential settling. Physical failure is not likely to occur if the liner is well designed and properly installed.

Some microbial organisms can damage the structural integrity of a liner and increase its permeability, thus inducing biological failure. Lyman et al.[26] found that: "The majority of synthetic polymers are rather inert towards biological enzymatic attack although, in principle, they are biodegradable. Generally, where attack has been observed, the rates of decomposition have been low because most synthetic polymers are attacked at the chain ends, which are often hidden in the polymeric matrix and are not, or only very slow accessible to enzymes."

Chemical degradation is perhaps the primary mechanism of synthetic liner failure. There are three types of chemical failure: swelling, extraction, and outdoor exposure. Swelling is considered the most serious chemical effect to polymeric liners, because it has the potential to cause a loss in strength, elongation, creep and flow, and puncture resistance. Chemical failure is most prevalent when caustic liquid waste comes into direct contact with the liner. Extraction of the plasticizer can result in embrittlement, shrinkage, and liner breakage. Liners which contain a large amount of monomeric plasticizer, such as polyvinyl chloride, are not susceptible to extraction. Outdoor exposure to heat, moisture, infrared and ultraviolet light (sunlight), oxygen, or ozone can also cause failure if the liner is not protected prior to installation. Chemical failure is not likely to occur if the liner is protected from reactive features such as sunlight and high strength liquid organic wastes.

Potential failure mechanisms of clay liners are numerous and varied. Seismic activity or subsidence can adversely affect the structural integrity of the landfill. Waste leachates can also modify clay liners. The main failure mechanisms associated with leachate-clay interactions are volume changes. Volume changes occur when there is a natural change in the water content of the clay. For any change of water content, the resulting clay volume change is dependent upon the clay mineral type, arrangement of clay particles, size of clay particles, surface area, and the nature of the compounds adsorbed to the clay.

The shrink-swell behavior of clay soils also contributes to both macro- and microstructural changes. Macrostructural changes resulting from bulk shrinkage can often be visually identified in clays by the presence of cracks, which enhance leakage rates by providing a conduit for the leachate to migrate. Interlayer shrinkages (or shrinkages in the space between the clay layer and the layer of sorbed fluid directly adjacent to it) are caused by micro structural alterations in the clay and, even though they are virtually undetectable, their influence on the permeability of a liner can be significant. Changes in interlayer spacing may effect the bulk volume, pore-size distribution, and permeability of the clay. Some of these changes could make the clay less rather than more permeable, however.

High concentrations of organic compounds can also have a substantial effect upon interlaying spacing. Although the interactions between clay surfaces and organic fluids are complex, the dielectric constant of an organic has been shown to be a controlling factor. If water is replaced

TABLE 7
EP Toxicity Test Maximum Allowable Concentrations (milligrams per liter)[30]

Contaminant	Level
Arsenic	5.0
Barium	100.0
Cadmium	1.0
Chromium	5.0
Lead	5.0
Mercury	0.2
Selenium	1.0
Silver	5.0

by an organic compound with a lower dielectric constant, the individual clay particles will contract as the result of a thinner interlayer space. This will make additional pore space for the clay particles to reorient themselves, and the clay layer will be less permeable. Hence, if the leachate is mostly water, the clay will be more permeable. However, exposure of the clay liner to leachate will only occur after the synthetic liner has failed.

Haxo et al.[27] estimated the permeability of a new synthetic liner to be 1×10^{-12} cm/s. Thus, an intact membrane would prevent the passage of most leachate. If the liner were to fail, the most likely consequence would be an increase in permeability rather than a total loss of protection. As an example, if the compacted clay under the liner has a permeability of 10^{-7} cm/s and there is a 10% failure of the liner, the resulting permeability of the composite liner system would be 10^{-8} cm/s, still less than the clay itself. If liquid flows through the membrane, it will encounter a layer of compacted clay. There must be sufficient flow through the membrane at this point to establish a saturated front. Unlike synthetic membranes, clay liners do not appear to be susceptible to degradation by dilute aqueous leachate.[28]

Thus, it is unlikely that liner failure will occur if certain conditions are met. First, the liner must be properly selected and installed. Second, high strength liquid organic waste should not be allowed to reach the liner. (The U.S. EPA has estimated that a concentration of 2 ppm of pure organic solvent will not degrade a synthetic liner.[29])

V. LEACHATE TEST PROCEDURES

There are several leaching tests that can be used to characterize ash. Current Resource Conservation and Recovery Act (RCRA) regulations[30] specify that if a leachate extracted using an EPA-approved extraction procedure contains any of the metals shown in Table 7 at concentrations equal to or greater than the stated limit, the waste is classified as toxic and, unless otherwise exempted, is subject to Subtitle C regulation. The maximum tolerated concentrations shown in Table 7 are 100 times the current Primary Drinking Water Standards (PDWS) established by the Safe Drinking Water Act for those compounds.

Waste extraction tests are used to predict the type and concentration of constituents that may leach from a waste disposal site under field conditions. Most extraction tests are conducted by mixing or washing a waste sample with a water-based solution for a specified length of time. The resulting leachate solution is then separated from the solids and chemical concentrations quantified. Although under current regulations only the Extraction Procedure (EP) toxicity test is used to determine whether a waste is EP toxic, the EPA has proposed a new procedure, the Toxic Characteristic Leaching Procedure (TCLP), to replace the EP toxicity test.[31] With the EP toxicity test procedure, samples are placed in a pH 5 solution buffered with acetic acid, and the solution is agitated for 24 h. The TCLP procedure is designed to be more waste-specific. Acidic wastes are placed in a pH 5 acetic acid solution with a sodium acetate buffer, while alkaline wastes are placed in an acetic acid solution only. A third extraction procedure, SW-924, differs from the EP and TCLP tests in that deionized or distilled water is used to extract the materials rather than a buffered solution.[3] Finally, the State of California has developed the Waste Extraction Test (WET), which extracts the waste under anoxic conditions in a sodium citrate solution at pH 5.[32]

It has been alleged that the EP test provides an inappropriate measure of leaching under field conditions, because it was developed to characterize leaching from municipal landfills. If ash is disposed of in a monofill, the only liquid that the ash will come in contact with is rainwater. Since rainwater does not contain organic acids, the results of either the EP or TCLP test tend to overestimate the leaching potential of ash. Thus, the ability of these extraction procedures to predict accurately concentrations in leachate is questionable. To address this issue, we compared the results of the various leaching tests.

A considerable amount of research has been conducted into the leaching of metals from ash. Rademaker and Young[33] analyzed extracts from ash produced at the Nashville massburn plant. They reached several conclusions, including:

- Leaching is a first order process (i.e., considerably higher concentrations are obtained early in the process).
- Most of the material leached is inert salts. Sulfate constituted 55.4% of the total dissolved solids leached, while chloride constituted 29.4%. For heavy metals, only 1.6×10^{-3} mg/g cadmium and 1.8×10^{-3} mg/g lead was leached.
- The presence of soil material generally inhibited the leaching of organics but not all metals.

In general, the results of the leaching tests depend upon the starting material, since the ash composition is directly correlated to the composition of the waste. Beyond this distinction, it is generally found that the EP and TCLP procedures are more aggressive in leaching inorganic constituents from ashes than the SW-924 procedure and that the EP and TCLP procedures yield comparable levels for inorganic constituents.[3] The WET procedure is the most aggressive with respect to the extraction of inorganics.[30] Only limited quantities of organic constituents tend to leach from fly ash, bottom ash, or a mixture of the two. In general, concentrations of PCDDs, PCDFs, PCBs, chlorinated benzenes, and other organics with the exception of chlorophenols, are below current detection limits.[3,34] Hence, only inorganic constituent analyses are discussed further.

Francis[35] examined leachate from four different plants, Chicago Northwest; Auburn, ME; Hampton, VA; and Gallatin, TN. The ashes were leached using four different extraction procedures and with MSW leachate. The analysis of the ash is presented in Table 8, and the analysis of the water ash leachate is presented in Table 9. These results show minimal leaching under normal test conditions. When MSW leachate was used instead of water, however, it was found that considerably higher concentrations of metals leached from the ash. This observation indicates that co-disposal of ash with MSW may result in a greater degree of metal leaching.

Fly ash, bottom ash, and a mixture of the two can be compared in the context of these three leaching tests. Inorganic constituents are more readily leachable from fly ash and to a lesser extent from the combined bottom and fly ash than from bottom ash.[3] Francis and White[5] reported that the WET method leached higher concentrations of the EP toxic inorganics (Table 7) than the other methods. Lead, and to a lesser extent, cadmium were extracted from the four ashes at levels greater than the EP toxicity maximum allowable limits (Table 7). NUS[3] found that the EP toxicity limit was exceeded frequently for cadmium, chromium, and lead.

TABLE 8
Elemental Concentrations[a] in the Ash from Four Resource Recovery Facilities (milligrams/kilogram)[35]

	Chicago	Sumner	Hampton	Auburn
Aluminum	53,000	60,000	54,000	45,000
Antimony	<250	<140	<260	<120
Arsenic	<1,200	<620	<1,200	<560
Barium	1,600	810	1,600	1,000
Beryllium	<2.4	<1.3	<2.4	<1.0
Cadmium	<17	<19	<17	<8
Calcium	85,000	79,000	68,000	42,000
Cobalt	<24	<19	<91	<11
Copper	550	2,000	650	1,700
Chromium	230	320	320	1,500
Iron	86,000	58,000	36,000	63,000
Lead	5,000	2,500	3,600	290
Magnesium	16,00	7,700	8,600	6,500
Manganese	1,200	1,600	1,000	1,100
Molybdenum	290	170	200	260
Nickel	<200	<110	<200	<96
Phosphorus	5,000	2,500	3,600	290
Potassium	11,000	5,800	12,000	5,500
Selenium	<780	<420	<780	<380
Silicon	84,000	61,000	120,000	47,000
Silver	<33	<17	<33	<16
Sodium	18,000	9,600	18,000	7,700
Strontium	270	380	300	220
Titanium	11,000	1,000	9,100	28,000
Vanadium	36	33	51	18

[a] Concentrations after drying at 100°C for 24 h determined by $LiBO_2$ fusion and analysis of the 3% wash by inductively coupled plasma spectroscopy.

Comparing the results of leaching tests with actual field data show that field measurements are routinely lower than leachate test results. Field generated leachate was found to meet the EP toxicity maximum allowable limits. This result was true for both co-disposal sites and monofills.[3] A comparison of the WET procedure results with actual field leachates found that with the exception of selenium, the field levels were below laboratory extracted concentrations.[5] The WET method overestimated lead concentrations by almost 800 times. Levels of the other metals were between 25 and 75 times higher in the WET method-produced leachate than in the field-generated leachate.

VI. OTHER CONSIDERATIONS IN ASH MANAGEMENT

Leachability of ash by standard tests also depends on the type of emissions control devices used. Sawell and co-workers[34] found that for cadmium, lead, and zinc, a greater percentage of the metal was leached from fabric filter ash than ash from either a dry or wet scrubber. For other metals, no trends were obvious. One recent example of air pollution control equipment design that is capable of producing an ash of low leachability was reported by Ferro and Parenteau.[16] At this facility, the flue gases at 350°F enter scrubbers where they react with a 15% lime slurry at 6 to 8 ppm. For 29 samples, the combined bottom and fly ash leachate was significantly below the EP toxicity regulatory limits.

In addition to the potential for leachate generation, the generation of fugitive dust during ash handling represents another potential pathway of human exposure. Fugitive dust can be

TABLE 9
Mean Elemental Concentrations in Water-Ash Leachates (milligrams per liter)[35]

	Control	Chicago	Sumner	Hampton	Auburn
Aluminum	<0.27[a]	<1.27	14.9	<8.34	<1.68
Antimony	<0.40	<0.40	<0.42	<0.42	<0.41
Arsenic	<2.04	<1.82	<1.9	<1.9	<2.01
Barium	<0.01	<0.05	<0.07	<0.05	<0.10
Beryllium	<0.01	<0.01	<0.01	<0.01	<0.01
Cadmium	<0.03	<0.03	<0.03	<0.03	<0.03
Calcium	<23.5	194	72	<60.9	53.3
Cobalt	<0.04	<0.04	<0.04	<0.04	<0.06
Copper	<0.29	<0.48	<0.29	<0.30	<0.29
Chromium	<0.63	<0.16	<0.08	<0.08	<0.55
Iron	<2.98	<0.82	<0.15	<0.07	<2.99
Lead	<0.76	<0.76	<0.80	<0.80	<0.78
Magnesium	2.64	19.08	<0.24	<0.77	2.46
Manganese	<0.57	<0.03	<0.01	<0.01	<0.23
Molybdenum	<0.10	<0.09	<0.09	<0.08	<0.22
Nickel	<0.41	<0.43	<0.33	<0.33	<1.29
Potassium	<22.6	<63.2	<56.4	<27.3	<30.2
Selenium	<1.24	<1.24	<1.30	<1.30	<1.27
Silver	<0.05	<0.05	<0.05	<0.05	<0.05
Sodium	<8.21	<82.0	<60.9	<17.5	<31.7
Strontium	0.07	0.70	0.88	0.34	<0.29
Zinc	<0.06	<0.07	<0.06	<0.06	<0.06

[a] The "<" signs indicates that at least one inductively coupled plasma analysis over the 103 d of leaching was below detection limit for that element and treatment.

generated when the ash is transferred between the points where it is generated and where it leaves the plant. Since ash is typically damp, containing up to 40% moisture, excessive exposures due to fugitive dust emissions are not expected. Mozzon et al.[36] measured inorganic materials in employee airspace at an RDF facility in Canada. They concluded that except for the boiler grate inspector and the precipitator cleaner at the plant, exposure to airborne metals was not inordinate. Fugitive emissions from ash handling have also been evaluated in a risk assessment.[37,38] It was found that the risks associated with ashfill emissions were low in comparison to the risks associated with stack emissions.

VII. CONCLUSIONS

Ash produced during the combustion of MSW is composed primarily of inorganic materials such as aluminum, calcium, iron, magnesium, silicon, and sodium. In addition to these relatively nontoxic materials, the ash may contain several potentially toxic organic compounds, such as PCDDs, PCDFs, and PAHs, and toxic heavy metals and metalloids such as arsenic, cadmium, chromium, nickel, and lead. The two types of ash, bottom ash and fly ash, generated during combustion are typically combined and disposed of in landfills. Sound management practices will reduce the likelihood of toxic material release to the environment. Environmentally unsound practices may result in the generation of leachate which can transport potentially hazardous materials from the landfill. The potential also exists for the generation of fugitive dust emissions during ash handling and transport, but since ash is generally moist, this potential is low.

There are three types of ash management techniques: chemical treatment, leachate confinement and control, and separation of ash from other types of waste. The principal method of chemical treatment is stabilization by a pozzolanic reaction. The pozzolanic process traps materials in the concrete-like matrix formed. The separation of ash from other types of waste is generally achieved through disposal in monofills. Disposing of ash in monofills helps to reduce the concentration of metals in ash leachate, since the types of materials that are present in leachate are limited to ash and infiltrating rainwater constituents. Leachate confinement and control is the primary method of ash management.

Leachate generated from co-disposal and monofill facilities has different characteristics. In general, co-disposal leachate contains higher concentrations of the EP toxic materials than does monofill leachate. The EP toxicity test can be used to predict leachate composition from co-disposal facilities although it generally overestimates the release of metals and metalloids. In general, actual field leachate generated from either co-disposal facilities or monofills meets the EP toxicity maximum allowable limits. Thus, with proper controls, the release of hazardous materials generated from MSW to the environment will be at or below allowable limits.

REFERENCES

1. **Eiceman, G. A., Clement, R. E., and Karasek, F. W.,** Analysis of fly ash from municipal incinerators for trace organic compounds, *Anal. Chem.*, 51, 2343,1979.
2. Engineering Science, Summary of MSW Incinerator Ash Survey, Contract No. 68-02-3888, Hazardous Waste Branch, U.S. Environmental Protection Agency, 1986.
3. NUS Corporation, Characterization of MWC Ashes and Leachates from MSW Landfills, Monofills, and Co-Disposal Sites — Summary, Vol. 1, EPA/530-SW-87-028A, Office Of Solid Waste, U.S. Environmental Protection Agency, Washington, D.C., 1987.
4. **Taylor, D. R., Tompkins, M. A., Kirton, S. E., Mauney, T., and Natusch, D. F. S.,** Analysis of fly ash produced from combustion of refuse-derived fuel and coal mixtures, *Environ. Sci. Technol.*, 16, 148, 1982.
5. **Francis, C. W. and White, G. H.,** Leaching of toxic metals from incinerator ashes, *J. Water Pollut. Control Fed.*, 59, 979, 1987.
6. **Vogg, H., Braun, H., Metzger, M., and Schneider, J.,** The specific role of cadmium and mercury in municipal solid waste incineration, *Waste Manage. Res.*, 4, 65, 1986.
7. **Greenberg, R. R., Zoller, W. H., and Gordon, G. E.,** Composition and size of distributions of particles released in refuse incineration, *Environ. Sci. Technol.*, 12, 566, 1978.
8. Study of Municipal Incineration Residue and its Designation as a Dangerous Waste, State of Washington, Department of Ecology, Seattle, 1986.
9. Coal Ash Disposal Manual, 2nd ed., Electric Power Research Institute, Menlo Park, CA, 1981.
10. **Furr, A. K., Parkinson, T. F., Hinrichs, R. A., Van Campen, D. R., Bache, C. A., Gutenmann, W. H., St. John, L. E., Jr., Pakkala, I. S., and Lisk, D. J.,** National survey of elements and radioactivity in fly ash, *Environ. Sci. Technol.*, 11, 1194, 1977.
11. **Furr, A. K., Parkinson, T. F., Wachs, T., Bache, C. A., Gutenmann, W. H., Wszolek, P. C., Pakkala, I. S., and Lisk, D. J.,** Multi-element analysis of municipal sewage sludge ashes, *Environ. Sci. Technol.*, 13, 1503, 1979.
12. **Norton, G. A., DeKalb, E. L., and Malaby, K. L.,** Elemental composition of suspended particulate matter from the combustion of coal and coal/refuse mixtures, *Environ. Sci. Technol.*, 20, 604, 1986.
13. **Hrudey, S. E. and Perry, R.,** Assessment of organic content of incinerator residues, *Environ. Sci. Technol.*, 7, 1140, 1973.
14. **Clement, R. E., Viau, A. C., and Karasek, F. W.,** Comparison of solvent extraction efficiencies for toxic organic compounds of flyash, *Can. J. Chem.*, 62, 2629, 1984.
15. **Surgi, R.,** Residues from Resource Recovery Facilities — Current Research, presented at State University of New York at Stony Brook, January 24, 1986.
16. **Ferraro, F. A. and Parenteau, R. J.,** Results of Emissions and Ash Testing at the Maine Energy Recovery Company Waste-to-Energy Plant, presented at 81st Annu. Meet. of the Air Pollut. Control Assoc., Dallas, June 1988.

17. **Kaiser, E. R.,** Physical-chemical character of municipal refuse, *Combustion J.*, 26, 28, 1977.
18. **Freeman, H.,** Pollutants from waste-to-energy conversion systems, *Environ. Sci. Technol.*, 12, 1252, 1978.
19. Guide to the Disposal of Chemically Stabilized and Solidified Waste, SW-872, Office of Solid Waste, U.S. Environmental Protection Agency, Washington, D.C., 1982.
20. Malcolm Pirnie, Inc., Sprout Brook Residue Disposal Facility Leachate Analytical Characterization Research, Summary of Results, Report to Westchester County, New York, Department of Public Works, 1987.
21. **Cundari, K. L. and Lauria, J. M.,** Ashfills and leachate, *Waste Age,* November, 82, 1986.
22. Emcon Associates, *Field Verification of Liners from Sanitary Landfills,* NTIS PB 83-21781, Municipal Environmental Research Laboratory, U.S. Environmental Protection Agency, Cincinnati, 1983.
23. **Haxo, H. G.,** Durability of liner materials for hazardous waste disposal facilities, in *Proc. 7th Annu. Res. Symp. on Land Disposal of Hazardous Waste,* EPA-600/9-81-002b, U.S. Environmental Protection Agency, Washington, D.C., 1981.
24. **Roberts, S. A., Nelson, N. A., and Haxo, H. E.,** Evaluation of a waste impoundment liner after long-term exposure, in *Proc. 9th Annu. Res. Symp. on Land Disposal of Hazardous Waste, EPA-600/9-83-019,* U.S. Environmental Protection Agency, Washington, D.C., 1983.
25. U.S. Environmental Protection Agency, Hazardous waste management system land disposal restrictions; proposed rule, *Fed. Reg.,* 51, 1725, 1986.
26. **Lyman, W. J., Sidman, K. R., Tratnyek, J. P., D'Amani, M., Ong, J. H., Schwope, A. D., and Bass, J. M.,** *Expected Life of Synthetic Liners and Caps. Draft Final Report,* Office of Water Regulations and Standards and Office of Research and Development, U.S. Environmental Protection Agency, Cincinnati, 1983.
27. **Haxo, H. E., Miedema, J. A., and Nelson, N. A.,** Permeability of polymeric membrane lining materials, in *Int. Conf. Geomembranes Proc.,* 1, 151, 1984.
28. **Ely, R. L., Jr., Kingsbury, G. L., Branscome, M. R., Goldman, L. J., Northeim, C. M., Turner, J. H., and Mixon, F. O., Jr.,** Final Report. Performance of Clay Caps and Liners for Disposal Facilities, Municipal Environmental Research Laboratory, U.S. Environmental Protection Agency, Cincinnati, 1983.
29. U.S. Environmental Protection Agency, Hazardous waste management system; land disposal restrictions; proposed rule, *Fed. Reg.,* 51, 160214, 1986.
30. Code of Federal Regulations, U.S. Government Printing Office, Washington, D.C., 40, 261.41, January 16, 1987.
31. Code of Federal Regulations, U.S. Government Printing Office, Washington, D.C., 40, 26, Appendix II, 1987.
32. State of California Administrative Code, *WET Test Procedure,* Sect. 66706, Chap. 30, Div. 4, Title 22, 1987.
33. **Rademaker, A. D. and Young, J. C.,** Leachate from solid waste resource recovery operations, *J. Energy Dev.,* 107, 17, 1981.
34. **Sawell, S. A., Bridle, T. R., and Constable, T. W.,** Leachability of Organic and Inorganic Contaminants in Ashes from Lime-Based Air Pollution Control Devices on a Municipal Waste Incinerator, presented at 80th Annu. Air Pollut. Control Assoc. Meet., New York, 1987.
35. **Franics, C. W.,** Leaching Characteristics of Resource Recovery Ash in Municipal Waste Landfills, DOE Project No. ERD-83-289, prepared for California Waste Management Board, Sacramento, 1984.
36. **Mozzon, D., Brown, D. A., and Smith, J. W.,** Occupational exposure to airborne dust, respirable quartz and metals arising from refuse handling, burning and landfilling, *Am. Ind. Hyg. Assoc. J.,* 48, 111, 1967.
37. **Kellermeyer, D. A. and Ziemer, S. E.,** What are the health risks of ash monofills, *Solid Waste Power,* August, 40, 1989.
38. **Wells, R. C.,** Fugitive Particulate Emission Estimates for Incinerator Ash Handling, presented at the 81st Annu. Air Pollut. Control Assoc. Meeting, Dallas, 1988.

Chapter 14

MUNICIPAL SOLID WASTE INCINERATION ASH MANAGEMENT: A STATE PERSPECTIVE

Richard J. Cook*

TABLE OF CONTENTS

* The author gratefully acknowledges useful comments and suggestions from Dr. Charles Cubbage and Dr. James Bedford from the TSCC staff and from Dr. Anthony Thomas and Dr. Jesse Goodwin, members of a Commission ad hoc committee on ash policy.

I. INTRODUCTION

Consistent with its legislatively mandated responsibility to review and recommend state policy on matters involving toxic or potentially toxic materials, the Toxic Substance Control Commission (TSCC) has investigated the issue of municipal solid waste (MSW) incinerator ash management. With one waste-to-energy (WTE) facility in operation, two major projects under construction, several others in the planning stage, and with ash being imported from out-of-state, a sound ash management plan is essential if Michigan is to avoid confusion, controversy, expense, and environmental degradation. Despite possible action and clarification at the federal level, the Commission has determined that Michigan should not rely on federal activity alone to develop a definitive ash management plan.

Incineration of solid waste with energy recovery is slated to play a significant role in Michigan's comprehensive solid waste management plan. Mass burn incinerators produce approximately one ton of ash from every three tons of mixed trash burned. In-state WTE facilities slated to be operating by 1990 are expected to generate some 2000 tons/d of residue. Ash generation could be over 4000 tons/d shortly after the turn of the century.[1] Furthermore, the importation of additional ash, such as that currently coming into southwestern Michigan from the Chicago area, may increase this load still further.[2]

Despite the projection that within 5 years nationwide some 50,000 to 55,000 tons of ash will be generated daily, ash management currently could be characterized as chaotic.[3] The fate of WTE incinerator ash has come under increased scrutiny by federal and state regulators because of concerns for protection of public health and the environment. Attention has also been focused on ash due to widespread and vigorous public resistance to siting landfills for the disposal of incinerator residue. Unfortunately, no regulatory leadership or clear guidance has emerged from the U.S. Environmental Protection Agency (EPA), leaving it up to operators or individual states to decide the immediate fate of incinerator ash.[4-7] Even the EPA assistant administrator for solid and hazardous waste has been quoted as saying that EPAs approach is not totally rational but follows federal law as the Agency has interpreted it.[8]

Significant variations in management practices exist from state to state and from facility to facility. At the present time, most units mix fly ash and bottom ash prior to disposal. Much of this material is co-disposed with unburned trash in normal landfills or is used as daily cover, although there is a trend toward dedicated landfills (so-called monofills), particularly for new facilities. Some materials have been used for a variety of other purposes, including building aggregate, roadbeds, and fill.

The federal Resource Conservation and Recovery Act (RCRA) specifically exempts household refuse from consideration as a hazardous waste. However, the matter of exemption of residues from the incineration of such waste, particularly in consideration of the usual residential-commercial mixed waste stream, has been confusing and controversial. Deciphering the U.S. EPAs position has been difficult because of conflicting signals and continuing delays on issuing regulations or guidance.[9,10] Industry and municipalities have exerted pressure for a hazardous waste exemption, arguing that the designation is unnecessary, would overload already strained hazardous waste landfill capacity, and would limit WTE facilities due to exorbitant disposal costs. Apparently, the EPA has decided to let Congress decide the issue through further legislation; several bills had been pending, but are currently inactive.[11,12] It seems likely that under any such legislation, ash would be classified as a "special waste" to be governed under Subtitle D of RCRA (solid waste) rather than the more stringent Subtitle C (hazardous waste). Failing timely legislation, the courts may well decide the issue, at least on an interim basis.

II. CONCERNS ABOUT WTE RESIDUES

Residues from WTE facilities fall into a number of categories depending in part on the design of individual units: (1) bottom or grate ash; (2) fly ash, (3) solids from dry or semi-dry scrubbers; (4) water from wet scrubbers, quenching, heat transfer, or other operations; and (5) bypass materials. This report will focus on (1), (2), and (3), comprising roughly 10, 1, and 1% by volume of the incinerator feed, respectively.[13]

Incinerator ash contains chemical compounds which are of concern from the standpoint of protecting public health and environment.[13-17] For example, the concentration of TCDD equivalents in fly ash and scrubber residue may exceed action guidelines for soil by manyfold. Concentrations of cadmium, lead, and other heavy metals exceed natural soil levels by orders of magnitude and generally exceed levels found in urban airborne particulates. In addition, alkalinity can be very high in cases involving acid scrubbing equipment, rendering direct contact with living organisms problematic, and potentially causing leachate to be caustic. Bottom ash, while generally lower in extractable organic and inorganic components by unit mass, comprises a much larger fraction of the residuals. Thus, from a mass loading standpoint, these materials may be problematic as well.

A study sponsored by the EPA found known inorganic and organic toxicants in incinerator residues, with significantly higher levels of both generally found in fly ash compared with bottom ash.[16] The EPA Extraction Procedure (EP) Toxicity Test (40 CFR, Part 261, Appendix II), used for a wide variety of industrial wastes, has been applied to incinerator ash. This acidic extraction procedure was designed to simulate leaching conditions in mixed solid waste landfills. Virtually all fly ash and about one-third of the bottom ash samples failed the standard EP toxicity test due to extracted lead and cadmium levels.[16] Except for possible statutory exemptions, in those cases the ash would be deemed hazardous under RCRA.

A study of several WTE facilities conducted by the State of Washington confirmed unacceptable levels of lead and cadmium in incinerator ash as measured by the EP toxicity procedure.[14,18] Furthermore, Washington's more extensive testing and classification system found that three of five fly ash samples tested caused acute fish toxicity and classified those samples as an "extremely hazardous waste". The remaining two were termed "dangerous waste" by virtue of the possible presence of International Agency for Research on Cancer (IARC) listed carcinogens and/or actual lead and cadmium leaching levels. A Swedish study reported that several heavy metals leached out of ash in greater amounts than in untreated municipal waste; furthermore, metals mobility was found to be greater in incinerator ash than in residues from the combustion of coal.[13] Chloride levels, particularly in cases involving dry flue gas cleaning residues, were also identified as a matter of concern.[13]

These and other studies have led to widespread agreement among researchers, regulators, and vendors that incinerator ash contains components which require control over exposure to people and the general environment. The extent to which measures need to be taken during the handling, storing, transporting, and disposing of ash to avoid immediate or eventual exposure is a matter of considerable debate.[19,20] Classifying the hazards associated with a waste, particularly one as complex and variable as incinerator residue, is an imperfect science. Present methods in this case generally deal with the extractability (leachability) of selected metals and organic compounds. While there is a growing consensus that the use of the EP toxicity test points to the desirability of isolating ash from untreated refuse (i.e., to use monofills), few seem satisfied that the procedure is a reliable predictor of either toxicity or mobility of potentially problematic constituents.

Some vendors, municipalities, and regulators have argued that conditions in an ash monofill, including pH of leachate and impermeability, are such that mobility and resulting risk are grossly overestimated by the EP test. Furthermore, they reason that increased use of dry scrubbers (or addition of lime) and mixing fly and bottom ash constitutes a form of treatment, further insuring immobilization. Their conclusion is that isolating mixed ash in a monofill is a sufficient safeguard to keep harmful components from reaching groundwater and surface water supplies. Some maintain that even this measure is unnecessary.

Data from the EP test as applied to ash do indicate serious questions about its reliability. Test results on identical samples during interlaboratory comparisons have shown poor reproducibility, with variations reported of comparable magnitude to that of the regulatory limit.[20] Adding to this problem is great variability in ash composition itself from facility to facility, and even from hour to hour in the same facility.[21] Nonetheless, conclusions have frequently been drawn and regulatory decisions made from EP test results that are totally lacking in statistical validity.[20]

A number of other tests have been developed. The EPAs Toxicity Characteristic Leaching Procedure (TCLP)), the Monofill Waste Extraction Procedure (MWEP), and Sequential Chemical Extraction (SCE) procedure have been suggested as improved predictors. All have been subject to criticism. At least two bills introduced in congressional committees call for the development of an improved testing and classification procedure by the EPA.[22,23] While improvements in testing might be made, it is unlikely that any test could completely account for high variability in the waste stream and actual long-term leaching conditions in co-disposal or monofill facilities.

The remainder of this section examines findings relevant specifically to the trend (in some cases voluntary, other cases required) toward the practice of mixed ash monofilling, which offers the potential of simpler and more predictable leachate composition, decreased permeability, pH control, and smaller total volumes. However, there are early indications that monofilling, with or without extra precautions, is no panacea.

Higher pH resulting from inherent alkalinity and from acid scrubber residues or added lime does retard the leaching of many metals. However, amphoteric metals, such as lead, are quite soluble at high pH levels. Lead is extracted quite efficiently with plain water from highly alkaline residues, which calls into question the common assumption that high alkalinity assures immobilization. Furthermore, chloride and other complexes have also been implicated to have higher than expected mobility.[13,24] Changes in pH over time and the complex variations in solubility of both inorganic and organic species as a function of pH, particle size, and metal speciation render reliable predictions of leachability from monofills over time difficult or impossible to make.[19,24-27] Work in Sweden and Switzerland indicates that maximum leachate levels of various substances do not appear for a relatively long time.[13,24] Moreover, adoption of flue gas cleaning devices and continuing modifications and improvements assure that ash will continue to change in character by comparison to the bulk of ash now buried.

A study on leachate from the Sprout Brook ash monofill in New York provides an example of results available from an operating facility.[28] The pH and concentrations of suspended and dissolved solids, chloride, manganese, zinc, and other parameters in the actual leachate from this combined ash monofill deviated widely from those predicted by simulated leaching studies. Although short-term leachates generally exhibited contaminant levels lower than those obtained in the EP toxicity test, the significance of this observation is unclear. There simply is little or no detailed information on the long-term leaching behavior from such facilities and the possible effects of extended low-level contamination of ground water.

III. FINDINGS

First and foremost, the TSCC would like to emphasize that deliberations at all levels about the characteristics and fate of incinerator residues have focused far too narrowly on a *disposal* approach. This is the same mindset which the Commission identified in its broader policy recommendations on incineration and other management options.[29] The TSCC urges state and federal regulators and elected officials to renew their commitment to the priorities set forth in RCRA in the following order: reduce, reuse/remanufacture, recycle, treat, incinerate with heat recovery, landfill.

Elemental constituents entering the combustion process must also exit that process — whether as carbon dioxide, water, acid gases, metal salts, organic compounds, or other trace components. Efforts to clean the flue gas result in higher concentrations of metals and other species in captured residue; furthermore, adjustments in combustion conditions to minimize one problem may well create others (e.g., destruction of organic compounds vis a vis particulate size, metals speciation, nitrogen oxide production, etc.). While efficient combustion and pollution control improve the incineration process, the inherent materials and energy inefficiency of the approach cannot be overcome by improved technology or regulation. It must be recognized that even state-of-the-art incineration represents a concession to the convenience of current mixed municipal solid waste disposal. Massburn facilities require little or no change in waste collection, do not cause economic dislocations due to produce reformulation or bans, require little citizen involvement other than paying the oft-hidden bill, and their design, construction and management can be handled by the private sector. The question of whether this concession on the scale projected is necessary or desirable goes beyond the scope of this chapter, but it must be made clear that the concession comes with serious short- and long-term costs, many of them hidden.

The TSCC recognizes the reality of WTE as part of the State's waste management strategy. However, the Commission urges that higher priority actions be taken *prior* to incineration and landfilling, specifically, separation of reusable and hazardous materials from the waste stream. Reduction of heavy metal content in the waste feed prior to combustion is much sounder technically than subsequent attempts to avoid dispersal. Deposits on problem items (such as batteries, tires, etc.), product reformulation (e.g., printing inks, plastics, etc.), and other measures to divert metals and other problematic materials from the waste stream are essential to sound resource management. The TSCC recognizes that such measures will not happen overnight, but it is frustrated that real efforts toward sound, long-range materials management have been largely shunted aside as a concession to perceived practicality.

Considering the composition and other properties of fly ash and scrubber residue, it is improbable that such materials will be suitable for productive reuse. However, if reduction/diversion measures for toxic components are implemented, the likelihood of useful applications of bottom ash will be enhanced. Because bottom ash accounts for the large majority of MSW incinerator residues, safe reuse could increase materials efficiency, preserve valuable landfill space, and reduce disposal costs. Bottom ash has been used on a limited basis as aggregate for asphalt and concrete, for fill, and for other applications. While this approach is consistent with the Commission's materials management vs. disposal approach, it is premature to promote reuse at this time. Further removal of selected heavy metals from the initial waste stream and additional review and study of the long-term mobility of metals and other components are necessary.

Landfilling of ash rather than reuse (with or without treatment) will be the norm in the near future. Increasing the level of containment in monofills through design, maintenance, and monitoring should decrease but will not eliminate the risk of releases during active, closure, and postclosure periods, However, the Commission wants to re-emphasize that all disposal methods are temporary in the long view and may eventually compromise the environment. Long after operators and regulators have left a site, substances will remain which are undesirable in water, soil, or air.

After evaluating various approaches and options, the TSCC finds the present system of ash management unsatisfactory. Wide variations in results from the EP toxicity test and serious questions about its relevance render its use as a criterion for disposal in normal solid waste disposal facilities a stop-gap measure. The practice of co-disposal of ash with unprocessed trash seems ill-advised under these circumstances.

It is the opinion of the TSCC that states should not wait for nor be limited to minimum standards which may eventually be forthcoming from federal agencies. Suggested policies and draft legislation may well be inadequate, even if they were to be enacted immediately. All indications are that any federal guidance or regulations will create a "special waste" category for incinerator ash. The somewhat arbitrary classification of solid waste into one of two categories (hazardous or nonhazardous) is not necessarily protective of public health and the environment, particularly if significant volumes of what may be marginally hazardous materials are mixed with other vastly different wastes in the most secure but severely limited hazardous waste landfills. Therefore, whether ash is designated as hazardous or nonhazardous in the conventional sense and regulated under Subtitle C or D is less important than a sensible, comprehensive plan to detoxify and manage these materials in an appropriate manner.

Mixing of bottom ash and fly ash (and scrubber residue if present) has obvious practical advantages for operators and vendors. While the claimed immobilization benefits of such mixing are promising, they have not been adequately substantiated, nor has the problem of increased mobility of certain constituents been solved. The Commission is aware that the major effect (if not the purpose) of the practice of combining ash to date has been the dilution of fly ash sufficiently to classify the mixture as nonhazardous by the EP toxicity test. This practice runs counter to the state's intention of avoiding delisting-through-dilution, renders a much larger volume of waste more highly contaminated, complicates interpretation of test results, minimizes the chances of metals recovery/detoxification, and eliminates the prospect of using less-contaminated bottom ash for productive purposes. While mixing lessens the concentration of contaminants, it does nothing to alter the amount of hazardous constituents present in repositories and potentially available for release into the environment. The TSCC is reluctant to endorse the practice of mixing waste streams of such different chemical and physical properties without substantial evidence that this practice is in the best long-term interest of sound ash management. Swedish and Canadian researchers among others have also expressed this view.[13,30,31]

Although the trend toward monofilling is quite clear, the level of containment necessary is not agreed upon by scientists, regulators, industry, and environmentalists. Liners and leachate collection seem to be emerging as a *minimum* requirement. A variety of other approaches has emerged which employ more stringent standards. For example, the State of Washington has taken an active, comprehensive approach to incinerator ash management.[32,33] The proposed regulations provide incentives for separate handling and solidification of fly ash and delineate a number of alternatives involving combinations of siting and engineering considerations (e.g., linkage of engineered component requirements with natural hydrogeological conditions). In addition, a comprehensive approach to handling, storage, testing, transportation, etc., is specified. While most attention nationally has been focused on groundwater protection, Washington's approach as been much more circumspect in examining and controlling other possible exposure routes as well (e.g., fugitive dust emissions).

The Commission also notes its concern for managing other ash-type residues, including coal ash. Exemptions have been granted for such residues largely on the basis of practical, economic, and technical grounds. The TSCC supports state legislation which would increase study of these materials and provide for increased regulation if necessary. An exclusive focus on incinerator ash to the exclusion of other virtually unregulated residues would not serve to adequately protect human health and the environment.

Finally, a word about costs. It is obvious that some, if not most, of the measures outlined here involve identifiable costs. Nonetheless, some painful lessons are being learned about the actual, long-term consequences of expedient and "low-cost" waste disposal practices. For example, we now realize that "cheap" methods of land disposal of the past were short-sighted bargains which will cost billions of dollars to correct. Societal costs of corrective actions, long-term monitoring and maintenance, and degraded resources have been largely externalized in traditional judgment of waste disposal options.

Modern landfills and incinerators only partially address waste disposal problems; they continue to rely on *containment* of problem components of waste and waste by-products. Isolation of materials from the general environment in perpetuity is something that is not only technically problematic but also of significant economic cost. It is long overdue that economic analysis and feasibility determinations take these costs explicitly into account and that they be made part of the up-front cost of waste disposal. When total and long-term costs are acknowledged and made part of user fees or surcharges, substantial funding becomes available to pay for source reduction, materials recovery, toxicant removal, and other measures which represent true solutions to waste problems rather than deferral.

IV. CONCLUSIONS

1. While efficient combustion and pollution control improve the incineration process, the inherent materials and energy inefficiency of the approach cannot be overcome by improved technology or regulation. Thus, incineration should be used selectively, only after reusable and problematic materials have been separated from the waste stream.
2. Deliberations at all levels about the characteristics and fate of incinerator residues have focused far too narrowly on a *disposal* approach. Too little attention has been given to reduction/removal of problem materials from the waste stream prior to incineration, to extraction/immobilization of residues, and to possible use of detoxified residues.
3. While careful site selection, design and operation are positive steps, landfilling options must be considered temporary in the long view. Solidification/immobilization measures are yet unproven but may aid in the isolation of residues containing high concentrations of problem components.
4. Current sampling and testing procedures for incinerator residues are inadequate. Inherent limitations due to waste stream heterogeneity and dependence of ash characteristics on incinerator operating conditions will render improvements difficult.
5. While expedient, the mixing of the fly ash with bottom ash may be ill-advised due to significant differences in their quality and quantity. Claims made for this practice as a treatment process have not been adequately substantiated, and the negative features could well outweigh the positive.
6. States should not wait for nor be limited to regulations or guidance which may eventually emerge at the federal level. With ash currently being generated within an increasing number of states and with ash frequently being transported across state boundaries, expeditious state regulations are needed.
7. Coal ash and similar residues have been granted certain regulatory exemptions based on practical, economic, and technical grounds. It is logical to subject such residues to further study and to provide increased regulation, if necessary.

V. RECOMMENDATIONS

In formulating state policy and regulations, the following recommendations should be considered:

1. First and foremost, *serious* attention should be given to keeping problematic substances out of incinerator feed. Incentives, separation and removal requirements, product reformulation, deposits, chemical and material composition audits, and other methods should be mandated as needed to remove toxic metals and other problematic materials from the waste stream.
2. Treatment of ash, particularly fly ash, through metals extraction or immobilization/solidification should be given priority over attempts which rely solely on isolating ash from the general environment through burial. Fly ash and scrubber residue should be managed separately from bottom ash unless and until it can be clearly shown that mixing constitutes a better rather than simply more expedient management method.
3. The study and development of safe, long-term uses of ash should be encouraged and supported.
4. Specific ash management plans should be required of all generators and submitted for review and approval.
5. The collection, storage, handling, and transportation of incinerator ash should be regulated to minimize human exposure and dispersal of contaminants into the air, soil, and surface water.
6. Separate and regular sampling and testing of incinerator fly and bottom ash should be required to increase the technical database for future regulatory improvements and for measuring progress in toxicant reduction in the waste stream.
7. A standard record-keeping system should be instituted which includes, at a minimum, amounts of waste incinerated, amounts of fly and bottom ash generated, the location and disposal/treatment methods used, and the test results on residues.
8. Design and siting standards for ash landfills should be specified. Single and double composite liner systems, leachate collection and treatment, leak detection, monitoring wells, cap design, and closure/postclosure requirements, etc. will need to be considered and requirements developed. Ash character, including separation, treatment, or preincineration toxicant reduction might beneficially be linked to siting criteria and/or design and monitoring requirements.
9. Some form of positive financial assurance should be required of generators/operators to assure proper operation, closure, and postclosure maintenance and monitoring.
10. Investigation of coal ash and similar exempted residues should be conducted simultaneously, and regulations developed if potential threats to the public health and environment are discovered.

REFERENCES

1. State of Michigan, Solid Waste Policy, Office of the Governor, June, 1988.
2. Michigan Waste Report, Michigan Waste Report, Inc., Grand Rapids, 8, 74, 1988.
3. Based upon figures provided by National Solid Waste Management Association, 1986, and from an industry survey conducted by *Newsday,* 1987.
4. Permit to Operate a Municipal Solid Waste Management Facility, Project 20-85-0306, New York Department of Environmental Conservation, Albany, 1988.

5. **Goldman, M.,** Waste-to-Energy Ash: Hazardous or Non-Hazardous Waste, presented at Air Pollution Control Association Annu. Meet., Dallas, June 20-24, 1988.
6. **May, R.,** Department of Environmental Protection, State of Connecticut, personal communication, 1988.
7. **Knudson, J.,** Department of Ecology, State of Washington, personal communication, 1988.
8. **Porter, J. W.,** The rush to burn — America's garbage gamble, special reprint from *Newsday*, p. 30, 1988.
9. **Tonetti, R.,** EPA's Views on Municipal Waste Combustion Ash Management, presented at 4th Annu. Conf. Solid Waste and Materials Policy, New York, January 28, 1988.
10. **Watson, E. B.,** Incineration of Municipal Solid Waste: Toxicity of Ash Residue, Legislative Science Office, Lansing, MI, 1987.
11. **McManus, F.,** *Resource Recovery Rep.*, 12, 1, 1988.
12. **Herz, M.,** Environmental Defense Fund, testimony before the House Committee on Energy and Commerce Subcommittee, April 13, 1988.
13. **Hartlen, J. and Elander, P.,** Residues from Waste Incineration—Chemical and Physical Properties, Swedish Geotechnical Institute, Linköping, 1986.
14. **Knudson, J. C.,** Study of Municipal Incineration Residue and Its Designation as a Dangerous Waste, State of Washington Department of Ecology, Olympia, 1986 (as revised, 1986, 1987).
15. Environment Canada, National Incinerator Testing and Evaluation Program, Burlington, Ontario, 1986-87.
16. U. S. Environmental Protection Agency, Office of Solid Waste, Characterization of Municipal Waste Combustor Residues, Volume V of Characterization of Municipal Waste Combustor Ashes and Leachates from Municipal Solid Waste Landfills, Monofills, and Codisposal Sites, U.S. Government Printing Office, Washington, D.C., 1987.
17. Environmental Defense Fund, Summary of Available Extraction Procedure Toxicity Test Data for Lead and Cadmium from MSW Incinerator Ash, Washington, D.C., 1988.
18. **Knudson, J. C.,** Metal Cations in Solid Waste Incinerator Residues, presented at GRCDA Solid Waste Symp., Seattle, April 23, 1987.
19. **Denison, R. A.,** Fundamental Objectives of Municipal Solid Waste Incinerator Ash Management, presented at Air Pollution Control Association Annu. Meet., Dallas, June 19-24, 1988.
20. **Shaub, W. M.,** Incinerator Ash—Some Things I Would Like to Know About and Why, presented at Air Pollution Control Association Annu. Meet., Dallas, June 19-24, 1988.
21. **Schmidt, J.,** New York State Department of Environmental Conservation, presented at Air Pollution Control Association Annu. Meet., New York, June 23, 1987.
22. **Luken, T.,** Analysis of Municipal Solid Waste Incinerator Act of 1988, House of Representatives, Washington, D.C.
23. Environmental Defense Fund, Ash Provisions of H.R. 4902, Washington, D.C., 1988.
24. **Brunner, P. H. and Baccini, P.,** The Generation of Hazardous Waste by MSW-Incineration Calls for New Concepts in Thermal Waste Treatment, presented at 2nd Int. Conf. New Frontiers for Hazardous Waste Management, Pittsburgh, September 27-30, 1987.
25. **Sawell, S. E., Bridle, T. R., and Constable, T. W.,** Leachability of Organic and Inorganic Contaminants in Ashes from Lime-Based Air Pollution Control Devices on a Municipal Waste Incinerator, presented at Air Pollution Control Association Annual Meeting, New York, June 21-26, 1987.
26. U.S. Environmental Protection Agency, Report of the Environmental Effects, Transport, and Fate Committee, Evaluation of Scientific Issues Related to Municipal Waste Combustion, Final Report, U.S. Government Printing Office, Washington, D.C., 1988.
27. **Karasek, F. W., Charbonneau, G. M., Revel, G. J., and Tong, H. Y.,** Determination of organic compounds leached from municipal incinerator fly ash by water at different pH levels, *Anal. Chem.*, 59, 1027, 1987.
28. **Malcolm Pirnie, Inc.,** Westchester County Sprout Brook Residue Disposal Facility, Leachate Physical and Chemical Parameters (Table 1), 1987.
29. Michigan Toxic Substance Control Commission, Policy Recommendations on Waste Incineration and Alternatives, Lansing, 1987.
30. **Constable, T. W.,** Environment Canada, personal communication, 1988.
31. **Sawell, S.,** Assessment of Ash Contaminant Leachability, NITEP, Phase II, Environment Canada, Burlington, Ontario, 1986.
32. Report on Ash Management Standards by the Department of Ecology to the State of Washington Legislature, March 1988.
33. Department of Ecology, State of Washington, Special Incinerator Ash Management Standards, WAC 173-306, Draft 4, 1988.

Chapter 15

COMPREHENSIVE MANAGEMENT OF MUNICIPAL SOLID WASTE INCINERATION: UNDERSTANDING THE RISKS

Richard A. Denison and Ellen K. Silbergeld

TABLE OF CONTENTS

I. INTRODUCTION

The management of municipal solid waste (MSW) — including both household and commercially generated waste — has until recently been a neglected subject in environmental management. As a result, current policies and practices often succeed only in shifting risks from one release point to another, and in some instances may actually increase the hazards associated with waste disposal.

For understandable reasons, national policy and debate has focused almost exclusively on redressing and preventing the mismanagement of the subset of wastes legally defined as hazardous. In the last decade, however, decisions about hazardous waste management, including its definition, have implicitly shaped the dimensions of the present debate over legally nonhazardous municipal waste.

Hazardous waste management policy centers on the recognition that both the inherent properties of an individual waste stream (e.g., toxic metal content or physical form) and its method of management determine its degree of hazard, and that addressing both aspects is critical to reducing such hazard. While these principles logically extend to MSW, the major methods currently used to manage this waste (landfilling and incineration) do not account for its heterogeneous nature, nor do they effectively reduce the hazards associated with its management.

Indeed, we are belatedly coming to realize that we have done little systematically to cope with managing MSW. City after city claims (sometimes legitimately, sometimes not) to be confronted with a management problem of crisis proportions, and decision making within these constraints is often considerably less than reasoned or deliberate. The perceived pressure to make immediate decisions, often in the face of diminishing landfill capacity, makes it increasingly difficult to persuade decision makers to consider long-term, efficient strategies of comprehensive management over admittedly attractive offers of short-term and incomplete (albeit often highly expensive) solutions to the immediate crisis.

The most frequent casualty in this atmosphere of charged debate is the appreciation of the nature and seriousness of the environmental and public health issues involved. Too often, one criticism is played off against another, and one proposed technological solution is arrayed against another without considering the alternative, indeed, the necessity, of selecting and combining options from a menu of feasible methods for waste management and disposal.

The major challenge is how to resolve issues of risk to the environment and public health while dealing effectively and expeditiously with MSW. It is the purpose of this chapter to demonstrate that a major impediment to this resolution has been a failure to comprehensively assess risks associated with various options, particularly with incineration as a method of waste processing.

Given the present lack of resolution of such major risk-related issues, the extent to which incineration can be deployed to serve waste management is still questionable. It will be unfortunate if we are unable to optimize use of incineration as a waste management tool, since it may provide useful advantages if implemented as part of a comprehensive waste management program.

Incineration, while increasingly adopted or proposed as the method of choice for dealing with MSW, is widely perceived as risky and remains highly controversial. For example, recent evidence indicates that use of incineration to manage the entire municipal waste stream frequently transforms it into an ash residue that possesses characteristics of a hazardous waste. Health and environmental risks, as well as the extraordinary expense, of properly managing this toxic by-product should give rise to caution in evaluating the utility of indiscriminate (or "massburn") incineration as a means of waste management.

Despite the clearly perceived risks of incineration, the major focus of the risk debate is only beginning to encompass the full range of risks which this technology can pose. Too frequently, public or regulatory concern has focused on halogenated dibenzodioxins and dibenzofurans to the exclusion of other toxic constituents of incinerator by-products, or on cancer to the exclusion of other adverse health effects. In addition, incinerators have primarily been characterized as stationary sources of toxic air pollutants, that is, with reference to their impact upon ambient air quality. Only recently have risk analyses for proposed incinerators begun to quantitatively assess pathways of exposure to air emissions in addition to direct inhalation. An additional significant limitation is the lack of a sufficient quantity and quality of information needed to support comprehensive risk assessments. Perhaps the most graphic illustration of the lack of data that would provide for a serious consideration of incinerator-associated risks from sources other than direct air emissions is the fact that, to date, a quantitative risk assessment of incinerator ash has never been conducted for a proposed incinerator project.

This scope of debate is misleading and dangerous. An inaccurate definition of risk encourages inadequate and misdirected strategies for control. The concept of incinerator-associated risks as limited to a few toxic air pollutants fails to comprehend the full range of risks, with respect to the panoply of toxic substances involved and the variety of relevant exposure routes, or our ability to moderate those risks through control over the role that incineration plays in overall waste management.

The perspective of this chapter is to discuss the serious and largely neglected public and environmental health considerations associated with MSW incineration, with an emphasis on the need to consider such risks in a comprehensive manner. Indeed, it is our contention that risk issues, not the apparent advantages of incineration as a tool for waste management, have emerged as the central focus of the debate concerning this technology.

Moreover, the appropriate consideration of risk issues includes a comprehensive assessment not only of the individual impacts of a specific new source such as a MSW incinerator, but more importantly, consideration of this source in the context of ongoing exposures to other environmental releases of pollutants. This perspective is particularly critical in evaluating incineration, because two of the major toxins released by incinerators, lead and dioxins, are persistent toxins with universal exposure. All persons have been exposed to and carry measurable levels of these pollutants in their bodies. In some populations, existing levels of exposure and body burdens are already in a range associated with detectable adverse impacts. Because of this, the evaluation of incremental inputs, even if they are apparently small when considered in isolation, must be of concern. Consideration of the cumulative nature of both exposure to and the health effects induced by many incinerator-associated pollutants may be critical to the siting of such facilities.

II. INCINERATION AS A STAGE IN WASTE MANAGEMENT

An MSW management system comprises a series of discrete activities that usually begins with the discard of used consumer goods and continues with their collection and transfer, recycling or recovery, processing, and, finally, disposal. Viewed in such a systems context, it is critical to recognize the limitations as well as advantages of incineration. It is altogether incorrect to consider incineration a method of waste disposal. Incineration is a processing technology, which provides the important benefit of reducing the amount (particularly the volume) of waste. Incineration does nothing, however, to alter the process of waste collection and transfer, and it leaves behind its own substantial burden of residues that must be managed and disposed of properly. Moreover, while incineration is well suited to certain types of materials, it is far less appropriate for handling without discrimination the mass of consumer and commercial products that comprise MSW. In particular, incineration can greatly enhance the mobility and bioavailability of toxic metals contributed to MSW by many consumer goods.

Consider in the broadest sense a MSW management system that includes incineration as a processing step. Waste enters the system from both producers and consumers of products that are distributed through the market system. The composition and physical forms of these goods are derived from chemical engineering and market decisions made at the level of primary production (e.g., mining or harvesting) and at intermediate stages of processing and packaging. Following their use, most of these goods are discarded, as only a small fraction is presently being retrieved and reused.

Materials not retrieved at this early stage require collection and delivery to central locations for further handling. These transfer stations may store waste or deliver it directly to interim processing or final disposal sites. Interim processing centers offer further opportunities for reclamation or separation, or can simply process the waste using other reduction technologies.

Following these opportunities for recycling, compaction and incineration are the only major options for further volume reduction. Compaction can be achieved during collection (through the operation of collection vehicles), during processing (through shredding, crushing, or baling methods), and during final disposal (through landfill compaction).

Incineration is a long-standing method of volume reduction. Incinerators may receive all or a fraction of the collected waste or a processed form of the waste (e.g., refuse-derived fuel, or RDF). It is critical to recognize that incineration can serve only an interim processing function and is not the final stage of MSW management.

The interaction of incineration with other strategies for waste management is of particular concern, since choices made at the waste processing stage may have a major impact on the options available at other points in the system. Without careful advanced planning, the use of incineration may limit other options for waste management (e.g., recycling or source separation) that are either less expensive overall or could optimize the efficient and safe use of incineration. Indeed, present or future opportunities to recycle or recover components of MSW, management options that are rapidly gaining in both political and economic acceptance, may directly conflict with the contractual arrangements and operating requirements of incinerators that were sized or designed without sufficiently accounting for such opportunities.

Many existing incinerator contracts require that a municipality or region guarantee a minimum tonnage of waste for delivery to the incinerator, often for the life of its operation. Clearly, incentives to reduce the amount of waste generated through changes in consumption patterns or to implement more aggressive recycling programs may be compromised or entirely eliminated by such long-term contractual obligations.

In addition, alteration of the waste input to an incinerator through recycling or other means can directly affect the efficiency, safety, and even the economy of incineration.[1] Most massburn incinerators have restricted operating ranges which delineate minimum waste quantities and calorific values under which the incinerator will operate optimally. Failure to meet these waste feed requirements can result in less efficient combustion and a greater likelihood of generating products of incomplete combustion such as dioxins. Alternatively, supplemental fuel such as oil or gas may be required to maintain proper combustion, which would turn these resource recovery facilities into resource requiring operations, and could significantly increase their operating costs.

These and other considerations discussed later suggest that incineration must be viewed as only one component in an integrated MSW management system. Decisions to employ incineration, including both the scale of its use and the particular form of incineration technology used, must be preceded by a thorough evaluation of all options for volume reduction, including methods presently available as well as those likely to become available during the life of the incinerator. These other volume reduction options — source reduction, separation, and recycling — should be viewed as serious waste management tools. Opportunities for their use must be maximized, not only to reclaim reusable materials, but also to reduce the amount of waste that must be incinerated or otherwise managed, and to increase the overall safety and efficiency of MSW management. Finally, risk considerations dictate that we examine alternatives to the use of toxic metals at the production stage as a critical element in designing an effective, long-term MSW management strategy.

III. PRODUCTS OF INCINERATION

Incineration yields products which take three forms: energy, gases, and solid residues. Energy (heat) can be recovered partially and used to produce steam for heating or electricity generation; however, this process is not related to the other products of incineration. In modern incinerators, gases exit almost exclusively via the stack, with or without prior conditioning before discharge into the atmosphere. The solid residue emerges from two points within most facilities: at the bottom on the grates, where it is called bottom ash, and at points beyond the combustion chamber, where it is called fly ash. Fly ash may fall out in the boiler or economizer or at points of gas conditioning, or it may be deliberately collected by particle trapping devices, such as electrostatic precipitators (ESPs), fabric filters, or scrubbers.

Incineration differs from other methods of waste management, such as compaction and direct landfilling, in that major portions of the waste stream are physically and chemically transformed during the combustion process. Moreover, the products of this process, both solids and gases, differ markedly from the original waste in their environmental and biological behavior as discussed later.

By way of illustrating these differences, consider the various points of potential release of incinerator by-products. At the facility site itself, opportunities for release of the solid products arise from the moment of their generation, and continue during onsite management, handling, storage, and transport of ash.[2] Further points of release may arise both during disposal (e.g., dust generation during landfilling; runoff or wind dispersal from uncovered ash) and after disposal (e.g., accidental or deliberate discharge of leachate or disposal of leachate treatment residues). Many or most of these pathways of exposure to waste products are unique to, or far more significant for, incineration than for other forms of waste management.

Consideration of potential release points demonstrates another fundamental point: impacts of incineration on air quality and the nature and amounts of solid residue generated are inextricably linked. In order to condition stack emissions to meet specific health and regulatory goals, modern incinerators are increasingly required to be equipped with such air pollution control technologies as fabric filters, high-efficiency ESPs, and scrubbers. Each of these control devices, however, increases the amount and changes the nature of the solids retained. The association between air quality controls and ash quality is shown in Table 1, taken from studies in the National Incinerator Testing and Evaluation Program (NITEP) in Canada.[3,4] These data show that as more effective controls are imposed upon stack emissions, the quality of fly ash changes markedly. In particular, both the concentration and leachability of several toxic metals in the fly ash increase.

Similarly, controls over the combustion process that increase combustion efficiency and reduce emissions of unburned substances can also affect ash quality. Recent data indicate that fly ash-bound dioxins become more bioavailable to fish as the organic carbon content of the ash decreases.[5] Hence, reducing dioxin air emissions through better combustion and stack controls may still yield an ash that poses a greater risk of dioxin exposure.

Failure to understand the essential linkage between air quality control and ash toxicity results in risk transfer, a practice which typically characterizes incinerator operations at present. It is not acceptable to purchase improvements in air emissions at incinerators at the cost of increasing risks associated with uncontrolled or insufficiently regulated disposal of ash.

IV. BEHAVIOR OF METALS IN INCINERATION

Quantitatively, the compounds of greatest concern for environmental and health impacts are the toxic heavy metals, specifically lead, cadmium, arsenic, mercury, selenium, and beryllium. Other metals which have been identified in incinerator emissions and ash are gallium, thallium, zinc, copper, aluminum, and manganese.[6-10] This discussion focuses primarily upon lead,

TABLE 1
Total and Leachable Metals in MSW Incineration Fly Ashes from Various Air Pollution Control Devices[a]

	Wet-dry scrubber ash	Dry scrubber ash	Fabric filter ash
Total metals (μg/g ash)[b]			
Cadmium	37	63	299
Copper	652	666	538
Lead	1919	3792	7055
Zinc	5154	9391	13695
Percent of total metals leachable in the short term[c]			
Cadmium	62%	68%	94%
Copper	5%	16%	41%
Lead	25%	56%	75%
Zinc	32%	62%	17%
Percent of total metals leachable in the long term[d]			
Cadmium	62%	83%	100%
Copper	30%	42%	69%
Lead	63%	80%	95%
Zinc	49%	73%	54%

[a] Ash from Quebec City municipal waste incinerator using a pilot-scale air pollution control plant.
[b] Average of six (wet-dry scrubber and fabric filter) or four (dry scrubber) samples. Units are micrograms per gram of ash (μg/g).[3]
[c] Sum of fractions A and B in the Sequential Chemical Extraction (SCE) Procedure.[4]
[d] Sum of fractions A, B and C in the SCE Procedure.[4]

cadmium, and mercury for two reasons: first, there is considerable information available on the health and ecotoxicological effects of these three metals, and second, their presence and behavior in incinerator air emissions and ash residues are exemplary of the problems posed by this technology.

Metals, being basic chemical elements, are neither created nor destroyed by incineration. Their amounts in the waste stream before incineration are identical to those found in air emissions and ash after incineration. However, the presence of toxic metals in the waste being combusted threatens our ability to safely employ incineration, since these metals are particularly difficult to manage once they have been transferred into incineration by-products. In contrast to the potential to reduce the organic component of these by-products through better combustion, there are no alterations of the incineration process itself (e.g., increasing temperature or changing incinerator design and operation) that can alter the output of metals.

What is critical for purposes of risk assessment is to understand that incineration is uniquely unsuited for managing metals. Incineration essentially destroys the bulky matrix of MSW (paper, plastics, or other materials) which contains metals and which acts to retard their entrance and dispersion into the environment (see, for example, Reference 11). In this respect, incinerators can be compared to secondary metal smelters: the burning of combustible materials liberates metals, which are subsequently released in air emissions or ash residues in a highly bioavailable form.

The mobilization and enhanced bioavailability of metals in incinerator waste products arises from several phenomena associated with combustion. First, several toxic metals volatilize during combustion and then condense onto the surface of fly ash particles. The concentrations of these metals on fly ash particulates increase with decreasing particle size.[12-16] (Any dioxins or other products of incomplete combustion formed during combustion will behave similarly.) A large fraction of these particles are of respirable size (less than 10 μm in diameter) and can also be readily ingested. Hence, they pose a risk of exposure whether they exit the stack and directly enter the ambient air, or are captured by particle control devices and managed in a manner that does not prevent subsequent entrainment into air.

Moreover, the small size of released particles actually promotes both short- and long-range airborne dispersion, as has been demonstrated for MSW incinerators[2,17] and is well known for metal-containing particles released by other stationary and mobile sources.[18-21]

Second, metals may also be released as fumes, depending upon exit gas temperature. As noted in Swedish and Canadian studies of the behavior of metals in MSW incinerators,[3,8,9] several metals, in particular, mercury, exist largely in the vapor form and escape out the stack unless gas scrubbing equipment is present and properly conditions the combustion gases prior to release. The transformation of metals into the vapor as well as particulate phase greatly increases the difficulty of their safe management.

Third, the physical and chemical states of metals sorbed onto fly ash particles enhance their leachability, for several reasons. Metals are present at or near the surface of such particles[13,22] and are therefore more available for leaching. In addition, the small particle size increases the surface area exposed to the leaching medium.[12] Finally, the high chlorine content of MSW results in the transformation of many toxic metals from their original chemical forms into metal chloride complexes,[8,23,24] which are generally much more soluble in water. A large and growing body of data employing several leaching tests, including the federally mandated extraction procedure (EP) toxicity test, demonstrate that toxic metals, lead and cadmium, in particular, leach from ash at levels that frequently exceed federal standards defining a hazardous waste. These data are discussed in the section on hazards of ash mismanagement.

V. HEALTH EFFECTS OF METALS MOBILIZED BY INCINERATION

Many of the heavy metals present in MSW incinerator by-products have well-defined health effects that have been demonstrated in numerous studies of exposed populations. The toxicity of heavy metals has been exhaustively reviewed, most recently by Friberg et al.[25] Many of the heavy metals are carcinogenic, but they can also exert a broad spectrum of devastating neurological, hepatic, renal, hematopoietic, and other adverse effects, both in humans and in other biota. Arsenic, cadmium, beryllium, and lead are carcinogenic metals; arsenic, lead, vanadium, cadmium, and mercury are neurotoxic; zinc, copper, and mercury are acutely toxic to aquatic life.[35,38]

Data on metal toxicology comes directly from a large body of clinical and epidemiological studies. Hence, objections often raised to extrapolation of risks (usually cancer risks only) from animal models to humans do not pertain to these cases. Indeed, risk assessments for these metals can be and have been done using human data exclusively (see, for example, Reference 26).

Because of their elemental (and therefore, permanent) nature, heavy metals accumulate both in environmental compartments and within the human body. Thus, long-term releases even at low levels have the potential to increase substantially metal levels in certain environmental compartments (e.g., surface dusts) and humans. The strong correlation in the U.S. between automobile lead emissions and body lead burdens[27] demonstrates how individually small but widely dispersed releases can have a significant impact on general population exposure.

Lead is a neurotoxin at very low doses. Its effects on prenatal neurological development are probably without a threshold, that is, the effects occur at any dose.[28-31] Moreover, given the continuing status of general population exposure to lead in the U.S. associated with past uses and failure to remediate particular instances of contamination, even marginal increments in exposure are of major public health concern.[32] Table 2, taken from national epidemiological data compiled between 1976 and 1980,[27,33] shows the distribution of lead levels in the U.S. population. It can be seen that a small increment in exposure, sufficient to raise median blood lead levels by 1 to 2 μg/dl, would drastically increase the prevalence of overt lead toxicity,[34]

TABLE 2[33]
Blood Lead Levels of Persons Six Months to 74 Years Old, with Mean, Standard Error, and Percent Distribution by Race and Age, U.S., 1976—1980[a]

	Estimated population (thousands) (1978)	Blood lead level (µg/100 ml)		Percent distribution				
		Mean	Standard error	<10	10-19	20-29	30-39	>39
All races								
All ages	203,554	13.9	0.24	22.1	62.9	13.0	1.6	0.3
6 mo—5 yr	16,852	16.0	0.42	12.2	63.3	20.5	3.6	0.4
6—17 yr	44,964	12.5	0.30	27.6	64.8	7.1	0.5	—
18—74 yr	141,728	14.2	0.25	21.2	62.3	14.3	1.8	0.4
White								
All ages	174,528	13.7	0.24	23.3	62.8	12.2	1.5	0.3
6 mo—5 yr	13,641	14.9	0.43	14.5	67.5	16.1	1.8	0.2
6—17 yr	37,530	12.1	0.30	30.4	63.4	5.8	0.4	—
18—74 yr	123,357	14.1	0.25	21.9	62.3	13.7	1.8	0.4
Black								
All ages	23,853	15.7	0.48	13.3	63.7	20.0	2.3	0.6
6 mo—5 yr	2,584	20.9	0.61	2.5	45.4	39.9	10.2	2.0
6—17 yr	6,529	14.8	0.53	12.8	70.9	15.6	0.7	—
18—74 yr	14,470	15.5	0.54	14.7	62.9	19.6	2.0	0.9

[a] Numbers may not add to total because of rounding.

especially in light of recent findings of effects in infants and young children with blood lead levels as low as 10 μg/dl.[28] Failure to understand the incremental nature of lead poisoning and current exposure levels will blind us to the real risks of a major new potential source of lead in the environment.

Cadmium is also neurotoxic and can cause lung damage (e.g., emphysema) and kidney damage as well; it is a carcinogen and mutagen, and may also be toxic to the fetus.[35] Exposures to cadmium are also widespread in the U.S., although documentation of clinical disease has as yet been confined to occupational populations. Lead and cadmium together may be synergistic, particularly as toxins to the bone.[36] Mercury in its organic forms is also neurotoxic, particularly to the fetus.[35] Since microorganisms in the environment can alkylate inorganic mercury species,[37] the release of mercury salts and oxides from incinerators must be evaluated in light of their potential conversion to methyl mercury.

Other metals present in incinerator by-products also pose significant human health risks.[35] Arsenic is a human and animal carcinogen, as well as a neurotoxin. Thallium is a systemic poison, and gallium has been identified as a potent reproductive toxin. Additionally, mercury, arsenic, cadmium, zinc, and copper are toxic to aquatic organisms.[35,38] Hence, in addition to human health concerns, the release of metals from incinerators and their accumulation in aquatic compartments pose substantial risk to the environment.

VI. RELEVANT ROUTES OF EXPOSURE

As noted above, it is critical to sound risk assessment to expand consideration of incinerators beyond their definition as stationary air pollution sources. While air emissions of a wide range of pollutants are at levels which require control using readily available technology,[6] comprehensive exposure assessment must go beyond calculations of direct inhalation exposure. Given the respirable size of fly ash captured by or escaping the particulate controls systems, direct inhalation and uptake of incinerator emissions by both pulmonary and nasal portals are obviously significant potential exposure routes. However, particularly for metals, it is clear that the majority of opportunities for human contact occur *after* metal-bearing particulates are deposited on soils, surface water, food, and dust, or accumulate in the buildings where people live or work. Postdeposition exposure to lead has been demonstrated to be 5 to 50 times more intense than exposure to metals in ambient air.[18,39-42] These studies demonstrate the importance of postdeposition contamination of dust and surface soils as vectors of exposure resulting in toxicity.[41-44] Contamination of surface soils by deposition of incinerator emissions has also been demonstrated. High levels of dioxins in soils were detected in the vicinity of an Italian MSW incinerator, resulting in its shutdown.[17]

The predominance of postdepositional exposure reflects two factors: (1) the continuing presence and accumulation of metals in environmental sinks following deposition, as compared to their relatively short lifetime suspended in air, and (2) the opportunities for intensive direct contact with deposited particles, particularly by children. Postdeposition exposure involves contact with dusts, surface soils, re-entrained particles, contaminated food, and drinking water. Risk assessments are only beginning to model human exposure for all these routes. Unfortunately, the tendency to consider exposure only via inhalation of ambient air results in an extremely limited and potentially misleading estimate of both human exposure and risk.

Only rarely have risk assessments for proposed MSW incinerators discussed the equally important and diverse pathways of exposure to ash residues and other incinerator waste products (e.g., quench water). For these wastes, exposure assessments must consider not only ultimate disposal, but all earlier steps: onsite handling from the time of generation, storage, transport, and handling and depositing at the landfill until time of final cover (see, for example, Reference 2). Moreover, postdisposal exposure can occur as a result of direct ground or surface water contamination by leachate — whether as a result of the deliberate discharge of leachate, failure of the leachate collection system, a breach in containment systems, or the lack of maintenance of such systems that will inevitably follow the end of any required postclosure period. Studies have documented that under a range of circumstances, ashborne metals can have significant mobility in soils,[45,46] emphasizing the need for thorough long-term containment of both ash and leachate.

Exposure assessment must also account for exposures to ashborne toxins that may occur long after initial disposal. The potential for erosion of the final cap over time or for transport of metals out of the landfill by plant uptake or other means must be considered as well. Several studies have demonstrated that metals in incinerator ash can be taken up from ash-amended soil by plants.[45,47] Anticipated end uses following closure of landfill sites (e.g., as recreational areas) may provide additional plausible exposure routes. Given the permanent nature of metals and the impermanence of all containment systems, exposure modeling must clearly be long-term in perspective.

EPA has begun to develop a reasonable methodology for estimating health risks from multiple-pathway exposure to incinerator air emissions[48] but has yet to consider the even larger set of pathways relevant in considering exposure to other incinerator waste products.

VII. HAZARDS POSED BY ASH MISMANAGEMENT

Foremost among the underexplored risks of MSW incineration are the hazards posed by the routine presence of high levels of several toxic metals in ash residues. Ironically, as previously discussed, the growing use of more efficient air pollution control devices on modern incinerators results in ash containing even higher levels of these metals in even more bioavailable forms. In addition to metals, highly toxic dioxins have been detected in all samples of incinerator fly ash tested, in some cases at levels that greatly exceed government standards.[10,78] While dioxins appear to be lower in fly ash from newer facilities, the ability of these facilities to consistently achieve acceptably low levels over their entire life span remains to be demonstrated. As is the case for metals, more efficient pollution control devices will act to increase the concentration of dioxins detected in ash residues.

Several measures of the risk posed by incinerator ash are germane to this discussion. The most fundamental and important measure of risk is the *total* toxic metal (and dioxin) content of ash, given the potential for direct inhalation and adsorption into the lung or direct ingestion of metal-laden ash particles. Indeed, a full accounting of the hazards of ash posed during all phases of its management requires knowledge of its total chemical composition.

Table 3 compares typical concentration ranges of several toxic metals in MSW incinerator fly ash to those found in natural soils, illustrating their extreme enrichment in this waste. Less extreme but still significant metal enrichment characterizes bottom ash. The total metal content of incinerator ash is comparable to other materials clearly regarded or classified as hazardous. For example, dusts and sludges from emissions control devices on secondary lead smelters, which are wastes listed as hazardous under federal regulations, exhibit concentrations of lead and cadmium similar to those found in incinerator fly ash.[49] A recent Washington State study[7] documented levels of several carcinogenic metals in both fly and bottom ash that were sufficiently high to classify the ash as dangerous or extremely hazardous waste under state regulations.

The leachability of metals present in incinerator ash is another measure of hazard. EPA regulations specify that metal leachability from solid wastes be assessed using the EP toxicity test.[50] The EP toxicity test is a screening procedure intended to identify wastes that pose hazards due to metal leachability and therefore require particularly careful management. The test is a short-term batch leaching procedure in which a waste is extracted with a slightly acidic fluid to roughly approximate the leaching conditions to which a waste would be subjected if disposed of in a typical sanitary landfill. If any of eight metals are present in the resulting waste extract at levels that exceed their respective drinking water standards by more than a factor of 100, the

TABLE 3
Metals Concentrations in Fly Ash and Natural Soil

	Concentration range (ppm)[a]	
Metal	**Fly ash[b]**	**Natural soil[c]**
Lead	2,300—50,000	10—13
Cadmium	100—2,000	0.1—0.2
Arsenic	10—750	2
Mercury	8—300	0.05—0.08

[a] Parts per million.
[b] NUS, 1987;[10] Environmental Defense Fund database.
[c] Vogg et al., 1986.[9]

waste is deemed to be a hazardous waste and must be managed in accordance with hazardous waste regulations.

Table 4 presents a summary of available data on ash from more than 45 U.S. incinerators tested for leaching using the federally mandated EP toxicity test. The data are drawn from federal, state, or local government and industry studies of ash collected from new and old facilities that employ a wide range of technologies. Together, they demonstrate that:

1. Virtually every sample of fly ash tested has exceeded federal standards defining a hazardous waste, usually for both lead and cadmium.
2. At almost half of the facilities where bottom or combined ash has been tested, the sample mean for that facility has also exceeded the standard for lead.

These results indicate that incinerator ash routinely exhibits the EP toxicity characteristic of a hazardous waste. Despite these findings, virtually no incinerator ash is currently managed as a hazardous waste. In fact, the majority of this ash is disposed of in sanitary landfills along with unburned waste,[10] exactly the disposal scenario which the EP test is designed to simulate.

Certain toxic metals exhibit another chemical property that is important in assessing their leachability from MSW incinerator ash. Some metals, most notably, lead, are amphoteric, that is, they are readily soluble in water under both acidic (low pH) and alkaline (high pH) conditions. This property becomes critical when evaluating the quality of ash generated by facilities equipped with acid gas scrubbers. These devices introduce a slurry or powder of lime into the combustion gases to neutralize acid components (sulfur dioxide and hydrogen chloride). Reaction products and excess lime become intimately mixed into fly ash, which forms a scrubber residue that is subsequently removed by downstream particulate control devices (e.g., electrostatic precipitators, fabric filters). Test data for the several U.S. facilities now in operation that possess such scrubbers indicate that the introduction of lime produces an ash residue, even when bottom and fly ash are combined, which is highly alkaline. pH values of 11 to 12 or higher are typical.[51,52]

The amphoteric behavior of lead in such highly alkaline ash is illustrated in Figure 1, in which the level of lead in leachates of combined fly and bottom ash from an operating incinerator[51] is plotted as a function of leachate pH. As shown in Figure 1, lead was found to leach from the ash at levels exceeding federal standards defining a hazardous waste at both low pH (using the EP toxicity test) and high pH (using distilled water). In tests of ash collected from several other U.S. incinerators[52-54] and a Canadian incinerator[12] that have acid gas scrubbers, lead was also found to readily leach from the ash, frequently in excess of hazardous waste standards, even when leached using distilled water.

In leaching studies using simulated rainwater, Swedish investigators[55] have also found that

TABLE 4
Summary of Available Extraction Procedure Toxicity Test Data for Lead and Cadmium from MSW Incinerator Ash

	Lead	Cadmium	Either
Fly ash from 23 facilities			
Number of samples analyzed	185	97	185
Number of samples over the EP limit	168	94	173
Percentage of samples over the EP limit	91%	97%	94%
Mean of all samples (mg/l)	22.0[a]	37.3[a]	—
Number of facilities over the EP limit[b]	20	21	22
Mean of facility means (mg/l)	23.4[a]	31.8[a]	—
Bottom ash from 22 facilities			
Number of samples analyzed	773	271	773
Number of samples over the EP limit	276	5	278
Percentage of samples over the EP limit	36%	2%	36%
Mean of all samples (mg/l)	6.39[a]	0.25[a]	—
Number of facilities over the EP limit[b]	9	1	9
Mean of facility means (mg/l)	7.18[a]	0.41[a]	—
Combined ash from 47 facilities			
Number of samples analyzed	933	806	933
Number of samples over the EP limit	373	115	390
Percentage of samples over the EP limit	40%	14%	42%
Mean of all samples (mg/l)	6.45[a]	0.59[a]	—
Number of facilities over the EP limit[b]	22	6	22
Mean of facility means (mg/l)	6.26[a]	0.48[a]	—

Note: Due to the large number of individual samples analyzed from certain facilities, the aggregate data tend to be skewed and overly dependent on the quality of ash from those few dominating facilities. Caution should be exercised in drawing conclusions about overall exceedance rates.

[a] Values exceed EP limits defining a hazardous waste. These values are 5.0 mg/l for lead and 1.0 mg/l for cadmium.
[b] Number of facilities for which mean of all available samples exceeds the EP limit.

Source: Environmental Defense Fund database. A full list of references is available upon request.

lime-based scrubber residues containing fly ash readily released large amounts of lead, as well as cadmium, mercury, copper, and zinc. The amounts of these metals, particularly lead and zinc, that leached from the scrubber residues were significantly higher than the amounts that leached from fly ash lacking lime, and far higher than the amounts that leached from bottom ash. The authors concluded that the enhanced leaching of toxic metals from such residues "is a major problem that will cause difficulty when it comes to disposal".[55]

The increased alkalinity of ash from facilities possessing acid gas scrubbers may also increase the leachability of organic chemicals present in the ash. A recent Canadian study found a marked increase in the solubility of a wide range of organic chemicals present in fly ash as pH increased.[56] These findings raise new concerns about the potential for acid gas scrubbers to enhance the leaching of dioxins or other toxic substances from ash that are normally not leachable due to their relative insolubility in water. This possibility has yet to be tested at any U.S. facility possessing acid gas scrubbers.

Significant controversy presently surrounds the use of the EP toxicity test as the sole means of determining the legal classification of incinerator ash. One obvious shortcoming is that the test only accounts for one of the many routes of exposure to ash: the leaching of metals into groundwater or surface water. In addition, the slightly acidic conditions of the leaching

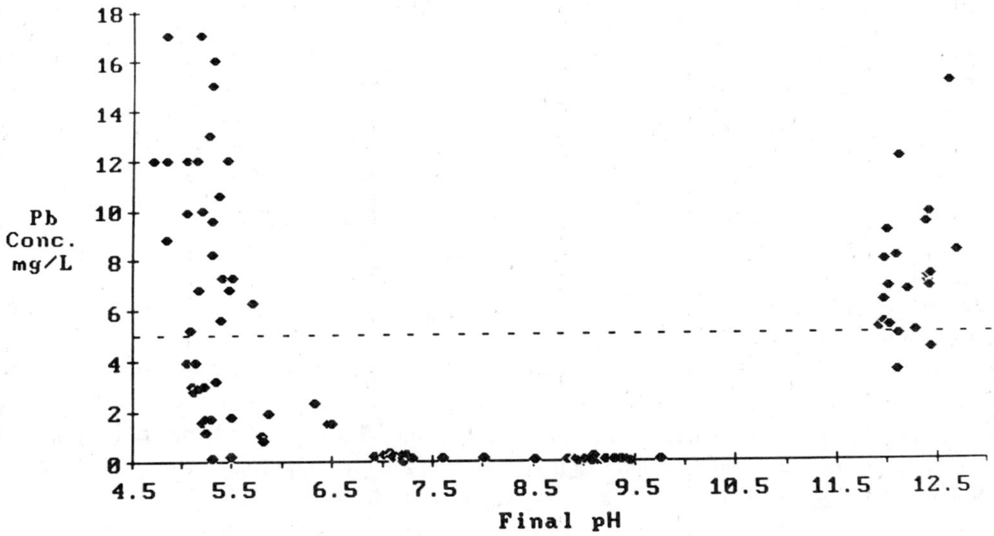

FIGURE 1. Concentration of lead in leachates of combined fly ash, scrubber residue, and bottom ash as a function of leachate pH. The dotted line represents the regulatory limit for lead in the EP Toxicity Test, 5.0 mg/l.[51]

procedure may not reasonably predict the leaching behavior under various disposal scenarios. For example, if ash is temporarily or permanently segregated from unburned wastes, leaching conditions may initially be determined primarily by the properties of the ash and the infiltrating precipitation. These conditions may be better approximated in the laboratory by extracting ash with distilled water or rainwater, rather than or in addition to the acidic fluid used in the EP toxicity test.

Finally, there has been considerable debate over the degree to which the EP test predicts or is even intended to predict actual metal concentrations in leachates collected in the field. The test includes certain conservative features (e.g., high liquid-to-solid ratio, continual agitation) that may be more aggressive than expected field conditions. In this respect, EPA intended for the procedure to predict the maximum concentrations of toxic constituents likely to migrate at one time, rather than average or cumulative releases over time.[57] Other features of the test (e.g., 24-h duration, single extraction, regulatory limits set at 100 times the drinking water standards) are potentially nonconservative. Indeed, in adopting a 100-fold factor rather than the 10-fold factor it had first proposed, EPA stated that it was "confident that anything which fails the EP at this factor has the potential to present a substantial hazard regardless of the attenuation mechanisms at play ... and no matter what leaching media it is actually exposed to".[57]

Few data are available to adequately characterize actual leachate quality from ash disposal sites. Very limited ash leachate data were reported in a recent EPA study of three landfills that contain only MSW incinerator ash.[10] The nine reported values varied over a wide range, but eight exceeded the drinking water standard for lead, and the average value exceeded the lead standard by more than 12-fold. Recent monitoring of leachate from a New York ash-only landfill during its first year of operation[58] found that average levels of many pollutants exceeded drinking water standards. Moreover, during this first year, pollutant levels in the leachate almost always exceeded (often dramatically) the highest levels predicted in laboratory simulations to occur during the first 25 years of operation.[59]

The potential for leachate contamination is not limited solely to the soluble fraction of metals or organic chemicals. Typically, leachate contains appreciable amounts of suspended solid material, which in the case of an ash-only disposal site consists of fine ash particles.[10] Given that metals and dioxins preferentially concentrate on smaller particles, leachate may be a significant vector even for very insoluble toxins. While few data addressing this issue are available, Ozvacic et al.[60] found significant dioxin contamination of an ash leachate sample and attributed the contamination to the suspended solid material in the sample.

The risks of ash must also be evaluated by direct bioavailability and toxicity testing, particularly with respect to the potential for ecosystem effects. Several studies demonstrate that toxic metals and dioxins present in ash are bioavailable to plants and animals[16,45,61-64] and toxic to organisms.[5,7,65,66] In addition to the potential for direct environmental damage, these data document the plausibility of human exposure through contamination of the food chain. These routes are particularly relevant in light of the permanent (metals) or persistent (dioxins) nature of the toxic constituents of the ash and the finite lifetimes of containment systems and treatment technologies.

The potential for improperly disposed ash to actually result in human exposure to toxic metals or dioxins is made more likely by the presently uncontrolled management of large amounts of incinerator ash by means such as open disposal, use as landfill cover or as fill material in marshy areas, reuse as construction aggregate, or use as deicing grit on winter roads. These uses clearly provide even greater opportunities for air- or water-borne dispersal of ash particles containing toxic metals or dioxins into the environment. Numerous studies on similar materials such as dusts from metal smelters directly demonstrate the significance of dispersal of metal-containing particulates, even long after initial disposal, as a route of human exposure.[19,42,67]

VIII. SOURCES OF UNCERTAINTY IN ESTIMATING INCINERATOR AIR EMISSIONS

Most incinerator risk assessments base their estimation of the magnitude of incinerator air emissions on sparse data derived from a handful of operating facilities. More often than not, these assessments purport to predict a planned facility's emissions to a remarkable degree of precision; rarely, however, do they account for many obvious factors that impede our ability to extrapolate from very limited emissions testing data to predict the routine emissions over the lifetime of a facility. Several such factors are briefly discussed below.

Emissions can vary tremendously over time due to waste heterogeneity. The EPA Science Advisory Board[68] has noted:

> "Because of the inherent variability of the MSW feedstock, it is difficult to predict the composition of the stack emissions that may result from combustion. Wide variations in feedstock composition can affect combustion conditions in the incinerator furnaces, and can cause cycles of poor combustion. Poor combustion conditions have a direct impact on emissions.... High concentrations of certain plastics, solvents, or other highly volatile materials will result in surges in the emissions of combustible gases."

Many of the existing data are skewed, since tests have usually been conducted on new facilities operating at peak performance, typically under the close scrutiny of highly trained personnel. Emissions are rarely measured under the full range of operating conditions that will be routinely encountered, including temperature fluctuations or start-up and shut down. For example, results of one of six dioxin emission tests recently conducted on a Massachusetts incinerator were excluded in reporting average emissions on the basis that the boiler had not stabilized in the 12 h since it was shut down to repair a broken grate; dioxin emissions during this test were a factor of ten higher than the average of the other tests.[69] Emissions tests at other incinerators have documented wide variations in emissions of dioxins and other pollutants.[70]

Emissions can reasonably be expected to increase significantly during upsets in combustion conditions or the operation of air pollution controls during periods of less-than-optimal performance resulting from lack of maintenance, or simply as a result of facility aging.[70] Yet data on the frequency of upsets are nonexistent, and such factors are rarely (if ever) modeled to "adjust" the source term. As a result, according to the SAB: "The capability for continued performance at such low emission levels under a variety of operating conditions remains to be demonstrated for full-scale municipal solid waste incinerators."[68] Indeed, an SAB review of the available data on emissions during poor or upset combustion conditions found that "total hydrocarbon levels have risen from a typical 1-5 parts per million (ppm) to 100 ppm and above".[68]

Another source of uncertainty in estimating routine emissions results from reliance on the measurement of surrogates (e.g., carbon monoxide, or CO) for emission components for which continuous monitoring is not possible. While low CO levels are a general indicator of good combustion efficiency, the quantitative correlation between CO levels and actual emissions can be rather poor, especially for chemicals such as dioxins that are likely created after combustion gases have left the furnace itself.[71-74]

IX. SUMMARY AND CONCLUSIONS

The risks associated with incineration require comprehensive analysis in order to develop appropriate guidelines for the use of this technology and to restrict its application to those portions of the waste stream where its benefits (e.g., volume reduction) are not overwhelmed by its risks. Sufficient information is available on the concentration and bioavailability of organic and inorganic compounds in both air emissions and ash residues to indicate the reality of these risks.

Techniques for comprehensive exposure assessment are still relatively new and unvalidated. Understandably, many risk assessments provided by incinerator vendors or municipalities do not adequately address all of the noninhalation or postdeposition routes of exposure, including estimations of accumulation of long-lived contaminants in slow-turnover environmental compartments such as surface soils and dusts.

The need to manage MSW more effectively without simply transferring risks, thereby continuing to incur long-term environmental and public health problems, requires that these risks be recognized and addressed. Both short- and long-term solutions need consideration. In the short term, appropriate technological and operating controls should be imposed on incinerators, including "best available control technology" (BACT) air pollution controls, maximally feasible on-line monitoring, periodic stack testing to ensure compliance with specific health-based emissions limits, and ash management under the "cradle-to-grave" rubric of the federal Resource Conservation and Recovery Act (RCRA).

The present massburn approach to incineration is propelled forward by the same myth that led to our disastrous reliance on mass landfilling, namely, that the municipal waste stream can be managed as a monolithic material using a single management technique. This approach does a disservice to the significant merits of incineration. Targeted removal of critical items which either compromise the efficiency of incineration or are identifiable as major sources of ash contamination can increase the safety and utility of incineration for the remainder of the waste stream.

Of course, the most efficient method of reducing the risks of incineration requires that we return to the broad systems view of MSW management with which we began this chapter, in which the use of incineration is only one component of a fully integrated waste management system. This view requires planners and managers to step back from the crisis mentality that now dictates management decisions, to consider all steps in the product cycle: production, use, discard, collection, recycling and recovery, processing, and ultimate disposal. Comprehensive planning of this type conducted on a regional or national basis can in turn affect available markets sufficiently to support a greatly enhanced recycling component of the MSW management system.

Critical attention needs to be directed at the earliest stages of the waste generation process: production and use. Such strategies of upstream intervention could serve to condition the waste stream by, for example, restricting certain production practices that introduce toxic materials into widely distributed household products. In several European countries, government initiatives have been directed toward reducing the use of certain metals (e.g., cadmium) in consumer products, particularly those (such as disposable plastic items) that are used and discarded in large amounts after limited use.[75-77] Given the use of toxic metals in the manufacture of a very broad range of consumer materials (e.g., printing inks, plastic stabilizing agents) as well as by more easily identified materials such as lead-acid batteries, only comprehensive source-based strategies are likely to prove successful in achieving significant reductions in the metal content and toxicity of MSW and incineration by-products.

Our present approach to MSW management has avoided solving difficult problems only by transferring them to later stages in the materials flow system. Many of the risks that characterize the final stage of the product cycle, disposal, result from decisions made at the earliest stages of production, packaging, and marketing. It is unreasonable to expect economics and technologies of waste disposal to efficiently manage such risks. Indeed, the most difficult and expensive method of reducing the risk of exposure to toxic metals in products is to delay action until after incineration has refined metals into a highly bioavailable and concentrated form. Both efficiency and the generally accepted goal of promoting prevention over remediation dictate that we examine alternatives to the use of such metals at the production stage. Such risk reduction strategies require expansion of the decision-making rubric beyond RCRA to include the tools for controlling product composition available under the Toxic Substances Control Act (TSCA).

Only in this way can we move beyond the widely perceived MSW management crisis toward collectively devised rational solutions.

REFERENCES

1. **Hershkowitz, A.,** Burning trash: how it could work, *Technol. Rev.,* 90, 26, 1987.
2. **Hutton, M., Wadge, A., and Milligan, P. J.,** Environmental levels of cadmium and lead in the vicinity of a major refuse incinerator, *Atmos. Environ.,* 22, 411, 1988.
3. Environment Canada, The National Incinerator Testing and Evaluation Program: Air Pollution Control Technology, Summary Report EPS 3/UP/2, Ottawa, Ontario, 1986.
4. **Sawell, S. E., Bridle, T. R., and Constable, T. W.,** Leachability of Organic and Inorganic Contaminants in Ashes from Lime-Based Air Pollution Control Devices on a Municipal Waste Incinerator, presented at Annu. Meet. Air Pollut. Control Assoc., New York, June 21-26, 1987.
5. **Kuehl, D. W., Cook, P. M., Batterman, A. R., Lothenbach, D. B., Butterworth, B. C., and Johnson, D. L.,** Bioavailability of 2,3,7,8-tetrachlorodibenzo-*p*-dioxin from municipal incinerator ash to freshwater fish, *Chemosphere,* 14, 427, 1985.
6. Environmental Protection Agency, Municipal Waste Combustion Study: Emission Data Base for Municipal Waste Combustors, EPA/530-SW-87-021b, U.S. Government Printing Office, Washington, D.C., 1987.
7. **Knudson, J. C.,** Study of Municipal Incineration Residue and Its Designation as a Dangerous Waste, Solid Waste Section, Department of Ecology, State of Washington, Olympia, 1986.
8. **Brunner, P. H. and Monch, H.,** The flux of metals through municipal solid waste incinerators, *Waste Manage. Res.,* 4, 105, 1986.
9. **Vogg, H., Braun, H., Metzger, M., and Schneider, J.,** The specific role of cadmium and mercury in municipal solid waste incineration, *Waste Manage. Res.,* 4, 65, 1986.
10. NUS Corporation, Characterization of Municipal Waste Combustor Ashes and Leachates from Municipal Solid Waste Landfills, Monofills, and Codisposal Sites, 7 Volumes, EPA Contract No. 68-01-7310, U.S. Environmental Protection Agency, U.S. Government Printing Office, Washington, D.C., October 1987.
11. **Wilson, D. C., Young, P. J., Hudson, B. C., and Baldwin, G.,** Leaching of cadmium from pigmented plastics in a landfill site, *Environ. Sci. Technol.,* 16, 560, 1982.

12. **Sawell, S. E., Bridle, T. R., and Constable, T. W.,** Assessment of Ash Contaminant Leachability, NITEP Phase II — Testing of the FLAKT Air Pollution Control Technology at the Quebec City Municipal Energy from Waste Facility, Industrial Programs Branch, Wastewater Technology Centre, Environment Canada, Burlington, Ontario, 1986.
13. **Norton, G. A., DeKalb, E. L., and Malaby, K. L.,** Elemental composition of suspended particulate matter from the combustion of coal and coal/refuse mixtures, *Environ. Sci. Technol.,* 20, 604, 1986.
14. **Carlsson, K.,** Heavy metals from "energy from waste" plants — comparison of gas cleaning systems, *Waste Manage. Res.,* 4, 15, 1986.
15. **Greenberg, R. R., Zoller, W. H., and Gordon, G. E.,** Composition and particle size in refuse incineration, *Environ. Sci. Technol.,* 12, 566, 1978.
16. **Wadge, A., Hutton, M., and Peterson, P. J.,** The concentrations and particle size relationships of selected trace elements in fly ashes from U.K. coal-fired power plants and a refuse incinerator, *Sci. Total Environ.,* 54, 13, 1986.
17. **Berlincioni, M. and di Domenico, A.,** Polychlorodibenzo-*p*-dioxins and polychlorodibenzofurans in the soil near the municipal incinerator of Florence, Italy, *Environ. Sci. Technol.,* 21, 1063, 1987.
18. Environmental Protection Agency, Air Quality Criteria for Lead, EPA-600/8-83028aF-dF, Research Triangle Park, NC, 1986.
19. **Harper, M., Sullivan, K. R., and Quinn, M. J.,** Wind dispersal of metals from smelter waste tips and their contribution to environmental contamination, *Environ. Sci. Technol.,* 21, 481, 1987.
20. **Nriagu, J. 0., Ed.,** *Changing Metal Cycles and Human Health,* Springer-Verlag, Berlin, 1984.
21. **Roberts, T. M., Hutchinson, T. C., Paciga, J., Chattopadhyay, A., Jervis, R. E., VanLoon, J., and Parkinson, D. K.,** Lead contamination around secondary smelters: estimation of dispersal and accumulation by humans, *Science,* 186, 1120, 1974.
22. **Wadge, A. and Hutton, M.,** The leachability and chemical speciation of selected trace elements in fly ash from coal combustion and refuse incineration, *Environ. Pollut.,* 48, 85, 1987.
23. **Bridle, T. R., Cote, P. L., Constable, T. W., and Fraser, J. L.,** Evaluation of heavy metal leachability from solid wastes, *Water Sci. Technol.,* 19, 1029, 1987.
24. **Brunner, P. H. and Baccini, P.,** The Generation of Hazardous Waste by MSW-Incineration Calls for New Concepts in Thermal Waste Treatment, presented at 2nd Conf. on New Frontiers for Hazardous Waste Management, Pittsburgh, September 27-30, 1987.
25. **Friberg, L., Nordberg, G. F., and Vouk, V. B.,** *Handbook of the Toxicology of Metals,* Vol. 1, Elsevier, Amsterdam, 1986.
26. Environmental Protection Agency, *Reducing Lead in Drinking Water: A Benefit Analysis,* EPA# 230/09-86-019, U.S. Government Printing Office, Washington, D.C., 1986.
27. **Annest, J. L., Pirkle, J. L., Makuc, D., Neese, J. W., Bayse, D. D., and Kovar, M. G.,** Chronological trend in blood lead levels between 1976-1980, *N. Engl. J. Med.,* 308, 1373, 1983.
28. **Bellinger, D., Leviton, A., Waterneaux, C., Needleman, H., and Rabinowitz, M.,** Longitudinal analyses of prenatal and postnatal lead exposure and early cognitive development, *N. Engl. J. Med.,* 316, 1037, 1987.
29. **Dietrich, K., Krafft, K., Bier, M., Succop, P. A., Berger, O., and Bornschein, R. L.,** Early effects of fetal lead exposure: neurobehavioral findings at 6 months, *Int. J. Biosocial Res.,* 8, 151, 1986.
30. **Dietrich, K. N., Krafft, K. M., Shukla, R., Bornschein, R. L., and Succop, P. A.,** Neurobehavioral effects of prenatal and early postnatal lead exposure, in *Toxic Substances and Mental Retardation: Neurobehavioral Toxicology and Teratology,* AAMD Monogr. Ser. No. 8, Schroeder, S. R., Ed., Washington, D.C., 1987, 71.
31. **Rutter, M. and Jones, R. R., Eds.,** *Lead Versus Health: Sources and Effects of Low Level Lead Exposure,* John Wiley & Sons, London, 1983.
32. Agency for Toxic Substances and Disease Registry, *The Nature and Extent of Lead Poisoning in Children in the United States: A Report to Congress,* U. S. Public Health Service, Atlanta, 1988.
33. **Annest, J. L., Mahaffey, K. R., Cox, D. H., and Roberts, J.,** Blood lead levels for persons 6 months-74 years of age: United States, 1976-1980, in *AdvanceData,* No. 79, National Center for Health Statistics, U. S. Public Health Service, Hyattsville, MD, 1982.
34. **Silbergeld, E. K.,** Neurotoxicology of lead, in *Neurotoxicology,* Blum, K. and Manzo, L., Eds., Marcel Dekker, New York, 1985, 299.
35. **Friberg, L., Nordberg, G. F., and Vouk, V. B.,** *Handbook of the Toxicology of Metals,* Vol. 2, Elsevier, Amsterdam, 1986.
36. **Silbergeld, E. K. and Schwartz, J.,** Lead and osteoporosis: mobilization of bone lead in postmenopausal women and possible etiologic role in bone demineralization, in *Proc. Int. Conf. Heavy Metals in the Environment,* Vol. 2, Lindberg, S. E. and Hutchinson, T. C., Eds., CEP Consultants, London, 1987, 360.
37. **Carson, B. L., Stockton, R. A., and Wilkinsdon, R. R.,** Organomercury, -lead, and -tin compounds in the environment and the potential for human exposure, in *Neurotoxicants and Neurobiological Function: Effects of Organoheavy Metals,* Tilson, H. A. and Sparber, S. B., Eds., John Wiley & Sons, New York, 1987, 1.

38. **Callahan, M., Slimak, M., Gabel, N., May, I., Fowler, C., Freed, J., Jennings, P., Durfie, R., Whitmore, F., Maestri, B., Maybey, W., Holt, B., and Gould, C.,** *Water Related Environmental Fate of 129 Priority Pollutants,* Report No. 440/4-79-0029, U.S. Environmental Protection Agency, Washington, D.C., 1979.
39. **Angle, C. R., Marcus, A., Cheng, I.-H., and McIntire, M. S.,** Omaha childhood blood lead and environmental lead: A linear total exposure model, *Environ. Res.*, 35, 160, 1984.
40. **Steenhout, A.,** Kinetic and epidemiologic approaches of the human organism response to lead exposure. A solution for the Pb-blood/Pb-air, or Pb-water, or Pb-dust ... relationships, in *Proc. Int. Conf. Heavy Metals in the Environment,* Vol. 2, Lindberg, S. E. and Hutchinson, T. C., Eds., CEP Consultants, London, 1987, 280.
41. **Davies, D. J. A., Thornton, I., Watt, J. M., Culbard, E. B., Harvey, P. G., Delves, H. T., Sherlock, J. C., Smart, G. A., Thomas, J. F. A., and Quinn, M. J.,** Relationship between blood lead and lead intake in two year old urban children in the UK, in *Proc. Int. Conf. Heavy Metals in the Environment,* Vol. 2, Lindberg, S. E. and Hutchinson, T. C., Eds., CEP Consultants, London 1987, 203.
42. **Roels, H. A., Buchet, J., Lauwerys, R. R., Bruaux, P., Claeys-Thoreau, F., Lafontaine, A., and Verduyn, G.,** Exposure to lead by the oral and pulmonary routes of children living in the vicinity of a primary lead smelter, *Environ. Res.*, 22, 81, 1980.
43. **Clark, C. S., Bornschein, R. L., Succop, P., Hammond, P. B., Peace, B., Krafft, K., and Dietrich, K.,** Pathways to elevated blood lead and their importance in control strategy development, in *Proc. Int. Conf. Heavy Metals in the Environment,* Vol. 1, Lindberg, S. E. and Hutchinson, T. C., Eds., CEP Consultants, London, 1987, 159.
44. **Mielke, H. W., Anderson, J. C., Berry, K. J., Mielke, P. W., Chaney, R. L., and Leech, M.,** Lead concentrations in inner-city soils as a factor in the child lead problem, *Am. J. Public Health,* 73, 1366, 1983.
45. **Giordano, P. M., Bethel, A. D., Lawrence, J. E., Solleau, J. M., and Bradford, B. N.,** Mobility in soil and plant availability of metals derived from incinerated municipal refuse, *Environ. Sci. Technol.*, 17, 193, 1983.
46. **Mika, J. S. and Feder, W. A.,** Resco Incinerator Residue Research Program: Results, Evaluations and Recommendations, Refuse-Energy Systems Company, University of Massachusetts, Waltham, 1985.
47. **Wadge, A. and Hutton, M.,** The uptake of cadmium, lead, and selenium by barley and cabbage grown on soils amended with refuse incinerator fly ash, *Plant Soil,* 96, 407, 1986.
48. Environmental Protection Agency, Methodology for the Assessment of Health Risks Associated with Multiple Pathway Exposure to Municipal Waste Combustor Emissions, Office of Air Quality Planning and Standards, U.S. Government Printing Office, Research Triangle Park, NC, 1986.
49. Environmental Protection Agency, Background Document, Resource Conservation and Recovery Act, Subtitle C — Hazardous Waste Management, Section 3001 — Identification and Listing of Hazardous Waste, Sections 261.31 and 261.32 —Listing of Hazardous Wastes, Book 11, Office of Solid Waste, U.S. Government Printing Office, Washington, D.C., 1980.
50. Code of Federal Regulations, 40, 261.24, 1987.
51. Resource Analysts, Inc., Extraction Procedure Toxicity Test results for combined fly and bottom ash from the New Hampshire-Vermont Solid Waste Project Incinerator, tests conducted for Signal Environmental Systems, Inc., Hampton, NH, unpublished data, reports dated 6 April, 1987, 30 May, 1987, 29 June, 1987.
52. **Shinn, C. H.,** Extraction Procedure Toxicity Characterization of Municipal Incinerator Ash from Ogden Martin, Brooks, Department of Environmental Quality, Laboratory Division, Portland, OR, 1987.
53. Roy F. Weston, Inc., Framingham Incinerator Ash Sampling and Analysis Report, prepared for ESI, Inc., Southborough, MA, Burlington, MA, 1987.
54. Commerce Refuse-to-Energy Authority, Ash Testing — Commerce Refuse to Energy Facility, letters to Dr. David Leu, California Department of Health Services, Sacramento, CA, Whittier, CA, dated 11 May, 21 July, 12 November, 13 November, 1987, 24 February, 16 March, 1988.
55. **Hartlen, J. and Elander, P.,** Residues from Waste Incineration: Chemical and Physical Properties, Swedish Geotechnical Insititute, Report No. 172, Linköping, Sweden, June 1986.
56. **Karasek, F. W., Charbonneau, G. M., Reuel, G. J., and Tong, H. Y.,** Determination of organic compounds leached from municipal incinerator fly ash by water at different pH levels, *Anal. Chem.*, 59, 1027, 1987.
57. Environmental Protection Agency, Background Document, Resource Conservation and Recovery Act, Subtitle C — Hazardous Waste Management, Section 3001 — Identification and Listing of Hazardous Waste, Section 261.24 — EP Toxicity Characteristic, Book 14, Office of Solid Waste, U.S. Government Printing Office, Washington, D.C., 1980.
58. Malcolm-Pirnie, Inc., Sprout Brook Residue Disposal Facility: Leachate Analytical Characterization Research, Summary of Results, letter to Calvin E. Weber, Deputy Commissioner of Solid Waste, Westchester County, 11 November, 1987.
59. **Cundari, K. L. and Lauria, J. M.,** The Laboratory Evaluation of Expected Leachate Quality from a Resource Recovery Ashfill, presented at 1986 Triangle Conference on Environmental Technology, Chapel Hill, NC, 1986.
60. **Ozvacic, V., Wong, G., Tosine, H., Clement, R. E., and Osborne, J.,** Emissions of chlorinated organics from two municipal incinerators in Ontario, *J. Air Pollut. Control Assoc.*, 35, 849, 1985.

61. **Van den Berg, M., Olie, K., and Hutzinger, O.,** Uptake and selective retention in rats of orally administered chlorinated dioxins and dibenzofurans from fly-ash and fly-ash extract, *Chemosphere,* 12, 537, 1983.
62. **Van den Berg, M., Van Greevenbroek, M., Olie, K., and Hutzinger, O.,** Bioavailability of polychlorinated dibenzo-*p*-dioxins and polychlorinated dibenzofurans on fly ash after semi-chronic oral ingestion by the rat, *Chemosphere,* 15, 509, 1986.
63. **Van den Berg, M., de Vroom, E., Olie, K., and Hutzinger, O.,** Bioavailability of polychlorinated dibenzo-*p*-dioxins and polychlorinated dibenzofurans on fly ash after semi-chronic oral ingestion by guinea pig and Syrian golden hamster, *Chemosphere,* 15, 519, 1986.
64. **Opperhuizen, A., Wagenaar, W. J., van der Wielen, F. W. M., van den Berg, M., Olie, K., and Gobas, F. A. P. C.,** Uptake and elimination of PCDD/PCDF congeners by fish after aqueous exposure to a fly-ash extract from a municipal incinerator, *Chemosphere,* 15, 2049, 1986.
65. **Kuehl, D. W., Cook, P. M., Batterman, A. R., and Butterworth, B. C.,** Isomer dependent bioavailability of polychlorinated dibenzo-*p*-dioxins and dibenzofurans from municipal incineration fly ash to carp, *Chemosphere,* 16, 657, 1987.
66. **Bronzetti, G., Bauer, C., Corsi, C., Del Carratore, R., Nieri, R., and Paolini, M.,** Mutagenicity study of TCDD and ashes from urban incinerator "in vitro" and "in vivo" using yeast D7 strain, *Chemosphere,* 12, 549, 1983.
67. **Landrigan, P. J., Gehlbach, S. H., Rosenblum, B. F., Shoults, J. M., Candelaria, R. M., Barthel, W. F., Liddle, J. A., Smrek, A. L., Staehling, N. W., and Sanders, J. F.,** Epidemic lead absorption near an ore smelter, *N. Engl. J. Med.,* 292, 123, 1975.
68. Science Advisory Board, Environmental Protection Agency, Evaluation of Scientific Issues Related to Municipal Waste Combustion, Report of the Environmental Effects, Transport, and Fate Committee, U.S. Government Printing Office, Washington, D.C., April 1988.
69. Radian Corporation, Final Emissions Test Report: Dioxin/Furans and Total Organic Chlorides Emissions Testing, Rust Corporation, Birmingham, AL, Research Triangle Park, NC, 1986.
70. **Goldfarb, T. D.,** Prefiled Testimony before the New York Department of Environmental Conservation Office of Administrative Law, in the Matter of the Application of Signal Environmental Systems, Inc., for permits to construct and operate the proposed Brooklyn Navy Yard Resource Recovery Facility, DEC Project No. 20-85-0306, 1988.
71. **Stieglitz, L., Zwick, G., Beck, H., Roth, W., and Vogg, H.,** On the de novo synthesis of PCDD/PCDF on fly ash of municipal waste incinerators, presented at Seventh Ann. Symp. on Chlorinated Dioxins and Related Compounds, Las Vegas, October 4-9, 1987.
72. **Ballschmiter, K., Niemczyk, R., Braunmiller, I., Ehmann, J., Duwel, U., and Nottrodt, A.,** Chemistry of formation and fate of PCDD/PCDF in the municipal waste incineration process, presented at 7th Annu. Symp. on Chlorinated Dioxins and Related Compounds, Las Vegas, October 4-9, 1987.
73. **Clarke, M. J.,** Improving Environmental Performance of MSW Incinerators, presented at Industrial Gas Cleaning Forum '88, Washington, D.C., November 3-4, 1988.
74. **Jenkins, A. C.,** Evaluation Test on a Hospital Refuse Incinerator at Saint Agnes Medical Center, Fresno, California, California Air Resources Board, Sacramento, CA, 1987.
75. **Bothen, M. and Fallenius, U.-B.,** Cadmium: Occurrence, Uses, and Stipulations, National Swedish Environmental Protection Board, Rep. No. SNV PM 1615, Solna, Sweden, 1982.
76. Ministry of Housing, Physical Planning, and Environment, Chemical Substances Act: Cadmium Decree, Government Gazette, 60, Leidschendam, The Netherlands, 1987.
77. Commission of the European Communities, *Environmental Pollution by Cadmium: Proposed Action Programme,* Report No. COM(87) 165, Brussels, Belgium, April 1987.
78. **Denison, R. A.,** unpublished data.

Chapter 16

COMPARING SOLID WASTE MANAGEMENT OPTIONS: A CASE STUDY

Bernd Franke

TABLE OF CONTENTS

I. INTRODUCTION

Historically, most U.S. cities selected landfilling as the solution of choice for disposing of their municipal solid waste (MSW).[1] As landfill space continues to dwindle, public officials are being forced to consider other waste management alternatives, including recycling, composting, and MSW incineration. Frequently, a particular waste disposal option is chosen without comparing the risks and benefits of all feasible options. To facilitate the comprehensive, concurrent evaluation of various waste disposal choices, the Institute for Energy and Environmental Research (IEER) developed a comprehensive and systematic approach for evaluating the costs, benefits, and environmental impacts of differing MSW handling systems. Application of this approach provides public officials with a detailed methodology for analyzing, comparing, and ranking MSW management alternatives.

This chapter outlines the IEER methodology and presents the results of a quantitative comparison conducted for Cologne, West Germany (population 966,000). Of the 973,000 metric tons of MSW generated per year, 98% is currently landfilled. Since future landfill capacity is limited to less than one third of the current amount, Cologne was forced to modify its existing MSW management concept. To develop a sound, integrated waste management strategy, Cologne hired the IEER to propose and compare feasible waste management options. The comparison of those alternatives, including information on costs, environmental impacts, energy savings, and future landfill space requirements, as well as conclusions and recommendations derived from this analysis, are presented in this chapter.[2]

II. CURRENT WASTE STREAM AND WASTE COMPOSITION

Table 1 summarizes the types and amounts of solid wastes which were generated in Cologne in 1987. About one third of the city's waste is from households, collected by the Department of Sanitation. With 20% of the total, the second largest component is commercial waste from several thousand small- and medium-sized companies. Sewage sludge is the major portion of the 111,309 metric tons/year of sludges for which the city has developed a separate utilization concept; this involves a combination of agricultural use of sludge compost, co-firing of dried sludge in industry, and landfilling of the remainder.

Table 2 gives an overview of the composition of the two major waste fractions, household and commercial waste. Paper and cardboard dominate with 27% of the commercial waste, whereas organics (food and yard waste) are the single largest waste fraction of household waste (26%).

III. SOURCE REDUCTION AND RECYCLING

A major part of all waste management options for Cologne require the extensive use of methods for source reduction. A thorough review of case studies, public relations activities, and legal actions by city administrations in the Federal Republic of Germany resulted in a catalog of source reduction activities which Cologne could pursue. A cornerstone was the support of backyard composting, as described by Platt et al.[3] Furthermore, a public relations campaign by the city was outlined which emphasized behavioral changes to be made by the consumer which the city of Heidelberg implemented.[4] Also recommended were changes in public procurement, as well as hiring waste reduction consultants to help businesses solve their specific problems. Of an overall potential of 20 to 25% source reduction (by weight), a reduction of about 7% of household and commercial wastes was estimated to be realistically achievable within a period of several years.

The possibilities for recycling and reuse were analyzed for each waste fraction. In regard to recycling of household wastes, two distinct options, "A" and "B", were found to be feasible

TABLE 1
Solid Waste Generation in Cologne

Type of waste	Amount (ton/year)	Percent of total (%)
Household waste	334,197	34.5
Commercial waste	183,072	18.8
Sludge	111,309	11.5
Clay, marl	104,328	10.8
Construction-site wastes	69,552	7.2
Production wastes	52,372	5.4
Foundry sands	25,009	2.6
Street sweepings	23,988	2.5
Bulky waste (private)	23,900	2.5
Market wastes	18,488	1.9
Leaves and yard waste	12,217	1.3
Hospital waste	5,928	0.6
Sand and silt	3,797	0.4
Other	1,427	0.1
Total	969,584	100

TABLE 2
Composition of Household and Commercial Waste

Waste fraction	Percent in household waste[a]	Percent in commercial waste[a]
Organics	25.7	10
Paper	4.3	27
Cardboard	16.4	
Ferrous metal	2.8	3
Nonferrous metal	0.4	
Composite materials	3.0	—
Glass	14.4	1
Plastics	5.2	9
Textiles	1.8	—
Inert material	1.8	12
Disposable diapers	1.9	—
Household toxics	0.3	—
Fine waste (up to 8 mm)	8.0	—
Medium size waste (8—40 mm)	14.2	—
Wood	—	11
Production wastes[b]	—	12
Other[c]	—	15

[a] Extrapolation from 1980/81 waste analysis performed in Cologne using data of the Federal household waste analysis to account for trends in waste composition.
[b] Rubber, rags, etc.
[c] Mainly a mixture of household-like wastes.

under the conditions in Cologne. These two options differ in basic strategies and in the extent of additional bins required for source separated dry recyclables and/or organics.

Recycling system A is a two-bin system targeting primarily the organic waste fraction for those households with enough space to locate at least two bins. In all 136,000 households (322,000 people) of the outlying suburban areas and in half of households of the inner city residential area (145,000 households, 300,000 people), food and yard waste are collected in a separate second bin. Paper and glass are collected not in a separate bin but with a dense network of drop-off containers throughout the entire city area.

Recycling system B is a combination of a two-bin/three-bin system targeting primarily mixed dry recyclables such as glass, paper, metal, and plastics. All 450,000 households in Cologne (966,000 people) would receive a second bin for recyclables. In addition to the bin for recyclables, all 136,000 households (322,000 people) of the outlying suburban areas would receive a third bin for separated food and yard waste. The limitation of the third bin system to the suburban areas is due to the high density of apartment buildings in inner city areas, which leaves little space for additional waste bins.

Both systems would yield substantial amounts of source separated organics that will be used to produce a high quality compost as reported elsewhere.[5] In the average of comparable programs, 91 kg/year of compostable organics are collected if an additional bin is provided.[5] Mixed recyclables collected from households (considered in system B) would be sorted in sorting plants. Currently, more than 60 of such sorting facilities operate successfully in West Germany.[6] Typically, about 40 kg of paper, 20 kg of glass, and 7 kg of metal are collected per person per year if an additional bin for mixed recyclables is provided.

Both systems have their specific advantages and disadvantages. Recycling system A requires less collection efforts than system B and is easier to implement. Also, the recycled glass and paper fraction collected in drop-off containers is of a higher quality than glass and paper collected co-mingled in a separate bin for subsequent sorting. System B, on the other hand, would yield a higher recovery rate and would result in a smaller rest for landfilling or incineration. Including source reduction, the first option would result in a 38% reduction in household waste, the second in a 43% reduction. This estimate was based on specific recovery rate achieved for each waste fraction in other West German communities. Given the residential structure of the city with its apartment buildings, recovery rates were estimated to be lower compared to areas dominated by single-family houses. The overall impact of the recycling systems (including source reduction) was estimated, taking into account rejects in organics and recyclables collection.

Given the fact that household wastes account for 34.5% of Cologne's total MSW, the difference in waste reduction relative to Cologne's overall MSW stream is less than 2% for the two systems, making the additional efforts associated with system B hardly cost effective. Hence, it was proposed to devote efforts to other waste fractions. Significant waste reductions, for example, can be achieved by recovering recyclables in commercial waste and from construction sites which should be recovered in sorting plants designed for this purpose. Since commercial waste is usually generated in large loads, separation at the source should be encouraged by lowering tipping fees for separated waste as opposed to unseparated waste. Organics in commercial and construction site waste should be separated and composted. In addition, separate collection and disposal of toxics in household and commercial waste ("problematic wastes") is recommended.

IV. MSW MANAGEMENT OPTIONS CONSIDERED FOR COLOGNE

All recycling systems result in a rest which currently cannot be recycled as a raw material. This rest could either be landfilled directly, anaerobically digested, or incinerated with or without pre-processing. The pre-processing system separates the combustible fraction (refuse derived fuel, or RDF) from the noncombustible fraction (fines, etc.) A further possibility for utilizing the rest was the ORFA (organic fibers from MSW) system whereby a fiber material is generated for use in paper production, construction, or as a fuel.

The recycling systems "A" and "B" for household waste were combined with a reasonable mixture of options for utilizing the remaining waste fraction. This resulted in eight MSW management options which were compared with the current system (option #0). Of course, these eight options are in themselves a selection of a great number of potential combinations. The eight options were:

#0 Current system (landfilling of 98% of the city's waste).
#1 Recycling system A for household waste, which would achieve a 38% reduction
Sorting plants for commercial and construction site waste
Landfilling of remaining waste fraction.
#2 Option #1 with subsequent processing of all suitable remaining waste with the ORFA system, requiring construction of a 340,000 metric tons/year ORFA plant
#3 Option #1 with subsequent incineration of all combustible remaining waste, requiring construction of a 387,000 metric tons/year incinerator
#4 Option #1 with only partial ORFA processing, requiring a #1 recycling system A for household waste, which would 130,000 metric tons/year ORFA plant, RDF processing of 257,000 metric tons/year, and incineration of the combustible waste fraction in a 154,000 metric tons/year incinerator
#5 Option #1 with RDF processing of 390,000 metric tons/year, and incineration of combustible waste fraction, requiring a 232,000 metric tons/year incinerator
#6 Option #5 with anaerobic digestion of the organic fraction from RDF sorting, requiring a 77,000 metric tons/year digestor
#7 Recycling system B for household waste, which would achieve a 38% reduction
Sorting plants for commercial and construction site waste
Landfilling of remaining waste fraction.
#8 Option #7 with anaerobic digestion of remaining household wastes, requiring a 147,000 metric tons/year digestor

For all options, flow charts were developed. Figure 1 shows option #5 as an example. These flow charts show the waste streams in the form of arrows and waste facilities in the form of boxes.

V. COMPARISON OF ALTERNATIVE OPTIONS

Each MSW management option was analyzed in detail regarding the associated landfill requirements, investment costs, specific costs (including collection), energy savings, amount of material recycling, and environmental impacts.

Investment costs include all costs for site selection, planning, and construction of the plants, as well as auxiliary facilities, plus investment costs for collection and hauling (additional containers and trucks).

Specific costs consist of the annual debt service divided by the number of tons of trash accepted by the system, the operation and maintenance costs per ton, and the costs for disposal of residues minus the revenues from sale of energy and/or recovered materials.

Landfill requirement is defined as the total landfill space required by the system. Special consideration needs to be given to the possibility that certain types of residues (fly ashes from incineration) might have to be disposed of in toxic waste or other designated landfills.

Energy savings result from production of steam and/or electricity in massburn incinerators and RDF, pyrolysis, or biogas plants. Energy savings also result from recycling of raw materials since they replace virgin materials the production of which is usually more energy intensive than use of secondary materials.

Environmental impacts can be of a positive or negative nature. Recycling will reduce emissions which result from production of virgin materials (e.g., aluminum from bauxite). A massburn incinerator might emit less SO_2 or NO_x than an outdated fossil fuel plant it replaces.

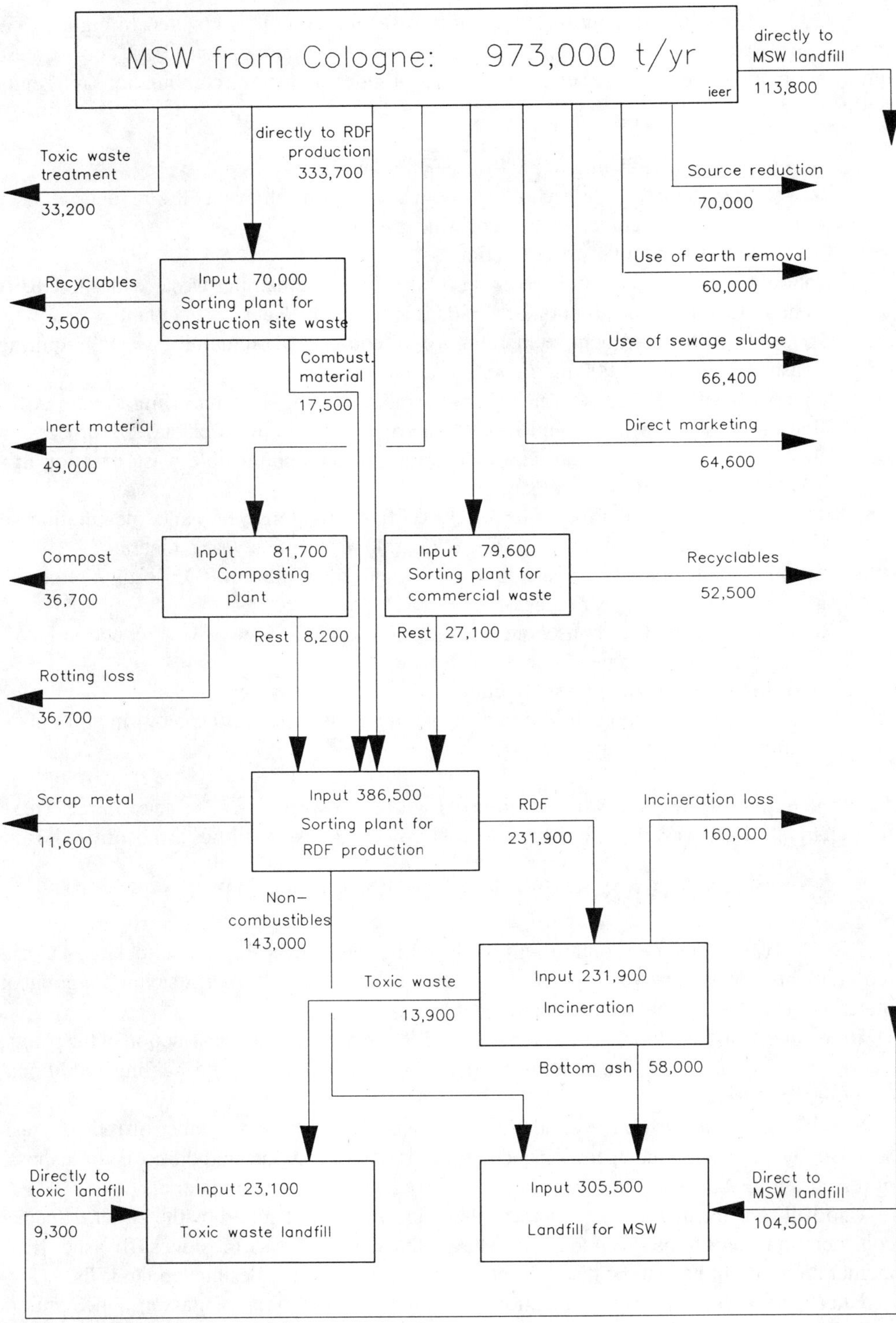

FIGURE 1. Mass flow diagram of MSW management option #5.

TABLE 3
Comparison of Planning Alternatives

	MSW management option							
	#1	#2	#3	#4	#5	#6	#7	#8
Landfill requirement	-	+	+	+	o	+	-	o
Security of disposal								
Reliability	+	-	+	-	+	o	+	-
Marketability	+	-	+	o	+	+	+	+
Citizen participation	+	+	+	+	+	+	+	+
Flexibility	+	-	o	o	+	o	+	-
Economy								
Investment costs	+	o	-	-	-	-	+	o
Specific costs	+	o	-	-	-	-	+	o
Energy balance	-	o	+	+	+	+	o	o
Recycling	o	+	+	+	+	+	+	+
Environmental impact								
Floor space required	+	o	o	-	-	-	+	o
Noise	o	o	o	o	o	o	-	-
Air	+	o	-	o	o	o	+	o
Water	+	+	+	+	+	+	+	o

+ = positive judgment; o = neutral judgment; - = negative judgment.

On the other hand, negative environmental impact will inadvertently result from any type of MSW handling system. For example, hauling waste leads to airborne pollution from vehicles, incinerators emit dioxins and furans, and the heavy metal content of MSW compost leads to soil pollution. Environmental impacts can also include long-term contamination of water supplies, endangerment of wildlife, etc. Theoretically, the impact from all actions and plants of the MSW management plan should be compared. Still, how to compare airborne emissions from an incinerator with water pollution from an anaerobic digester? Or how to compare delayed groundwater pollution from landfills and immediate impact from truck noise? A comparative assessment would be a major effort and was not possible within the scope of the study. Environmental impacts are therefore discussed on a semiquantitative basis, including a comparison of emission rates of airborne pollutants for the options involving incineration. A detailed environmental impact assessment will be performed once the size and sites for the various plants are finalized.

Another major factor to be considered is the acceptance of a specific system by the general public, industry, and the government officials involved. The problem of long-term security is evaluated on the basis whether the method selected provides a solution to the waste problem over a long time, which depends on the reliability of the technology and the amount of landfill space required. Flexibility is defined as a qualitative measure of adapting the system to varying external conditions, such as changes in the composition of the waste stream or in environmental regulations.

Table 3 and Figures 2 through 6 demonstrate the results of a comparative assessment which considered landfill requirements, investment costs, specific costs, energy savings, and incinerator emissions.

Option #2, the ORFA system in combination with intensive recycling of household and commercial waste (system A), would result in the lowest landfill requirement. However, the marketability of the ORFA fiber material is considered to be a problem. Although the fiber could be used as a fuel, it would be a very expensive RDF plant in the possible case that direct use in the paper industry is not feasible. Options #1 and #7, intensive recycling of household and

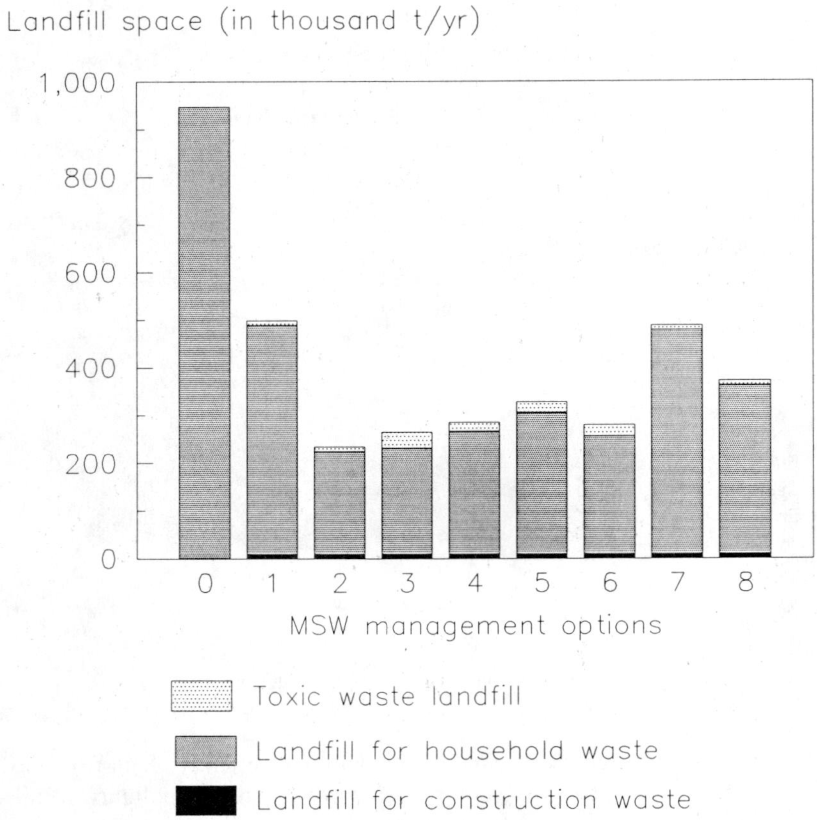

FIGURE 2. Landfill requirement for eight MSW management options.

commercial waste (systems A and B), would result in landfilling of about 500,000 ton/year — about half of the current amount. The largest amount of toxic waste would be generated with option #3, due to the amount of fly ash from the 387,000 ton/year incinerator.

Investment costs for the whole array of plants were evaluated on the basis of preliminary competitive bids from companies which have built or operated similar facilities. Investment and specific costs are low for options #1 and #7, which include recycling systems A and B with subsequent landfilling of the remaining waste. The cost for investment in additional containers and collection vehicles is about twice the cost for recycling system B. Since additional investment costs are rather small compared to the current collection costs (see Figure 4), this difference is not very significant.

The single largest investment would be an incinerator for 270 million Deustche Mark (DM) in option #3. Investment costs for sorting and composting plants are rather small with 59 million DM for options #1 through #6, and would increase to 84 million DM for options #7 and #8 due to the larger amount of recyclables to be sorted.

It is often overlooked that the dominating factor in specific costs per ton of household waste for all options is the cost for collection and transport — more than two thirds of the costs the household pays for solid waste disposal. Discussions about the costs for a recycling plant, for example, are quickly put into the right perspective when the total cost of the system is presented. Also, the landfill tipping fee for Cologne is very low (49 DM/ton) and does not adequately reflect the scarcity of landfill capacity. Hence, the fraction of specific costs due to disposal is very low. Costs for toxic waste treatment and disposal were estimated with 1000 DM/ton. Therefore,

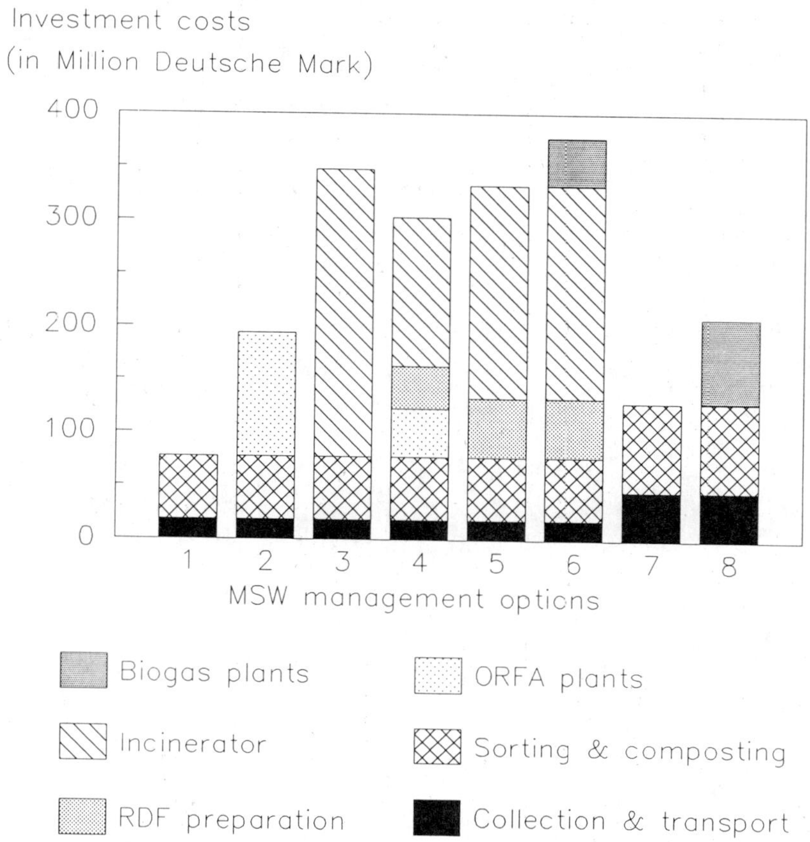

FIGURE 3. Investment costs for eight MSW management options.

despite the fact that the amount of toxic waste generated is small compared to the overall MSW generated, costs for toxic waste treatment and disposal are larger than for landfilling of the nontoxic waste.

Overall, the specific costs for all systems except #1 and #7 do not differ very much given the many uncertainties associated with such cost estimates.

Energy savings from recycling can be larger than the energy production from burning the combustible remaining waste fraction. This conclusion is illustrated in Figure 5. Whereas burning of waste would generate energy in Cologne, energy savings from recycling operations would result at those locations where secondary materials are used. The highest amount of potential energy savings was found for the ORFA system (option #2). This, however, assumes that the system would generate a fiber product with a long-term marketability.

A detailed investigation of environmental impacts is only possible after a site has been selected and plant emission data are available. In this first stage of the concept, the MSW management options were compared on a semiquantitative basis (see Table 3). Such a comparison is by its very nature subjective and should be regarded as a tool to clarify the decision making process rather than as a final environmental assessment.

In order to provide at least an impression of the environmental impacts of the incineration process, the annual emissions of particulates, CO, SO_2, HF, and HCl, were compared for three assumptions: (1) that emissions will merely meet the German Federal Clean Air Standards, (2) that emissions will meet current licensing requirements, or that (3) emissions will be controlled

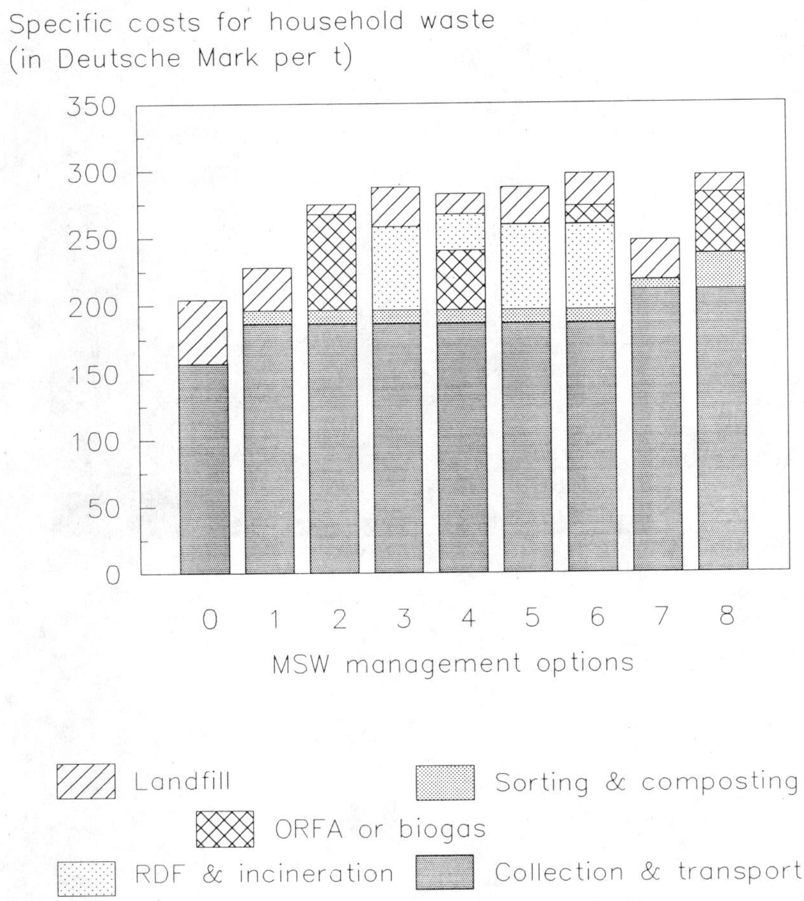

FIGURE 4. Specific costs for household waste for eight MSW management options.

by the best available control technology (BACT). As can be seen from Figure 6, the impact of air pollution control is much larger than the difference between the MSW management options which determine the overall amount of waste to be incinerated.

VI. CONCLUSIONS AND RECOMMENDATIONS

A comprehensive MSW management program is a complex system with many components. Each component is associated with different environmental, societal, and financial risks. The risks and benefits from all MSW management options must be evaluated and compared. The method presented here is a first step to help decision makers analyze, compare, and rank MSW management alternatives.

Landfill space is severely limited or virtually nonexistent in many cities. For many cities, the amount of waste to be landfilled must be decreased by 70%. Since MSW incinerators emit many potentially toxic pollutants, such as heavy metals and dioxins, the amount of waste incinerated should be minimized, and the incinerator waste stream should contain as little hazardous materials as possible. Three overall objectives were pursued in making the recommendation: (1) landfill as little as possible, (2) incinerate as little as possible, and (3) provide a viable MSW management system based on collection systems and plant technologies which have been

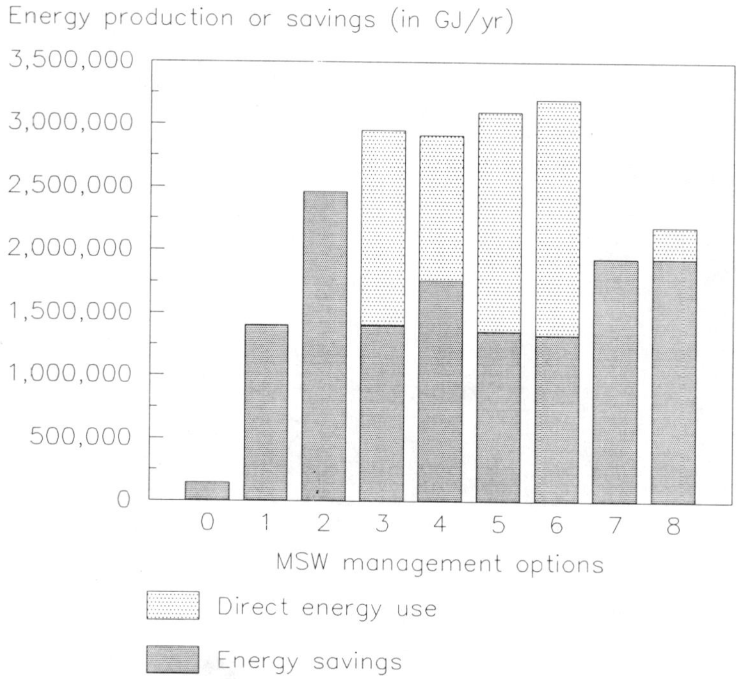

FIGURE 5. Energy savings for eight MSW management options.

reliably operated on a large scale over long periods of time. To achieve these three objectives for Cologne, extensive source reduction and recycling combined with reprocessing the rest and incineration of the combustible fraction appears to be the most feasible solution. Hence, we recommended option #5 (see Figure 1). If sufficient landfill space were available to bury about 500,000 tons of MSW per year for the next two decades, options #7 and #8 would be most feasible.

Of the total of 973,000 tons of MSW generated per year, option #5 requires incineration of only 231,900 ton/year, or 24% of the total waste. RDF combustion in a state-of-the-art MSW incinerator with air pollution control equipment of the highest standard (dual stage wet scrubber, baghouse, denox) was recommended rather than co-firing of RDF in existing boilers. The incinerator will co-generate electricity and heat to be distributed by the city's district heating network. Option #5 requires landfill space for about 300,000 ton/year. Use of the bottom ash was not recommended due to significant environmental contamination from leachate. Furthermore, bottom ash use would compete with material from construction waste recycling which then would have to be landfilled. The suggested system is flexible, allows expansion of the sorting process for RDF production in the future, and has been successfully applied on a large scale. The investment costs for the city were estimated with 332 million DM, of which 18 million DM are for additional containers and trucks, 59 million DM are for sorting and composting plants, 55 million DM are for pre-processing of the waste, and 140 million DM are for the incinerator. The specific costs for household waste collection and disposal will increase 35% from currently 213 DM/metric ton to 287 DM/metric ton.

Other options such as the ORFA system or biogasification systems still have to prove their marketability. If proven to be feasible, they could be added to the overall system (e.g., option #4 or #6), further decreasing the amount of waste that has to be combusted and/or landfilled.

The environmental impacts of any option selected need to be carefully addressed in an

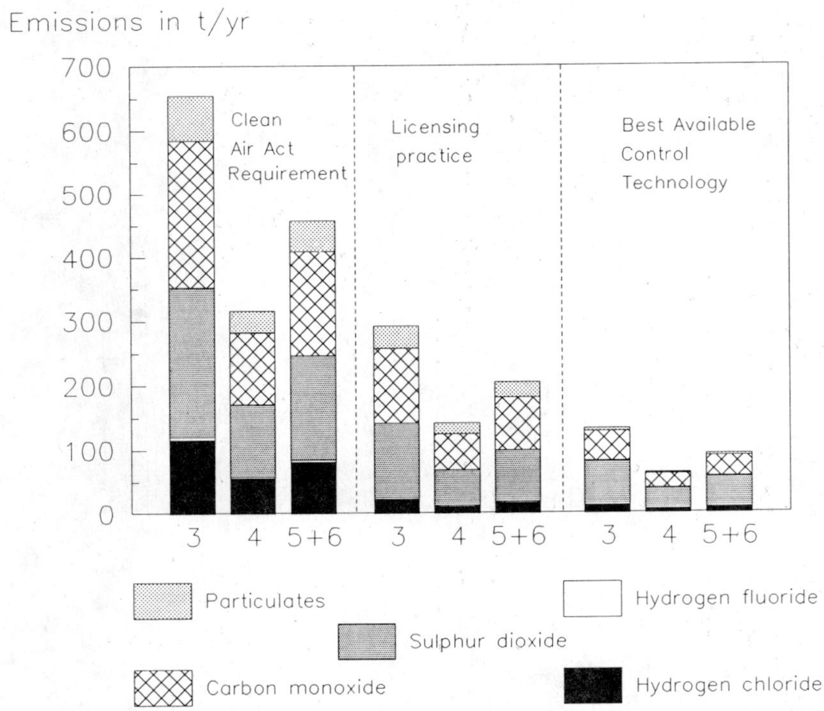

FIGURE 6. Emissions from incineration for four MSW management options involving incineration.

environmental impact analysis. The matrix presented here allows elected officials and the city administration to review the advantages and disadvantages associated with waste management options.

REFERENCES

1. Environmental Protection Agency, Municipal Waste Combustion Study: Characterization of the Municipal Waste Combustion Industry, EPA/530-SW-87-021h, U.S. Government Printing Office, Washington, D.C., 1987.
2. Institute for Energy and Environmental Research, Solid Waste Management Concept for the City of Cologne, Part I, Source Reduction and Use of Waste, Prepared for the City of Cologne, by IEER/IFEU, Heidelberg, West Germany, unpublished data, 1988.
3. **Platt, B. A., Seldman, N., Franke, B., and Mayer, B.,** *Garbage in Europe: Technologies, Economics, and Trends*, Institute for Local Self-Reliance, Washington, D.C., 1988.
4. **Franke, B.,** New Trends in Composting Municipal Solid Waste in West Germany, presented at Pennsylvania Recycling Conference, Harrisburg, May 3 to 5, 1987.
5. **Franke, B.,** Composting source separated organics, *Biocycle*, July 1987, p. 40.
6. **Bilitewski, B.,** Recycling facilities for household and commercial waste, *Beih. Müll Abfall*, 21, 1, 1985.

Chapter 17

THE POTENTIAL OF MATERIALS RECOVERY AND RECYCLING IN WASTE MANAGEMENT

Charles Papke

TABLE OF CONTENTS

I. INTRODUCTION

Integrated solid waste management has gained recent appeal as the waste disposal crisis grows more acute around the U.S. Many state and local governments have discarded their search for a single disposal alternative, such as waste-to-energy, and instead have developed a hierarchy of disposal alternatives, with waste reduction and recycling at the top. As part of waste management planning, recycling goals have been adopted that define the level of waste reduction to be achieved for their community. Setting these goals involves evaluating the performance of recycling technologies and estimating their maximum potential for reducing the amount of solid waste being landfilled. This chapter explores some of the factors involved in this evaluation process.

In 1988, the U.S. Environmental Protection Agency (EPA) adopted a goal of 25% reduction of waste going to landfills within the next 4 years in its solid waste agenda.[1] Some states, such as Michigan, have not only ranked disposal options, but have also projected the role of each method for handling the State's waste.[2] They are

1. Waste reduction and re-use — 12 to 18%
2. Recycling and composting — 28 to 42%
3. Waste-to-energy — 25 to 45%
4. Landfilling — remainder

The question of whether waste reduction and recycling can reduce a substantial portion of the waste stream often arises when a particular percentage goal for recycling is being set. Broad variations in recycling rates reported by different communities adds to the difficulty in determining appropriate goals for waste reduction activities. For example, residents of Wilton, NH (population 10,000) recycle 45% of their trash with a regional recycling program,[3] while citizens of Palo Alto, CA (population 57,200) recover only 7% of the total waste generated,[4] despite the fact that they have one of the oldest and most comprehensive recycling programs in the U.S. Other, usually larger cities with aggressive curbside recycling programs have thus far recovered even smaller percentages, while cities like Woodbury, NJ (population 10,700) boast a much higher recovery rate of 55%.[5]

Why are there so many variables in recovery rates, and what, if anything does this indicate about the potential of recycling to divert municipal solid waste (MSW) from landfills? Some advocates claim that vigorous recycling programs can solve our nation's waste disposal problems by achieving waste reduction rates as high as 80%,[6] but other recycling experts feel that more realistic expectations range from 20 to 30%.[7] The EPA estimates that approximately 5 to 10% of municipal wastes were recycled in 1988 using all methods of recovery.[8]

Both groups site specific community programs to support their contentions, but the real potential for recycling in waste reduction is difficult to determine. Closer examination of existing recycling programs around the U.S. can provide some clues, but many factors need to be considered when analyzing the potential impact of recycling on the waste stream. Four factors discussed in this chapter are

- Municipal solid waste: its quantity and composition for a given community
- Selection and use of appropriate recycling technologies to maximize recovery
- Keys to successful recycling programs: program design, promotion, and management
- Unbiased determination of recycling's cost-effectiveness as a waste management alternative

Accurate analysis of these factors can better predict how recycling can alleviate a community's waste disposal problems. External conditions, such as markets for recovered materials that are

TABLE 1
Comparison of Waste Composition Data from Selected American Communities

Recyclable material	Essex, NJ[a] Resid.[b]	Essex, NJ[a] Comm.[c]	Percent of total MSW: Northern California[d]	Marblehead MA[e]	National average[f]
Paper					
Newspaper	6.27	2.96	6.70	15.1	6.7
Cardboard	11.60	23.21	6.8	1.5	8.9
Misc. paper	NA[g]	NA	26.5	22.1	21.5
Total paper	NA	NA	40.0	38.7	37.1
Metals					
Ferrous metals	6.33	6.96	8.0	4.3	8.3
Nonferrous	1.36	0.74	2.0	0.6	1.3
Total metals	7.69	7.70	10.0	4.9	9.6
Glass	9.02	3.63	8.0	17.6	9.6
Plastics	6.24	4.35	2.5	NA	7.2
Yard wastes	NA	NA	3.0	NA	17.9
Remainder	59.12	58.15	36.5	48.8	18.5

[a] Reference 26.
[b] Residential waste.
[c] Commercial waste.
[d] Reference 30.
[e] Reference 21.
[f] Reference 31.

beyond the proper design and operation of a program, can limit the potential of recycling and its waste reduction impact. However, awareness of these and other constraints, and the possibilities for overcoming them are part of proper program management which is discussed in Section VI of this chapter.

II. MUNICIPAL SOLID WASTE: QUANTITY AND COMPOSITION

The source and composition of MSW are as important to the success of recycling as they are for incineration projects. The BTU value of the waste stream, available quantities, and relative amounts of combustible materials are important to incineration projects. Similarly, source separation recycling projects require more information about the amounts and ingredients of waste components as well as quantity variations per route, season, neighborhood, and waste generation sources. Detailed knowledge of these factors, however, is not as important for nonsource separation projects, such as disposal site recycling, as they are for source separation programs that require extensive collection procedures for their operation.

MSW is generated from two sources: the residential and commercial sectors of the community. The methods of collection usually vary for each source. Since most community recycling programs target only residential wastes with drop-off centers and curbside collection programs, the diversion impact on the total MSW generated is influenced by the relative proportion of residential solid waste in the entire waste stream.

The quantity of material recovered in residentially oriented recycling programs also depends upon the composition of MSW for a particular community. Table 1 compares the results of waste characterization studies for three areas of the country and indicates how quantities of different recyclable materials can vary. Newspaper, for example, comprises 15.1% of the MSW in Marblehead, MA, a suburban residential town, but comprises only 6.27% of the residential MSW in Essex County, New Jersey, a more urbanized community. The amount of other waste materials, such as glass and plastics, can also vary widely (Table 1).

Therefore, when comparing the effectiveness of residential recycling programs, the amount of each material that is available in the waste stream of each community should be noted. A residentially oriented program that recovers 50% of household newsprint, glass, and cans may divert over 30% of the total MSW generated in a rural community, but only 5% in a more industrialized urban area. Furthermore, when these figures are calculated for all of the materials targeted by a typical recycling program, the total waste diverted can vary even more for any given program. This example can explain some apparent differences in the impact of recycling on waste disposal.

Historically, community recycling programs have targeted specific materials in the residential waste stream using drop-off centers, buy-back centers, or curbside collection techniques. Only a few materials, primarily office paper, have been targeted in commercial wastes by organized programs that operate regardless of current market prices. Market-oriented recyclers collect mainly cardboard and other easily recoverable materials from local businesses and industry. For the most part, their efforts tend to fluctuate with the strength of the market. As a result, entrepreneurial collectors do not have a consistent waste reduction impact and usually remove only the higher market-valued materials. In this sense, these entrepreneurial efforts can adversely impact other recycling programs by reducing revenues that would otherwise be generated from the sale of these materials.

A. TYPES OF SCRAP MATERIALS RECYCLED

Certain large cities without extensive curbside recycling programs claim recovery rates over 20%, while other cities, such as San Jose, CA, with America's largest curbside recycling program, serving over 700,000 residents, are recovering less than 5% (Table 2).[9] Discrepancies like this are better understood by examining the definitions of recycling and the type of waste materials being recovered.

Postconsumer scrap, or waste materials commonly collected in community recycling programs, such as newspapers, glass bottles, and aluminum cans, is generated by consumers after products and packaging materials have been used and discarded. Most recycling collection programs track these materials in their recycling statistics, however, other materials are more difficult to track and can generate city-to-city recycling rate discrepancies between 10 and 20%.

One type of scrap material that rarely ends up in the waste stream is called preconsumer scrap. This secondary scrap is recycled "in-house" and is put back into the manufacturing process without being discarded. For instance, Bell Telephone recycles wire and cable scrap that are generated when workers replace or repair telephone lines and equipment. Copper scrap is delivered from nationwide Bell collection depots to the company's own North Carolina reclamation facility where it is processed into new copper rods.[10] Most preconsumer scrap is not usually landfilled or included in the statistics for recycling programs. With growing pressure to achieve high recycling rates, these materials are occasionally incorporated into recovery rate statistics.

Between these two types of wastes are scrap materials that traditionally have been recycled by scrap metal and paper stock dealers and have, therefore, not been part of the waste stream. These scrap materials are generated by converters, fabricators, and other retail manufacturers who usually save their scrap for recycling. To boost their recycling rates some cities have included these materials in their recovery totals. However, other cities include only the strict postconsumer scrap in their statistics, because they are focusing on methods to reduce their current waste volumes.

B. WASTE SOURCES

Another reason for discrepancies in recycling rates involves the geographical and categorical sources of MSW. Most scrap dealers do not note the location where a particular load of scrap originated, and some brokers buy and combine scrap loads from different dealers or generators.

TABLE 2
Comparison of Recovery Rates for Curbside Recycling Programs[a]

Location of program	Population[b]	Recycling goal (%)	Recycling rate[c]	Percent of MSW diverted
Ann Arbor, MI	40,000	25	58	10
Austin, TX	88,000	15	47	3
Boulder, CO	22,000	50	20	4
Davis, CA	9,500	NRG[d]	40	6
El Cerrito, CA	7,500	25	22	29
Groton, CT	11,000	25	15	10
Islip, NY	73,500	50	8	4
Madison, WI	177,700	NRG	52	26
Marblehead, MA	20,100	NRG	33	5
Marin Co., CA	40,000	30	30	5
Minneapolis, MN	12,000	16	11	1
Montclair, NJ	14,500	25	114	45
Palo Alto, CA	15,800	25	25	7
St. Louis Park, MN	16,700	16	55	4
San Jose, CA	20,700	28	8	5
Santa Rosa, CA	29,000	NRG	7	4
Springfield Township, PA	6,600	NRG	8	12
Sunnyvale, CA	129,000	25	19	3
Wilton, NH	109,000	NRG	8	18
Woodbury, NJ	8,600	85	74	90

[a] Some curbside programs also use drop-off, buy-back, composting, and other recycling options.
[b] Population directly served by the curbside recycling program.
[c] Recycling rate is in pounds per month per person.
[d] The local agency had not adopted a specific recycling goal as such but may have program goals.

From Resource Management Associates, 1989.

Since tracking of scrap materials is not a general practice, it is difficult to avoid including outside scrap materials in a city's recovery rate calculations. In addition, because surveys usually include data from collectors as well as scrap dealers, brokers, and end-users, it is possible to double or triple count some of the same scrap materials. Thus, surveys that do not carefully identify the geographical source of scrap materials and track them through the recycling system may be generating artificially high recovery rates.

MSW has been defined as solid wastes generated by the residential and commercial sectors of a community. There are other categories of waste sources that are sometimes included in MSW, such as industrial, construction and demolition debris, institutional, and occasionally sewage or agricultural wastes. Industrial and institutional wastes can substantially increase the total amount of MSW generated if they are included in the definition of MSW, particularly for urban and industrial areas. Some jurisdictions may include any solid waste that will be landfilled, while others restrict the definition of MSW to that which is hauled by traditional garbage packer trucks.

These differences in the definition of MSW can also lead to wide variation in recycling rates. Since calculations of the percent of waste diverted from landfill by recycling are based on the total MSW for the community, recycling rates are lowered as more waste source categories are included in the MSW definition.

In conclusion, there are several factors related to the quantities, composition, sources, and categories of MSW that affect the calculations for recycling rates. Consistent definitions of these waste and scrap categories as well as consistent methodologies for calculating waste reduction rates are needed to better evaluate the performance and potential of recycling upon waste.

Option	Waste Diversion	Recycling Convenience	Use Elsewhere	Program Costs
DROP-OFF CENTERS	Typical target materials: newspaper, cans, glass bottles. Recovery rates less than 2% of municipal wastes. Volumes generally 10-50 tons per month per center.	Hauling materials to centers lowers convenience. Can provide only opportunity for recycling in rural communities. Added convenience with 24 hr. service for more types of waste.	Most commonly provided recycling service - thousands in U.S. Innumerable single material centers, especially newspaper.	Can be the least expensive option. Fully staffed center may cost $40 per ton; lower with volunteers.
BUY-BACK CENTERS	Waste diversion of 1 to 5% for large multi-material sites. Limited diversion for single-material such as aluminum cans.	Requries generator to transport material. Provides financial incentive.	Commonly in use throughout the country. Mostly private operations, both for-profit and non-profit.	Start-up costs similar to drop-off centers; operational costs higher. Generally a profit generating operation.
CURBSIDE RECYCLING COLLECTION	Recovery rates up to 10-25% of municipal wastes. Target materials - newspaper, cans and glass containers usually.	Most convenient option for materials recycled. Weekly collection increases recovery rate.	500-1000 newspaper only collection programs nationwide. 400-600 multi-material programs nationally; number increasing.	Full program costs can be less than solid waste collection and disposal. Net program cost typically $20-50 per ton, or 30-90 cents per household per month.

	Can recovery 200-400 tons per month per 100,000 population.		Large portion of programs require citizen participation.	Mandatory partici-pation can reduce costs.
DISPOSAL SITE RECYCLING	Target materials include newspaper, cardboard, glass, cans, plastics, yard waste, plus ferrous metals, reusable items and others. Target materials compose 10-45% of the waste system.	Properly designed service can be conven-ient to facility users.	Practice being re-established with modern equipment and procedures. Some facilities also employ manual sorting from mixed wastes.	Highly variable depending on type of operation. Large-volume process-ing sites have high capital and operation costs. Programs can be designed to generate profits.

FIGURE 1. Recycling option overview.

Option	Waste Diversion	Recycling Convenience	Use Elsewhere	Program Costs
YARD WASTE COMPOSTING	Yard waste is generally between 8 and 20% of the waste stream. Drop-off centers can capture 20% of the yard waste. Curbside collection can capture essentially all yard waste.	Drop-off center programs require users to self-haul. Curbside collection adds convenience to residents.	Mixed success with past programs. Yard waste compositng on a large scale is relatively new.	Start-up and operation costs similar to curbside collection programs. Net costs comparable to total solid waste disposal costs.
OFFICE PAPER RECOVERY	Comprises 3-5% of the waste stream. Typical recovery rates: up to 50% of available materials per location	Office separation systems convenient for office staff. Recovery by sorting mixed office waste increases staff convenience and program costs.	Hundreds of successful programs, several operating from mid-1970.	Start-up and operation costs moderate. Costs generally covered by avoided disposal fees and/or sale of materials. Large-volume programs can generate revenues.
COMMERCIAL WASTES COLLECTION	Recovery rates from 10-60% of material, depending on design. Target materials	Medium convenience-requires some material preparation before collection.	Some collection activitity in all major metropolitan areas. Regular collection	Generator's costs can be offset by avoided disposal costs. Collection costs

	typically: cardboard, some glass, plastics and metals.		programs are rare.	generally less than revenues from sale of materials.
SPECIAL MATERIALS RECYCLING: PLASTICS, TIRES & OIL	Plastics comprise 6-9% of solid waste; tires and oils less than 1%. Existing recycling recovers small portion of available waste materials. Removal of these materials can make waste disposal easier.	Tires and oil must be taken to appropriate collection sites. Plastics recycling opportunity least available.	Except for large generators, recycling level is low for all materials. PET & PE plastics most common types recycled. Retreading most common form of tire recycling; some tire-to-energy use.	Curbside collection of some plastics and oil adds small cost to program. For tire generators, recycling costs lower than disposal where available.

FIGURE 1 (continued).

III. RECYCLING TECHNOLOGIES

Another important factor to consider when comparing recycling programs in different communities is the recycling method used. A drop-off center approach, or even a network of centers, is not as effective as a weekly curbside collection program, and neither system alone is as effective as concurrent implementation of both methods. Thus, a comprehensive recycling program using a combined approach can recover a significant portion of the waste stream as illustrated in Figure 1, which compares eight different programs. The ideal type of integrated program would include collection of yard wastes (for composting), plastics, commercial wastes, office papers, wastes from multifamily housing, and other special refuse, which together target nearly all recycling materials in the commercial and residential waste stream. As might be expected, comprehensive recycling has the best chance of achieving the high goals adopted by many jurisdictions. By surveying municipal programs, it is possible to estimate the impacts of a single recycling method. Since no one city has yet combined all technologies, it is difficult to predict the effectiveness of a truly comprehensive recycling program.

Of the eight recycling methods shown in Figure 1, disposal site recycling (DSR) differs from the other recycling technologies because it is not a source separation method. Source separation methods require recyclable materials to be separated from other wastes at the source of their generation, which necessitates more public cooperation and promotion to attain high recovery levels. Conversely, DSR systems remove recyclables from mixed wastes after they have been collected and transported to centralized disposal facilities, such as transfer stations or landfills. DSR systems, such as those used by Marin County, California, employ a combination of mechanical and manual means to pick through selected loads of refuse that contain high concentrations of recyclable materials, particularly corrugated cardboard and wood wastes. MSW has been processed via this method in some European cities,[17] but the technology is not as well developed in the U.S.

With respect to waste diversion potential, three options rank highest: (1) curbside collection; (2) yard waste composting; and (3) DSR. Each option can remove 10 to 20% of the waste stream depending upon actual program design and waste composition. However, no community has implemented all three options at once. The usual practice for implementing an intensive recycling program is to initiate a residential curbside program to collect multiple types of recyclable materials (called "multimaterial collection"). In a second phase, a composting program may be added, first for yard waste, brush and trimmings, and later for other organics such as wood and food wastes. So far, multimaterial DSR programs have not been fully implemented, but two multifaceted recycling programs, one serving Marin County, California, and one serving San Jose, CA, are described below to show how different recycling options can be integrated.

A. MARIN COUNTY, CALIFORNIA, CASE STUDY

About 138,000 of the 240,000 Marin County residents have had a weekly curbside recycling service since 1980. The program is operated by a consortium of local garbage companies, Marin Recycling, who received $575,000 in state funds to buy trucks and processing equipment for the program. After a 1-year pilot program, the company required additional funding to continue operation. The County Board of Supervisors approved a request by the company to assess a monthly household surcharge of $0.60 to maintain its collection service. The surcharge dropped annually thereafter to a low of about $0.40, but increased abruptly in 1987 to $1.00 to cover the costs of constructing and operating a $9 million Materials Recovery Facility (MRF). The new plant currently processes largely commercial waste that has high concentrations of cardboard, metals, glass, wood, and yard debris.

Marin Recycling gathers and processes source-separated materials that are collected in a combined residential and commercial program incorporating curbside collection and other recycling options. It has constructed a facility for separating the mixed (commingled) recyclables and processing them for market use. This materials processing facility (MPF) also receives some recyclables that have been recovered from mixed refuse in the MRF, which is located next to a nearby refuse transfer station. To recover the recyclables, the MRF primarily relies on manual picking from wastes that have been dumped on a floor and run over a series of conveyors and other mechanical devises. A wider variety of materials are recovered from the trash including cardboard, yard debris, wood waste, metal scrap, and plastics when markets are available.

Marin County's combined curbside, drop-off, and buy-back center recycling program collected over 18,000 tons of recyclable materials in 1986 requiring that only about 8.2% of the County's waste stream be landfilled.[11] The source-separated tonnage represents about 18% of the residential MSW generated by households in the service area. With the recovered materials in the MRF and expanded collection of materials from the commercial sector, total recovery reached nearly 22,000 tons/year by 1988, achieving levels of 22% waste diversion for the total recycling program.[12]

B. SAN JOSE, CALIFORNIA, CASE STUDY

South of San Francisco lies San Jose, which currently oversees the nation's largest curbside recycling program, serving over 700,000 residents. In 1985, the City contracted with a private garbage company to operate a pilot curbside collection service for about 21,000 households. The pilot area served two separate parts of the city, Berryessa and Willow Glen, which had similar demographic and geographic characteristics.

Similar to Marin County, the City had several drop-off and buy-back recycling centers, newspaper drives, and other recycling activities before city officials decided to expand the program's scope. Unlike Marin County, the City Council was unable to obtain outside funding for their pilot program and was forced to allocate $100,000 from its refuse disposal fund to conduct a 1-year pilot study. Based upon the recommendations of an earlier feasibility study,[13] the City decided to determine the impact of providing households with curbside containers for storing their cans, glass, and newspaper. Specially designed three-part stackable containers were given to every household in the Berryessa area while those in the Willow Glen portion were given no special containers.

The participation data in Table 3 show that the Berryessa recovery levels were about double those in Willow Glen. Further, when containers were later given to the Willow Glen residents, participation increased to a level nearly matching Berryessa.[14] Other factors, such as publicity, education efforts, and the type and level of service remained nearly identical for both neighborhoods.

The City conducted a thorough evaluation of the pilot program,[15] and decided to expand curbside collection city-wide in 1987. A more recent study showed that while participation and recovery rates are high, total waste diversion remains less than 5% of the City's total MSW stream.[16] Since residential wastes comprise nearly half of the total MSW, the curbside efforts are removing only about 10% of the residential waste.

Again, definitions play a role in judging the merits of various recycling methodologies. Perceived effectiveness of any given method will depend upon how effectiveness is measured. For example, an aggressive office paper recycling program can be virtually 100% effective in removing all types of paper targeted, but still have a negligible impact on reducing the *total* waste stream from the office building or the City. Many curbside recycling programs boast 60 to 70% public participation, yet they divert less than 10% of the total residential MSW (as seen in San Jose's program). Although participation levels are not necessarily indicative of a program's waste diversion effectiveness, a few programs measure their effectiveness in these terms. As a result, high participation of tonnage figures can be misleading in determining a program's true

TABLE 3
Material Recovery Rates for Selected Curbside Recycling Programs

Location	No. of households	Tons/year	Percent participation	Recovery rate
Austin, TX	88,000	8,347	25	15.8
Davis, CA	9,400	1,998	65	35.4
Palo Alto, CA	15,800	3,520	65	34.9
El Cerrito, CA	7,600	1,140	45	25.0
San Jose, CA				
Berryessa	9,500	1,445	58	25.4
Willow Glen	11,200	832	25	12.4
Marin Co., CA	40,000	9,900	60	41.2
Ann Arbor, MI	40,000	1,764	21	7.4
Montclair, NJ	14,500	3,382	80	38.9
Islip, NY	73,500	14,400	20	32.7
Boulder, CO	22,000	3,500	35	26.5
Springfield Township, PA	6,630	1,659	60	41.7
Naperville, IL	22,000	2,784	40	21.1
St. Louis Park, MN	16,700	1,800	28	17.9

waste diversion impact. As might be expected, this approach also adds to the confusion regarding diversion rates.

Because participation rates are difficult to measure accurately and can give misleading information about the effectiveness of a program, another procedure for measuring program performance is gaining wider use. This performance index measures the total pounds of MSW collected each month per household in the service area regardless of whether the household actually participated in the recycling program. The formula for the calculation is based on program data primarily obtained by the operator. The recovery rate formula is

$$\text{Recovery rate (pounds/HH/month)} = \text{(total TPM) (2000)/HH served} \quad (1)$$

where TPM = total tons of all recyclables collected by the program per month and HH = total number of households served by the program.

The recovery rate calculation correlates to participation since higher participation increases recovery rates. However, the correlation may not be linear, since households may participate by recycling unequal amounts of materials.

The recovery rate measures more directly the volume of materials removed from the waste stream and therefore provides a better figure for comparing the effectiveness of similarly designed recycling programs. Table 4 lists a number of curbside recycling programs serving cities across the U.S. and compares both the participation rates claimed by the program operator and the recovery rates calculated by using the above formula. Comparing the two sets of measurements shows that there is not a strong correlation between participation and recovery rates. However, since the goal of recycling is to divert materials from the waste stream, the recovery rate measurement more accurately estimates the effectiveness of a recycling program.

IV. KEYS TO SUCCESSFUL RECYCLING

Three key factors for high recycling performance are well known but not always emphasized. Convenience to the public, community awareness, and sound program management must be considered when designing any integrated program.

TABLE 4
Mandatory Ordinance Impact on Participation in Curbside Recycling Programs

Programs with Mandatory Ordinances			
Weekly collection		**Biweekly/monthly collection**	
Location	**Percent participation[a]**	**Location**	**Percent participation[a]**
Berlin, NJ	90	Barrington, RI	35
Dover, NJ	7	Monitowoc Co., WI	30
Groton, CT	85	Monroe Township, NJ	25
Haddonfield, NJ	60	Montclair, NJ	70
Islip, NY	40	Roxbury, NJ	85
Longmeadow, MA	80	St. Cloud, MN	44
Montgomery Co., MD	50		
Woodbury, NJ	85		
Hamburg, NY	98		
Program with Voluntary Participation			
Albany, OR	26	Bellingham, WA	19
Ann Arbor, MI	30	Bend, OR	4
Austin, TX	50	Berkeley, CA	30
Beaverton, OR	10	Boscobel, WI	60
Corvallis, OR	40	Burbank, CA	50
Davis, CA	80	El Paso, TX	5
El Cerrito, CA	50	Grand Rapids, MI	10
Hyde Park, IL	24	Los Altos, CA	65
Kitchener, ON	80	Minneapolis, MN	22
Madison, WI	25	Naperville, IL	28
Marin Co., CA	50	Port Townsend, WA	13
Mecklenbury Co., NC	18	Ritchfield, MN	30
Monroe Co., PA	52	St. Louis Park, MN	40
Palo Alto, CA	65	Santa Monica, CA	35
San Jose, CA	57		
Springfield, PA	65		
Sunnyvale, CA	58		

[a] Program participation rates are those claimed by the program operator.

From Resource Management Associates, 1989.

A. PROGRAM DESIGN

Recycling opportunities, whether they are voluntary or mandatory, must be convenient. Drop-off centers located at or near disposal sites where most residents take their trash attain higher recovery rates than those in hard to find or out of the way residential locations. Likewise, curbside programs with weekly collection are far more effective waste diverters than programs with less frequent or irregular collection schedules.

Design is also important for maximizing the efficiency of materials handling and processing. Providing household recycling containers to residents can dramatically increase recovery rates and participation.[18] Careful integration of recycling collection with refuse collection and disposal can also improve the impact of a program. The public is more likely to participate in recycling when it parallels with an existing garbage schedule.

Overlooking these design factors can severely limit the effectiveness of a recycling program and burden the effort with excessive cost and labor requirements. Some less effective programs handle and process materials more than necessary. Others, which use en route sorting in curbside collection, for example, could save labor and vehicle time by introducing simple processing steps for partially comingled materials. Also, expensive glass crushers are sometimes used when more efficient methods of handling and transportation of glass are available.

Another element in the design of a curbside recycling program is mandating public participation through an ordinance from the local or state government. In past years, most curbside programs did not require participation and depended upon education and promotion activities to encourage voluntary participation. However, with the growing crisis in locating new landfills, many communities and some states, such as Rhode Island, New Jersey, and Connecticut, began requiring residents to regularly separate certain recyclables from their trash.

While passing a law mandating recycling participation can have some impact on recovery rates,[19] it is clear that the degree of enforcement plays a larger role. When towns with and without mandatory ordinances are compared as shown in Table 5, the impact of participation is inconclusive. More data currently being gathered in states with mandatory recycling laws may clarify the role of enforcement in achieving higher participation and recovery rates. Even in communities with mandatory participation laws, making residents aware of laws and enforcing them is still necessary for achieving high participation levels. In many cities with mandatory ordinances, less than half the residents participate in their curbside recycling programs.

B. PUBLIC AWARENESS

Promotion has perhaps been the most important, single factor in determining the success of a program. In order to be effective, the promotion of a recycling program must involve all sectors of the community and use a wide range of educational materials and activities. In addition, these efforts must be continuous.[20] Many programs have devoted thousands of dollars to start-up promotion and then found participation declining when follow-up activities were dropped. In fact, most voluntary programs that have achieved high participation levels use full- or part-time staff to distribute materials and continue educational activities.

Most recycling programs, particularly those previously initiated, have not used sophisticated educational and promotional activities to increase awareness of the residents being served by the program. Even fewer programs have attempted to evaluate the effectiveness of the materials and events used to enhance public awareness. These aspects were evaluated for three programs, Marblehead and Somerville, MA,[21] and Palo Alto, CA.[22] Table 5 summarizes the results of the Palo Alto case study. The results show that two of the most effective activities were a letter from the mayor or city officials urging residents to participate in the program and personal communication through door-to-door solicitation. This case study also revealed that promotional activities need to be continuous for maximum impact on public support and, hence, materials recovery.

C. PROGRAM MANAGEMENT

One of the harder to define factors in evaluating effective recycling programs is the need for devoted and qualified managers. While many recycling programs have failed over the years due to poor market conditions, careful planning and management can increase operating efficiency and the ability to endure bad market conditions. Creative research can sometimes uncover new marketing opportunities for improving program revenues.

The causes of recycling program failure have received little attention in the past, although one case study of an established program in Fresno, CA, did point out some of the problems that are likely to occur when a program is poorly managed.[23] The City of Fresno's monthly collection program served nearly 300,000 residents but suffered from low participation and recovery rates. In addition, promotion was poor and sporadic, and operating costs were higher than normal due

TABLE 5
Public Awareness Activities for Palo Alto Recycling Program

Activity	Used in program?[a]	Materials cost[b]	Rated impact[c]
1. General approach			
a. Strategy, planning, and consulting	Yes	7,250	A
b. Logo/strategy	Yes	300	A
2. Generating active support			
a. City endorsement and letter	Yes	2,300	B
b. Community advisory committee	Yes	0	A
c. Special group support and planning	Yes	0	B
3. Media, printed and electronic			
a. Press releases	Yes	N/A[d]	A
b. Feature stores and editorials	Yes	0	A
c. PSAs, public announcements, etc.	Yes	0	A
d. Community newsletter and programs	Yes	0	C
e. Telephone book cover	Yes	0	A
4. Printer materials			
a. Brochure and fact sheet	Yes	N/A	A
b. Posters	Yes	900	C
c. Bumper stickers and decals	Yes	260	D
d. Displays	Yes	750	A
e. Slide show	Yes	4,200	A
f. Miscellaneous graphics and printing	Yes	3,290	—
5. Community outreach programs			
a. Public survey	Yes	0	C
b. Door-to-door canvassing	Yes	N/A	A
c. Billing inserts	Yes	N/A	B
d. Schools, churches, and civic group involvement	Yes	N/A	C
e. Workshops and seminars	Yes	N/A	B
f. Program newsletter	No	—	—
g. Mobile exhibit	Yes	2,844	A
h. Theatricals	Yes	N/A	B
6. Paid advertising			
a. Curbside emphasis	Yes	400	C
b. Buy-back emphasis	N/A	—	—
7. Other activities			
a. Fund raising by community groups	No	—	—
b. Special events (e.g., parades, celebrations, etc.)	Yes	—	—
Total costs		$22,194	

[a] Activities used during or immediately prior to grant period.
[b] Cost does not include labor or in-kind contributions.
[c] Impact rated after discussion with the program director with decreasing impact going from A to E.
[d] Specific cost figures not available.

From Resource Management Associates, Case Study of Curbside Recycling Program for the City of Palo Alto, California, 1981.

primarily to an excessively large staff. Program supervisors with direct knowledge of the daily activities were allowed little authority in program management. Since these programs were left uncorrected, the City Council eventually voted to end the program.

While this study and others[24] did not thoroughly analyze program management problems, the results indicate the importance of the program manager in being able to identify problems and take the necessary steps to correct and overcome them. Other constraints and resources can limit the flexibility and choices of management, but the success of most recycling programs can usually be attributed to one or two key management personnel who have been able to tackle obstacles and maintain efficient and cost-effective programs.

Recycling collection activities are still developing and have, in most cases, not yet been fully integrated into municipal waste management practice. In these circumstances, most successful recycling programs have required strong leadership and dedication from one or more staff members to initiate programs and develop their recovery potential. As recycling becomes more institutionalized and integrated into waste management, the need for ardent managerial effort may not be as great.

V. RECYCLING COST-EFFECTIVENESS

Two constraints to wider implementation of municipal recycling programs are high start-up and operation costs coupled with the expectation that recycling needs to pay for itself. Burdened with the requirement of self-sufficiency, it has been difficult to justify the cost-effectiveness of a program or to implement broad recovery programs in cities where commercial recycling projects already exist. However, this economic view ignores the updated role of recycling as a waste management tool designed to work together with reuse, composting, incineration, and landfilling.

Residents can reduce the amount of waste which must be disposed of by changing their buying habits and by recovering materials from the waste stream through recycling. Avoiding products with excessive packaging where possible is the most cost-effective way to reduce waste. However, items saved for recycling no longer carry the costs of disposal, and this avoided cost often becomes the second most cost-effective manner of waste disposal.

While there is cost for collecting and processing recyclables, these costs are offset by the revenues generated from the sale of recovered materials. Another economic advantage of recycling that is often overlooked is the fact that recovering recyclable materials from the waste stream cuts down on the volume of waste that would otherwise have to be collected, transported, and disposed. Thus, the net cost of recycling materials is often lower than the net cost of disposing of those materials in landfills or through incineration. The savings generated by avoiding the costs of disposal operations can be credited to recycling efforts. In many cases these avoided disposal credits are not included in the economic analysis of recycling, and as a result, the net cost of the program is overestimated. Furthermore, the chance of including recycling in a community's waste management plan is reduced because other disposal options appear to be more economically attractive.

However, a growing number of communities that have adopted full and balanced accounting for their recycling programs view recycling as cost-effective. They justify recycling not only in terms of the short-range cost savings described earlier, but also in terms of the long-range savings associated with having to dispose of less waste in the future. Less waste disposal translates into lower overall disposal costs. Figure 2 illustrates this concept by comparing costs associated with recycling, landfill, and incineration.

VI. OVERCOMING LIMITATIONS TO RECYCLING

Nearly everyone acknowledges that all of our waste cannot be recycled, but what factors really limit the potential of recycling in relation to total waste reduction? Answers relate to the technical and social aspects of recycling. Technically, not all waste materials are easily separable or have sufficient value to warrant the cost of separation. Also, the lack of adequate resell markets can limit recycling of some waste materials. Even when technical obstacles are cleared, social values play a role in the effectiveness of a program.

The social aspects of recycling include public education programs to increase awareness, remove biases, and change behavior patterns and public willingness to accept regulations that

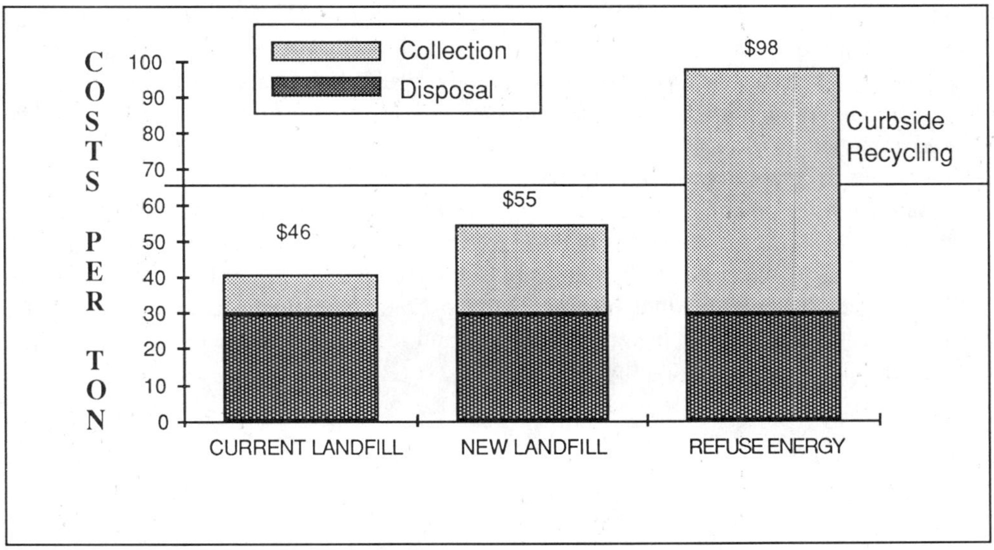

Note:

Excluded from these estimates are transfer hauling costs. Additional costs would be incurred to haul the waste to facilities further away than a city's current disposal site.

FIGURE 2. Summary economics of solid waste alternatives.

encourage and mandate recycling work in tandem with public support to establish economic incentives for developing technologies and markets for secondary materials.

Each of these issues contributes to the potential effectiveness of recycling, but perhaps the most important factor is the lack of public knowledge about the problems of waste disposal. A growing awareness of the Earth's limited resources has given recycling avid ecological support. Furthermore, communities are more pressed than ever to adopt recycling programs because of problems associated with locating new disposal methods or sites. A recent survey of state landfill capacity revealed that 14 states have 5 years or less left of remaining space. Moreover, all but five states claimed having less than 20 years of remaining capacity.[25]

The prospect of searching for new landfill sites coupled with building new MSW incinerators has triggered many communities to investigate ways of reducing the volume of waste generated. Less garbage means lower disposal costs in the long run. It also can result in lower capital costs for construction of waste-to-energy plants, as shown by the example in Essex County, New Jersey.[26]

There is renewed interest in overcoming some of these barriers to recycling. The EPAs new agenda[8] has identified several areas for increasing the effectiveness of recycling. While there have been studies investigating technological barriers to recycling, such as better processing methods and equipment for glass and metals,[27] more research is needed. Other studies are underway to further integrate recycling into solid waste handling procedures.

Market fluctuations for collected recyclables have triggered problems and even caused the failure of some recycling centers and programs. Inadequate market demands and material gluts can limit waste collection and strain a program financially. Materials such as tin-plated steel cans and plastics which do not benefit from established markets can present the same types of problems as an overabundance of recyclables collected due to inadequate market demands. For example, in 1986, Wellesley, MA, witnessed a 24% drop in recovered newspaper revenue despite a 25% increase in citizen participation of recycling newspaper supplies.[28]

Mandatory use of renewable resources such as recycled newsprint by publishing companies is a growing area of debate.[29] Besides market resistance, more mid-level processing facilities are needed to bridge the gap between raw recycled and processed recyclables more easily used by manufacturers. As additional recovery programs emerge, disruption of existing markets can also be a severe problem. More research and development of new market opportunities is needed to overcome limits imposed by lack of markets.

In summary, the potential for recycling to help solve MSW disposal problems promises to have a significant impact. However, to do so, recycling programs need to be comprehensive, target a full range of materials, including yard and commercial wastes, and be well designed and well managed. When fully integrated into the overall waste management strategy of a community, waste reduction, materials recovery, and recycling promise to be cost-effective tools in the future management of MSW.

REFERENCES

1. **Salimando, J.,** EPA'S Porter sets goal: 25% recycling by 1992, *Waste Age,* March 1988.
2. Community Assistance Division of the Michigan Department of Natural Resources, Resource Recovery: A Statewide Waste Management Strategy, Michigan Department of Natural Resources, Lansing, 1983.
3. **Moore, P.,** Success is 45% recycling, *Biocycle,* May/June 1987.
4. City of Palo Alto, The Palo Alto Materials Recovery Program, A Summary Report, City of Palo Alto, CA, 1986.
5. **Freeze, J. and Hood, H.,** Survey of U.S. Recycling and Waste Diversion Potential, January, personal interviews conducted by Resource Management Associates, 1988.
6. Environmental Defense Fund, Coming Full Circle, Successful Recycling Today, EDF Environmental Information Exchange, Washington, D.C., 1988.
7. Creating false recycling markets unwise, *Refuse News,* May, 1988.
8. U.S. Environmental Protection Agency, The Solid Waste Dilemma: An Agenda for Action, Office of Solid Waste, Washington, D.C., 1989.
9. **Liss, G. B.,** Personal interview by Chuck Papke, Resource Management Associates, February 1989.
10. National Association for Recycling Industries, *Recycling Copper and Brass,* 1980.
11. **Mattheis, A. H.,** Going back to the future, *Waste Age,* July 1988.
12. **Johnson, B.,** Recycling center is part of private hauler's plan, *Waste Age,* February 1988.
13. Resource Management Associates, Curbside Recycling Report: Implementation Plan for Curbside Recycling in the City of San Jose, CA, Phase I, San Jose, October 1984.
14. Resource Management Associates, San Jose Pilot Curbside Recycling Program: Evaluation Report of First Six Months Operation, San Jose, CA, December 1985.
15. **Gee, B. Y. and Papke, C. R.,** Economics of Curbside Recycling in the City of San Jose, California Energy Commission, Sacramento, 1987.
16. California Energy Commission, Economics of Curbside Recycling in the City of San Jose, Final Report, Sacramento, 1987.
17. **Albert, J.,** Waste recovery in Europe, *Waste Age,* October 1987.
18. **Peters, A. and Grogan, P.,** Curbside recycling: what is working best?, *Biocycle,* May/June 1988.
19. **Cohen, D. W.,** A National Survey of Separate Collection Programs, U.S. Environmental Protection Agency, Washington, D.C., 1979.
20. **Buckwalter, C.,** Public acceptance for source separation, *Biocycle,* October 1988.
21. U.S. Environmental Protection Agency, Multimaterial Source Separation in Marblehead and Summerville, Massachusetts, Vol. 2-5, Washington, D.C., 1979.
22. Resource Management Associates, Case Reviews of Curbside Recycling Programs in California, Part I: The City of Palo Alto and Part I: The City of Downey, Solid Waste Management Board, December 1981.
23. Resource Management Associates, Case Reviews of Curbside Recycling Programs in California, Part III: The City of Fresno, California, Solid Waste Management Board, December 1981.
24. Resource Management Associates, Case Reviews of Curbside Recycling Programs in California, Part I: The City of Palo Alto and Part II: The City of Downey, Solid Waste Management Board, December 1981.

25. **Glenn, J. and Riggle, D.,** Biocycle survey: where does the waste go?, *Biocycle,* April 1989.
26. Department of Planning and Economic Development, Division of Solid Waste Management, The Integration of Energy and Material Recovery in the Essex County Solid Waste Management Program, A Report Prepared for Essex County, April 1983.
27. Michigan Department of Natural Resources, Options to Overcome Barriers to Recycling, The Clean Michigan Fund, February 1987.
28. **Grady, J.,** Recyclers in Saabs and Volvos, *Waste Age,* February 1987.
29. **Glenn, J.,** PURPA legislation for recyclables, *Biocycle,* October 1988.
30. California Refuse Removal Council, Waste Generation and Disposal in California, State Solid Waste Management Board, Sacramento, 1982.
31. Franklin Associates, Characterization of Municipal Solid Wastes, U.S. Environmental Protection Agency, Washington, D.C., 1988.

Chapter 18

A PERSPECTIVE ON MUNICIPAL WASTE COMBUSTORS AS A SOURCE OF ENVIRONMENTAL DIOXIN

Holly A. Hattemer-Frey and Curtis C. Travis

TABLE OF CONTENTS

* Research was sponsored by the U.S. Environmental Protection Agency under Interagency Agreement number 1564-1564-A1 applicable under Martin Marietta Energy Systems, Inc., Contract No. DE-AC05-84OR21400 with the U.S. Department of Energy.

I. INTRODUCTION

Because emissions from municipal solid waste (MSW) incinerators are known to contain polychlorinated dibenzo-*p*-dioxins (PCDDs) and polychlorinated dibenzofurans (PCDFs), including the toxic 2,3,7,8-tetrachlorodibenzo-*p*-dioxin (TCDD),[1] much concern and debate has arisen about human exposure to dioxin emitted from these facilities. As the controversy surrounding the incineration of municipal solid waste (MSW) intensifies, government agencies, public officials, and the public question whether MSW incineration is an acceptable waste management option. In this atmosphere of intense political debate, decision makers are being presented with massive amounts of information and pressured to address public skepticism and concern about the safety of these facilities. It has become a formidable challenge for any one individual to critically evaluate the myriad of information available. Given the sensitivity and importance of this issue, we have attempted such an evaluation.

This chapter does not intend to resolve all aspects of the incineration debate but to discuss some key issues concerning risk assessment for human exposure to PCDDs and PCDFs. Since public fear about dioxin and cancer risks have heightened the controversy surrounding the combustion of MSW, we evaluated the following three issues concerning human exposure to facility-emitted dioxins and furans:

1. Are MSW incinerators the major source of dioxin input into the environment?
2. Are environmental concentrations of PCDDs/PCDFs around operating MSW incinerators substantially elevated?
3. Are MSW incinerators the major source of human exposure to PCDDs and PCDFs?

Many other issues, such as the health risks associated with ash management, are not addressed in this chapter and require further investigation.

II. ENVIRONMENTAL PARTITIONING OF DIOXIN

Various measurement and predictive techniques can be used to evaluate the movement and transfer of chemicals within and between environmental media. Since organic chemicals tend to accumulate in the media in which they are most soluble, a few basic physicochemical properties can be used to predict the behavior and fate of chemicals released into the environment. Multimedia transport models estimate the concentration of pollutant in various environmental media. One such model, Mackay et al.'s Level III Fugacity model treats the environment as a "unit world" divided into six homogenous compartments: air, water, soil, bottom sediment, suspended sediment (in water), and aquatic biota (fish).[2,3] While multimedia transport models should not be viewed as exact replicas of the environment, they are considered acceptable for predicting the equilibrium partitioning of nonparticulate chemicals.[2-7] Preliminary attempts to validate fugacity models show excellent agreement with measured environmental concentrations and estimates of human exposure.[2,8,9]

III. ESTIMATING THE MAGNITUDE OF DIOXIN EMISSIONS INTO THE U.S.

The public has been concerned about dioxin exposure from such diverse sources as MSW incinerators, pulp and paper mills, and contaminated soil and fish. There is a general perception among the public that dioxin contamination is a localized problem and control of a few isolated sources, such as MSW incinerators, will reduce human exposure to acceptable levels. The fact that dioxin has been measured in virtually *all* media, including air, soil, sediment, suspended

TABLE 1
Estimated Magnitude of Emissions into the U.S. Environment for Selected Organic Compounds (kg/year)

Chemical	Range of model-predicted emissions values (kg/year)	Best estimate of emissions (kg/year)	Reported emissions or production (kg/year)
TCDD	25 — 120	80	—
PCB (Arachlor 1254)	1.1×10^6 — 1.5×10^7	8.5×10^6	3.9×10^6 [a]
			2.8×10^6 [b]
Pentachlorophenol (PCP)	4.5×10^6 — 2.2×10^8	6.3×10^7	2.3×10^7 [c]
Hexachlorobenzene (HCB)	2.5×10^5 — 6.4×10^6	7.2×10^5	2.2×10^6 [d]
			2.7×10^6 [e]
Trichloroethylene (TCE)	4.8×10^8 — 1.4×10^9	4.9×10^8	2.8×10^8 [f]
			1.5×10^8 [g]
Benzene	5.8×10^9 — 3.3×10^{10}	8.3×10^9	8.5×10^9 [e]

[a] Estimated emissions into U.S., 1972—1977.[52]
[b] Total U.S. sales of Araclor 1254, 1974.[53]
[c] Annual U.S. production.[54]
[d] Estimated U.S. peak production, 1979.[57]
[e] Total input from fungicidal and industrial production.[58]
[f] Total input from industrial production 1972.[55]
[g] Total U.S. production, 1987.[56]

sediment, fish, cow milk, other food items, and human biological samples indicates that dioxin emissions are widespread.[10-16]

PCDDs and PCDFs are formed during many combustion processes.[17] Principal environmental sources of PCDDs and PCDFs are suspected to be (1) discharges from metal processing and treatment plants, pulp and paper mills, and copper smelting plants;[1,13,18-20] (2) motor vehicle emissions;[21-23] (3) chemical manufacturing processes;[24] (4) chemical, industrial, and hospital waste incinerators;[24-29] and (5) forest fires and residential wood burning.[27,30,31] Data concerning the relative contribution of each of these sources and their collective contribution to TCDD total environmental load of dioxin, however, are limited.

Given that the general population is being continuously exposed to small amounts of TCDD, the question arises as to the magnitude of total TCDD emissions into the environment and the relative contribution of various dioxin sources to total TCDD input. Mackay et al.'s Fugacity model[3] was used to determine the rate of input necessary to maintain measured background levels of TCDD in air[16] and soil,[14] assuming continuous input. The model was modified to represent the contiguous U.S. by adjusting compartment volumes and advective flow rates.[32,33] Our best estimate of TCDD emissions into the contiguous U.S. is 80 kg/year with a range of 25 to 120 kg/year as the assumed half-life of TCDD in soil varies from infinity to 20 years.

To validate this source term estimate for TCDD, we used the modified (U.S. scale) Fugacity model to estimate the magnitude of emissions into the contiguous U.S. for pentachlorophenol (PCP), polychlorinated biphenyls (PCBs as Araclor 1254), hexachlorobenzene (HCB), trichloroethylene, and benzene. Table 1 shows that total emissions predicted by the model for these compounds agree relatively well with reported emissions or production data. Nevertheless, these results represent *preliminary* attempts to estimate the magnitude of chemical input into the U.S. environment and should not be viewed as definitive values.

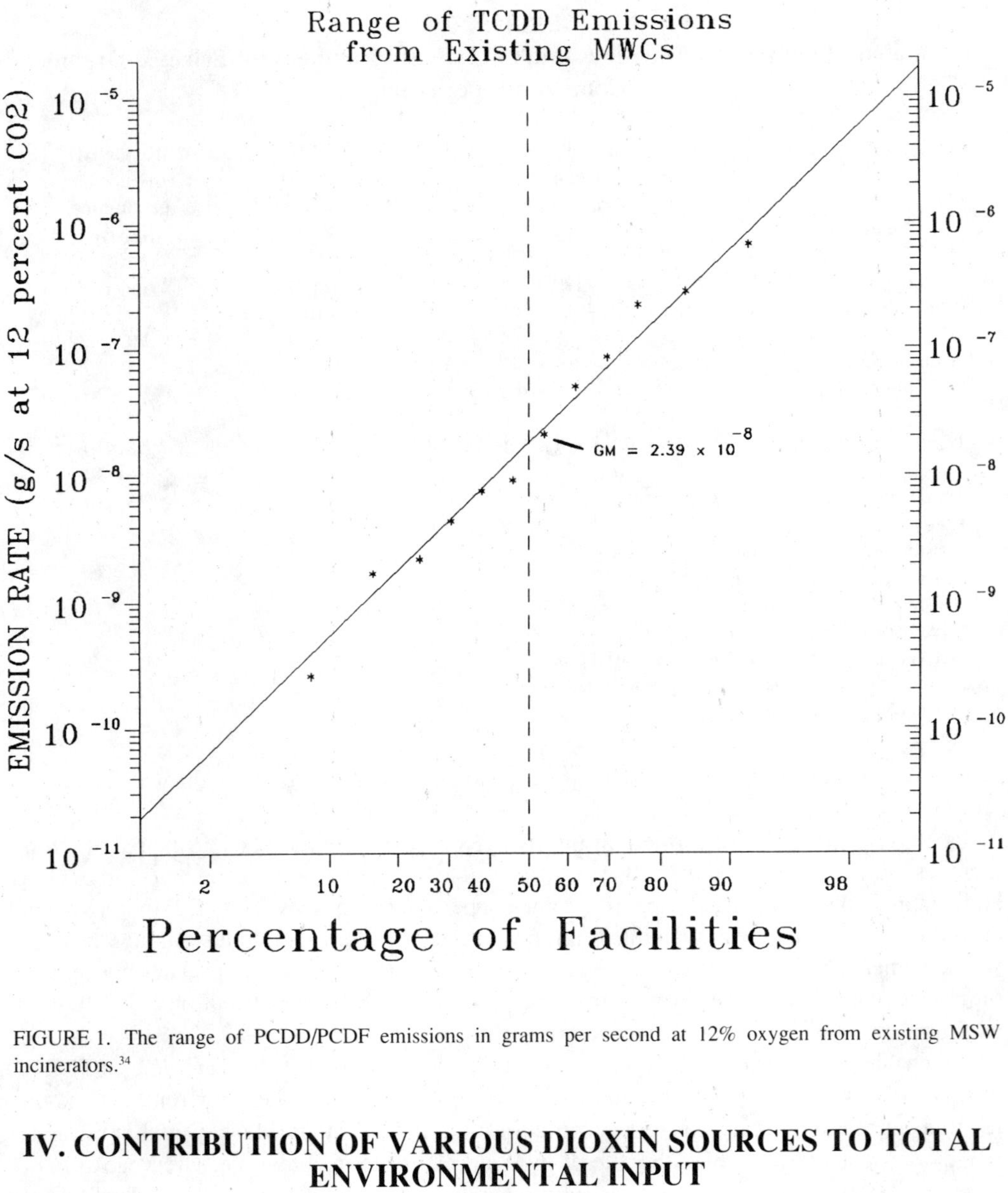

FIGURE 1. The range of PCDD/PCDF emissions in grams per second at 12% oxygen from existing MSW incinerators.[34]

IV. CONTRIBUTION OF VARIOUS DIOXIN SOURCES TO TOTAL ENVIRONMENTAL INPUT

Having estimated the magnitude of TCDD emissions into the environment, it becomes possible to evaluate the relative contribution of known and suspected sources to total TCDD input.

The five TCDD sources evaluated in the following sections were selected because emissions data were readily available. These values relates below are approximations that are limited by accuracy and availability of reported data. Since it is difficult to estimate the uncertainty of the existing data, it is equally difficult to estimate the uncertainty associated with our estimates.

A. MUNICIPAL SOLID WASTE INCINERATORS

The range of TCDD emissions in grams per second (g/s) at 12% oxygen from existing MSW incinerators is presented in Figure 1, which shows that emissions from MSW incinerators are lognormally distributed with a mean emission rate (50th percentile value) of 2.4×10^{-8} g/s.[34] Assuming that there are currently 130 operating MSW incinerators in the U.S.,[35] total emissions from this source are approximately 0.1 kg/year or less than 1% of total TCDD input. Assuming a worst-case scenario (i.e., all operating U.S. MSW incinerators emit TCDD at the maximum rate emitted by an existing facility, 7.0×10^{-7} g/s), then emissions from MSW incinerators into the U.S. would account for 2.9 kg/year, or about 4% of total TCDD input. These data suggest that MSW incinerators are not a major source of TCDD input in the U.S.

B. MOTOR VEHICLES

Direct and indirect evidence indicates that motor vehicles are a prominent source of environmental dioxin.[21-23] Although direct measurements of TCDD emission rates from motor vehicles are scarce, Marklund et al.[22] reported that cars run on leaded gasoline in Sweden emitted <2.0 to 12 ng of TCDD per mile driven. TCDD emissions from cars run on unleaded gasoline were nondetectable. Assuming that 42 million cars in the U.S. used leaded gasoline in 1985 and that the typical car was driven 7900 mi/year,[36] emissions of TCDD from vehicles run on leaded gasoline would range from <0.7 to 4.0 kg/year, which would account for up to 5% of total TCDD input. Past emissions could have been higher, since more vehicles used leaded fuel.

In 1969, the 87 million passenger vehicles run on leaded gasoline in the U.S. averaged 9650 mi/year.[37] TCDD emissions from automobiles in 1969 can thus be estimated to have been <1.7 to 10.1 kg/year, which could account for as much as 13% of TCDD input into the environment. These data indicate that motor vehicle emissions may have been a significant source of environmental dioxin in the past, although modifications to gasoline products instituted in the early 1970s have reduced dioxin emissions from automobiles.

C. HOSPITAL WASTE INCINERATORS

The California Air Resources Board (CARB) quantified pollutant emissions from hospital waste incinerators in California. TCDD emissions at the Cedars Sinai Hospital in Los Angeles ranged from 0.2 to 0.5 ng/s, while emissions from the Sutter General Hospital in Sacramento ranged from <0.11 to 0.24 ng/s.[25,26] Assuming that 6185 U.S. hospitals have an incinerator on site,[29] total TCDD emissions from all hospital incinerators in the U.S. appear to account for <1% of total TCDD input.

D. RESIDENTIAL WOOD BURNING

Nestrick and Lamparski[30] surveyed residential wood combustion (RWC) units and found that TCDD levels ranged from nondetectable (<1.1) to 20 pg/g. Assuming that in 1980/1981 U.S. households burned 42 million cords of wood and that 0.677 cords equal 1 ton,[38] total TCDD emissions from RWC units range from <0.1 to 1.1 kg/year or about 1% of total TCDD input. Thus, residential wood burning appears to contribute more TCDD than hospital incinerators but less than motor vehicles and MSW incinerators.

E. ISOMERIC PATTERN

Rappe et al.[1] and Rappe and Kjeller[16] reported that the isomeric pattern of PCDDs and PCDFs observed in air, sediment, and soil samples is virtually identical to the pattern found in several incineration sources including MSW incinerators, steel mills, copper smelting plants, and motor vehicle exhausts, which indicates that environmental concentrations cannot be linked to any one combustion source.

Cow and human milk samples, however, tend to exhibit a different isomeric pattern than that observed for combustion sources.[1] Biological samples tend to contain only 2,3,7,8-substituted congeners. The number of congeners present in emissions from combustion facilities tend to be much higher than the number of congeners present in the milk sampled from cows grazing near such facilities. A possible explanation of this finding is variations in bioaccumulation and/or biodegradation.[1] Interestingly, Beck et al.[11] found that the isomeric pattern of PCDD/PCDF congeners in food items was not similar to either the pattern observed for most combustion sources or the pattern exhibited by biological samples, although 2,3,7,8-substituted congeners did dominate. This indicates that bioaccumulation and/or biodegradation pathways for vegetation may be different than for animals.

F. CONCLUSIONS

The fact that the five dioxin sources characterized above (sources suspected to be major contributors) account for only 1 to 13% of total annual TCDD input into the U.S. environment is surprising and indicates that: (1) our source term estimate for TCDD may be too high; (2) some as yet unidentified major source(s) of dioxin contamination exists; or (3) there are multiple environmental sources of dioxin, and no one source dominates total input. Another potential source of TCDD is the widespread, trace-level contamination of industrial products. Given the omnipresence of dioxin in the environment, it is imperative that the major sources of TCDD be determined and controlled, so that background environmental concentrations are not further elevated.

V. ARE ENVIRONMENTAL LEVELS OF PCDDs AND PCDFs NEAR MSW INCINERATORS ELEVATED?

To determine if environmental PCDD/PCDF concentrations around operating MSW incinerators are elevated, they must be evaluated in the context of background environmental concentrations to which individuals are continuously exposed. Measured data on the concentration of PCDDs and PCDFs in air, soil, and animal tissues obtained from sites near operating MSW incinerators are scarce. The existing data, however, provide little evidence that PCDD/PCDF concentrations around these facilities are substantially elevated relative to background levels.

A. AMBIENT AIR

Background PCDD/PCDF concentrations have been measured in urban air in the U.S. (Indiana, New York, California, and Ohio), West Germany, New Zealand, and Japan and in air near operating MSW incinerators in The Netherlands, West Germany, and the U.S. (Table 2). Table 2 shows that background air concentrations of PCDDs and PCDFs are remarkably similar in urban areas of various industrialized countries. We could locate only two studies, however, that reported both the background urban air concentration and the concentration of PCDDs and PCDFs in air near an operating MSW incinerator.

Rappe et al.[23] found that measured concentrations of PCDDs and PCDFs in air 1.5 km from an MSW incinerator in West Germany were about four times higher than measured background air levels in West Germany. However, the same study found that PCDD and PCDF levels in air (32 and 36 pg/m^3) 100 m from a motorway and 600 m from a copper smelting plant were 11 to 12 times higher than measured background air levels and about three times higher than levels measured around the incinerator.[23] Christmann et al.[39] reported that mean PCDD/PCDF levels 300 m from a copper smelting plant in Austria (96 pg/m^3) were seven times higher than mean background levels in West Berlin (14 pg/m^3). These data support the theory that numerous combustion sources (not just MSW incinerators) are contributing to elevated levels of PCDDs and PCDFs measured in the ambient air of industrialized countries.

Edgerton et al.[40] found that the measured concentration of PCDDs and PCDFs in air 2 km downwind of an MSW incinerator located in an urban area (6 pg/m^3) was less than two times higher than PCDD/PCDF levels in the ambient air from an area with no known source of dioxin contamination (4 pg/m^3). Table 2 shows that mean PCDD/PCDF levels around MSW incinerators (12 pg/m^3) are about two times higher than mean urban air levels (6 pg/m^3). These data suggest that PCDD/PCDF levels 2 km away from MSW incinerators should be indistinguishable from background urban air levels.

TABLE 2
PCDD/PCDF Concentrations in Background Urban Air and in Air Near Operating MSW Incinerators (pg/m³ of air)

Location	Background levels in urban air	Levels near MSW incinerators	Ref.
California	16.1	NA	59
California (rural)	3.5	NA	59
West Germany (rural)	0.7	NA	23
West Germany (urban)	3.0	13.0	23
West Germany	8.7	NA	60
West Germany	14.3	NA	39
Ohio	3.6	5.9	40
West Germany	57.0	NA	16
New Zealand	6.6	NA	61
New York	2.0	NA	62
Indiana	2.2	NA	63
Netherlands	NA	21.0	64
Ohio	2.2	NA	65
Geometric mean	6.2[a]	11.7	

[a] Rural values not included in the geometric mean.

B. SOIL

PCDD/PCDF levels measured in background urban soil and in soil around MSW incinerators are presented in Table 3. Dow Chemical Company[30] measured PCDDs and PCDFs in surface soil samples at 15 urban sites in the midwest. These samples contained mostly HCDDs and OCDDs, a pattern which is similar to that observed in other media samples taken from sites with no known direct source of contamination.[41] Total PCDD/PCDF concentration in suburban and rural samples ranged from 0.47 to 9.1 ng/g, with a geometric mean concentration of 1.0 ng/g, which is about two times higher than the mean world concentration of PCDDs and PCDFs around incinerators.

Yasuhara et al.[42] measured PCDD/PCDF levels in soil near four MSW incinerators in Japan. (Incinerator type, age, and distance from the incinerator at which samples were taken were not specified.) The geometric mean concentration near these incinerators (5.3 ng/g) is about 18 times higher than the mean world background concentration but only about four times greater than the U.S./Canada mean concentration (1.3 ng/g). Since background concentrations of PCDDs/PCDFs in Japanese soil were not reported, it is impossible to determine if soil concentrations near these incinerators were elevated relative to background levels in that country.

McLaughlin et al.[43] measured soil concentrations of PCDDs and PCDFs at 11 sites 0.1 and 2.1 km away from the SWARU MSW incinerator, which has been operating in Hamilton, Ontario, Canada, since 1973, at two suburban sites (4.5 and 5.6 km away), and at one remote site (22 km away). The geometric mean concentration for the 11 test sites near the incinerator (0.4 ng/g) was about five times lower than the geometric mean concentration for the two suburban sites (1.8 ng/g) and about two times lower than the remote site concentration (0.8 ng/g).

Berlincioni and di Domenico[44] assayed soil samples collected at five sites within a 1-km radius of an MSW incinerator that has been operating in Florence, Italy, since 1973. Background levels were determined at two sites 8 km from the incinerator. The geometric mean PCDD/

TABLE 3
PCDD/PCDF Concentrations in Background Urban Soil and in Soil Near MSW Incinerators (ng/g of soil)

Location	Background levels in urban soil	Levels near MSW incinerators	Ref.
U.S. (urban)	0.96	NA[a]	30
Japan	NA	5.30	42
Canada (urban)	1.82	0.38	43
Canada (remote)	0.81	NA	43
Italy	0.01	0.25	44
Denmark	0.29	0.21	45
U.K.	0.45	NA	46
Geometric mean	0.35	0.57	

[a] Measurement not available.

PCDF soil concentration near the incinerator (0.25 ng/g) is 25 times higher than Italian background levels (0.01 ng/g) but virtually identical to mean world background levels (0.35 ng/g).

Finally, Nielsen and Lokke[45] examined soil samples from three "relatively unpolluted sites" 0.1 to 0.5 km away from a small MSW incinerator and found that PCDD/PCDF concentrations near that facility were not elevated relative to background concentrations. Depending on release point and plume characteristics, however, maximum soil concentrations may have occurred further away from the incinerator.

Creaser et al.[46] sampled 77 sites across the U.K. to quantify typical environmental levels of PCDDs and PCDFs in British soil. Total PCDD/PCDF levels ranged from <0.05 to 2.3 ng/g with a mean concentration of 0.45 ng/g. U.K. soil levels appear to be about half as high as U.S. levels but virtually identical to mean levels near MSW incinerators (0.57 ng/g).

These data are not sufficient to make definitive statements about whether the concentrations of PCDDs and PCDFs around operating MSW incinerators are substantially elevated. For example, if the Japanese data are included in the geometric mean, levels of PCDDs and PCDFs in soil around MSW incinerators appear to be about two times higher than background concentrations. Conversely, if the Japanese data are not included in the geometric mean, PCDD/PCDF levels near MSW incinerators (0.27 ng/g) are virtually identical to background urban levels (0.35 ng/g).

C. COW MILK

PCDD/PCDF levels measured in commercial milk samples and samples obtained from cows grazing on farmland near operating MSW incinerators are presented in Table 4.

Rappe et al.[47] analyzed single milk samples from cows that grazed on farmland located: (1) 16 km from a small MSW incinerator and 20 km from a large MSW incinerator; (2) 0.2 to 1 km from a typical medium-sized MSW incinerator; (3) 0.3 to 1 km from the same incinerator in (2); and (4) 1 km from a production site for chlorinated products. Two commercial milk samples represented background samples. PCDD/PCDF concentrations (whole milk basis) for the three incinerator samples ranged from 0.84 to 3.2 ng/kg, with a geometric mean concentration of 2.1 ng/kg. PCDD/PCDF concentrations in the two control samples were 0.95 and 0.33 ng/kg, with a geometric mean concentration of 0.56 ng/kg. The concentration of PCDDs and PCDFs in the sample collected from cows grazing near the chlorinated products production site was 2.7 ng/kg, which is similar to levels in milk from cows grazing near MSW incinerators and about four

TABLE 4
PCDD/PCDF Concentrations in Commercial Milk Samples and Samples from Cows Grazing on Farmland Near MSW Incinerators (ng/kg whole milk basis)

Location	Levels in commercial samples	Levels in samples near MSW incinerators	Ref.
Switzerland	0.56	2.1	47
West Germany	0.81	NA[a]	10
Japan	0.99	NA	48
Canada	3.10	NA	49
West Germany	1.46	NA	50
U.K.	0.52	NA	51
Geometric mean	1.01		

[a] Measurement not available.

times higher than background levels. These data indicate that other processes (not just MSW incineration) are capable of contributing to elevated levels of PCDDs and PCDFs in cow milk.

Beck et al.[10] analyzed eight cow milk samples collected from different dairies in West Germany not located near MSW incinerators and found that the mean concentration of PCDDs and PCDFs was 0.81 ng/kg. Ogaki et al.[48] analyzed two commercial milk samples from Japan and found PCDD/PCDF levels of 0.99 ng/kg. The Ontario Ministry of the Environment (OME)[49] analyzed concentrated extracts of composite milk samples pooled from four regions in Ontario and found that PCDD/PCDF levels in whole milk were 3.1 ng/kg. In West Germany, Fürst et al.[50] tested 10 cow milk samples for the presence of dioxins and furans and found that mean PCDD/PCDF levels ranged from 0.37 to 4.1 ng/kg with a mean total concentration of 1.5 ng/kg. Finally Startin et al.[51] collected milk samples from seven farms located in rural areas of England and Wales. Total PCDD/PCDF levels ranged from 0.33 to 0.72 ng/kg with a mean concentration of 0.52 ng/kg.

Thus, the single measured concentration of PCDDs and PCDFs in milk samples from cows grazing near MSW incinerators (2.1 ng/kg) is about two times higher than mean measured world background concentrations (1.0 ng/kg). Data on PCDD/PCDF levels in commercial milk samples and samples from cows near operating MSW incinerators, however, are not sufficient to make definitive statements. Given that dairy products are a major pathway of human exposure to background levels of dioxin,[9] it is imperative that more measurements be taken.

VI. CONCLUSIONS

We acknowledge that many people are quite concerned about dioxin emissions from MSW incinerators. Unfortunately, existing data on environmental concentrations of PCDDs and PCDFs around operating MSW incinerators, especially near newer facilities that have improved design and state-of-the-art pollution control equipment, are not sufficient to make definitive statements about the extent of human exposure to facility-emitted dioxin. Our analysis, however, indicates that: (1) MSW incinerators are not the major source of environmental dioxin; (2) environmental levels of PCDDs and PCDFs around MSW incinerators are not substantially elevated relative to background levels; and (3) MSW incinerators are not the major source of human exposure to PCDDs and PCDFs. Thus, individuals expressing concern over dioxin emissions from MSW incinerators need to produce data and analyses showing why that concern is warranted. While we recognize that there are limitations inherent to extrapolating data from a small sample size, until the potential adverse human health effects from continued, low-level exposure to TCDD are fully characterized, we argue for a more focused effort to identify the major source(s) of TCDD input, so that background levels and human exposure can be minimized.

Given the relatively small contribution by MSW incinerators to total environmental dioxin, it is imperative that future research efforts focus on characterizing the sources of PCDD/PCDF input into the environment. Research efforts need to stress:

- Determination of PCDD/PCDF levels in air, soil, vegetation, and cow milk samples from sites near MSW incinerators
- Determining background levels of PCDDs and PCDFs in cow milk and food samples
- Further quantifying PCDD/PCDF emissions from vehicles that use unleaded, leaded, and diesel fuels
- Quantifying the magnitude of PCDD/PCDF emissions from other facilities known or suspected to emit dioxins and furans (e.g., steel mills, copper smelting plants, pulp and paper mills, etc.), so that the major contributors to the high background levels can be identified

Given the public concern over dioxin, it is surprising that the U.S. Environmental Protection Agency has yet to undertake one of these research objectives.

REFERENCES

1. **Rappe, C., Andersson, R., Bergqvist, P.-A., Brohede, C., Hansson, M., Kjeller, L.-O., Lindstrom, G., Marklund, S., Nygren, M., Swanson, S. E., Tysklind, M., and Wiberg, K.,** Overview of environmental fate of chlorinated dioxins and dibenzofurans. Sources, levels and isomeric pattern in various matrices, *Chemosphere*, 16, 1603, 1987.
2. **Mackay, D., Paterson, S., and Cheung, B.,** Evaluating the environmental fate of chemicals: the fugacity level III approach as applied to 2,3,7,8-TCDD, *Chemosphere*, 14, 859, 1985.
3. **Mackay, D., Paterson, S., Cheung, B., and Neely, W. B.,** Evaluating the environmental behavior of chemicals with a level III fugacity model, *Chemosphere*, 14, 335, 1985.
4. **Cohen, Y. and Ryan, P. A.,** Multimedia modeling of environmental transport: trichloroethylene test case, *Environ. Sci. Technol.*, 19, 412, 1985.
5. **Eberhardt, L. L., Gilbert, R. O., Hollister, H. L., and Thomas, J. M.,** Sampling for contaminants in ecological systems, *Environ. Sci. Technol.*, 10, 917, 1986.
6. **Hushon, J. M., Klein, A. W., Strachan, W. J. M., and Schmidt-Bleek, F.,** Use of OECD premarket data in environmental exposure analysis for new chemicals, *Chemosphere*, 12, 887, 1983.
7. **Mackay, D. and Paterson, S.,** Calculating fugacity, *Environ. Sci. Technol.*, 15, 1006, 1982.
8. **Hattemer-Frey, H. A. and Travis, C. C.,** Pentachlorophenol: environmental partitioning and human exposure, *Arch. Environ. Contam. Toxicol.*, 18, 482, 1989.
9. **Travis, C. C. and Hattemer-Frey, H. A.,** Human exposure to 2,3,7,8-TCDD, *Chemosphere*, 16, 2331, 1987.
10. **Beck, H., Eckart, K., Kellert, M., Mathar, W., Ruhl, Ch.S., and Wittowski, R.,** Levels of PCDDs and PCDFs in samples of human origin and food in the Federal Republic of Germany, *Chemosphere*, 16, 1977, 1987.
11. **Beck, H., Eckart, K., Mathar, W., and Wittowski, R.,** PCDD and PCDF body burden from food intake in the Federal Republic of Germany, *Chemosphere*, 18, 417, 1989.
12. **Czuczwa, J. W., McVeety, B. D., and Hites, R. A.,** Polychlorinated dibenzo-*p*-dioxins in sediments from Siskiwit Lake, Isle Royale, *Science*, 226, 568, 1984.

13. The National Dioxin Study: Tiers 3,5,6, and 7, EPA 440/4-87-003, Office of Water Regulation and Standards, U.S. Environmental Protection Agency, Washington, D.C., 1987.
14. **Nestrick, T. J., Lamparski, L. L., Frawley, N. N., Hummel, R. A., Kocher,** C. W., Mahle, N. H., McCoy, J. W., Miller, D. L., Peters, T. L., Pillepich, **J. L., Smith, W. E., and Tobey, S. W.,** Perspectives of a large scale environmental survey for chlorinated dioxins: overview and soil data, *Chemosphere,* 15, 1453, 1986.
15. **Patterson, D. G., Jr., Hoffman, R. E., Needham, L. L., Roberts, D. W., Bagby, J. R., Pirkle, J. L., Falk, H., Sampson, E. J., and Houk, V. N.,** 2,3,7,8-Tetrachlorodibenzo-*p*-dioxin levels in adipose tissue of exposed and control persons in Missouri, *JAMA,* 256, 2683, 1986.
16. **Rappe, C. and Kjeller, L.-O.,** PCDDs and PCDFs in environmental samples air, particulates, sediments and soil, *Chemosphere,* 16, 1775, 1987.
17. **Bumb, R. R., Crummett, W. B., Cutie, S. S., Gledhill, J. R., Hummel, R. H., Kagel, R. O., Lamparski, L. L., Luoma, E. V., Miller, D. L., Nestrick, T. J., Shadoff, L. A., Stehl, R. H., and Woods, J. S.,** Trace chemistries of fire: a source of chlorinated dioxins, *Science,* 210, 385, 1980.
18. **Amendola, G., Barna, D., Blosser, R., LaFleur, L., McBride, A., Thomas, F., Tiernan, T., and Whittmore, R.,** The occurrence and fate of PCDDs and PCDFs in five bleached kraft pulp and paper mills, *Chemosphere,* 18, 1181, 1989.
19. **Swanson, S. E., Rappe, C., Malstrom, J., and Kringstad, K. P.,** Emissions of PCDDs and PCDFs from the pulp industry, *Chemosphere,* 17, 681, 1988.
20. **Marklund, S., Kjeller, L.-O., Hansson, M., Tysklind, M., Rappe, C., Ryan, C., Collazo, H., and Dougherty, R.,** Determination of PCDDs and PCDFs in incineration samples and pyrolytic products, in *Chlorinated Dioxins and Dibenzofurans in Perspective,* Rappe, C., Choudhary, G., and Keith L., Eds., Lewis Publishers, Chelsea, MI, 1986.
21. **Ballschmiter, K., Buchert, H., Niemczyk, R., Munder, A., and Swerev, M.,** Automobile exhausts versus municipal waste incineration as sources of the polychloro-dibenzodioxins (PCDD) and -furans (PCDF) found in the environment, *Chemosphere,* 15, 901, 1986.
22. **Marklund, S., Rappe, C., Tysklind, M., and Egeback, K.-E.,** Identification of polychlorinated dibenzofurans and dioxins in exhausts from cars run on leaded gasoline, *Chemosphere,* 16, 29, 1987.
23. **Rappe, C., Kjeller, L.-O., Bruckmann, P., and Hackhe, K.-H.,** Identification and quantification of PCDDs and PCDFs in urban air, *Chemosphere,* 17, 3, 1988.
24. **Hutzinger, O., Blumich, M. J., van den Berg, M., and Olie, K.,** Sources and fates of PCDDs and PCDFs: an overview, *Chemosphere,* 14, 581, 1985.
25. Evaluation Retest on a Hospital Refuse Incinerator at Sutter General Hospital, Rep. No. ARB/ML-88-026, California Air Resources Board, Sacramento, 1988.
26. Evaluation Test on a Hospital Refuse Incinerator at Cedars Sinai Medical Center, Rep. No. ARB/SS-87-11, California Air Resources Board, Sacramento, 1987.
27. Dioxins in the Environment: The Inside Story, in *Proc. Semin. on Dioxins and Furans,* Ontario Ministry of the Environment, Rexdale, October 22, 1985.
28. **Weerasinghe, N. C. A. and Gross, M. L.,** Origins of polychlorodibenzo-*p*-dioxins (PCDD) and polychlorodibenzofurans (PCDF) in the environment, in *Dioxins in the Environment,* Kamrin, M. A. and Rodgers, P. W., Eds., Hemisphere Publishing, Washington, D.C., 1985, 133.
29. Hospital Waste Combustion Study Data Gathering Phase Final Report, EPA-450/3-88-017, Office of Air Quality Planning and Standards, U.S. Environmental Protection Agency, Research Triangle Park, NC, 1988.
30. **Nestrick, T. J. and Lamparski, L. L.,** Isomer-specific determination of chlorinated dioxins for assessment of formation and potential environmental emission from wood combustion, *Anal. Chem.,* 54, 2292, 1982.
31. **Sheffield, A.,** Sources and releases of PCDD's and PCDF's to the Canadian environment, *Chemosphere,* 14, 811, 1985.
32. **Mackay, D. and Paterson, S.,** Evaluation of the regional multimedia fate of organic chemicals: a level III Fugacity model, *Environ. Sci. Technol.,* in press.
33. **Hattemer-Frey, H. A. and Travis, C. C.,** Estimating the magnitude of TCDD emissions into the U.S., *J. Air Waste Manage. Assoc.,* submitted, 1989.
34. Municipal Waste Combustion Study: Characterization of the Municipal Waste Combustion Industry, EPA/530-SW-87-021H, Pollutant Assessment Branch, U.S. Environmental Protection Agency, Research Triangle Park, NC, 1987.
35. Resource Recovery Focus, Waste-to-Energy Capacity Soars, Survey Shows, Institute of Resource Recovery, 1, 3, 1989.
36. Statistics of the United States 1988, U.S. Department of Commerce, Washington, D.C., 1989.
37. Historical Statistics of the United States Colonial Times to 1970, U.S. Department of Commerce (USDC), Washington, D.C., 1975, p. 716.
38. **Skog, K. E. and Watterson, I. A.,** Residential Fuelwood Use in the United States: 1980-81, Rep. No. ADA-131724, Forest Products Laboratory, U.S. Department of Agriculture, Madison, WI, 1983.

39. **Christmann, W., Klöppel, K. D., Partscht, H., and Rotard, W.,** Determination of PCDD/PCDF in ambient air, *Chemosphere,* 19, 521, 1989.
40. **Edgerton, S. A., Czuczwa, J. M., Rench, J. D., Hodanbosi, R. F., and Koval, P. J.,** Ambient air concentrations of polychlorinated dibenzo-*p*-dioxins and dibenzofurans in Ohio: sources and health risk assessment, *Chemosphere,* 18, 1713, 1989.
41. Risk Assessment for Dioxin Contamination Midland, Michigan, EPA 905/4-88-005, U.S. Environmental Protection Agency, Chicago, 1988.
42. **Yasuhara, A., Ito, H., and Morita, M.,** Isomer-specific determination of polychlorinated dibenzo-*p*-dioxins and dibenzofurans in incinerator-related samples, *Environ. Sci. Technol.,* 21, 971, 1987.
43. **McLaughlin, D. L., Pearson, R. G., and Clement, R. E.,** Concentrations of chlorinated dibenzo-*p*-dioxins (CDD) and dibenzofurans (CDF) in soil from the vicinity of a large refuse incinerator in Hamilton, Ontario, *Chemosphere,* 18, 851, 1989.
44. **Berlincioni, M. and di Domenico, A.,** Polychlorodibenzo-*p*-dioxins and polychlorodibenzofurans in the soil near the municipal incinerator of Florence, Italy, *Environ. Sci. Technol.,* 21, 1063, 1987.
45. **Nielsen, P. G. and Lokke, H.,** PCDDs, PCDFs, and metals in selected Danish soils, *Ecotoxicol. Environ. Safety,* 14, 147, 1987.
46. **Creaser, C. S., Fernandes, A. R., Al-Haddad, A., Harrad, S. J., Homer, R. B., Skett, P. W., and Cox, E. A.,** Survey of background levels of PCDDs and PCDFs in UK soils, *Chemosphere,* 19, 767, 1989.
47. **Rappe, C., Nygren, M., Lindstrom, G., Buser, H. R., Blaser, O., and Wuthrich, C.,** Polychlorinated dibenzofurans and dibenzo-*p*-dioxins and other chlorinated contaminants in cow milk from various locations in Switzerland, *Environ. Sci. Technol.,* 21, 964, 1987.
48. **Ogaki, J., Takayama, K., Miyata, H., and Kashimoto, T.,** Levels of PCDDs and PCDFs in human tissues and various foodstuffs in Japan, *Chemosphere,* 16, 2047, 1987.
49. Polychlorinated Dibenzo-*p*-Dioxins and Polychlorinated Dibenzofurans and Other Organochlorine Contaminants in Food, Ontario Ministry of the Environment, Toronto, 1988.
50. **Fürst, P., Fürst, C., and Groebel, W.,** Level of PCDDs and PCDFs in Food-Stuffs from the Federal Republic of Germany, presented at 9th Int. Symp. on Chlorinated Dioxins and Related Compounds, September 17-21, Toronto, 1989.
51. **Startin, J. R., Rose, M., Wright, C., and Gilbert, J.,** Surveillance of British Foods for PCDDs and PCDFs, presented at 9th Int. Symp. on Chlorinated Dioxins and Related Compounds, September 17-21, Toronto, 1989.
52. **Holton, G. A., Travis, C. C., and Etnier, E. L.,** A comparison of PCB exposures to human populations from oceanic and terrestrial incineration, *Haz. Waste Haz. Material,* 2, 453, 1985.
53. *IARC Monographs on the Evaluation of the Carcinogenic Risk of Chemicals to Humans: Some Halogenated Hydrocarbons,* Vol. 18, International Agency for Cancer Research, Lyon, 1978.
54. **Cirelli, D. P.,** Pentachlorophenol position document I, *Fed. Reg.,* 43, 48446, 1978.
55. Assessment of trichloroethylene as a potentially toxic air pollutant, *Fed. Reg.,* 50, 52422, 1985.
56. Directory of Chemical Producers U.S., SRI International, 1987.
57. Ambient Water Quality Criteria for Chlorinated Benzenes, EPA 440/5-80-028, Office of Water Regulation and Standards, U.S. Environmental Protection Agency, Washington, D.C., 1980.
58. **Courtney, K. D.,** Hexachlorobenzene: a review, *Environ. Res.,* 20, 225, 1979.
59. Ambient Concentrations of PCDDs/PCDFs in the South Coast Air Basin, Rep. No. A6-100-32, California Air Resources Board, Sacramento, 1988.
60. **Buck, M. and Kirschmer, P.,** *Emission Measurements of Polychlorinated Dibenzo-p-Dioxins and Dibenzofurans in the North Rhein-Westphalin State,* LIS-Berichte No. 62, Essen, West Germany, 1986.
61. **Bingham, A. G., Edmunds, C. J., Graham, B. W. L., and Jones, M. T.,** Determination of PCDDs and PCDFs in car exhaust, *Chemosphere,* 19, 669, 1989.
62. **Smith, R. M., O'Keefe, P. W., Hilker, D. R., Aldous, K. M., Mo, S. H., and Stelle, R. M.,** Ambient air and incinerator testing for chlorinated dibenzofurans and dioxins by low resolution mass spectrometry, *Chemosphere,* 18, 585, 1989.
63. **Eitzer, B. D. and Hites, R. A.,** Dioxins and furans in the ambient atmosphere: a baseline study, *Chemosphere,* 18, 593, 1989.
64. **Olie, K., van den Berg, M., and Hutzinger, O.,** Formation and fate of PCDD and PCDF from combustion processes, *Chemosphere,* 12, 627, 1983.
65. **Tiernan, T. O., Wagel, D. J., Vanness, G. F., Garrett, J. H., Solch, J. G., and Harden, L. A.,** PCDD/PCDF in the ambient air of a metropolitan area in the U.S., *Chemosphere,* 19, 541,1989.

Chapter 19

CHARACTERIZING THE EXTENT OF HUMAN EXPOSURE TO PCDDS AND PCDFS EMITTED FROM MUNICIPAL WASTE INCINERATORS

Holly A. Hattemer-Frey and Curtis C. Travis

TABLE OF CONTENTS

* Research was sponsored by the U.S. Environmental Protection Agency under an Interagency Agreement applicable under Martin Marietta Energy Systems, Inc., Contract No. DE-AC05-84OR21400 with the U.S. Department of Energy.

I. INTRODUCTION

Although sanitary landfills are currently the primary method of municipal solid waste (MSW) management, the U.S. Environmental Protection Agency[1] (EPA) estimates that within 10 years half of all municipalities will lack adequate landfill space. Consequently, numerous cities and counties are considering MSW combustion as an alternative waste management option.[2] Because combustion is widely recognized as a source of polychlorinated dibenzo-*p*-dioxins (PCDDs) and polychlorinated dibenzofurans (PCDFs),[3-9] including the highly toxic 2,3,7,8-substituted isomers,[9] considerable local and national opposition to MSW combustion has arisen. Many people living MSW incinerators fear that they will be exposed to high levels of these compounds and subsequently develop cancer.

The fact that the general population is being continuously exposed to small amounts of dioxin, as exemplified by the fact that virtually *all* human adipose tissue samples contain measurable levels of dioxins and furans,[10] is often overlooked in this debate. The widespread background contamination of PCDDs and PCDFs suggests that a more prevalent source of dioxins and furans than MSW incinerators must exist.[9,11] This chapter provides a singular perspective on municipal waste combustion as a source of human exposure to PCDDs and PCDFs.

II. RELATIVE POTENCIES OF DIOXINS AND FURANS

Although there are actually 75 chemical compounds (congeners) in the PCDD family and 135 PCDFs, 2,3,7,8-tetrachlorodibenzo-*p*-dioxin (TCDD) is commonly referred to as dioxin. PCDDs and PCDFs are organized into congener groups distinguished by the degree of chlorination. Within each congener group, the compounds are further speciated into isomers according to the position each chlorine atom occupies on the ring structure. Since no cancer bioassays have been reported for furans and only three PCDDs have been tested, dioxin and furan toxicities are often estimated by reference to the known toxicity of TCDD.[12] Dioxin and furan subgroups are assigned a potency value relative to the known potency of TCDD. This potency value is referred to as a toxicity equivalence factor (TEF). By multiplying the concentration of each congener present in stack emissions by its corresponding TEF value, the approximate toxicity of the mixture can be estimated. Thus, a mixture of PCDDs and PCDFs emitted at the rate of 1 g of TCDD-toxic equivalents per second is assumed to have the same toxic effect as a release of 1 g of TCDD per second.

III. CURRENT RISK ASSESSMENT PRACTICES

To assess the extent of human exposure to facility-emitted dioxins and furans, we examined 12 risk assessments prepared for the following proposed MSW incinerators designed to use sophisticated, efficient pollution control equipment: (1) Los Angeles;[13] (2) San Diego;[14] (3) Philadelphia;[15] (4) Irwindale, CA;[16] (5) Brooklyn Navy Yard, NY;[17] (6) Stanislaus County, California;[18] (7) North Hempstead, NY;[19] (8) Milliken, CA;[20] (9) Boston;[21] (10) Bloomington, IN;[22] (11) Pomona, CA;[23] and (12) Montgomery County, Maryland.[24] Although these reports were produced by industry groups and were not subjected to an independent peer review, most have undergone extensive regulatory review and intense public scrutiny. As such, they probably represent current industry practices.

Another important factor to note is that most health risk assessments, such as the ones evaluated for this chapter, use deterministic models to derive single-value human exposure estimates. Relying on single-value estimates, which are often worst-case estimates, can yield a misleading impression of the accuracy of such estimates and the likelihood that individuals living near an MSW incinerator might actually experience the estimated exposure. Finally, we

TABLE 1
Predicted Maximum Emission Rates, Air Concentrations, and Deposition Velocities of PCDDs and PCDFs (expressed in toxic equivalents) for 12 Proposed MSW Incinerators

Facility	Max annual av. emission rate (ng/s)	Max air conc (fg/m^3)	Max deposition rate (ng/m^2/year)
Irwindale	640	2.1	1.32
San Diego	1700	48.0	9.84
Los Angeles	19	2.4	0.18
Philadelphia	390	17.0	0.70
Brooklyn	638	20.3	8.93
Stanislaus	1700	44.0	2.77
North Hempstead	130	8.1	0.31
Milliken	160	6.1	0.31
Boston	470	15.0	1.98
Bloomington	80	20.0	3.72
Pomona	110	40.0	4.28
Montgomery	367	1.3	0.60
Typical facility[a]	284	10.9	1.43

[a] Geometric mean of the 12 facilities evaluated.

acknowledge that data on PCDD/PCDF emissions from operating MSW incinerators and environmental levels near existing facilities are scarce and that definitive statements about resulting human exposures cannot be made.

IV. THE MAGNITUDE OF MSW INCINERATOR EMISSIONS

Table 1 lists the magnitude of predicted emissions, air concentrations, and deposition velocities of dioxins and furans (expressed in TCDD-toxic equivalents) for each facility. (Since the values presented in Table 1 were taken directly from the risk assessments, weighing schemes used to estimate values in TCDD-toxic equivalents may not be consistent.) Emission rate estimates are important because they are used to predict air concentrations and subsequent inhalation exposures to facility-emitted contaminants. PCDD/PCDF emissions estimates for proposed MSW incinerators are typically derived from short-term source test data collected from existing facilities that are similar in design and have similar combustion and air pollution control systems. Long-term operating data are virtually nonexistent.[25]

The absence of a correlation between emission rates and maximum air concentrations can be attributed to differences in stack height, facility design, the type and extent of pollution control equipment installed, and the quality of source test data available at the time the risk assessment was prepared. Higher stacks cause combustor emissions to be dispersed over a greater area, which results in lower air concentrations in the immediate vicinity of the facility. Similarly, as more MSW incinerators become operational, the quantity and quality of source test data improve.[26] New facilities generally use more sophisticated, efficient pollution control equipment and are expected to have lower overall emissions. Table 1 shows that PCDD/PCDF emissions (expressed in TCDD-toxic equivalents) from a typical, modern MSW incinerator equal 284 ng/s.

Evaluating the magnitude of human exposure to PCDDs and PCDFs emitted from MSW incinerators is complicated, since various PCDD/PCDF isomers are emitted. The relative quantity of each isomer formed during combustion may vary slightly from facility to facility according to design, operating conditions, and waste feed. Knowledge of which compounds are

TABLE 2
Concentration of PCDDs and PCDFs in Background Ambient Air

Location	Conc (fg/m³)[a]	Ref.
California	305	51
West Germany	204	27
	245	52
	153	53
Ohio	103	29
	17	54
New Zealand	71	34
New York	50	32
Indiana	17	31
Geometric mean	86	

[a] Concentrations are expressed in toxic equivalents using the International Toxic Equivalent Factor (I-TEF) weighing scheme[12] and assuming a perfect isomeric distribution if values for 2,3,7,8-substituted congeners were not reported.

present in MSW incinerator emissions is crucial in assessing risk, because the potency factors for the different isomers vary widely.[12]

The incremental air concentration (expressed in TCDD-toxic equivalents) predicted to occur at the point of maximum individual exposure near a typical, modern MSW incinerator is 11 fg/m^3 (Table 1). *Measured* background urban air concentrations of PCDDs and PCDFs (also in TCDD-toxic equivalents) in the U.S., West Germany, and New Zealand range from 17 to 305 fg/m^3, with a geometric mean concentration of 86 fg/m^3 (Table 2). [The relative toxicity of measured background dioxin and furan urban air concentrations was calculated according to the International Toxic Equivalent Factor (I-TEF) weighing scheme,[12] assuming a perfect isomeric distribution if values for 2,3,7,8-substituted congeners were not reported.] Thus, the *maximum* incremental air concentration of dioxins and furans estimated to occur around a typical, modern MSW incinerator (11 fg/m^3) is about eight times lower than mean background levels of PCDDs and PCDFs measured in urban air (86 fg/m^3).

Measurements of atmospheric levels of PCDDs and PCDFs (in toxic equivalents) near operating MSW incinerators are scarce. Rappe et al.[27] reported that the mean concentration of PCDDs and PCDFs measured 1.5 km downwind from an MSW incinerator near Hamburg, West Germany, was 154 fg/m^3. Olie et al.[28] found that the concentration of PCDDs and PCDFs in air 1.5 km from an MSW incinerator near Amsterdam was 133 fg/m^3 (the value of 133 fg/m^3 represents the geometric mean of the concentrations measured at 1 and 2 km downwind from the incinerator). Edgerton et al.[29] measured the concentration of PCDDs and PCDFs in air 2 km downwind from an MSW incinerator located in an urban area and found that PCDD/PCDF levels equaled 106 fg/m^3. The latter study is the only one to report both measured air concentrations around an operating facility and background levels. Their results show that levels near the MSW incinerator (106 fg/m^3) are virtually identical to background concentrations (103 fg/m^3).[29] Thus, the geometric mean measured air concentration around operating MSW incinerators (129 fg/m^3) appears to be slightly higher than the mean background urban air concentration of PCDDs and PCDFs (86 fg/m^3). (Note that incinerator age and the type of air pollution control equipment used at these facilities were not specified.)

Deposition velocities (Table 1) define the rate at which airborne particulates settle from the atmosphere onto soil, water, and vegetation surfaces. Deposition of airborne dioxins and furans can contribute to human exposure by contaminating food items. For example, the deposition of facility-emitted contaminants onto pastures and grasslands, accumulation of contaminants in beef and cow milk fat, and ingestion of contaminated beef and dairy products can be a significant pathway of human exposure to environmentally released pollutants. Travis and Hattemer-Frey[30] showed that this pathway is the major source of human exposure to background levels of TCDD. Similarly, chemicals deposited onto bodies of water can be accumulated by fish and transferred to humans through the aquatic food chain.

V. ASSESSING HUMAN EXPOSURE TO FACILITY-EMITTED PCDDs AND PCDFs

Human exposure to incinerator-emitted contaminants can occur via the following pathways: (1) inhalation; (2) ingestion of soil and dust, fish, drinking water, backyard garden produce, and milk and meat products; (3) infant consumption of mother's milk; and (4) dermal absorption. Table 3 presents conservative estimates of daily intake (expressed in TCDD-toxic equivalents) from exposure to facility-emitted dioxins and furans. (Again, values in Table 2 were taken directly from the 12 risk assessments evaluated, and weighing schemes used to estimate toxic equivalents may not be consistent.)

Inhalation exposures were calculated assuming constant exposure to the maximum ambient air concentration for 24 h/d. Exposure from the ingestion of dairy and meat products was not considered in most cases due to the absence of commercial dairies and slaughterhouses in the immediate vicinity of the proposed facilities. Since commercial dairies and slaughterhouses are present within the Stanislaus, Milliken, and Bloomington study areas (i.e., the area surrounding the facility that would be directly affected by facility emissions), deposition of incinerator-emitted contaminants onto livestock feed and soil and subsequent human exposure were evaluated in the risk assessments prepared for those incinerators.

Ingestion of contaminated drinking water was considered a negligible source of human exposure to incinerator-emitted contaminants, because it is obtained from underground supplies that are usually not contaminated by incinerator emissions. Exposure via consumption of home-grown produce contaminated by deposition of particulates onto outer plant surfaces, root uptake of pollutants from contaminated soil, and uptake of organics vapors by aerial plant parts is frequently a major pathway of human exposure to facility-emitted contaminants. The Los Angeles, Stanislaus, and North Hempstead assessments assumed that 25% of contaminants deposited onto vegetative surfaces is removed during cleaning and food preparation. Ingestion of contaminated fish was considered only if fishable bodies of water existed within a facility's study area. All soil, fish, drinking water, and home-grown produce exposure estimates were based on the percent of soil, fish, water, and produce consumed that was produced or grown within the study area and, hence, contaminated by facility emissions. Consumption of mother's milk was corrected for the average amount of breast milk consumed and the fraction of life (typically $^{1}/_{70}$) during which breast milk is ingested.

VI. BACKGROUND EXPOSURE TO PCDDs AND PCDFs

Background levels of PCDDs and PCDFs were measured in air, soil, sediment, suspended sediment, fish, cow milk, human adipose tissue, and human breast milk samples.[10,27,29,31-40] Because dioxins and furans tend to bioaccumulate in the food chain, even the small concentrations found in vegetation, fish, beef, and dairy products result in human exposure. For example, Travis and Hattemer-Frey[30] showed that the food chain, especially beef and dairy products, accounts for 98% of human exposure to background levels of TCDD. Inhalation and ingestion of contaminated water are not major pathways of human exposure to TCDD.[30,41]

Beck et al.,[42] Ono et al.,[43] Birmingham et al.,[44] and Fürst et al.[45] analyzed randomly collected food samples of the human diet and estimated the average, background daily intake of PCDDs and PCDFs (in TCDD-toxic equivalents) by the general population to be 90, 60, 92 and 89 pg/

TABLE 3
Conservative Estimates of Total Daily Intake of PCDDs and PCDFs (expressed in toxic equivalents) from MSW Incinerators

	Facility[a]											
	1	2	3	4	5	6	7	8	9	10	11	12
	Daily intake (fg/d)											
Inhalation	23	28	20	210	345.8	930	250	370	380	190	410	26.4
Total ingestion	2.6	180	15,000	2.1	506.1	96	25	1,300	3.9	2,000	1,400	224.6
Soil and dust	0.61	0.61	140	2.1	122.6	0.90	0.39	1.3	3.9	1,700	25	NA[b]
Garden produce	0.77	180	3,100	NA	182.7	74	25	1,100	NA	NA	1,300	12.5
Drinking water	NA	0.00005	NA	NA	5.5	NA	NA	NA	NA	0.0041	NA	NA
Fish	1.2	0.43	12,000	NA	195.3	0.64	NA	120	NA	30	35	0.003
Milk and beef	NA	NA	NA	NA	NA	20	NA	41	NA	290	NA	212.1
Mother's milk	6.3	100	NA	14	232.4	25	NA	460	NA	NA	320	NA
Dermal absorption	0.29	9.0	200	NA	41.4	0.057	0.34	0.91	0.29	NA	51	0.034
Total intake[c]	32	320	15,000	230	1,126	1,100	280	2,100	380	2,200	2,200	251

[a] 1 = Los Angeles; 2 = San Diego; 3 = Philadelphia; 4 = Irwindale; 5 = Brooklyn Navy Yard; 6 = Stanislaus; 7 = North Hempstead; 8 = Milliken; 9 = Boston; 10 = Bloomington; 11 = Pomona; 12 = Montgomery County.

[b] NA = Intake value for that pathway was not calculated.

[c] Total intake values were obtained by summing the inhalation, total ingestion, mother's milk, and dermal absorption values.

TABLE 4
Total Daily Intake of PCDDs and PCDFs (expressed in toxic equivalents) by the Maximally Exposed Individual Living Near a Typical, Modern MSW Incinerator

Source	Daily intake (pg/d)	% of total daily intake	Ref.
Background	92.1[a]	99.2	42—45
Incinerators	0.7	0.8	This chapter
Total	92.8	100	

[a] Geometric mean of daily intake estimates reported by Beck et al.,[42] Ono et al.,[43] Birmingham et al.,[44] and Fürst et al.[45]

d, respectively. The Japanese and Canadian intake estimates, however, reflect the dietary consumption of an individual weighing 50 and 60 kg, respectively. If these estimates are adjusted for an individual weighing 70 kg, then the average daily intake of PCDDs and PCDFs is 84 and 107 pg/d, respectively. The predicted daily intake of PCDDs and PCDFs by the maximally exposed individual living near a typical MSW incinerator (0.7 pg/d) was calculated by taking the geometric mean of the 12 total daily intake estimates presented in Table 3.

Table 4 shows that human exposure to PCDDs and PCDFs emitted from a typical, state-of-the-art MSW incinerator is dominated by background environmental contamination. One might question why, if measured environmental concentrations of PCDDs and PCDFs in air around operating MSW incinerators are similar to background levels, the total daily intake of PCDDs and PCDFs by the maximally exposed individual living near a typical, modern MSW incinerators is so small? One reason is that risk assessments prepared for proposed MSW incinerators generally assume that only a small percentage of food is locally produced and consumed. For example, only about 5% of meat (except poultry), 21% of vegetables, and 8% of fruits consumed by all households nationwide are home grown.[46] Those percentages are even lower for individuals living in urban areas. Since the food chain is the primary pathway of human exposure to a large class of organics including DDT, TCDD, and most pesticides,[30,47-50] this assumption can substantially reduce exposure estimates. In areas of high agricultural production, the 0.7 pg daily intake estimate for combustors could be low.

VII. CONCLUSIONS

As the fear of dioxin spreads, government agencies struggle to decide if combusting our MSW is an acceptable waste management alternative. Since the incineration issue has become so emotional and political, decision makers are being forced to address public skepticism and concerns about the safety of these facilities. Although the acceptance of an MSW incinerator must be based on community needs, health risks, costs, and waste management benefits, the evaluation of potential risks and benefits should be based on the best available scientific evidence and be made in a rational, logical manner.

We do not conclude from this analysis that the risks associated with facility-emitted dioxins and furans are negligible or that MSW incinerators are safe. We are not advocating nor criticizing MSW combustion as a waste management tool. Furthermore, this chapter evaluates only the risks associated with exposure to dioxins and furans. Many other health-related issues concerning combustion of MSW remain unresolved. Current scientific data, however, indicate that modern, well-operated MSW incinerators are not a major source of human exposure to PCDDs and PCDFs relative to background levels of dioxins and furans. Despite the slightly elevated air concentrations measured around two existing MSW incinerators, background intake of PCDDs and PCDFs overwhelms exposure to facility-emitted dioxins and furans. A disturbing aspect of this analysis is the high background levels of dioxins in the environment. The relatively small contribution by MSW incinerators to total human exposure suggests that some as yet unidentified source(s) are contributing to background levels of PCDDs and PCDFs. It is imperative that further research efforts focus on characterizing the major source(s) of environmental PCDDs and PCDFs, because collectively they may elevate background concentrations to levels that may pose a far greater threat to human health than MSW incinerators.

There is a general perception that dioxin contamination is a localized problem and that control of a few isolated sources, such as municipal solid waste incinerators, will reduce human exposure to acceptable levels. The reality is that dioxin is everywhere and that virtually every man, woman, and child in the U.S. is being exposed to dioxin daily. PCDDs and PCDFs have been measured in virtually all media, including air, soil, meat, milk, fish, vegetation, fruits, and human biological samples. The EPAs regulatory focus, unfortunately, has been on sources rather than exposures. The EPA continues to function as if control of dioxin emissions from a few localized sources will avert human exposure. Until the potential adverse human health effects from continued exposure to PCDDs and PCDFs are fully characterized, we argue that a more focused effort should be made to identify the major source(s) of PCDD/PCDF input, so that background levels and human body burdens can be controlled.

REFERENCES

1. Municipal Waste Combustion Study: Characterization of the Municipal Waste Combustion Industry, EPA/530-SW-87-021H, Pollutant Assessment Branch, U.S. Environmental Protection Agency, Research Triangle Park, NC, 1987.
2. **Petty, J. and Rodgers, P.,** Source, distribution, and fate workshop summary, in *Dioxins in the Environment,* Kamrin, M. A. and Rodgers, P. W., Eds., Hemisphere Publishing, Washington, D.C., 1985.
3. **Olie, K., Lustenhouwer, J. W. A., and Hutzinger, O.,** Polychlorinated dibenzo-*p*-dioxins and related compounds in incinerator effluents, in *Chlorinated Dioxins and Related Compounds: Impact on the Environment,* Hutzinger, O., Frei, R. W., Merian, E., and Pocchiari, F., Eds., Pergamon Press, Oxford, 1982, 227.
4. **Lao, R. C., Thomas, R. S., Chiu, C., Li, K., and Lockwood, J.,** Analysis of PCDD-PCDF in environmental samples, in *Chlorinated Dioxins and Dibenzofurans in the Total Environment II,* Keith, L. H., Rappe, C., and Choudhary, G., Eds., Butterworths, Boston, 1983, 65.
5. **Tiernan, T. O., Taylor, M. L., Garrett, J. H., van Ness, G. F., Solch, J. G., Deis, D. A., and Wagel, D. J.,** Chlorodibenzodioxins, chlorodibenzofurans and related compounds in the effluents from combustion processes, *Chemosphere,* 12, 595, 1983.
6. **Hutzinger, O., Blumich, M. J., van den Berg, M., and Olie, K.,** Sources and fate of PCDDs and PCDFs: an overview, *Chemosphere,* 14, 581, 1985.
7. **Weerasinghe, N. C. A. and Gross, M. L.,** Origins of polychlorodibenzo-*p*-dioxins (PCDD) and polychlorodibenzofurans (PCDF) in the environment, in *Dioxins in the Environment,* Kamrin, M. A. and Rodgers, P. W., Eds., Hemisphere Publishing, Washington, D.C., 1985, 133.
8. **Marklund, S., Kjeller, L.-O., Hansson, M., Tysklind, M., Rappe, C., Ryan, C., Collazo, H., and Dougherty, R.,** Determination of PCDDs and PCDFs in incineration samples and pyrolytic products, in *Chlorinated Dioxins and Dibenzofurans in Perspective,* Rappe, C., Choudhary, G., and Keith, L., Eds., Lewis Publishers, Chelsea, MI, 1986.
9. **Rappe, C., Andersson, R., Bergqvist, P.-A., Brohede, C., Hansson, M., Kjeller, L.-O., Lindstrom, G., Marklund, S., Nygren, M., Swanson, S. E., Tysklind, M., and Wiberg, K.,** Overview on environmental fate of chlorinated dioxins and dibenzofurans: sources, levels and isomeric pattern in various matrices, *Chemosphere,* 16, 1603, 1987.
10. Analysis for Polychlorinated Dibenzo-*p*-Dioxins (OCDD) and Dibenzofurans (PCDF) in Human Adipose Tissue: Method Evaluation Study EPA-560/5-86-020, Office of Toxic Substances, U.S. Environmental Protection Agency, Washington, D.C., 1986.
11. **Ballschmiter, K., Buchert, H., Niemczyk, R., Munder, A., and Swerev, M.,** Automobile exhausts versus municipal waste incineration as sources of the polychloro-dibenzodioxins (PCDD) and -furans (PCDF) found in the environment, *Chemosphere,* 15, 901, 1986.

12. Pilot Study on International Information Exchange on Dioxins and Related Compounds: International Toxicity Equivalency Factor (I-TEF) Method of Risk Assessment for Complex Mixtures of Dioxins and Related Compounds, NATO/CCMS Rep. No. 176, U.S. Environmental Protection Agency Contr. No. 68-02-4254, North Atlantic Treaty Organization, Committee on the Challenges to Modern Society, Washington, DC, 1988.
13. **Smith, A. H.,** *Health Risk Assessment of the Los Angeles City Energy Recovery (LANCER) Project, Vol. 1,* Health Risk Associates, Berkeley, CA, 1987.
14. **Eschenroeder, A., Petito, C., and Wolff, S.,** *A Public Health Risk Assessment for the Proposed SANDER Facility in San Diego, California,* (85-AFC-4), Alanova Corporation, Lincoln, MA, 1987.
15. *Risk Assessment for the Proposed Trash-to-Steam Municipal Solid Waste Incinerator at the U.S. Naval Base in Philadelphia,* ICF-Clement Associates, Fairfax, VA, 1987.
16. **Eschenroeder, A., Guldberg, P., Hahn, J., Kellermeyer, D., Smith, A., Wolff, S., and Ziemer, S.,** *An Analysis of Health Risks from the Irwindale Resource Recovery Facility,* Alanova Corporation, Lincoln, MA, 1986.
17. **Lipsky, D., Boldt, K. R., Manto, M. S., and Bobal, R. R.,** *Assessment of Potential Public Health Impacts Associated with Predicted Emissions of Polychlorinated Dibenzo-Dioxins and Polychorinated Dibenzo-Furans from the Brooklyn Navy Yard Resource Recovery Facility,* Fred C. Hart Associates, New York, 1984.
18. Stanislaus Waste-to-Energy Facility Health Risk Assessment, Radian Corporation, Sacramento, 1986.
19. *Draft Site and Technology Specific Impact Addendum to the Generic Environmental Impact Statement for the North Hempstead Solid Waste Management Facility Project,* Vol. 2, *Health Risk Assessment,* Malcolm Pirnie, White Plains, NY, 1987.
20. Milliken Waste-to-Energy Project Health Risk Assessment, Radian Corporation, Sacramento, 1987.
21. Environmental Research and Technology, Inc. and Meta Systems, Inc., *Proposed City of Boston Waste-to-Energy Facility South Bay, Suffolk County, Massachusetts,* CSI Resource Systems, Boston, 1986.
22. Bloomington Incinerator Project Risk Assessment, Westinghouse Electric Corporation, Pittsburgh, 1987.
23. Assessment of Health Risks Associated with Trace Element and Organic Emissions from the Proposed SPADRA Refuse-to-Energy Project, Pomona, California, HDR Techserv, Inc., Santa Barbara, CA, 1987.
24. Health Risk Assessment for a Resource Recovery Facility in Montgomery County, Maryland, Weston, Inc., Seattle, 1988.
25. **Lowe, J., Alberts, M., and Dietrich, W.,** The Stanislaus Waste-to-Energy Project: A Health Risk Assessment Case Study in California, presented at 80th Ann. Meet. Air Pollution Control Association, New York, June 21-26, 1987.
26. **Jones, K. H., Walsh, J., and Alston, D.,** The statistical properties of available worldwide MSW combustion dioxin/furan emissions data as they apply to the conduct of risk assessments, *Chemosphere,* 16, 2183, 1987.
27. **Rappe, C., Kjeller, L.-O., Bruckmann, P., and Hackhe, K.-H.,** Identification and quantification of PCDDs and PCDFs in urban air, *Chemosphere,* 17, 3, 1988.
28. **Olie, K., van den Berg, M., and Hutzinger, O.,** Formation and fate of PCDD and PCDF from combustion processes. *Chemosphere,* 12, 627, 1983.
29. **Edgerton, S. A., Czuczwa, J. M., Rench, J. D., Hodanbosi, R. F., and Koval, P. J.,** Ambient air concentrations of polychlorinated dibenzo-*p*-dioxins and dibenzofurans in Ohio: sources and health risk assessment, *Chemosphere,* 18, 1713, 1989.
30. **Travis, C. C. and Hattemer-Frey, H. A.,** Human exposure to 2,3,7,8-TCDD, *Chemosphere,* 16, 2331, 1987.
31. **Eitzer, B. D. and Hites, R. A.,** Dioxins and furans in the ambient atmosphere: a baseline study, *Chemosphere,* 18, 593, 1989.
32. **Smith, R. M., O'Keefe, P. W., Hilker, D. R., Aldous, K. M., Mo, S. H., and Stelle, R. M.,** Ambient air and incinerator testing for chlorinated dibenzofurans and dioxins by low resolution mass spectrometry, *Chemosphere,* 18, 585, 1989.
33. **Rappe, C. and Kjeller, L.-O.,** PCDDs and PCDFs in environmental samples air, particulates, sediments and soil, *Chemosphere,* 16, 1775, 1987.
34. **Bingham, A. G., Edmunds, C. J., Graham, B. W. L., and Jones, M. T.,** Determination of PCDDs and PCDFs in car exhaust, *Chemosphere,* 19, 669, 1989.
35. **Beck, H., Eckart, K., Kellert, M., Mathar, W., Ruhl, Ch.-S., and Wittowski, R.,** Levels of PCDFs and PCDDs in samples of human origin and food in the Federal Republic of Germany, *Chemosphere,* 16, 1977, 1987.
36. **Nestrick, T. J., Lamparski, L. L., Frawley, N. N., Hummel, R. A., Kocher, C. W., Mahle, N. H., McCoy, J. W., Miller, D. L., Peters, T. L., Pillepich, J. L., Smith, W. E., and Tobey, S. W.,** Perspectives of a large scale environmental survey for chlorinated dioxins: overview and soil data, *Chemosphere,* 15, 1453, 1986.
37. **Czuczwa, J. M. and Hites, R. A.,** Dioxins and dibenzofurans in air, soil and water, in *Dioxins in the Environment,* Kamrin, M. A. and Rodgers, P. W., Eds., Hemisphere Publishing, Washington, D.C., 1985, 85.
38. The National Dioxin Study: Tiers 3,5,6, and 7 EPA 440/4-87-003, Office of Water Regulation and Standards, U.S. Environmental Protection Agency, Washington, D.C., 1987.
39. **van den Berg, M., van der Wielen, F. W. M., Olie, K., and van Boxtel, C. J.,** The presence of PCDDs and PCDFs in human breast milk from the Netherlands, *Chemosphere,* 15, 693, 1986.

40. **Rappe, C., Nygren, M., Lindstrom, G., Buser, H. R., Blaser, O., and Wuthrich, C.,** Polychlorinated dibenzofurans, dibenzo-*p*-dioxins and other chlorinated contaminants in cow milk from various locations in Switzerland, *Environ. Sci. Technol.,* 21, 964, 1987.
41. **Oehme, M., Mano, S., Mikalsen, A., and Kirschmer, P.,** Quantitative method for the determination of fentogram amounts of polychlorinated dibenzo-*p*-dioxins and dibenzofurans in outdoor air, *Chemosphere,* 15, 607, 1986.
42. **Beck, H., Eckart, K., Mathar, W., and Wittowski, R.,** PCDD and PCDF body burden from food intake in the Federal Republic of Germany, *Chemosphere,* 18, 587, 1989.
43. **Ono, M., Kashima, Y., Wakimoto, T., and Tatsukawa, R.,** Daily intake of PCDDs and PCDFs by Japanese through food, *Chemosphere,* 16, 1823, 1986.
44. **Birmingham, B., Thorpe, B., Frank, R., Clement, R. E., Tosine, H., Fleming, G., Ashman, J., Wheeler, J., Ripley, B. D., and Ryan, J. J.,** Dietary intake of PCDD and PCDF from food in Ontario, Canada, *Chemosphere,* 19, 507, 1989.
45. **Fürst, P., Fürst, C., and Grobel, W.,** Levels of PCDDs and PCDFs in Food-Stuffs from the Federal Republic of Germany, presented at the 9th Int. Symp. on Chlorinated Dioxins and Related Compounds, September 17-21, Toronto, 1989.
46. Household Food Consumption Survey, 1965-1966, Rep. No. 12, Food Consumption of Households in the United States, Seasons and Year, 1965-1966, U.S. Department of Agriculture, Washington, D.C., 1966.
47. **Travis, C. C. and Arms, A. D.,** The food chain as a source of toxics exposure, in *Toxic Chemicals, Health and the Environment,* Lave, L. B. and Upton, A. C., Eds., Johns Hopkins University Press, Baltimore, 1987.
48. **Stevens, J. B. and Gerbec, E. N.,** Dioxin in the agricultural food chain, *Risk Anal.,* 8, 329, 1988.
49. **Lioy, P. L., Harkov, R., Waldman, J. L., Pietarinen, C., and Greenberg, A.,** The total human environmental exposure survey (THEES) to benzo(a)pyrene: comparison of the inhalation and food pathways, *Arch. Environ. Health,* 43, 304, 1988.
50. **Hattemer-Frey, H. A. and Travis, C. C.,** Pentachlorophenol: environmental partitioning and human exposure, *Arch. Environ. Contam. Toxicol.,* 18, 482, 1989.
51. Ambient Concentrations of PCDDs/PCDFs in the South Coast Air Basin, California Air Resources Board, Sacramento, 1988.
52. **Christmann, W., Klöppel, T., Partscht, H., and Rotard, W.,** Determination of PCDD/PCDF in ambient air, *Chemosphere,* 19, 521, 1989.
53. **Buck, M. and Kirschmer. P.,** Emission Measurements of Polychlorinated Dibenzo-*p*-Dioxins and Dibenzofurans in the North Rhein-Westphalin State, LIS-Berichte Nr. 62, Stuttgart, 1986.
54. **Tiernan, T. O., Wagel, D. J., Vanness, G. F., Garrett, J. H., Solch, J. G., and Harden, L. A.,** PCDD/PCDF in the ambient air of a metropolitan area in the U.S., *Chemosphere,* 19, 541, 1989.

Chapter 20

USEFUL COMMUNICATION OF RISKS TO DECISION MAKERS

Carolyn S. Konheim

TABLE OF CONTENTS

I. FACING A RISKY DECISION

The caustic observation, "Any politician would prefer a dead body to a frightened voter",[1] expresses the cynicism that many citizens attribute to elected officials whom they feel are covering up the risks of modern society. People today harbor a heightened fear of technology. An authority on risk perception notes: "The dominant perception for most Americans (and one that contrasts sharply with the views of professional risk assessors) is that they face more risk today than in the past and that future risk will be even greater than today's."[2] Nowhere is this more evident than in the numerous battles being waged across the U.S. over public approval for building resource recovery, or refuse-to-energy incineration plants.

Normally, an elected official has a well-established standard to which an exposure can be compared. The standard provides a basis against which to measure whether conditions such as the quality of the water or the air are safe. Individuals also know when they are exposing themselves to risks by exceeding a speed limit or choosing to smoke. However, in circumstances where no standards exist, the growing practice is to estimate the probability of risk, usually cancer risk, by a tool widely used in making engineering or regulatory choices, a health risk assessment.

The health risk assessment methodology is so well understood by its practitioners that they feel very comfortable with upper-bound results expressed in risks of 10^{-6}. They forget the admonition by Crouch and Wilson: "No one is born with an intuitive understanding of one in a million. It is an acquisition which can only be made by comparison."[1]

With this in mind, calculated risks associated with a proposed activity are sometimes compared to other calculated risks of cancer, such as the lifetime risk from consuming 30 diet sodas or 40 tablespoons of peanut butter. Instead of achieving perspective, however, the authors of risk assessments find the public hostile to what the public perceives to be a trivialization of their genuine concerns.

To the public, the very act of undertaking a "risk" assessment seems to legitimize their initial concerns about the seriousness of the threat. If so many Ph.D.s are being paid so many thousands of dollars to calculate the effects of so many potential pathways of exposure in so many numbers after the decimal place, something very serious must be afoot! The scientifically responsible approaches — accounting for all factors by assigning them a quantitative value, and recognizing that zero risk does not exist — seem to obfuscate, rather than answer, the fundamental, straightforward question in the minds of the public and their elected officials: "Is the project safe?" Nor does the typical risk assessment procedure even attempt to address what most people mean by risk, i.e., "What will happen if something goes wrong?"

Since few public policies are decided by rational assessments of the risks, especially in the public arena, the whole concept is foreign to most people. They are much more familiar with an ordering of public priorities according to popular demand, irrespective of its scientific basis.

II. IS RISK ASSESSMENT THE BEST TOOL?

The misunderstandings between the public and the risk analysis have led many scientists to conclude there is a communication problem when it comes to the concept of risk. In response, practitioners try to better explain the contents of their toolbox and to describe their carefully constructed worst case scenarios in more relevant terms. It is arguable, though, that they need to question whether they are using the right tools to evaluate a specific proposed project for public decision makers.

Risk assessment is a tool developed and ever-refined by regulators to decide whether an exposure to a substance or to a proposed action is of sufficient risk to merit investing significant resources to control it. The results are sometimes weighed against the benefits to society of permitting the exposure, or they are compared to risks of other exposures or actions which may be more deserving of control.

It is the absence of standards for many pollutants that has caused officials to reach out for the only tool available to try to determine the acceptability of very site-specific proposals. This was exactly the case in 1981 when state and local officials were trying to determine whether a Hempstead, NY, resource recovery plant could safely reopen. The facility had been closed largely for economic reasons and due to odors from its failing water-based material separation technology, but it was believed by most people to have been closed because dioxin was discovered in the stack gases by the Environmental Protection Agency (EPA). Somewhat disingenuously, local opponents of reopening the plant made the reasonable demand that the state or federal government should first establish a standard by which they could judge whether trace emissions of dioxin were safe. When governmental agencies, for a number of reasons, declined to take the initiative, the facility owners asked two Cornell University professors, Neil Orloff and John Babish, to ask all the "what if" questions and produce a worst-case response. An attempt to report the results of the careful analysis to the public was lost, however, on the crowd of committed opponents who filled a high school gymnasium. It was an event that gave Dr. Barry Commoner his first platform on the subject of dioxin from waste-to-energy (WTE). Since then, the scene has been replicated in dozens of school auditoriums across the country.

The pioneering Hempstead risk analysis was crude by today's standards. Although there has been little reexamination of the assumptions of toxicity of pollutants to man, which are based largely on extrapolation from animals, these assumptions have been applied to more and more potential pathways of human exposure, especially in agricultural areas. The predictable result is a higher calculated risk for the hypothetical "most exposed person", whether such a person exists or is ever likely to exist. The refinement of the precision of the calculations carries the implication that the calculations are, in actuality, very accurate, but this is more true mathematically than in "the real world".

III. PUBLIC EXPECTATIONS OF THE RISK ASSESSMENT PROCESS

The public and their elected officials have a more pragmatic view of risk assessment. Basically, they seek the guidance of professionals they can trust on the safety of a proposed project. A citizens advisory committee on the proposed Brooklyn Navy Yard (New York) Resource Recovery Facility was more comforted by the qualitative conclusions of consultants they had selected — two physicians from the Mt. Sinai School of Medicine known for their public health advocacy — that the potential health effects resulting from the plant emissions would be "minimal" and "nondetectable",[3] than they were by the quantitative prediction of the pioneering, peer-reviewed Fred C. Hart study in 1984, which concluded that the risks associated with dioxins emitted from the plant would be "in the range of 0.24 to 5.9×10^{-6}".[4]

As part of the environmental review of a proposed Broome County, New York, resource recovery facility, a citizens' advisory committee (CAC) of unusually technically qualified people was appointed by the county executive to oversee the health risk assessment. In meetings held approximately every 6 weeks over a period of 1 year, the member engineers, biologists, physicians, and an agricultural expert, some of whom were from the host community, came to understand the methodology and critique the assumptions of the study.

In the environmental reviews of over 20 resource recovery facilities, the recurrent questions that citizens have expressed and expect to have answered by the process of risk assessment are

1. What are the specific risks compared to the benefits of this project? Is the risk to each group worth the benefit gained? What are the benefits and risks of alternative solutions? What are the benefits and risks of taking no action?
2. How did you calculate the risk? Is there one standard way of doing it, or are there several? Is there a prevailing consensus in the scientific community on the basic premises of the analysis, or are there distinguished dissidents?
3. Did you base your calculation on data from facilities already in operation, or is the database theoretical?
4. If you based the data on already-operating facilities, were they similar to the proposed facility? If not, how would their differences alter the analyses?
5. Does the design of the facility make the risk as low as possible? Can the facility be updated later if new ways are found to lower the risk?
6. Who in the community bears the burden of risk? Are older, younger, and sick people more at risk?
7. What is the chance of a serious accident? If one occurred, what would be the worst possible impact? How often do accidents happen in currently operating facilities? Will their likelihood increase over time? What is their magnitude? Would the effects of an accident be irreversible? What provisions have been made to handle accidents?
8. Will risks be identifiable? Who will monitor the performance of the plant? Can the risk be reduced?
9. Can the public influence how the facility is designed and operated?
10. Does approving the project mean foreclosing future, potentially less risky, options?

Since such practical questions are rarely even addressed in the standard risk assessment process, citizens feel shortchanged, and even deceived, by what appears to them to be an unresponsive process. Thus, communicating risk in "the real world" may require a reexamination of the conventional risk assessment process and augment, if not substantially revise, it for project-specific applications. What appears to be needed is a risk/benefit analysis, not merely a prediction of upper-bound risks.

Risk assessors are generally adamant about drawing a firm line between quantitative risk assessment and risk management. A former president of the Society of Risk Analysis has written emphatically about the "ideal separation of the risk assessor and the risk manager", warning that the penetration by either into the other's domain can introduce bias or undermine the risk assessment principles established by the National Academy of Sciences.[5] The reluctance of the professional analyst to superimpose his or her values on the affected public or the decision maker is understandable, but it is unacceptable unless the decision maker is given enough information to make an informed decision. In presenting conclusions to people unaccustomed to the risk assessment process, one cannot simply express the risk of the proposed action as a single number and walk away from it. As explained earlier, this works only when the risk can easily be compared to other risks, e.g., one chemical vs. another, by regulators or others familiar with the context. In evaluating a controversial project, the public and the decision maker also need to know the risks of alternative actions and of no action. Without such information, speculation on low or no risk alternatives proposed by project opponents fills the void and handicaps the decision maker.

Introducing a comparison to benefits is tricky, because one is faced with results of noncomparable values, e.g., the possibility of developing cancer vs. reducing disposal costs to the taxpayer. Costs and benefits are relevant only if there is a certainty that reduced disposal costs will actually result in improving the quality of life of the affected population.

It is also inconsistent with other scientific evaluations for risk assessors of proposed projects to abdicate any expression of professional judgment. A risk assessment is often a component of an environmental impact statement (EIS). Throughout such documents, professionals carry out complex analyses to arrive at quantitative results, such as concentrations of air pollutants, increments of noise levels, and traffic levels of service. Since any of these conclusions would have no meaning to readers untrained in the specific disciplines, the environmental specialist typically interprets the impact as being either "significant" or "insignificant". As with an EIS, it would not jeopardize the integrity of the risk assessment process for the assessor to explain whether he or she considers the risk to be significant and why. Doing so is not making a judgment about acceptability of the risk, nor is the risk manager deprived of his or her decision-making role on the acceptability of the risk in view of many other practical and political considerations.

IV. LIMITATIONS OF SITE-SPECIFIC RISK ASSESSMENTS

The controversies in risk assessments are related to lack of engineering and scientific agreement on emission factors, deposition models and values of toxicity, uptake, bioavailability, and consumption. These differences of opinion have been relentlessly played out in the public arena. Citizens get drawn into methodological arguments among experts for which they are poorly prepared. There is no authoritative reference document on the risk of state-of-the-art resource recovery, although a Radian Corporation study for the EPA presents some ranges of risk of generic facilities.[6] Indeed, the only real differences from one site to another are related to local plume dispersion characteristics and the number of potential pathways of exposure. Under present practices, due to the importance of deposition of emissions and ingestion pathways, an impact on a forested hillside in the vicinity of where cows graze is likely to result in a higher hypothetical individual risk than a densely populated, flat, urban area. This is due to the assumption that the hillside may one day be farmed and the farm family might derive its entire subsistence from the immediate locale, consuming local water, vegetables, beef, poultry, fish, and milk that may be contaminated by incinerator-emitted pollutants. These exposure routes are added to the dust and inhalation exposures (and a limited diet of backyard grown vegetables) that are the main routes affecting urban dwellers. While the 19th century ideal of farm life is rare today, it is even more unlikely to be established in the future on undeveloped, hilly terrain. Nevertheless, it provides a fertile field for generating hypothetical routes of exposure.

Two approaches attempt to identify a more typical risk to an entire population, but these cannot account for the prevalence of exposure pathways. The EPAs Human Exposure Model calculates the inhalation exposure of census populations at modeled receptor points. Another model, developed for Broome County, New York, shows that the land area within each isopleth of modeled ground level concentrations and deposition is calculated to determine the average exposure. Both approaches could become more realistic by developing an average profile of pathways for the general area as well.

Addressing the underlying limitations in the risk assessment process (the focus on the most exposed individual, extreme assumptions about exposure and toxicities, failure to quantify risks of alternatives) might produce a result that could be more easily communicated. In the absence of such guidance, it is important to address how best to communicate the general methodology most often used.

V. IMPROVING THE CREDIBILITY OF THE PROCESS

The Brooklyn CACs firm trust in the individual physicians advising them greatly influenced their acceptance of the physicians' conclusion regarding the proposed Brooklyn Navy Yard Resource Recovery Facility. The two operative words are "trust" and "individual". The CAC selected the Mt. Sinai physicians themselves through a competitive process. Second, the physicians became familiar to CAC members through several presentations made by one individual physician. Trust in institutions, firms, and public agencies is at an all-time low in America, particularly when the topic concerns the effects of new technology. The experience in numerous reviews of resource recovery projects confirms that the expertise of the individual assessor and his or her conscientiousness in responding to public concerns builds more credibility than the impressive credentials or extensive experience of the firm with which the individual is associated. The visible continuity of that individual in the process also helps to build trust.

Even with this, people trust their neighbors even more than experts. If a few civic leaders take the time and effort to master complex analytical material, their reasoned conclusions are likely to be accepted by the broader community. A way of formalizing this process is to establish a CAC. In the case of the Brooklyn CAC, elected officials appointed a cross section of community and environmental activists, including two chemistry professors, who met monthly for nearly 4 years to review numerous iterations of an environmental impact statement and a dioxin risk assessment. The CAC was funded by the New York City Department of Sanitation to retain their own technical consultants who helped develop an informed group that was capable of recommending specific measures to limit risk and monitor plant operations. As a result, the CAC felt a degree of control over the risk — an important aspect of tolerance of risk. (It is this characteristic of personal control which has been used to explain why the average person accepts the risk of driving his or her own car more than flying in an airplane.)[7] Indeed, providing means to exercise control over the risk, such as monitoring a facility's operations, is considered a way of managing risk.

In the case of Broome County, New York, the familiarity that developed between the advisory panel and the risk assessors through one year of dialogue helped to build trust in the findings of the study. While the CAC as a whole never took a formal position on the outcome, individual members expressed to the media their endorsement of the fairness and thoroughness of the process. In subsequent public hearings on the EIS, local opponents to the project focused on aspects of the project other than the health risk analysis.[8]

In the City of Milwaukee, a large Advisory Task Force on Solid Waste representing a broad cross section of interests was created by City legislation. The 18 voting members included environmental activists, recycling business owners, landfill operators, business and labor leaders, engineering professors, public health specialists, and a physician. The 13 nonvoting members represented municipal and regional agencies. The 3 to 4 hour meetings were videotaped for cable television. At the onset of the planning process, the Advisory Task Force was successful in causing the City officials to reorient the scope of work drafted by consultants from focusing primarily on a WTE facility to a balanced investment of resources in developing solid waste alternatives. An aspect of the study was a generic environmental and health impact assessment which contained qualitative evaluations of all the alternatives under study. The Task Force was frustrated and somewhat incredulous that the lack of data on landfills, material separation and reprocessing, residue disposal, and other solid waste management activities prevented predicting a quantitative risk for all options equal to the precision they found in analyzing the effects of emissions from a WTE plant.

A CAC usually consists of individuals concerned enough about the public interest to donate a substantial amount of their own time to learning about the evaluation process so they can feel confident about the findings. Such groups require more than just a rudimentary explanation of the methodology and the assumptions a health risk assessment. Complex information must be presented in increments, allowing sufficient time for the material to be comprehended and questioned. Fundamental concepts, like the significance of dosage, rather than the inherent toxicity of a substance, must be mastered. Similarly, the advisory group must agree to the appropriateness of the pathways being studied and the pathways that are dismissed. Since these issues are of concern to both citizens and professional peer reviewers who may be retained, interaction of the two groups may be beneficial. In the case of the Broome County and Milwaukee groups, the participation of scientifically trained members has enabled the dialogue to be conducted on a very substantive level.

Most important, any advisory group, whether it is composed of citizens or scientific peers, must fully understand and agree to the scope of work at the onset. The undermining criticism of the LANCER (Los Angeles County) resource recovery project's health risk assessment by its advisory committee was that the study did not examine certain questions, such as effects of Clean Air Act-regulated pollutants. The fact is that these questions were never reflected in the initial scope. As noted above, the public's expectation of what will be addressed often differs greatly from the professionals'. That is why it is so important for the scope to be fully described at the onset.

Not every advisory group succeeds in building public confidence. An observer of the stalemated Philadelphia resource recovery process has described the shortfalls so vividly, it is worth quoting her extensively:[9]

> Environmental and community groups rejected the analysis for two reasons. First, these groups were excluded from the analysis. ... The physicians appointed by the mayor to lead the study had no background in environmental protection or community participation. None were from South Philadelphia. The consulting firm hired by the physicians was not well known in Philadelphia. Second, the mayor's charge to the risk assessment team was to consider only one disposal option — the South Philadelphia plant. The risk assessment examined health effects associated with emissions under one set of operating conditions — those considered appropriate by the plant's manufacturer. (While) ... a wide array of alternatives ... was still available ... the risk assessment did not consider how refuse recycling might reduce emissions from the plant, or how health impacts might change as the plant grew older. If Mayor Goode had publicly recognized the role of project opponents in the decision to commission the risk assessment, these groups might not have rejected the study at the outset. If environmental and community leaders had served as full participants in the risk assessment effort, helping in the selection of the risk assessment team and the consulting firm, and in decisions about the scope of the study, the pollutants of concern, the emissions estimates, and the level of acceptable risk, they would have found the final analysis more credible. The risk assessment might have helped all concerned to see disposal choices more clearly.

VI. PUTTING THE FINDINGS INTO PERSPECTIVE

Even if all the recommended steps had been followed in Philadelphia, participants there, as elsewhere, would still need a frame of reference for the results, an answer to: "What does this calculated risk mean?"

First, it needs to be conveyed that the risk number is an index of probability, not a prediction of actual effects in a given population. The entire notion of probability eludes many people. They envision the lottery jackpot, not the odds. The fact that the odds of winning worsen as the jackpot gets bigger and more people play, does not discourage most people. In predicting adversity, the opposite is the case. The severity of the risk is more important to the public than its probability. If the risk is a dreaded risk, such as cancer, the actual likelihood of occurrence is less important than the general fear associated with the disease.

Comparing predicted risk of cancer with actual incidences is tempting to put the effect in perspective. According to the National Cancer Institute, the statistical chance of dying from cancer in 1985 for Americans was 171,000 in 1 million.[10] However, it is inappropriate to compare the actual incidence of cancer in a population with a calculated upper-bound risk. The calculated risk is an index of probability, not a prediction of actual cancers in a given population.

Only if one makes this distinction very clear, is it excusable to project the incremental risk to the actual rate of cancer in the study area. In Broome County, at the urging of the CAC, the added average countywide cancer case risk of 0.11 additional cancers every 70 years from the proposed WTE plant was applied to the expected annual County rate of 60,000. It translated into one additional cancer every 600 years. To the public, this was a powerful perspective.

More properly, the results of a risk assessment for a proposed resource recovery facility can only be compared to other risk estimates that use similar assumptions. The most comparable are risks that people experience in everyday life associated with background exposures to pollutants in the environment. Such pollutants may be man made or naturally occurring, but for the most part, the general population is unaware of them. For example, it was relevant in the Broome County study to note, for purposes of comparison to the risks from the proposed resource recovery facility, that the New York State Department of Health measured radon in 172 homes in Broome County and found that mean levels translated into an increased lifetime risk of lung cancer of 7000 to 30,000 chances in 1 million.[11] A far less recognized risk was one calculated by the National Research Council of the National Academy of Sciences. They found that 80% of the American dietary risk of cancer comes from fungicide, herbicide, and insecticide residues found on fresh foods.[12] Using the standard risk assessment approach of consuming a typical diet of vegetables grown commercially for 70 years and the potency factors for each contaminant extrapolated from animal experiments, the calculated risk associated with eating fresh vegetables is 4640 in 1 million.[12]

Many people find that comparing the risks associated with exposure to resource recovery emissions to risks from eating peanuts offensive, because eating peanuts seems so absurdly trivial. They are unaware that peanuts and peanut products, such as peanut butter, contain a mycotoxin mold called aflatoxin, which is as potent an animal carcinogen as dioxin.[1] A child who eats one peanut butter sandwich per week for 15 years has a lifetime cancer risk of 250 chances in 1 million.[13] However, trying to explain the parallels may be lost on a skeptical audience.

Three experts on risk communication call the peanut butter risk a "last choice — rarely acceptable" comparative risk to be "used with extreme caution". They explain: "People do not widely recognize eating peanut butter as risky, and they see even less analogy with plant emissions. A plant manager who tells an audience of mothers that the air toxicity is less dangerous than the peanut butter they feed their children is likely to be seen as patronizing and not believable."[14]

Similarly, the legitimate comparison to the daily accepted risk from driving in an automobile (16,800/million)[1] irritates people, because they feel they have more control over the auto risk than they do with involuntary exposure to refuse burning emissions. Comparisons to other involuntary, but naturally occurring risks, such as exposure to natural background radiation (1400/million) fare no better.[1] Such risks are not willfully imposed by other individuals.

The risk/benefit concept makes more sense to people in the context of the choices they regularly make concerning their own health. People accept the use of X-rays, for instance, as a part of their medical treatment, even though each X-ray carries about a 7 in 1 million risk.[1] They drink chlorinated water to prevent disease, even though such water meeting federal Drinking Water Standards can carry a 460 in 1 million risk[15] — an amount which is increased 1- to 6-fold from dermal absorption and inhalation of chloroform when showering.[15] In the case of Broome County, the press was able to convey that the average risk from the resource recovery facility, 1.7 in 1 million, was less than the risk from drinking and showering each day with the local chlorinated public water supply.[1]

However, Covello and his colleagues[14] warn against comparisons to unrelated risks. This insight was reflected by members of the Milwaukee Advisory Task Force, some of whom said at a scoping meeting: "We are only interested in comparisons of risks to alternative waste disposal methods, not to risks of driving a car. And don't tell us about peanut butter!"[16]

To most people, comparative risks mean the risks of alternative courses of action. The public frequently assumes that there are no environmental or health risks associated with recycling and that continued reliance on landfilling is a more acceptable, near-term alternative than building anything with a smokestack. These alternatives must be examined if the risk assessment is to be useful to decision makers. In the Broome County study, the potential for toxic air and water emissions from the reprocessing of recycled materials and the vast array of carcinogens and other regulated pollutants measured from sanitary landfills provides evidence that each option carries some risk.[13]

Covello and his associates[14] seriously address the important task of putting a quantitative risk number into perspective. Problems arise, they say, since "Comparisons aimed at minimizing risk irritate the public because they are seen and resented as an effort to persuade people that the risk is smaller than they thought."

In view of this, Covello and his co-authors recommend that it is best to acknowledge explicitly the intention to convince the audience the risk is tolerably small. One should not expect to be trusted but should point to independent sources of information the audience can evaluate. They feel that the key to communicating a risk is not to seem to be prejudging the acceptability of the risk without accepting the community's right to participate fully in that decision and their right to base that decision on factors other than the magnitude of the risk.[14]

VII. RANKING RISK COMPARISONS

The Covello team rank risk comparisons according to the following framework.

A. MOST ACCEPTABLE

1. Comparisons of the Same Risk at Two Different Times

In the solid waste arena, this might be a comparison of the risk due to emissions of older, poorly controlled municipal incinerators to the risk of the emissions from proposed state-of-the-art facilities.

2. Comparisons with a Standard

Clearly, having an official health-based reference is ideal. It is the absence of such a standard in the case of dioxins and furans that created public skepticism in the first place. In some cases, as with lead, the existing standard is not viewed as sufficiently protective by some public health advocates.[17]

3. Comparisons with Different Estimates of Risk

Citing the range of risk estimates from many different parties helps to demonstrate that even if some detractors are right, and the risk is, say, ten times higher than the calculated risk, even that risk is likely to be extremely small. This occurred in Broome County, where the risk assessor retained by the host community of the resource recovery facility said that with some change of assumptions, he might come up with a risk twice as high as that of the County's assessor. When he used different emissions data, the risk would be as high as 440 in 1 million, but he pointed out that: "The convolution of emissions biases, modeling biases, and most of all, dose-response biases forces the [cancer risk estimate from dioxins and furans] to be several hundred times higher than the *best* estimate."[18]

B. LESS DESIRABLE

1. Comparisons of the Risk of Doing Nothing and Doing Something

This is especially applicable to solid waste disposal solutions that have been precipitated by a disposal crisis in environmental or economic terms. Doing nothing for some communities means continued landfilling; for others, it means changing to hauling to a distant landfill. Rarely are the risks of these options analyzed quantitatively. Data on emissions of more than a dozen known carcinogens from landfills were reported in the Broome County study[13] and, in the future, can be augmented by data from a statewide landfill emission testing program in California.[19] While it is accepted that leachate can contaminate ground water in some locales, the actual exposures to contaminants of water or air are not usually translated into risks. Even more rare is any attempt to calculate the environmental and health risks from added truck emissions or the potential for accidents due to extra hauling miles. While truck-related risks are familiar, rather than dreaded, they help to put the proposed action into a legitimate perspective.

2. Comparisons of Alternative Solutions to the Same Problem

The public's desire for information on risks of alternative disposal methods has been discussed elsewhere. It is the responsibility of the risk assessor to consider these alternative risks and of the risk communicator to make them known. This allows the risk manager to decide which option poses an acceptable risk.

3. Comparisons with the Same Risk as Experienced in Other Places

In its generic risk assessment on municipal waste combustion, the EPA[6] compared the risk of emissions from existing refuse burning facilities to facilities expected to be built by 1995, as well as making comparisons of risks from different combustion technologies.

C. EVEN LESS DESIRABLE

1. Comparisons of Average Risk with Peak Risk at a Particular Time or Location

Estimating the risk for an average exposed individual was a recommendation of a task force of the American Society of Mechanical Engineers (ASME) Dioxin Committee in its comments on the EPAs proposed methodology for risk assessment of municipal waste combustors. Examples of doing this were described earlier in the chapter.

2. Comparison of the Risk from One Source of a Particular Adverse Effect with the Risk from All Sources of the Same Adverse Effect

This principle applies to resource recovery risk assessments in two ways. Covello et al.[14] suggest comparing the predicted chance of cancer to the incidence of cancer in a given population from all causes. The advantages and disadvantages of doing so are discussed above. One might also compare the risk from a particular pollutant to the risk from exposure to the same pollutant from other sources. Of particular relevance is the growing inventory of sources of dioxin and furans in the environment, including the most pervasive combustion source, automobiles.[20,21]

Another means of providing perspective on risk is to convey that some of the same pollutants under study for a resource recovery project are also present in the environment from background sources. One example is PCBs. It was reported in the Broome County study that the average concentration of PCBs measured in 103 kitchens in New York State is about 10,000 times greater than the maximum ground level concentration predicted to occur from the proposed refuse-to-energy facility.[13] Similar conditions can be identified for background levels of metals and numerous organics. A predictable public response, one often voiced by the Environmental Defense Fund with respect to lead, is that no increment of exposure is acceptable.

D. MARGINALLY ACCEPTABLE

1. Comparisons of Risk with Cost or with a Cost/Risk Ratio

Comparing the huge cost to taxpayers and other public programs of adding emission control equipment to reduce risk when the size of the risk is, indeed, standard practice for elected officials seeking approval for resource recovery projects. The New York City Sanitation Deputy Commissioner in 1985 compared the cost of adding scrubber emission controls to the Brooklyn

Navy Yard and subsequent facilities to the cost of hiring 1000 additional police officers.[22] As Covello predicted, the comparison was marginally effective, probably because the people concerned about the risk of emissions from the resource recovery plant were not the same people who place a high priority on clean streets.

2. Comparisons of Risks with Benefits

The desirability of a risk/benefit analysis to enable decision makers to make an informed decision has been stated earlier. It is, in reality, what a decision maker has to do. However, since the risk assessor usually fails to apply the same analytic rigor to an evaluation of benefits, the comparison of risk to benefits is often dismissed.

3. Comparisons of Occupational With Environmental Risks

Some risk assessors addressing dioxin exposure point to numerous studies of workers exposed to massive amounts of dioxin that have found no long-lasting effects other than chloracne to put into perspective the minute ambient exposures to dioxin emissions caused by a modern resource recovery plant. When audiences are already captured by the notion that "dioxin is the most potent man-made carcinogen", the comparison defies credulity.

4. Comparisons with Other Risks from the Same Source or the Same Risk Agent

A resource recovery example would be a comparison of the risk of death from stack emissions to the risk of death from truck emissions generated by travel to the facility or from accidents caused by the trucks. Since the latter two are familiar risks, while the former is a dreaded risk, even though the end result, death, is the same, the two risks are not likely to be viewed with the same importance.

5. Comparisons with Other Specific Causes of the Same Disease, Illness, or Injury

The examples cited earlier about the risk of cancer from chlorinated drinking water or X-rays are often used and, with some groups, make sense.

E. RARELY ACCEPTABLE: COMPARISONS OF UNRELATED RISK

(Warning: using these comparisons may severely damage your credibility)

The peanut butter example provided earlier exemplifies this risk. Other frequently invoked comparisons which seem equally absurd to the average citizen include being struck by lightning and getting skin cancer from being a frequent flier or from living in Denver. While the low risks associated with those events are acknowledged, the causes seem too unconnected to the project at hand. For the same reason, the use of a high risk activity, such as death by driving an automobile, makes the speaker appear insensitive to the genuine concerns of people who may be affected by a project.

The value of understanding the psychology of the listeners of the message should provoke all risk assessors to take stock of their standard procedures and consider which is the most useful approach. As Crouch and Wilson teach, "A risk assessment cannot be used if it cannot be understood."[1]

VIII. ACKNOWLEDGING THE POTENTIAL FOR ERROR

All this information may be greeted with a skeptical, "Now, tell us what's really going to happen?" The media coverage of incidents like Three Mile Island have caused people to have heightened fear of technology running amuck. Defining the probability of such occurrences and their consequences is a generally unchartered area of risk assessments, but one that must be pursued. The potential for mechanical or human error in all aspects of operating a facility seems particularly great to the public based on the unfortunate record of waste disposal and poor performance of many older, manually controlled plants. It is insufficient to outline the safeguards that will be established without an accompanying "fault tree" analysis and an estimate of the potential risk from activities such as transporting ash.

At the same time, one needs to be able to answer: "By how much could your estimate be off?" An uncertainty analysis that predicts the upper and lower 95th confidence interval may be used. One should be able to postulate, for example: "If people consume more locally grown crops than we assumed, if our emissions are too optimistic, and if we adjusted all other factors upward, the actual risk may be twice as high as we calcûlated; on the other hand, it may actually be lower by about one third."

Another approach is to produce a range of risk estimates, as called for by California guidelines, "with an explanation of the reliability of the data that support estimates at different points on the range".[23] Unless one can put boundaries on the uncertainty, the mere acknowledgment of uncertainty will produce anxiety and unbridled speculation.

IX. MAKING THE RESULTS MORE USEFUL

The purpose of a risk assessment is to assist decision makers. It cannot be used, however, if it cannot be understood.[1] Indeed, most people, even the most interested, will never read more than the executive summary of an extensive risk assessment. A very terse executive summary which fails to convey the conservatism of the assumptions and thoroughness of the process provides no basis for the reader to assess the accuracy of the conclusions. The executive summary should be written for a nontechnical readership and be able to be read independently of the full report.

However, the reality is that even the executive summary will be read by only the most dedicated citizens and elected officials. Authors have to recognize that the decision makers for whom the carefully executed analysis was prepared have little patience for more than a 5- or 10-min presentation of the results. Many local officials never attempt to read even an executive summary. In fact, they often get their information from the local newspaper or television station. Thus, communicating risk in "the real world" demands that the authors, or someone closely identified with the study, have the skill and talent to make an authoritative, crisp verbal presentation of the methodology and findings. The absence of such a spokesperson in so many cases explains why local officials often rely more on the conclusions of their appointed volunteer advisors than those of their well-paid consultants. At the same time, regulatory agencies scrutinize risk assessments with a rigor that is rarely appreciated by the public. Public confidence in the regulators' eventual endorsement of most studies would be enhanced if the extent of regulatory review of preliminary versions was more widely known.

The need for a spokesperson who is comfortable explaining the many disciplines that were brought to bear on the assessment is also evident in the inevitable debates that ensue between self-appointed "experts". These "experts" are generally masters of rhetoric who do not feel constrained to limit their remarks to their particular areas of expertise. They will challenge anything and everything, including the reliability of emission estimates, the appropriateness of the models, the failure to include all exposure pathways, the biological uptake, the assumptions concerning toxicity, the consideration of alternative waste disposal options, and the objectiveness of the preparers of the study. To respond to such "experts", a multidisciplined spokesperson is needed who can defend the work without sounding defensive.

X. COMMUNICATING THROUGH THE MEDIA

Most members of the public and their elected representatives learn of the results of a risk assessment through the media. The press cannot be expected to master mountains of reports nor to convey the care that went into the analysis without a series of background briefings over the course of the study. Since electronic media rarely make a long-term commitment to any news issue, an even more concise summary of the study is indispensible to assure reasonably accurate communication of risk assessment information on radio and television.

XI. CONCLUSION

The problems of risk communication lie in the limitations of applying the standard health risk assessment methodology to site-specific proposals. The conventional assessment generally fails to address the public's concept of risk, and its numerical results for one course of action are not very helpful to decision makers who have to weigh risks against benefits. Ideally, a risk/benefit approach will be introduced that will incorporate evaluation of alternative risks, examine the potential for catastrophic circumstances, and assess ways that enable those affected to control the risks in some meaningful way. Until such an approach is developed, there are steps to improve the communication of the results obtained from a standard risk assessment.

- The first step is to build trust in the individuals who are undertaking the analysis and in their assumptions, preferably through a citizens' oversight mechanism.
- The second step is to put the results into perspective through discussions of alternative courses of action and use of meaningful comparative risks.
- A third measure is to ensure that the study is accompanied by an executive summary that will enable the nontechnical reader to follow the assumptions and methodology of the study as well as its conclusions.
- A fourth is to assure that the media is well briefed and provided with concise materials.
- Finally, the term "risk assessment" should be jettisoned in site-specific applications in favor of "health impact statement" or some other title with less provocative connotations.

REFERENCES

1. **Crouch, E. A. C. and Wilson, R.,** *Risk/Benefit Analysis,* Ballinger Publishing, Cambridge, MA, 1982, 75.
2. **Slovic, P.,** Perception of risk, *Science,* 236, 1987.
3. **Fischbein, A. and Levin, S.,** Comments on Health Risk Posed by Emissions from the Proposed Resource Recovery Facility at Brooklyn Navy Yard for the Citizens Advisory Committee on Resource Recovery in Brooklyn, Environmental Science Laboratory, Mt. Sinai School of Medicine, New York, 1985.
4. **Hart, F. C.,** *Assessment of Potential Public Health Impacts Associated with Predicted Emissions of Polychlorinated Dibenzo-Dioxins and Polychlorinated Dibenzo-Furans from the Brooklyn Navy Yard Resource Recovery Facility,* 1984.
5. **Deisler, P. F., Jr.,** The risk management-risk assessment interface, *Environ. Sci. Technol.,* 22, 15, 1988.
6. Assessment of Health Risks Associated with Municipal Waste Combustion Emissions, EPA/530-SW-87-021g, U.S. Environmental Protection Agency, Washington, D.C., 1987, 2-9.
7. **Conn, W. D. and Feimer, N. R.,** Communicating with the public in environmental risk: integrating research and policy, *Environ. Prof.,* 7, 39, 1985.

8. **Wood, D.,** Broome County, New York, Resource Recovery Project Manager, personal communication, 1987.
9. **Nash, J.,** Assessing the health risks from municipal waste incineration: an example from Philadelphia, *Environ. Impact Assess. Rev.*, 7, 249, 1987.
10. *Cancer Facts and Figures,* 88-500M, Rev.2/88-No.-5008-LE, American Cancer Society, New York, 1988.
11. U.S. Environmental Protection Agency, A Citizen's Guide to Radon, Office of Air and Radiation, U.S. Government Printing Office, Washington, D.C., 1986, 10.
12. National Research Council/National Academy of Sciences, *Regulating Pesticides in Foods,* National Academy Press, Washington, D.C., 1987, 69.
13. Broome County Resource Recovery Project, Draft Health Risk Assessment, 11-3, 1987.
14. **Covello, V., Sandman, P. M., and Slovic, P.,** *Risk Communication, Risk Statistics and Risk Comparisons: A Manual for Plant Managers,* Chemical Manufacturers Association, Washington, D.C., 1988, 21.
15. **Wallace, L.,** Chloroform: available on tap, *Health Environ. Dig.*, 2, 3, 1988.
16. Minutes of the Third Meeting of the Milwaukee Citizens Advisory Task Force on Solid Waste Disposal, April 14, 1988.
17. Testimony of the Environmental Defense Fund on H.R. 4939 and the Significance of Lead in Drinking Water, House Subcommittee on Health and the Environment, Washington, D.C., 1988.
18. **Eshenroeder, A. and Wolff, S.,** A Reassessment of Potential Health Risk Effects of the Broome County Resource Recovery Stack Emissions, prepared for the Town of Kirkwood, NY, Alanova, Inc., January 1988.
19. California Air Resources Board, Engineering Evaluation Branch, Evaluation Tests of Landfill Emissions, Sacramento, 1985-present.
20. **Jones, K. H. and Konheim, C.,** A Comparative Analysis of Leaded Gasoline Auto Exhaust Versus MSW Incinerators as Sources of U.S. Population Dioxin/Furan Exposure, presented at 24th Annu. Meet. Government Refuse Collection and Disposal Association, Reno, NV, 1986.
21. U.S. Environmental Protection Agency, National Dioxin Study, Office of Solid Waste and Emergency Response, EPA/530-SW-87-025, U.S. Government Printing Office, Washington D.C., 1987.
22. **Casowitz, P.,** Comments to Brooklyn Navy Yard Citizens Advisory Committee, February 1985.
23. California Department of Health Resources, Guidelines for Chemical Carcinogen Risk Assessments and their Scientific Rationale, Sacramento, 1985.

Chapter 21

ADDRESSING PUBLIC CONCERNS IN SITING WASTE-TO-ENERGY FACILITIES

Lillie Craig Trimble and Albert Iannacone

TABLE OF CONTENTS

I. INTRODUCTION

The current waste management problem did not develop overnight. The perception of seemingly endless space for garbage disposal coupled with industrial expansion has contributed to the production of 200 million tons/year of household and commercial waste,[1] plus an unknown amount of industrial waste. Concurrently, mass media coverage of problems associated with improper waste management, such as the Mobro 4000 "garbage barge",[2] has heightened public awareness of the waste disposal issue. One result of increased awareness about waste management problems is that the public is taking a more active role in deciding how to dispose of waste. Unfortunately, as various waste management options come to the forefront, citizens are uneasy about many of them, partly because of the stigma attached to waste disposal and partly because they are often left uninformed.

After years of virtual total dependence on open dumps and sanitary landfills for the disposal of our municipal solid waste (MSW), many U.S. cities are evaluating alternative waste management plans, including waste minimization, recycling, incineration, and landfilling as future disposal options. The goal of maximum recovery of materials and energy and minimum use of land seems to be accepted by the public, but they are confused about which option or combination of options will best achieve that goal. When a waste-to-energy (WTE) incinerator is proposed as part of an integrated waste management plan, public opposition swells. Most citizens do not want WTE facilities located in their neighborhood.[3] Addressing some of the public's concerns associated with WTE incineration, however, may make it easier to site these facilities.

II. COMPREHENSIVE WASTE MANAGEMENT

Source reduction, recycling, and incineration are resource recovery techniques which are rapidly gaining public attention as components of a comprehensive waste management strategy.[3] The target most often mentioned for source reduction is packaging. Rice[1] reported that throwaway packaging alone has increased by 80% since 1960 and now accounts for one third of all household trash. Because of the many facets of this multibillion-dollar industry, and society's obsession for the convenience of disposable goods, source reduction must be considered a "long-term" solution that will require much education of the buying public concerning the additional product and disposal costs attributable to packaging.

A more feasible, short-term approach is recycling. The EPA announced a goal of 25% recycling by 1992,[4] and some members of Congress have proposed establishing a goal of 50% reduction in solid waste by the year 2000.[5] Unless the public makes drastic changes in their purchasing and disposal habits, however, these goals are not likely to be achieved. Although most citizens favor the development of recycling programs, recycling is no panacea, because not everyone cooperates. A National Solid Waste Management Association survey[6] indicated that 70% of U.S. citizens believe that recycling can solve much of our country's disposal problem, while only 15% would be willing to pay a tax if the program needed to be subsidized.

Most people are not aware of the costs and limitations associated with recycling.[7,8] The long-term success of any recycling program depends largely on the demand for transforming recyclables into useful products. If there is no market demand for recycled materials, they must be disposed or stored until market conditions change. Stockpiling recyclables merely replaces landfilling with warehousing. As market prices shift, some collection centers have gone out of business and materials intended for reuse have been landfilled. Any realistic appraisal for recycling must differentiate between what can be physically or chemically recycled and what is economically feasible to recycle. An analysis of the benefits and costs of recycling must include an evaluation of: (1) waste stream composition; (2) potential markets for recycled materials; (3) the costs and feasibility of collecting separated materials; (4) energy requirements; and (5) the amount of pollution generated by recycling.[9]

Recycling can be accomplished through the combination of source separation and materials transformation.[10] Source separation provides for recovery of the cleanest, most valuable recyclables, such as paper, glass, and aluminum products. Separation of materials before disposal also reduces possible contamination from other substances present in the waste stream. Not all materials can be recycled, however. Disposable diapers, oily rags, expended ballpoint pens, and tennis shoes, for example, represent products that require expensive equipment for separation and are items that do not readily lend themselves to recycling.[6,11] Other products that contain various oils, glues, laminates, or layers of different materials are more difficult and expensive to recycle.[12]

Some idealistic individuals greatly overestimate the ability of recycling to solve the disposal problem.[2,3,7,11] In Japan, where one of the most effective recycling programs in the world exists, only 50% of the country's household and commercial waste is recycled, 34% is incinerated, and the remaining 16% is landfilled.[13] (In the U.S. 86% of our MSW is currently being landfilled.[1]) Japanese residents separate their waste into seven categories, and waste collection of different categories occurs on alternating days.[13] While, Machida City, Japan, with one of the most successful recycling programs in the world, achieves a 65% reduction,[13] most U.S. metropolitan cities with effective recycling programs achieve reductions of only 25 to 30%.[13]

One reason that this fact is true is that most recycling programs in the U.S. target the residential waste stream, which represents 25 to 90% of the total amount of waste generated.[14] When commercial and industrial wastes are not targeted, recovery rates drop off dramatically.[14] A recycling program oriented to the residential sector that recovers 50% of household newsprint, glass, and cans may divert over 30% of the total waste stream in a rural community but only 5% of the total waste stream in an industrialized urban area. A study conducted in Kansas City, MO showed residential areas of the city generated 518 tons of MSW per day or 24% of the 2163 tons of MSW generated daily in the city.[15] If 100% of the residential community participated, and a 25% recycling rate were achieved, 2033 tons of MSW would still remain for disposal, which corresponds to a 6% reduction in the total daily waste stream. If 25% of the residential, commercial and industrial waste generated were recovered, 1622 tons (or 75%) of the total amount of MSW generated would still remain for disposal. Even if 40% of the waste stream were removed for reutilization, the estimated volume of waste to be disposed in the U.S. in the year 2000 will be 109.3 million tons.[9]

A balanced program involving both recycling and WTE incineration to burn the nonrecoverable portion of the waste stream could provide the best solution to many cities' waste management problem.[10] Because WTE incineration eliminates 85 to 90% of the total waste volume, it is frequently considered an integral component of a comprehensive waste management strategy for metropolitan areas. On the other hand, due to the increasing costs of landfilling and the high costs of constructing WTE facilities, reducing the volume of waste that must be disposed makes economic sense. When a WTE plant is being planned, it should be sized to be compatible with a strong recycling program.[16] Noncombustibles, such as batteries, bottles, and metal cans, should be recycled even if there is a marginal or no net economic benefit, because these materials can create problems in the plant during the combustion process and contaminate the ash residue.[3,17]

III. FACTORS THAT AFFECT THE SITING OF WTE FACILITIES

A. COMMUNICATING WITH THE PUBLIC

Acceptance of WTE facilities by an often-skeptical public can be traced to open communication between elected officials, project developers, the media, and the public. The public wants to be, and will be, a part of any siting process. Citizens must be involved early.[18,19] If facility proponents do not communicate with the public early, project opponents will. By the time a formal public hearing is scheduled, citizens' views may have already crystallized. Project proponents should communicate beyond elected officials and community leaders. They should contact organizations that are active in the area and exchange information with as many diverse groups as possible, such as schools, churches, garden clubs, Scouts, the League of Women Voters, and civic organizations.

Most individuals have limited knowledge about WTE incineration and find it arduous to judge claims made by proponents or opponents.[18,19] The technical issues associated with any proposed facility should be explained to the public in simple, uncomplicated terms.[20] Complex, technical language and vague answers may actually increase rather than decrease public concerns. Conversely, if spokespersons appear uncaring or unresponsive, they will probably not be trusted no matter how technically sound the plan is.

The public wants to know the specific amounts of various pollutants to be emitted from the incinerator and question whether these amounts will be harmful. Audio-visual presentations and printed handouts used at public meetings can help explain technical points. These aids should be prepared with the same level of professionalism that would be used in a presentation to a major client, so that the public can sense the importance placed on their concerns.

B. CITIZEN ADVISORY COMMITTEES

A balanced citizen advisory committee composed of individuals that represent the government, business leaders, the waste industry, teachers, civic leaders, environmentalists, and citizens can sort through information and work with project developers to get information to the general public.[18-21] Members should be knowledgeable about waste management issues, as they will face hard questions during the decision-making and implementation phases.[19] Marion County, Oregon,[21] and San Marcos, CA,[22] represent two case studies where well-informed citizen advisory committees were involved throughout the process.

In 1967, Marion County and four surrounding counties were faced with a waste disposal problem after a state agency informed them they could no longer use the only landfill in the area because of pollution problems.[21] As other sites for a landfill were being evaluated, public opposition to landfilling surfaced. As a result, in 1974 government officials decided to build a WTE facility instead of a new landfill. This plan, however, also met with intense public opposition when none of the five counties wanted the incinerator sited in their county. When Marion County's waste disposal problems intensified in 1979, officials initiated a citizen involvement program. Officials organized a 200-member Solid Waste Advisory Council (SWAC) with representation from all five counties. In 1980, the SWAC recommended that Marion County build a WTE facility instead of another landfill. The SWAC organized a public education campaign after project opponents placed a state referendum on the ballot to stop the project. The referendum failed, and a 550 ton/d WTE facility opened in 1986.

In 1982, after San Marcos officials awarded a developer the contract to construct a WTE facility, the company launched a public education program which involved circulating newsletters, issuing monthly press updates, conducting public opinion surveys, and working with an independent citizens' group.[22] The citizens' group sent letters to the editors of local newspapers and gave testimony at public hearings. The city drafted a ballot measure that accurately explained the project and its benefits, which included funding for a citywide curbside recycling program and road improvements, as well as a redevelopment project around the site to ensure that most of the property tax generated would be retained by the city and local school district. The citizens' group became the foundation for a successful WTE project.

C. RISK ASSESSMENTS

A health risk assessment can be a powerful tool for communicating potential risks to the public.[23] However, most risk assessments represent a worst-case scenario, not the most likely or plausible case.[9] Cancer risk estimates usually assume that the maximally exposed individual lives at the point of maximum impact for 70 years (an extremely unlikely scenario) to ensure that no risk assessment underestimates the potential magnitude of human exposure.[24] Excessively conservative assumptions are generally made to compensate for uncertainties inherent to the risk assessment process,[24] the result of which is virtually always an overestimation of cancer risk.[25] As the risk assessment process for WTE facilities improves, newer methodologies are emerging that do not rely as heavily on worst-case assumptions and allow more realistic estimates of human exposure to be made.

Equally important, risk assessments should provide a realistic estimate of the probability that the estimated risk will occur.[9] Probability distributions can be used to select the estimated health risk that is more reasonable rather than using the estimate which represents the worst-case scenario. Comparing the risks of a WTE facility to a voluntary activity, such as cigarette smoking, only serves to antagonize the public. Relating the relative risk of health or environmental damage from a controlled WTE facility with the risk from other familiar facilities, such as the local utility (which is probably exempt from emission control regulations) or automobiles, may advance the public's understanding of risk-related issues.[25] Examples similar to the following may help the public gain a perspective on risk-related issues. Nearly 70% of Florida's sulfur dioxide emissions originate from utility plants, 40% of nitrogen oxide emissions come from cars, buses, trains, and trucks, while WTE facilities contributed less than 1% of total sulfur dioxide and nitrogen oxide emissions.[26]

Comparative risk assessments for all waste management options, including recycling, composting, incineration, and landfilling are needed so that the regulatory agencies, governmental bodies, and the public can critically evaluate the risks associated with various options.[9,26,27]

D. "EXPERT" VS. "EXPERT"

The public hearing process is a necessary step in the development of a WTE facility, because it allows citizens to express their concerns and ask questions. It has become a formidable challenge for any one individual to critically evaluate the myriad of information available on issues associated with WTE incineration.[28] Because the majority of the general public and segments of the scientific community have a limited knowledge of some key issues concerning WTE incineration, the public hearing process has turned into an arena for an "expert vs. expert" debate.

Experts can honestly disagree about the interpretation of data. To many, a Ph.D. bestows instant credibility, but higher degrees are no guarantee that the views presented by a well-educated individual represent the consensus of the scientific community. One issue on which experts often do not agree is dioxin. Dioxin has been labeled the "most toxic chemical known to man", and WTE facilities have been described as "dioxin factories".[19] Despite the fact that dioxin causes cancer in laboratory rodents, no human deaths have ever been linked to dioxin exposure.[19] Hoffman et al.[29] reported that although human health effects associated with long-term exposure to dioxin are not well known, individuals exposed to dioxin for an average of 2.8 years did not experience serious health problems when compared with an unexposed group.

In this situation where experts may reach diametrically opposite conclusions, decision makers are being forced to address public skepticism and concern regarding the safety of these facilities.[30] Until the scientific community reaches a consensus concerning the risks of WTE incineration, and the public is educated about those risks, fear of the unknown will remain a major influence shaping public opinion.

E. FACILITY OPERATIONS AND CONTROLS

1. Emissions Control

In recent years significant advances have been made in emission monitoring technology in response to concerns over potential health and environmental effects of facility emissions.[31-33] Although most new facilities use sophisticated pollution control equipment, such as scrubbers or baghouses, to greatly reduce the volume of pollutants released,[34] the public is not convinced that emissions from existing and proposed WTE incinerators will not cause serious human health effects. To lessen public fears, older facilities should be tested and, if necessary, modern pollution control equipment installed to meet emission standards. Alternatively, plant operators could demonstrate that emission standards at existing facilities are being met by maintaining good operating procedures, including controlling the moisture content of the waste, maintaining stable temperatures and sufficient air supplies in the combustion chamber, and removing lead batteries and metals from the waste stream before combustion occurs.[35] Tests in Gallatin, TN, and Salem, VA, showed that removal of inert, noncombustible materials (aluminum, steel, glass, and batteries) can significantly reduce the metals content of emissions. One recent Japanese study showed that the addition of several mercury dry cell batteries to the waste stream increased mercury emissions by almost tenfold,[35] while another Japanese study indicated that it is possible to remove mercury vapors from emissions using a variety of packed-bed adsorbents.[36] Continuing research in these areas holds promise for developing more practical means of removing other metallic pollutants from the emissions as well.[37,38]

Even modern, well-designed facilities that use state-of-the-art, sophisticated pollution control equipment can emit unacceptably high levels of toxic pollutants if they are poorly operated. Monitoring programs, however, can be implemented to ensure that emissions of pollutants are kept reasonably low and in compliance with state and/or federal permit levels. A monitoring program designed to provide evidence that the facility is operating properly is part of good management practices at any facility. Once compiled, emissions data should be accessible to the public. If a company or agency makes it difficult for individuals to review such information, the public is likely to perceive that a serious problem exists. Conversely, facility operators could build a position of credibility and trust by initiating monitoring programs and explaining the results to interested persons.

2. Operator Certification

Because of high costs of building and operating facilities and the complex technical equipment involved, it is important that all employees be thoroughly trained and certified.[39,40] Several major companies train employees at existing plants before placing them in newer facilities.[41] However, public fears about improper operation of WTE facilities could be reduced with the establishment of a national standard for operator certification. Certification should consist of written exams on operation of generic and specific plants. Exams should include knowledge of (1) environmental regulations, (2) incinerator/combustion controls, (3) optimum combustion practices, (4) operation of emissions control equipment, and (5) operation of continuous emissions monitoring equipment.[39] Training programs and certification could aid facility owners in assuring worker protection, public safety, and investment protection.[39]

F. MANAGING THE ASH RESIDUE

After incineration is complete, approximately 10% of the total volume of waste remains as a relatively inert ash.[42] For example, in Marion County, Oregon, about one truck of ash residue leaves the facility for every ten trucks that enter the plant. Opponents to WTE incineration often claim that 30% of the MSW remains as ash after combustion. This 30% value reflects the weight of the ash residue, not the volume. The heaviness of the remaining ash is due mainly to the presence of dense, noncombustible items, such as metals and glass slag. Since project proponents must estimate the volume of landfill space required to dispose of the remaining ash, we believe that it is more appropriate to characterize the amount of ash remaining after combustion in terms of volume, not mass. Understanding the volume of ash that must be disposed after incineration is important if the public is to appreciate the role that WTE incineration can play in an integrated waste management plan.

The health risk related to this residue is one of the latest concerns expressed by the public. Until the question of ash toxicity is resolved, the ash should be disposed of in a lined monofill or processed into an environmentally safe product. Lined monofills should also have a leachate collection system and monitoring wells. Such monofills should not be located near floodplains or drinking water supplies.

Research to find product uses for ash is ongoing and several options appear feasible. When a cement-like substance is mixed with ash, it forms an aggregate compound that can be used for building blocks as well as construction and roadbed material.[42-47] Additional research and development into safe methods of ash utilization may reduce a community's future landfill requirements.[3]

G. PROPERTY VALUES

Another factor affecting the siting of WTE facilities is the fear that property values will decline.[18,48] For most people, the largest single investment ever made is the purchase of a house. Any threat to property values may be seen as a very real threat to the economic stability of the community.[18]

Early WTE plants in the U.S. were largely offensive and had tall, exposed stacks.[48,49] Today, architectural aesthetics, including designing facilities with landscaping to complement the building and mute sounds, are being used to gain and maintain public support. For example, the Hillsboro County, Florida, facility is constructed of brick and glass and has the stacks enclosed in a "bell tower". The Baltimore Bresco plant has round windows and a sloped roof, while the Brooklyn Navy Yard plant is designed with a curved roof, banks of windows, and a 500-ft stack which has been camouflaged to look like a slender skyscraper with a lighted top.[49]

Citizens want facilities that are designed to minimize negative impacts, both economic and ecological. More aesthetically pleasing facilities may reduce the stigma attached to garbage disposal.[49] Residential and commercial development surrounding the WTE facilities in Hillsboro and Pinellas Counties, Florida, indicate that property values may actually increase with the construction of an aesthetically pleasing and properly sited WTE facility.[50,51] Facility proponents should cite examples of economic development around existing WTE facilities, or arrange for local representatives to tour an existing plant comparable to the proposed facility.[22]

H. SITE CRITERIA

Consistent, federal regulations governing the siting of waste management facilities are imperative to avoid long distance transportation of waste being used as a solution to local siting problems.[52-55] In the interim, some U.S. counties have established reasonable siting criteria.[48,50,51] Hillsboro County, Florida, successfully sited a WTE plant by establishing a County Resource Recovery Program, whose major task was site selection.[50] A comprehensive procedure was outlined to objectively review three broad categories of siting criteria: engineering, institutional, and environmental concerns. Engineering considerations included identifying drainage and floodplain areas, geologic constraints, proximity to utilities, accessibility, and comparative costs. Institutional considerations were zoning and land use, permitting, legal considerations, and site ownership, while environmental parameters included air and water quality and noise and odor reduction.

Twenty-three sites were evaluated using a weighted numbering scale for each criteria. Engineering criteria were multiplied by a factor of three, environmental criteria by a factor of two, and institutional criteria by a factor of one. Point totals for each site were calculated, and the four site with the highest point total were compared on the basis of cost to develop. Transportation costs of hauling waste to the site, construction cost differences due to site selection, and operation costs were contrasted. This process narrowed the choice to one site. The county then established specific standards for site development, so that the facility was compatible with other buildings in the area. These requirements included a maximum building height, specific landscaping and architectural parameters, and guidelines for proper facility operation. This procedure could serve as a model for other areas.

I. INCENTIVES

Traditionally, incentives were not discussed when trying to site a facility. Many companies, however, find it easier to site a facility if there are incentives or benefits for the host community.[22] These benefits range from an improved fire department, more emergency response equipment and personnel, better roads, and school courses to educate students for jobs at the facility. Incentives can also be proposed to offset the costs a community or state incurs which are directly related to or increased because of a waste management facility operating in the area. Tonnage taxes, for example, that accrue to the community can help to compensate it for the extra expense involved in operating the facility and monitoring emissions. By offering to pay a $0.75 to $1.00/ton fee to local governments, a waste management company not only eliminated opposition to a landfill but also had two counties bidding against each other for the facility.[54]

In addition, burning MSW to generate energy could lessen our demand on foreign oil and help to keep energy prices down.[56] Several states, including Connecticut, New York, and Oregon,[27,57] have laws or regulations requiring utilities to purchase energy from a WTE plant at rates above avoided costs. In Missouri, for example, if any portion of a WTE plant is owned by a governing body, the utility must purchase energy from the plant at the same rate that the utility charges the governing body.[58]

J. LOSS OF FAITH IN GOVERNMENT AGENCIES

Facility noncompliance and inconsistent enforcement by government agencies has led to public mistrust of governmental ability or desire to provide adequate monitoring of waste management facilities. To reestablish public trust, it is important that agency personnel ensure that existing laws are enforced objectively and that communications with citizens are open, frank, and clear.

K. EDUCATION AND INFORMATION

Presently, there is little waste management educational material available. No comprehensive program has evolved that teaches young people about the seriousness of wasting resources and potential problems associated with waste disposal.[59,60] The public must be made aware that prudent use of resources and proper disposal of waste is their responsibility, not the responsibility of future generations.

Role of the Media: One aspect of the information process that has been overlooked by most people is an informed media. The same ratio of misinformation regarding waste management options generally, and WTE facilities specifically, exists among the media as with the general public. In opposition to WTE facilities, a small but very vocal minority asserts that we could solve all of the waste management problems through recycling if we just tried hard enough.[3] Repeated often enough, without information to the contrary, such claims become fact to the media and the general public.

In the past, the WTE industry has only communicated with the media when a facility was proposed.[61] An educated media could be a starting point toward an educated public. Companies, professional groups, scientists, and government officials must take a more active role in disseminating valid, scientific information to the media. Opposition to the siting of a facility makes good headlines, but most of the media will cover all sides of an issue if they receive balanced information.[62]

IV. CONCLUSIONS

Unless public attitudes change drastically, the volume of waste generated in the U.S. will continue to increase dramatically.[9] Changing public opinion is a long-term endeavor. In the interim, the most efficient and environmentally safe means of waste disposal must be found. The optimal waste disposal strategy will vary from community to community. Recycling and WTE incineration are not either/or options; both are complementary aspects of a comprehensive waste management plan.[3,16] WTE incineration can be used to dispose of the waste that cannot be recycled. A combination of a strong recycling program that achieves a 25 to 40% reduction in MSW, and a WTE facility to reduce the remaining volume of combustible waste by 85 to 90%, would extend the life of existing landfills, and change our disposal emphasis from landfilling to resource recovery.[3]

Major public concerns associated with WTE incineration include environmental and human health risks as well as economic effects, both positive (incentives) and negative (property values). Risk assessments comparing potential hazards of various waste management options should be conducted for every proposed facility. Well-designed and imaginatively landscaped facilities that blend into the community may help abate concerns over property values. Facilities should be sized to allow for strong recycling programs, and batteries and other noncombustible items should be removed from the waste stream to reduce toxic emissions and lessen ash contamination. Modern emission control equipment should be installed at existing and proposed facilities, and ash residue should be disposed of in a lined monofill.

Having little or no information on waste management options leaves the public open to scare tactics when a facility is proposed for their area.[61] The EPA, state and local agencies, waste management professionals, and environmentalists must work together to get accurate information to the public before the siting process begins.[61] Until greater public awareness of waste issues is achieved, funding to deal with the problem will likely remain inadequate. Rather than planning rationally for future needs, an uneducated public will continue to be involved only when facilities are proposed for their community or neighborhood.

It is becoming increasingly difficult to site WTE facilities without full public support. The public will be involved in the siting process for WTE facilities, and must be communicated with early, clearly, and often. In the absence of information on all options, the public reacts to fear of the unknown. Unanswered inaccuracies become facts in the public's view. Obtaining public support necessitates initiating an education program, so that the need for a facility will be apparent and citizens, elected officials, and project proponents can work together to solve our waste management problem.

REFERENCES

1. **Rice, F.,** Where will we put our garbage?, *Fortune*, April 11, p. 100, 1988.
2. **Salimando, J.,** Barge; positive or negative?, *Waste Age*, 18(7), 28, 1987.
3. **Birks, D. W.,** Choosing all of the above, *Solid Waste Power*, 2(3), 32, 1988.
4. **Anon.,** EPA's Porter sets goal: 25% recycling by 1992, *Waste Age*, 19(3), 22, 1988.
5. **Meade, K.,** Congress to tackle a myriad of issues, *Waste Age*, 20(1), 14, 1989.
6. *At A Glance: Recycling And Solid Waste*, National Solid Waste Management Association, Washington, D.C., 1988.
7. **Ogg, J.,** Recycling: does it work? Is it worth it?, *Solid Waste Power*, 1(4), 20, 1987.
8. **Shotwell, R.,** Recycling is great: but there's a limit, *Waste Alternatives*, 1(1), 12, 1988.
9. **Shaub, W. M.,** Incineration, Some Environmental Perspectives, presented at Symp. Emissions from Combustion Processing: Origin, Measurement and Control, Toronto, June 5 to 10, 1988.
10. *Materials and Energy from Waste, Final Draft*, Office of Technology Assessment, Washington, D.C., 1978.
11. **Johnson, C. A.,** "Total recycling" test ends, *Waste Age*, 19(2), 30, 1988.
12. *Coming Full Circle: Successful Recycling Today*, Environmental Defense Fund, Washington, D.C., 1988.
13. **Hershkowitz, A. J. and Salerni, E.,** *Garbage Management in Japan: Leading the Way*, 1st ed., INFORM, New York, 1987.
14. **Papke, C.,** Getting real with recycling, *Solid Waste Power*, 2(2), 28, 1988.

15. **Cerwick, J., Brown, T., and Robinson, R.,** *Solid Waste Management Plan, District Heating and Cooling Plan and Waste-To-Energy Feasibility Study,* ES-4, Barnes and McDonnell Engineering Company, Kansas City, MO, 1988.
16. **Kiser, J.,** Waste-to-energy and recycling, a compatible solution, *Waste Alternatives,* 1(1), 21, 1988.
17. **Clarke, M.,** How plant operators can minimize emissions, *Waste Age,* 18(12), 156, 1987.
18. **Konheim, C., Reiss, N., and Hasselriis, F.,** Get citizens involved in siting and do it early!, *Waste Age,* 19(3), 37, 1988.
19. **Eldred, B.,** The spector of dioxin, *American City County,* August, 32, 1986.
20. **Nabor, T.,** Nanograms are not the answer to "will I be hurt?" questions, *Waste Age,* 19(3), 44, 1988.
21. **Shotwell, R.,** State-of-the-art in Oregon, *Waste Alternatives,* 1(2), 24, 1988.
22. **Shepherd, T.,** Blueprint for a rare California siting victory, *Waste Age,* 19(9), 97, 1988.
23. **Fontana, P. H.,** What a risk assessment is and how it can help in siting, *Waste Age,* 19(4), 331, 1988.
24. **Anderson, E. L., Chrostowski, P. C., and Foster, S.,** Calculating the risks, *Solid Waste Power,* 2(3), 40, 1988.
25. **Michaels, R. A.,** Health risk assessment: WTE vs. peanut butter, *Solid Waste Power,* 2(5), 22, 1988.
26. **Anon.,** How much will plant retrofits cost?, *Waste Age,* 19(3), 119, 1988.
27. **Musello, D.,** Red light, green light, factors that can make or break a project, *Solid Waste Power,* 1(4), 1, 1987.
28. **Travis, C. C. and Hattemer-Frey, H. A.,** A perspective on dioxin emissions from municipal solid waste incinerators, *Risk Anal.,* 9(1), 91, 1989.
29. **Hoffman, R. E., Stehr-Green, P. A., and Webb, K. B.,** Health effects of long-term exposure to 2,3,7,8-tetrachlorodibenzo-*p*-dioxin, *JAMA,* 225, 2031, 1986.
30. **Travis, C. C. and Hattemer-Frey, H. A.,** Human exposure to dioxin from municipal solid waste incinerators, *Waste Manage.,* in press, 1989.
31. **Clarke, M.,** Emissions control: a never ending quest, *Waste Age,* 19(1), 83, 1988.
32. **Kiser, J. V. L.,** Continuous emission monitoring: a primer, *Waste Age,* 19(5), 64, 1988.
33. **Kiser, J. V. L.,** More on continuous emission monitoring, *Waste Age,* 19(6), 119, 1988.
34. **Couppis, E. C.,** Will smaller plants get scrubbers?, *Waste Age,* 19(3), 119, 1988.
35. **Clarke, M.,** How plant operators can minimize emissions, *Waste Age,* 18(11), 156, 1987.
36. **Otani, Y., Emi, H., Kanaoka, C., Uchijima, I., and Nishino, H.,** Removal of mercury vapor from air with sulfur-impregnated adsorbents, *Environ. Sci. Technol.,* 22, 708, 1988.
37. **Carlson, K.,** Heavy metals from energy from waste plants comparison of gas cleaning system, *Waste Manage. Res.,* 4, 15, 1986.
38. **Clarke, M.,** Air pollution control status report, *Waste Age,* 18(11), 102, 1987.
39. **Grady, J. C.,** How will plant operators be certified?, *Waste Age,* 18(11), 32, 1987.
40. **Anon.,** Operator certification standard is coming, *Waste Age,* 19(11), 95, 1988.
41. **Grady, J.,** Saugus: plant as university, *Waste Age,* 18(11), 6, 1987.
42. **Wingerter, E.,** An integral part of solid waste management, *Waste Alternatives,* 1(2), 21, 1988.
43. **Mills, G.,** Ogden Martin, Inc., personal correspondence, 1989.
44. **Anon.,** Ash vitrification may solve residue problem, *Solid Waste Power,* 1(4), 29, 1987.
45. **Anon.,** Tetrapods, *World Wastes,* 31(10), 20, 1988.
46. **Anon.,** An idea for ash use from the barge man!, *Waste Age,* 19(2), 16, 1988.
47. **Anon.,** What's new in ash disposal?, *Waste Age,* 19(11), 28, 1988.
48. **Rogoff, M. J. and Berry, P. V.,** Can resource recovery facilities be good neighbors?, *Solid Waste Power,* June, p. 24, 1987.
49. **Burch, G.,** Bringing beauty to the beast, *Solid Waste Power,* 1(4), 10, 1987.
50. **Hauser, R., Jr.,** Siting the Hillsboro County Solid Waste Energy Recovery Facility, presented at Int. Symp. Resource Recovery, Tampa, FL, February 18 to 21, 1986.
51. **Zier, R. E.,** Put siting first: period, *Waste Age,* 19(3), 57, 1988.
52. **Hastreiter, J.,** Rail may make connections with distant disposal sites, *Waste Wastes,* 31(3), 28, 1988.
53. **Voell, P. and Voell, A.,** Shrinking Northwest fills force long distance hauls, *World Wastes,* 31(3), 32, 1988.
54. **Johnson, B.,** Portland: first in the West to send waste long distance, *World Wastes,* 31(10), 21, 1988.
55. **Adler, C. A.,** Moving wastes on rail may help contain the crisis, *World Wastes,* 30(12), 20, 1987.
56. **Salimando, J.,** Someday we might need some energy, *Waste Age,* 19(3), 6, 1988.
57. **Kiser, J.,** New utility rules threaten refuse-to-energy, *Waste Age,* 19(10), 87, 1988.
58. 260.800 RSMo., *Waste to Energy Facilities,* 1986.
59. Michigan Resource Recovery Strategy Ad Hoc Committee, Educational Materials for Schools, Policy Statement B-4, Policy Development Forum, Detroit, November 6, unpublished data, 1983.
60. **Browne, E. R.,** Teaching kids to teach others, *Waste Age,* 19(4), 154, 1988.
61. **Salimando, J.,** A president for a time of transition, *Waste Age,* 19(6), 57, 1988.
62. **Magnusson, J.,** Selling the technology, *Solid Waste Power,* 2(3), 8, 1988.

Chapter 22

THE ROLE OF PUBLIC PARTICIPATION IN THE RISK ASSESSMENT PROCESS FOR MUNICIPAL SOLID WASTE INCINERATORS

James F. Walsh and Kay H. Jones

TABLE OF CONTENTS

I. INTRODUCTION

Landfilling has historically been the solution of choice to the nation's municipal solid waste (MSW) problems. However, as available landfill space continues to dwindle, it has become clear that alternatives to landfilling are needed. At the present time, the incineration of MSW and subsequent utilization of the energy created from this combustion process is one alternative to continued landfilling. Waste reduction and recycling are the other alternatives. While successful programs can reduce the volume of waste which must be disposed of by about 20 to 30% if they can muster significant public support and participation,[1] waste reduction and recycling alone cannot solve the continuing and growing waste management crisis.[2] Thus, as part of an integrated solid waste management program, many municipalities are turning to incineration.[3,4]

As an increasing number of MSW incineration projects are being planned and developed, a concomitant increase in organized public opposition has also arisen. This chapter discusses some of the reasons behind this public opposition and proposes ways to effectively address public concerns about potential health risks associated with emissions from these facilities. Several case studies illustrating these concepts are also presented.

II. CAUSES OF OPPOSITION

There are several reasons why individuals might be opposed to the siting of an MSW incinerator at a certain location. It should be noted that some of these issues were interrelated on past projects and will likely be interrelated on future projects. [The following discussion is not based on scientific research but is derived from the authors' direct experience on a number MSW incineration permitting, risk assessment, and siting projects. Despite this, we believe that individuals involved with MSW incineration issues may benefit from knowledge gained during our experiences.]

A. NONHEALTH-RELATED ISSUES

1. Solid Waste Management Policy Concerns

Probably the single most significant nonhealth-related reason for the recent but growing public opposition to MSW incinerators is public concern about the overall solid waste management policy of the particular jurisdiction responsible for waste management (typically the city or county). The focal point of public concern is usually that the county is relying too heavily on incineration and devoting insufficient time and money to developing other alternatives, such as waste reduction and recycling. Incorporating decisive reduction and recycling programs into the overall integrated waste management plan typically reduces this type of opposition. Contrary to public opinion, most developers of MSW incinerators are not opposed to recycling programs.[2]

2. Socioeconomic Issues

The siting of an MSW incinerator in a poor or industrialized section of the city or county often raises questions about inequities and the fair distribution of misery. This problem is often complicated by the fact that poor neighborhoods, especially in older cities, are frequently located near prime industrial areas. Such areas are often excellent locations for an MSW incinerator when zoning, land availability, proximity to other industry, transportation access, and development costs are considered. Citizens living in these areas, however, often believe that they had already paid their fair share to the community by having other major industrial facilities in their neighborhood.

3. Outside Waste Issues

Another concern often raised is that a proposed MSW incinerator is oversized (i.e., the proposed size necessitates accepting waste from other cities or counties). Local citizens are reluctant to accept the responsibility of disposing of MSW from the other areas, again, largely because they perceive that such transactions are inequitable. Local residents often have a difficult time accepting the problems associated with disposing of their own trash. Being faced with accepting trash from other areas seems to compound their frustration. On the other hand, building an MSW incinerator large enough to service several cities and counties might be the most economical way for all parties involved to dispose of their trash.

4. Traffic Concerns

The increased number of trucks traveling through a neighborhood and the anticipated congestion resulting from this increased truck traffic is a typical concern of local residents. Many residents believe that the additional truck traffic will reduce the quality of life and increase traffic hazards in their neighborhood.

5. Recommendations

Although there are other nonhealth-related issues that might be of concern to individuals or groups opposed to a particular project, the above-mentioned items are the problems that have occurred most often in the authors' experience. Dealing with these issues can be very difficult. One method commonly employed is the establishment of a "host community agreement" by the vendor, county, or other entity that is proposing the project or stands to gain economically from it. This type of agreement is a fairly straightforward attempt to recognize the inequities of siting an incinerator in a particular area by offering some reward or other form of compensation to the host community for allowing the incinerator to be built in their neighborhood. However, this type of agreement may be inappropriate or unworkable in some cases. Regardless, it is important for project proponents to recognize that numerous nonhealth-related causes of public opposition exist and that these issues can greatly complicate the entire project.

B. HEALTH-RELATED ISSUES

A great deal of attention has been focused on potential human health impacts of MSW incineration. Numerous claims and counterclaims have been made about the dangers of various pollutants, especially dioxins and furans, which have resulted in much confusion on the part of the public and regulatory agencies. Amid this confusion, many concerned citizens groups have begun to question and, in some cases, to strongly oppose MSW incineration projects based on the potential for adverse health effects. Public fears about potential adverse health effects associated with exposure to facility-emitted pollutants have been heightened by increased media attention of recent environmental catastrophes, such as the Union Carbide chemical plant explosion in Bhopal, India, and by the incipient distrust of both industry and government that comes in the wake of such disasters.

C. MISTAKES MADE BY PROJECT PROPONENTS

Some common mistakes made by project proponents (e.g., local solid waste authorities, consultants, or vendors) can also fuel public distrust and opposition. For example, project proponents who fail to understand the seriousness and depth of public concern, who tend to treat concerned individuals with paternalism, or who offer unsubstantiated assurances instead of involving the public in the overall process, run the risk of seriously alienating the public from the outset. Such complacency and lack of understanding is often fueled by early impressions that the level of interest is low and that any attempt to involve the public would only generate unwelcome conflict. This is a dangerous strategy since opposition can occur at any time. Furthermore, project proponents who make bold statements to the public about the ultimate safety of the facility before any environmental risk assessment has been conducted or who conduct inadequate siting studies or environmental risk assessments (which are subsequently subjected to close scrutiny) run the risk of losing their credibility. Once lost, credibility and trust are extremely difficult to regain.

D. SUMMARY

Although there are other issues of concern to individuals opposed to the siting of an MSW incinerator in their neighborhood, the above-mentioned items are ones that have arisen most often in our experience. It is important for project proponents to recognize that numerous causes of opposition exist and can greatly complicate the siting process. It is vital that the public be dealt with in an open, honest, and straightforward manner, so that they can obtain a genuine understanding of the issues involved. In other words, public participation in the overall process, not just public relations, is required.

III. PUBLIC PARTICIPATION

The public is demanding to be included in major local policy decisions, such as the siting of MSW incinerators, in an increasingly forceful manner. Regardless of the motivation behind this involvement, local concerned citizen groups are influencing decisions concerning the final disposition of these projects. Therefore, it is vital that the public be included in the process as early as possible, preferably during the siting studies, so that their concerns can be addressed quickly and thoroughly and to allow meaningful public input before final decisions are made. In many cases, failure to include the public early in the process has cost project proponents a great deal of time and money and, in several cases, has resulted in the ultimate cancellation of the project.

The earlier the public is brought into the process, the greater the likelihood of public acceptance. In many cases, this involvement can best be accomplished by organizing a citizens' advisory committee (CAC) to review and evaluate key issues of concern. This committee should be viewed by project proponents as a legitimate part of the decision making process. Figure 1 illustrates the composition of a model CAC. Although it may not be feasible to include individuals from all categories listed in Figure 1, and in some cases it may be appropriate to include other individuals not identified here, it is advisable to have as many of these types of groups represented as possible.

A. PUBLIC PARTICIPATION IN THE RISK ASSESSMENT PROCESS

It is extremely important to include some type of CAC in the risk assessment process from the outset. If the CAC is not involved until after the site has been selected and the risk assessment completed, they may feel that their review is merely a "rubber stamp" and that they have been denied a meaningful role in the entire process.

In a well-conceived program (assuming that a site has already been selected), a series of meetings with the CAC should be organized to discuss and evaluate health-related issues as the risk assessment is being developed. Initial meetings should discuss the project scope with subsequent meetings examining all of the various technical components of the risk assessment process. For example, a typical series of meetings should cover the following broad risk assessment topics:

1. Project description and scope
2. Identifying pollutants of concern

FIGURE 1. Organization of a formal citizens' advisory committee (CAC).

3. Determining the appropriate emissions data, deposition models, exposure assessment methodologies, and toxicological data to be used in the analysis
4. Performing the exposure assessment
5. Estimating the magnitude of risk to humans

This process accomplishes several objectives. First, it allows citizens to participate in the entire risk assessment process. Second, it helps to ensure that citizens' concerns will be taken seriously. Finally, it provides a platform for addressing key concerns early in the process before public opposition has solidified. Involving the public throughout the entire risk assessment process improves the likelihood that the majority of the public will accept the final results of the study, no matter what they are.

This process also provides a reasonable alternative for including and educating members of the media. Media personnel are more likely to report speculative claims about potential health impacts in the absence of clear information to the contrary. Involving members of the press in the risk assessment process, however, offers them the opportunity to acquire and report balanced information concerning the risks associated with MSW incineration.

B. RISK MANAGEMENT

Throughout the process, it is important that the distinction between risk assessment and risk management be made clear. The objective of the risk assessment effort is to evaluate risks from the proposed facility. Risk management is the process in which individuals determine if the estimated risks are acceptable. Maintaining this distinction is important, because it allows the risk management process to focus on the technical aspects of the risk assessment without becoming encumbered by all of the other, nonhealth-related issues that can often dominate the process.

C. RISK COMMUNICATION

The inherent theme behind public participation is risk communication. The risk assessment document can be a technically accurate and scientifically credible document, but if the public perceives it as an attempt to obfuscate their concerns, it will fail to communicate information on the risks associated with municipal waste combustion. Risk assessments, by necessity, are complicated documents that include many complex calculations. However, if they are written in a simple, straightforward, well-organized manner and are presented to the public committee over a series of meetings during which their questions and concerns can be answered, the public will be more likely to accept the results and the opinions of the committee.

IV. CASE STUDIES

The following case studies represent projects where CACs were formed and CAC members participated in the risk assessment process. These case studies illustrate that including the public in the risk assessment and siting process does not, in itself, necessarily facilitate risk communication.

A. MONTGOMERY COUNTY, MARYLAND

The preparation of a risk assessment for a proposed 1200 ton/d (TPD) resource recovery facility in suburban Washington, D.C., was initiated in 1984. This project was one of the first projects in which a formal review committee consisting of local, concerned citizens was established. The committee was composed of ten technically competent individuals, a number of whom had health-related backgrounds. The two most ardent opponents to the project were either members of the committee or ad hoc advisors to the committee. Meetings were held almost once a month during the first year of the project, because of the inherent difficulty associated with the risk assessment process.

At the end of the first phase of the project, which lasts more than 1 year, a minority report was submitted to the County government highlighting the concerns of those still opposed to the project. However, the risk assessment and the public participation process undertaken were considered successful by the County government, which supported proceeding with the project.

As other issues were considered by the County government, additional phases were added to the original risk assessment to address the following concerns:

- Comparing a rural site vs. the original suburban site
- Examining the risks of co-firing refuse-derived fuel (RDF) at an existing, rural coal-fired power plant
- Determining the risks associated with substituting a dedicated RDF boiler for the proposed massburn system at both the rural and suburban sites
- Performing a multipathway health risk assessment (i.e., evaluating both direct and indirect pathways of human exposure) for the final rural site selected

The community review committee remained active and involved throughout all phases of the risk assessment process even though the composition of the committee changed to include residents that lived near the selected rural site. In fact, the overall process continued after the preliminary risk assessment had been completed, so that some additional issues relating to the siting of the resource recovery facility close to an existing power plant could be evaluated.

B. BROOME COUNTY, NEW YORK

In the authors' opinion, this case study offers an excellent example of an appropriate way to include the public in the risk assessment/risk management process. Before the risk assessment process was initiated in late 1986, a CAC had already been formed. This committee was comprised of a diverse group of local residents who had expertise in engineering, biology, agricultural, medicine, and chemistry and several individuals from the community in which the preferred site existed.

Initial meetings were aimed at ensuring skeptical members that the risk assessment process would be impartial and valid, since it was clear some members of the CAC believed that the overall process would be biased. However, during the first year as technical questions raised by committee members were answered, the level of trust in the entire process began to increase.

Originally, meetings were held to discuss the overall risk assessment approach. Significant input from committee members on other risk-related issues, however, led to the investigation of additional matters outside the original project scope. Subsequent meetings focused on evaluation of the proposed technology, pollutants of concern, emission factors, dispersion and deposition modeling, exposure assessment, toxicological criteria, and the site-specific landfill risk assessment. Final meetings concentrated on the prediction of risk and the comparison of risks from incineration to risks associated with other waste management options.

There was never complete agreement within the committee on every point under discussion within the committee. For example, some committee members were concerned that the County was not devoting sufficient resources and attention to recycling, while other CAC members believed that the proposed facility was too large and that fly and bottom ash should be separated before disposal. By the time the risk assessment was completed, however, most CAC members believed that the process had been thorough and unbiased, and they were confident that the report adequately characterized an upper-bound level of risk based on the assumptions made in the risk assessment process. In fact, several committee members were quoted in the local newspaper as saying that they were comfortable with the process and that they believed that the risks were low.

In effect, as confidence in the overall process heightened, risks from the facility began to diminish as the primary concern. Although other concerns did not disappear, these concerns were not risk-related but were public policy and risk management concerns.

In our opinion, this effort was very successful. The public was brought into the process early, dealt with in an open and honest manner, and given all the facts and information they needed to evaluate the proposed facility from a risk assessment perspective.

C. DELAWARE COUNTY, PENNSYLVANIA

For this project, work on the risk assessment began in 1987 as part of the permitting process. Most concerns revolved around political, economic, and social issues. Nevertheless, a CAC was established in an attempt to focus attention on the health risk assessment. Members of the CAC were individuals with diverse backgrounds who lived in the affected community, including a college professor (engineering), a medical doctor, two schoolteachers, and a minister. The committee was remarkably united in pursuing their main goal of evaluating health risks. Other, nonhealth-related concerns were aggressively ignored by the committee.

After a series of meetings similar to those described in the Broome County case, the committee wrote a report summarizing their review of the existing risk assessment, which had been completed earlier as part of the permit submission effort. (In this case, it took so long for the CAC to be formed that it was necessary to complete the risk assessment document without direct input from the committee. This is not the preferred sequence.) The CACs report supported the findings of the risk assessment, for the most part, and made a number of appropriate points about the relative nature of the facility compared to other industrial developments (i.e., the facility creates a certain amount of pollution and yields a certain number of jobs). The report further stated that the risks were not unacceptable. The committee report was released at a public hearing which only a handful of local citizens attended. This lack of public participation supports argument that the health risks were not the major issue of concern.

There were so many nonrisk-related issues encumbering the overall process in this case, that public participation in the risk assessment process, while successful, seemed to have very little impact on settling existing conflicts. In reality, many concerns raised about the risks associated with the proposed facility appeared to be, in our opinion, simply efforts to obscure other concerns and terminate the project. This case shows that even if the public is involved in the risk assessment process, other concerns may outweigh health concerns. (*Note:* This project has received regulatory approval and is currently under construction.)

D. OTHER EXAMPLES

There are many instances where the public was not included in the process. In many of these cases, substantial public concerns caused major project delays and cancellations. For example, in Montgomery County, Maryland, the first opportunity for public involvement came during the public hearing for the permit application. The crowd was hostile and distrustful of the project proponents. Black balloons, negative signs, and hostile chants were evident throughout the proponents' presentation. Understandably, the results of the risk assessment were not accepted as realistic but rather were perceived as mere propaganda put forth to benefit project proponents. The permit, which was eventually granted by the state agency, is still under review over 1 year later by the state environmental review judge as a result of legal actions taken on behalf of the concerned citizens.

In another instance, a large U.S. city made virtually no attempt to explain or communicate potential health risks to the public early in the process. By the time that city officials realized their mistake, intense public opposition had coalesced. This project is no longer under active consideration.

V. CONCLUSIONS

MSW incinerators are part of the solution to the growing waste disposal crisis facing the U.S., and the public is determined to be involved in waste management decisions. Public concerns about potential health impacts from these MSW incinerators cannot be ignored or given minimal consideration. The public must be brought into the process early and included in the risk assessment effort if they are to gain trust and confidence in the results of these complex documents. Risk assessments can be used to educate the public about potential health risks associated with facility-emitted pollutants. Involving the public in the risk assessment process facilitates public understanding and acceptance of information contained within these reports and allows the public to make more informed decisions concerning the acceptability of such risks.

REFERENCES

1. **Papke, C.,** Getting real with recycling, *Solid Waste Power,* 2(2), 28, 1988.
2. **Kiser, J.,** Waste-to-energy and recycling, a compatible solution, *Waste Alternatives,* 1, 21, 1988.
3. **Thompson, J. R.,** The Illinois solid waste management program, *Solid Waste Power,* 3(1), 42, 1989.
4. **Birks, D. W.,** Choosing all of the above, *Solid Waste Power,* 2, 32, 1988.
5. **Huff, W. J. and Smith, G. A.,** Managing the wastes of Babylon, *Solid Waste Power,* 2(2), 24, 1988.

INDEX

A

B

C

G

H

I

J

K

L

M

N

O

P

Q

R

S

T

U

V

W

Y

Z